CLIMATE CHANGE

The IPCC Scientific Assessment

WORLD METEOROLOGICAL ORGANIZATION / UNITED NATIONS ENVIRONMENT PROGRAMME

INTERGOVERNMENTAL PANEL ON CLIMATE CHANGE

This is the final Report of Working Group 1 of the Intergovernmental Panel on Climate Change, which is sponsored jointly by the World Meteorological Organization and the United Nations Environment Programme. The report considers the scientific assessment of climate change. Several hundred working scientists from 25 countries have participated in the preparation and review of the scientific data. The result is the most authoritative and strongly supported statement on climate change that has ever been made by the international scientific community. The issues confronted with full rigour include: global warming, greenhouse gasses, the greenhouse effect, sea level changes, forcing of climate, and the history of Earth's changing climate. The information presented here is of the highest quality. It will inform the necessary scientific, political and economic debates and negotiations that can be expected in the immediate future. Appropriate strategies in response to the issue of climate change can now be firmly based on the scientific foundation that the Report provides. The Report is, therefore, an essential reference for all who are concerned with climate change and its consequences.

Climate Change

THE IPCC SCIENTIFIC ASSESSMENT

Climate Change

The IPCC Scientific Assessment

Edited by

J. T. HOUGHTON, G. J. JENKINS and J. J. EPHRAUMS

The right of the
University of Cambridge
to print and sell
all manner of books
was granted by
Henry VIII in 1534.
The University has printed
and published continuously
since 1584.

Published for the

Intergovernmental Panel on Climate Change

CAMBRIDGE UNIVERSITY PRESS

Cambridge

New York Port Chester Melbourne Sydney

Published by the Press Syndicate of the University of Cambridge
The Pitt Building, Trumpington Street, Cambridge CB2 1RP
40 West 20th Street, New York, NY 10011, USA
10 Stamford Road, Oakleigh, Melbourne 3166, Australia

First published 1990

Printed in Great Britain at the University Press, Cambridge

British Library cataloguing in publication data
Intergovernmental Panel on Climate Change
    Climate Change.
    1. Climate Changes
    I. Title   II. Houghton, John T. (John Theodore) *1931–*
    III. Jenkins, G. J.   IV. Ephraums, J. J.
    551.6

Library of Congress cataloguing in publication data available

    ISBN 0 521 40360 X hardback
    ISBN 0 521 40720 6 paperback

INTERGOVERNMENTAL PANEL ON CLIMATE CHANGE

CLIMATE CHANGE
The IPCC Scientific Assessment

Report Prepared for IPCC by Working Group 1

Edited by J.T.Houghton, G.J.Jenkins and J.J.Ephraums

(Meteorological Office, Bracknell, United Kingdom)

WMO

UNEP

Contents

Preface

The Intergovernmental Panel on Climate Change (IPCC) was jointly established by our two organisations in 1988. Under the chairmanship of Professor Bert Bolin, the Panel was charged with:

(i) assessing the scientific information that is related to the various components of the climate change issue, such as emissions of major greenhouse gases and modification of the Earth's radiation balance resulting therefrom, and that needed to enable the environmental and socio-economic consequences of climate change to be evaluated.

(ii) formulating realistic response strategies for the management of the climate change issue.

The Panel began its task by establishing Working Groups I, II and III respectively to:

(a) assess available scientific information on climate change.

(b) assess environmental and socio-economic impacts of climate change.

(c) formulate response strategies.

It also established a Special Committee on the Participation of Developing Countries to promote, as quickly as possible, the full participation of developing countries in its activities.

This volume is based upon the findings of Working Group I, and should be read in conjunction with the rest of the IPCC first assessment report; the latter consists of the reports and policymakers summaries of the 3 Working Groups and the Special Committee, and the IPCC overview and conclusions.

The Chairman of Working Group I, Dr John Houghton, and his Secretariat, have succeeded beyond measure in mobilizing the co-operation and enthusiasm of hundreds of scientists from all over the world. They have produced a volume of remarkable depth and breadth, and a Policymakers Summary which translates these complex scientific issues into language which is understandable to the non-specialist.

We take this opportunity to congratulate and thank the Chairman for a job well done.

G.O.P. Obasi
Secretary-General
World Meteorological Organization

M.K. Tolba
Executive Director
United Nations Environment Programme

Foreword

Many previous reports have addressed the question of climate change which might arise as a result of man's activities. In preparing this Scientific Assessment, Working Group I [1] has built on these, in particular the SCOPE 29 report of 1986 [2], taking into account significant work undertaken and published since then. Particular attention is paid to what is known regarding the detail of climate change on a regional level.

In the preparation of the main Assessment most of the active scientists working in the field have been involved. One hundred and seventy scientists from 25 countries have contributed to it, either through participation in the twelve international workshops organised specially for the purpose or through written contributions. A further 200 scientists have been involved in the peer review of the draft report. Although, as in any developing scientific topic, there is a minority of opinions which we have not been able to accommodate, the peer review has helped to ensure a high degree of consensus amongst authors and reviewers regarding the results presented. Thus the Assessment is an authoritative statement of the views of the international scientific community at this time.

The accompanying Policymakers' Summary, based closely on the conclusions of the Assessment, has been prepared particularly to meet the needs of those without a strong background in science who need a clear statement of the present status of scientific knowledge and the associated uncertainties.

The First Draft of the Policymakers Summary was presented to the meeting of the Lead Authors of the Assessment (Edinburgh, February 1990), and the Second Draft which emanated from that meeting was sent for the same wide peer review as the main report, including nationally designated experts and the committees of relevant international scientific programmes. A Third Draft incorporating a large number of changes suggested by peer-reviewers was tabled at the final plenary meeting of Working Group I (Windsor, May 1990) at which the Lead Authors of the main report were present, and the final version was agreed at that meeting.

It gives me pleasure to acknowledge the contributions of so many, in particular the Lead Authors, who have given freely of their expertise and time to the preparation of this report. All the modelling centres must be thanked for providing data so readily for the model intercomparison. I also acknowledge the contribution of the core team at the Meteorological Office in Bracknell who were responsible for organising most of the workshops and preparing the report. Members of the team were Professor Cao Hong-Xing from China, Dr Reindert Haarsma from The Netherlands, Dr Robert Watson from the USA, and Dr John Mitchell, Dr Peter Rowntree, Dr Terry Callaghan, Chris Folland, Jim Ephraums, Shelagh Varney, Andrew Gilchrist and Aileen Foreman from the UK. Particular acknowledgment is due to Dr Geoff Jenkins, the Coordinator of Working Group I who led the team. Thanks

---

[1] Organisational details of IPCC and Working Group I are shown in Appendix 2.

[2] The Greenhouse Effect, Climate Change and Ecosystems, SCOPE 29; Bolin, B., B. Döös, J. Jäger and R. A. Warrick (Eds.), John Wiley and Sons, Chichester, 1986.

are also due to the Dr Sundararaman and the IPCC Secretariat in Geneva. Financial support for the Bracknell core team was provided by the Departments of the Environment and Energy in the UK.

I am confident that the Assessment and its Summary will provide the necessary firm scientific foundation for the forthcoming discussions and negotiations on the appropriate strategy for response and action regarding the issue of climate change. It is thus, I believe, a significant step forward in meeting what is potentially the greatest global environmental challenge facing mankind.

Dr John Houghton
Chairman, IPCC Working Group I

Meteorological Office
Bracknell
July 1990

Policymakers Summary

Prepared by IPCC Working Group I

CONTENTS

EXECUTIVE SUMMARY

We are certain of the following:

- there is a natural greenhouse effect which already keeps the Earth warmer than it would otherwise be.

- emissions resulting from human activities are substantially increasing the atmospheric concentrations of the greenhouse gases: carbon dioxide, methane, chlorofluorocarbons (CFCs) and nitrous oxide. These increases will enhance the greenhouse effect, resulting on average in an additional warming of the Earth's surface. The main greenhouse gas, water vapour, will increase in response to global warming and further enhance it.

We calculate with confidence that:

- some gases are potentially more effective than others at changing climate, and their relative effectiveness can be estimated. Carbon dioxide has been responsible for over half the enhanced greenhouse effect in the past, and is likely to remain so in the future.

- atmospheric concentrations of the long-lived gases (carbon dioxide, nitrous oxide and the CFCs) adjust only slowly to changes in emissions. Continued emissions of these gases at present rates would commit us to increased concentrations for centuries ahead. The longer emissions continue to increase at present day rates, the greater reductions would have to be for concentrations to stabilise at a given level.

- the long-lived gases would require immediate reductions in emissions from human activities of over 60% to stabilise their concentrations at today's levels; methane would require a 15-20% reduction.

Based on current model results, we predict:

- under the IPCC Business-as-Usual (Scenario A) emissions of greenhouse gases, a rate of increase of global mean temperature during the next century of about 0.3°C per decade (with an uncertainty range of 0.2°C to 0.5°C per decade); this is greater than that seen over the past 10,000 years. This will result in a likely increase in global mean temperature of about 1°C above the present value by 2025 and 3°C before the end of the next century. The rise will not be steady because of the influence of other factors.

- under the other IPCC emission scenarios which assume progressively increasing levels of controls, rates of increase in global mean temperature of about 0.2°C per decade (Scenario B), just above 0.1°C per decade (Scenario C) and about 0.1°C per decade (Scenario D).

- that land surfaces warm more rapidly than the ocean, and high northern latitudes warm more than the global mean in winter.

- regional climate changes different from the global mean, although our confidence in the prediction of the detail of regional changes is low. For example, temperature increases in Southern Europe and central North America are predicted to be higher than the global mean, accompanied on average by reduced summer precipitation and soil moisture. There are less consistent predictions for the tropics and the Southern Hemisphere.

- under the IPCC Business as Usual emissions scenario, an average rate of global mean sea level rise of about 6cm per decade over the next century (with an uncertainty range of 3 - 10cm per decade), mainly due to thermal expansion of the oceans and the melting of some land ice. The predicted rise is about 20cm in global mean sea level by 2030, and 65cm by the end of the next century. There will be significant regional variations.

There are many uncertainties in our predictions particularly with regard to the timing, magnitude and regional patterns of climate change, due to our incomplete understanding of:

- sources and sinks of greenhouse gases, which affect predictions of future concentrations.

- clouds, which strongly influence the magnitude of climate change.

- oceans, which influence the timing and patterns of climate change.

- polar ice sheets which affect predictions of sea level rise.

These processes are already partially understood, and we are confident that the uncertainties can be reduced by further research. However, the complexity of the system means that we cannot rule out surprises.

Our judgement is that:

- Global - mean surface air temperature has increased by 0.3°C to 0.6°C over the last 100 years, with the five global-average warmest years being in the 1980s. Over the same period global sea level has increased by 10-20cm. These increases have not been smooth with time, nor uniform over the globe.

- The size of this warming is broadly consistent with predictions of climate models, but it is also of the same magnitude as natural climate variability. Thus the observed increase could be largely due to this natural variability; alternatively this variability and other human factors could have offset a still larger human-induced greenhouse warming. The uneq-

uivocal detection of the enhanced greenhouse effect from observations is not likely for a decade or more.

- There is no firm evidence that climate has become more variable over the last few decades. However, with an increase in the mean temperature, episodes of high temperatures will most likely become more frequent in the future, and cold episodes less frequent.

- Ecosystems affect climate, and will be affected by a changing climate and by increasing carbon dioxide concentrations. Rapid changes in climate will change the composition of ecosystems; some species will benefit while others will be unable to migrate or adapt fast enough and may become extinct. Enhanced levels of carbon dioxide may increase productivity and efficiency of water use of vegetation. The effect of warming on biological processes, although poorly understood, may increase the atmospheric concentrations of natural greenhouse gases.

To improve our predictive capability, we need:

- to **understand** better the various climate-related processes, particularly those associated with clouds, oceans and the carbon cycle.

- to **improve** the systematic observation of climate-related variables on a global basis, and further investigate changes which took place in the past.

- to **develop** improved models of the Earth's climate system.

- to **increase** support for national and international climate research activities, especially in developing countries.

- to **facilitate** international exchange of climate data.

Introduction: what is the issue ?

There is concern that human activities may be inadvertently changing the climate of the globe through the enhanced greenhouse effect, by past and continuing emissions of carbon dioxide and other gases which will cause the temperature of the Earth's surface to increase - popularly termed the "global warming". If this occurs, consequent changes may have a significant impact on society.

The purpose of the Working Group I report, as determined by the first meeting of IPCC, is to provide a scientific assessment of:

1) the factors which may affect climate change during the next century, especially those which are due to human activity.
2) the responses of the atmosphere - ocean - land - ice system.
3) current capabilities of modelling global and regional climate changes and their predictability.
4) the past climate record and presently observed climate anomalies.

On the basis of this assessment, the report presents current knowledge regarding predictions of climate change (including sea level rise and the effects on ecosystems) over the next century, the timing of changes together with an assessment of the uncertainties associated with these predictions.

This Policymakers Summary aims to bring out those elements of the main report which have the greatest relevance to policy formulation, in answering the following questions:

- What factors determine global climate?
- What are the greenhouse gases, and how and why are they increasing?
- Which gases are the most important?
- How much do we expect the climate to change?
- How much confidence do we have in our predictions?
- Will the climate of the future be very different ?
- Have human activities already begun to change global climate?
- How much will sea level rise?
- What will be the effects on ecosystems?
- What should be done to reduce uncertainties, and how long will this take?

This report is intended to respond to the practical needs of the policymaker. It is neither an academic review, nor a plan for a new research programme. Uncertainties attach to almost every aspect of the issue, yet policymakers are looking for clear guidance from scientists; **hence authors have been asked to provide their best-estimates wherever possible**, together with an assessment of the uncertainties.

This report is a summary of our understanding in 1990. Although continuing research will deepen this understanding and require the report to be updated at frequent intervals, basic conclusions concerning the reality of the enhanced greenhouse effect and its potential to alter global climate are unlikely to change significantly. Nevertheless, the complexity of the system may give rise to surprises.

What factors determine global climate ?

There are many factors, both natural and of human origin, that determine the climate of the earth. We look first at those which are natural, and then see how human activities might contribute.

What natural factors are important?

The driving energy for weather and climate comes from the Sun. The Earth intercepts solar radiation (including that in the short-wave, visible, part of the spectrum); about a third of it is reflected, the rest is absorbed by the different components (atmosphere, ocean, ice, land and biota) of the climate system. The energy absorbed from solar radiation is balanced (in the long term) by outgoing radiation from the Earth and atmosphere; this terrestrial radiation takes the form of long-wave invisible infrared energy, and its magnitude is determined by the temperature of the Earth-atmosphere system.

There are several natural factors which can change the balance between the energy absorbed by the Earth and that emitted by it in the form of longwave infrared radiation; these factors cause the **radiative forcing** on climate. The most obvious of these is a change in the output of energy from the Sun. There is direct evidence of such variability over the 11-year solar cycle, and longer period changes may also occur. Slow variations in the Earth's orbit affect the seasonal and latitudinal distribution of solar radiation; these were probably responsible for initiating the ice ages.

One of the most important factors is the **greenhouse effect**; a simplified explanation of which is as follows. Short-wave solar radiation can pass through the clear atmosphere relatively unimpeded. But long-wave terrestrial radiation emitted by the warm surface of the Earth is partially absorbed and then re-emitted by a number of trace gases in the cooler atmosphere above. Since, on average, the outgoing long-wave radiation balances the incoming solar radiation, both the atmosphere and the surface will be warmer than they would be without the greenhouse gases.

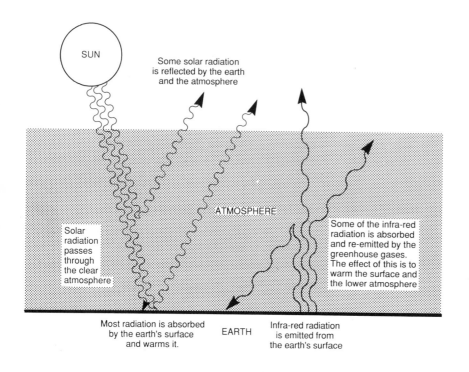

Figure 1: A simplified diagram illustrating the greenhouse effect.

The main natural greenhouse gases are not the major constituents, nitrogen and oxygen, but water vapour (the biggest contributor), carbon dioxide, methane, nitrous oxide, and ozone in the troposphere (the lowest 10-15km of the atmosphere) and stratosphere.

Aerosols (small particles) in the atmosphere can also affect climate because they can reflect and absorb radiation. The most important natural perturbations result from explosive volcanic eruptions which affect con-centrations in the lower stratosphere. Lastly, the climate has its own **natural variability** on all timescales and changes occur without any external influence.

How do we know that the natural greenhouse effect is real?

The greenhouse effect is real; it is a well understood effect, based on established scientific principles. We know that the greenhouse effect works in practice, for several reasons.

Firstly, the mean temperature of the Earth's surface is already warmer by about 33°C (assuming the same reflectivity of the earth) than it would be if the natural greenhouse gases were not present. Satellite observations of the radiation emitted from the Earth's surface and through the atmosphere demonstrate the effect of the greenhouse gases.

Secondly, we know the composition of the atmospheres of Venus, Earth and Mars are very different, and their surface temperatures are in general agreement with greenhouse theory.

Thirdly, measurements from ice cores going back 160,000 years show that the Earth's temperature closely paralleled the amount of carbon dioxide and methane in the atmosphere (see Figure 2). Although we do not know the details of cause and effect, calculations indicate that changes in these greenhouse gases were part, but not all, of the reason for the large (5-7°C) global temperature swings between ice ages and interglacial periods.

How might human activities change global climate?

Naturally occurring greenhouse gases keep the Earth warm enough to be habitable. By increasing their concentrations, and by adding new greenhouse gases like chloro-fluorocarbons (CFCs), humankind is capable of raising the global-average annual-mean surface-air temperature (which, for simplicity, is referred to as the "global temperature"), although we are uncertain about the rate at which this will occur. Strictly, this is an **enhanced** greenhouse effect - above that occurring due to natural greenhouse gas concentrations; the word "enhanced" is usually omitted, but it should not be forgotten. Other changes in climate are expected to result, for example changes in precipitation, and a global warming will cause sea levels to rise; these are discussed in more detail later.

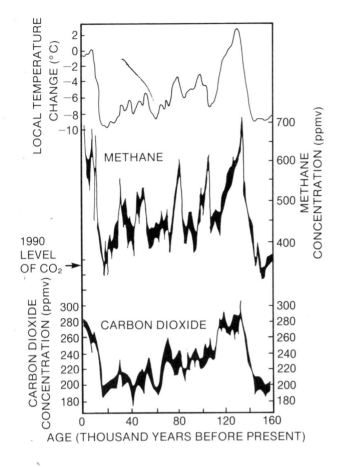

Figure 2: Analysis of air trapped in Antarctic ice cores shows that methane and carbon dioxide concentrations were closely correlated with the local temperature over the last 160,000 years. Present day concentrations of carbon dioxide are indicated

There are other human activities which have the potential to affect climate. A change in the albedo (reflectivity) of the land, brought about by **desertification or deforestation** affects the amount of solar energy absorbed at the Earth's surface. Human-made **aerosols**, from sulphur emitted largely in fossil fuel combustion, can modify clouds and this may act to lower temperatures. Lastly, changes in **ozone in the stratosphere** due to CFCs may also influence climate.

What are the greenhouse gases and why are they increasing?

We are certain that the concentrations of greenhouse gases in the atmosphere have changed naturally on ice-age time-scales, and have been increasing since pre-industrial times due to human activities. Table 1 summarizes the present and pre-industrial abundances, current rates of change and present atmospheric lifetimes of greenhouse gases influenced by human activities. Carbon dioxide, methane,

and nitrous oxide all have significant natural and human sources, while the chlorofluorocarbons are only produced industrially.

Two important greenhouse gases, water vapour and ozone, are not included in this table . Water vapour has the largest greenhouse effect, but its concentration in the troposphere is determined internally within the climate system, and, on a global scale, is not affected by human sources and sinks. Water vapour will increase in response to global warming and further enhance it; this process is included in climate models. The concentration of ozone is changing both in the stratosphere and the troposphere due to human activities, but it is difficult to quantify the changes from present observations.

For a thousand years prior to the industrial revolution, abundances of the greenhouse gases were relatively constant. However, as the world's population increased, as the world became more industrialized and as agriculture developed, the abundances of the greenhouse gases increased markedly. Figure 3 illustrates this for carbon dioxide, methane, nitrous oxide and CFC-11.

Since the industrial revolution the combustion of fossil fuels and deforestation have led to an increase of 26% in carbon dioxide concentration in the atmosphere. We know the magnitude of the present day fossil-fuel source, but the input from deforestation cannot be estimated accurately. In addition, although about half of the emitted carbon dioxide stays in the atmosphere, we do not know well how much of the remainder is absorbed by the oceans and how much by terrestrial biota. Emissions of chlorofluorocarbons, used as aerosol propellants, solvents, refrigerants and foam blowing agents, are also well known; they were not present in the atmosphere before their invention in the 1930s.

The sources of methane and nitrous oxide are less well known. Methane concentrations have more than doubled because of rice production, cattle rearing, biomass burning, coal mining and ventilation of natural gas; also, fossil fuel combustion may have also contributed through chemical reactions in the atmosphere which reduce the rate of removal of methane. Nitrous oxide has increased by about 8% since pre-industrial times, presumably due to human activities; we are unable to specify the sources, but it is likely that agriculture plays a part.

The effect of ozone on climate is strongest in the upper troposphere and lower stratosphere. Model calculations indicate that ozone in the upper troposphere should have increased due to human-made emissions of nitrogen oxides, hydrocarbons and carbon monoxide. While at ground level ozone has increased in the Northern Hemisphere in response to these emissions, observations are insufficient to confirm the expected increase in the upper troposphere. The lack of adequate observations prevents us from accurately quantifying the climatic effect of changes in tropospheric ozone.

Table 1: *Summary of Key Greenhouse Gases Affected by Human Activities*

	Carbon Dioxide	Methane	CFC-11	CFC-12	Nitrous Oxide
Atmospheric concentration	ppmv	ppmv	pptv	pptv	ppbv
Pre-industrial (1750-1800)	280	0.8	0	0	288
Present day (1990)	353	1.72	280	484	310
Current rate of change per year	1.8 (0.5%)	0.015 (0.9%)	9.5 (4%)	17 (4%)	0.8 (0.25%)
Atmospheric lifetime (years)	(50-200)†	10	65	130	150

ppmv = parts per million by volume;

ppbv = parts per billion (thousand million) by volume;

pptv = parts per trillion (million million) by volume.

† The way in which CO_2 is absorbed by the oceans and biosphere is not simple and a single value cannot be given; refer to the main report for further discussion.

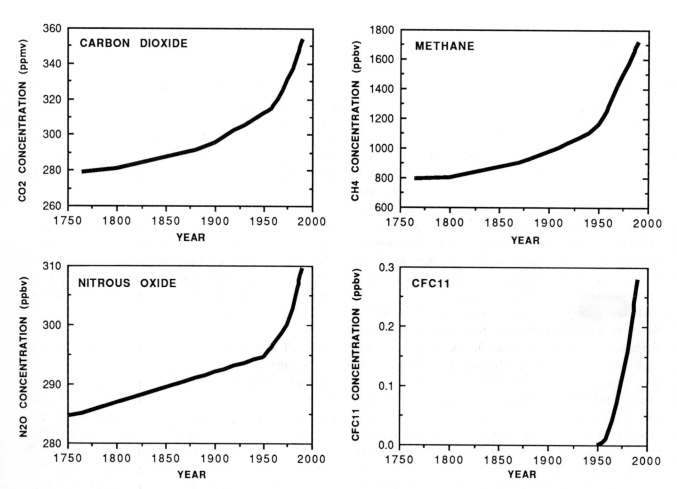

Figure 3: Concentrations of carbon dioxide and methane after remaining relatively constant up to the 18th century, have risen sharply since then due to man's activities. Concentrations of nitrous oxide have increased since the mid-18th century, especially in the last few decades. CFCs were not present in the atmosphere before the 1930s.

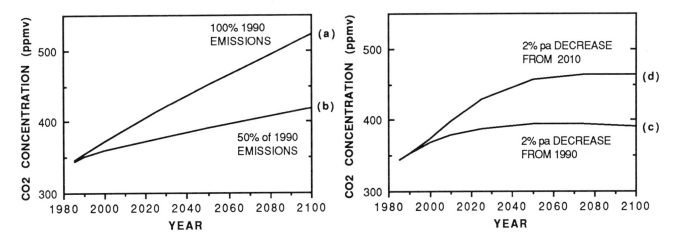

Figure 4: The relationship between hypothetical fossil fuel emissions of carbon dioxide and its concentration in the atmosphere is shown in the case where (a) emissions continue at 1990 levels, (b) emissions are reduced by 50% in 1990 and continue at that level, (c) emissions are reduced by 2% pa from 1990, and (d) emissions, after increasing by 2% pa until 2010, are then reduced by 2% pa thereafter.

In the lower stratosphere at high southern latitudes ozone has decreased considerably due to the effects of CFCs, and there are indications of a global-scale decrease which, while not understood, may also be due to CFCs. These observed decreases should act to cool the earth's surface, thus providing a small offset to the predicted warming produced by the other greenhouse gases. Further reductions in lower stratospheric ozone are possible during the next few decades as the atmospheric abundances of CFCs continue to increase.

Concentrations, lifetimes and stabilisation of the gases

In order to calculate the atmospheric concentrations of carbon dioxide which will result from human-made emissions we use computer models which incorporate details of the emissions and which include representations of the transfer of carbon dioxide between the atmosphere, oceans and terrestrial biosphere. For the other greenhouse gases, models which incorporate the effects of chemical reactions in the atmosphere are employed.

The atmospheric lifetimes of the gases are determined by their sources and sinks in the oceans, atmosphere and biosphere. Carbon dioxide, chlorofluorocarbons and nitrous oxide are removed only slowly from the atmosphere and hence, following a change in emissions, their atmospheric concentrations take decades to centuries to adjust fully. Even if all human-made emissions of carbon dioxide were halted in the year 1990, about half of the increase in carbon dioxide concentration caused by human activities would still be evident by the year 2100.

In contrast, some of the CFC substitutes and methane have relatively short atmospheric lifetimes so that their atmospheric concentrations respond fully to emission changes within a few decades.

To illustrate the emission-concentration relationship clearly, the effect of hypothetical changes in carbon dioxide fossil fuel emissions is shown in Figure 4: (a) continuing global emissions at 1990 levels; (b) halving of emissions in 1990; (c) reductions in emissions of 2% per year (pa) from 1990 and (d) a 2% pa increase from 1990-2010 followed by a 2% pa decrease from 2010.

Continuation of present day emissions are committing us to increased future concentrations, and the longer emissions continue to increase, the greater would reductions have to be to stabilise at a given level. If there are critical concentration levels that should not be exceeded, then the earlier emission reductions are made the more effective they are.

The term **"atmospheric stabilisation"** is often used to describe the limiting of the concentration of the greenhouse gases at a certain level. The amount by which human-made emissions of a greenhouse gas must be reduced in order to stabilise at present day concentrations, for example, is shown in Table 2. For most gases the reductions would have to be substantial.

How will greenhouse gas abundances change in the future?

We need to know future greenhouse gas concentrations in order to estimate future climate change. As already mentioned, these concentrations depend upon the magnitude of human-made emissions and on how changes in climate and other environmental conditions may influence the biospheric processes that control the exchange of natural greenhouse gases, including carbon dioxide and methane, between the atmosphere, oceans and terrestrial biosphere - the greenhouse gas "feedbacks".

Table 2: *Stabilisation of Atmospheric Concentrations. Reductions in the human-made emissions of greenhouse gases required to stabilise concentrations at present day levels:*

Greenhouse Gas	Reduction Required
Carbon Dioxide	>60%
Methane	15 - 20%
Nitrous Oxide	70 - 80%
CFC-11	70 - 75%
CFC-12	75 - 85%
HCFC-22	40 - 50%

Note that the stabilisation of each of these gases would have different effects on climate, as explained in the next section.

Four scenarios of future human-made emissions were developed by Working Group III. The first of these assumes that few or no steps are taken to limit greenhouse gas emissions, and this is therefore termed Business-as-Usual (BaU). (It should be noted that an aggregation of national forecasts of emissions of carbon dioxide and methane to the year 2025 undertaken by Working Group III resulted in global emissions 10-20% higher than in the BaU scenario). The other three scenarios assume that progressively increasing levels of controls reduce the growth of emissions; these are referred to as scenarios B, C, and D. They are briefly described in the Annex to this summary. Future concentrations of some of the greenhouse gases which would arise from these emissions are shown in Figure 5.

Greenhouse gas feedbacks

Some of the possible feedbacks which could significantly modify future greenhouse gas concentrations in a warmer world are discussed in the following paragraphs.

The net emissions of carbon dioxide from terrestrial ecosystems will be elevated if higher temperatures increase respiration at a faster rate than photosynthesis, or if plant populations, particularly large forests, cannot adjust rapidly enough to changes in climate.

A net flux of carbon dioxide to the atmosphere may be particularly evident in warmer conditions in tundra and boreal regions where there are large stores of carbon. The opposite is true if higher abundances of carbon dioxide in the atmosphere enhance the productivity of natural ecosystems, or if there is an increase in soil moisture which can be expected to stimulate plant growth in dry ecosystems and to increase the storage of carbon in tundra peat. The extent to which ecosystems can sequester increasing atmospheric carbon dioxide remains to be quantified.

If the oceans become warmer, their net uptake of carbon dioxide may decrease because of changes in (i) the chemistry of carbon dioxide in seawater, (ii) biological activity in surface waters, and (iii) the rate of exchange of carbon dioxide between the surface layers and the deep ocean. This last depends upon the rate of formation of deep water in the ocean which, in the North Atlantic for example, might decrease if the salinity decreases as a result of a change in climate.

Methane emissions from natural wetlands and rice paddies are particularly sensitive to temperature and soil moisture. Emissions are significantly larger at higher temperatures and with increased soil moisture; conversely, a decrease in soil moisture would result in smaller emissions. Higher temperatures could increase the emissions of methane at high northern latitudes from decomposable organic matter trapped in permafrost and methane hydrates.

As illustrated earlier, ice core records show that methane and carbon dioxide concentrations changed in a similar sense to temperature between ice ages and interglacials.

Although many of these feedback processes are poorly understood, it seems likely that, overall, they will act to increase, rather than decrease, greenhouse gas concentrations in a warmer world.

Which gases are the most important?

We are certain that increased greenhouse gas concentrations increase radiative forcing. We can calculate the forcing with much more confidence than the climate change that results because the former avoids the need to evaluate a number of poorly understood atmospheric responses. We then have a base from which to calculate the relative effect on climate of an increase in **concentration** of each gas in the present-day atmosphere, both in absolute terms and relative to carbon dioxide. These relative effects span a wide range; methane is about 21 times more effective, molecule-for-molecule, than carbon dioxide, and CFC-11 about 12,000 times more effective. On a kilogram-

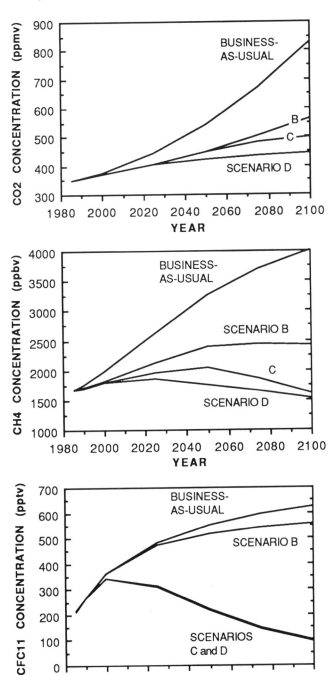

Figure 5: Atmospheric concentrations of carbon dioxide, methane and CFC-11 resulting from the four IPCC emissions scenarios

how this quantity has changed in the past (based on observations of greenhouse gases) and how it might change in the future (based on the four IPCC emissions scenarios). For simplicity, we can express total forcing in terms of the amount of carbon dioxide which would give that forcing; this is termed the **equivalent carbon dioxide concentration**. Greenhouse gases have increased since pre-industrial times (the mid-18th century) by an amount that is radiatively equivalent to about a 50% increase in carbon dioxide, although carbon dioxide itself has risen by only 26%; other gases have made up the rest.

The contributions of the various gases to the total increase in climate forcing during the 1980s is shown as a pie diagram in Figure 7; carbon dioxide is responsible for about half the decadal increase. (Ozone, the effects of which may be significant, is not included)

How can we evaluate the effect of different greenhouse gases?

To evaluate possible policy options, it is useful to know the relative radiative effect (and, hence, potential climate effect) of equal emissions of each of the greenhouse gases. The concept of relative **Global Warming Potentials (GWP)** has been developed to take into account the differing times that gases remain in the atmosphere.

This index defines the time-integrated warming effect due to an instantaneous release of unit mass (1 kg) of a given greenhouse gas in today's atmosphere, relative to that of carbon dioxide. The relative importances will change in the future as atmospheric composition changes because, although radiative forcing increases in direct proportion to the concentration of CFCs, changes in the other greenhouse gases (particularly carbon dioxide) have an effect on forcing which is much less than proportional.

The GWPs in Table 3 are shown for three time horizons, reflecting the need to consider the cumulative effects on climate over various time scales. The longer time horizon is appropriate for the cumulative effect; the shorter timescale will indicate the response to emission changes in the short term. There are a number of practical difficulties in devising and calculating the values of the GWPs, and the values given here should be considered as preliminary. In addition to these direct effects, there are indirect effects of human-made emissions arising from chemical reactions between the various constituents. The indirect effects on stratospheric water vapour, carbon dioxide and tropospheric ozone have been included in these estimates.

Table 3 indicates, for example, that the effectiveness of methane in influencing climate will be greater in the first few decades after release, whereas emission of the longer-lived nitrous oxide will affect climate for a much longer time. The lifetimes of the proposed CFC replacements range from 1 to 40 years; the longer lived replacements are

per-kilogram basis, the equivalent values are 58 for methane and about 4,000 for CFC-11, both relative to carbon dioxide. Values for other greenhouse gases are to be found in Section 2.

The total radiative forcing at any time is the sum of those from the individual greenhouse gases. We show in Figure 6

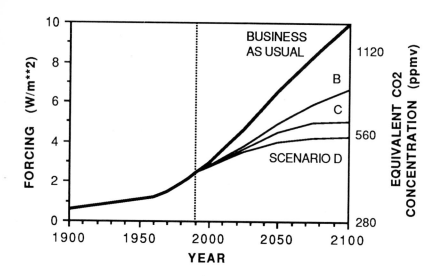

Figure 6: Increase in radiative forcing since the mid-18th century, and predicted to result from the four IPCC emissions scenarios, also expressed as equivalent carbon dioxide concentrations.

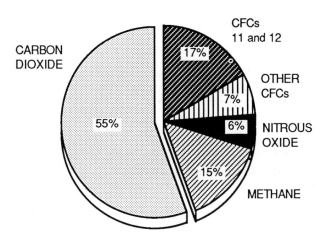

Figure 7: The contribution from each of the human-made greenhouse gases to the change in radiative forcing from 1980 to 1990. The contribution from ozone may also be significant, but cannot be quantified at present.

still potentially effective as agents of climate change. One example of this, HCFC-22 (with a 15 year lifetime), has a similar effect (when released in the same amount) as CFC-11 on a 20 year time-scale; but less over a 500 year time-scale.

Table 3 shows carbon dioxide to be the least effective greenhouse gas per kilogramme emitted, but its contribution to global warming, which depends on the product of the GWP and the amount emitted, is largest. In the example in Table 4, the effect over 100 years of emissions

of greenhouse gases in 1990 are shown relative to carbon dioxide. This is illustrative; to compare the effect of different emission projections we have to sum the effect of emissions made in future years.

There are other technical criteria which may help policymakers to decide, in the event of emissions reductions being deemed necessary, which gases should be considered. Does the gas contribute in a major way to current, and future, climate forcing? Does it have a long lifetime, so earlier reductions in emissions would be more effective than those made later? And are its sources and sinks well enough known to decide which could be controlled in practice? Table 5 illustrates these factors.

How much do we expect climate to change?

It is relatively easy to determine the direct effect of the increased radiative forcing due to increases in greenhouse gases. However, as climate begins to warm, various processes act to amplify (through positive feedbacks) or reduce (through negative feedbacks) the warming. The main feedbacks which have been identified are due to changes in water vapour, sea-ice, clouds and the oceans.

The best tools we have which take the above feedbacks into account (but do not include greenhouse gas feedbacks) are three-dimensional mathematical models of the climate system (atmosphere-ocean-ice-land), known as General Circulation Models (GCMs). They synthesise our knowledge of the physical and dynamical processes in the overall system and allow for the complex interactions between the various components. However, in their current state of development, the descriptions of many of the processes involved are comparatively crude. Because of this, considerable uncertainty is attached to these

Table 3 *Global Warming Potentials. The warming effect of an emission of 1kg of each gas relative to that of CO$_2$*
These figures are best estimates calculated on the basis of the present day atmospheric composition

	TIME HORIZON		
	20 yr	100 yr	500 yr
Carbon dioxide	1	1	1
Methane (including indirect)	63	21	9
Nitrous oxide	270	290	190
CFC-11	4500	3500	1500
CFC-12	7100	7300	4500
HCFC-22	4100	1500	510

Global Warming Potentials for a range of CFCs and potential replacements are given in the full text.

Table 4 *The Relative Cumulative Climate Effect of 1990 Man-Made Emissions*

	GWP (100yr horizon)	1990 emissions (Tg)	Relative contribution over 100yr
Carbon dioxide	1	26000†	61%
Methane*	21	300	15%
Nitrous oxide	290	6	4%
CFCs	Various	0.9	11%
HCFC-22	1500	0.1	0.5%
Others*	Various		8.5%

* These values include the indirect effect of these emissions on other greenhouse gases via chemical reactions in the atmosphere. Such estimates are highly model dependent and should be considered preliminary and subject to change. The estimated effect of ozone is included under "others". The gases included under "others" are given in the full report.
† 26 000 Tg (teragrams) of carbon dioxide = 7 000 Tg (=7 Gt) of carbon

Table 5 *Characteristics of Greenhouse Gases*

GAS	MAJOR CONTRIBUTOR?	LONG LIFETIME?	SOURCES KNOWN?
Carbon dioxide	yes	yes	yes
Methane	yes	no	semi-quantitatively
Nitrous oxide	not at present	yes	qualitatively
CFCs	yes	yes	yes
HCFCs, etc	not at present	mainly no	yes
Ozone	possibly	no	qualitatively

predictions of climate change, which is reflected in the range of values given; further details are given in a later section.

The estimates of climate change presented here are based on

i)   the "best-estimate" of equilibrium climate sensitivity (i.e the equilibrium temperature change due to a doubling of carbon dioxide in the atmosphere) obtained from model simulations, feedback analyses and observational considerations (see later box: "What tools do we use?")

ii) a "box-diffusion-upwelling" ocean-atmosphere climate model which translates the greenhouse forcing into the evolution of the temperature response for the prescribed climate sensitivity. (This simple model has been calibrated against more complex atm-osphere-ocean coupled GCMs for situations where the more complex models have been run).

How quickly will global climate change?

a. If emissions follow a Business-as-Usual pattern

Under the IPCC Business-as-Usual (Scenario A) emissions of greenhouse gases, the average rate of increase of global mean temperature during the next century is estimated to be about 0.3°C per decade (with an uncertainty range of 0.2°C to 0.5°C). This will result in a likely increase in global mean temperature of about 1°C above the present value (about 2°C above that in the pre-industrial period) by 2025 and 3°C above today's (about 4°C above pre-industrial) before the end of the next century.

The projected temperature rise out to the year 2100, with high, low and best-estimate climate responses, is shown in Figure 8. Because of other factors which influence climate, we would not expect the rise to be a steady one.

The temperature rises shown above are **realised** temperatures; at any time we would also be **committed** to a further temperature rise toward the equilibrium temperature (see box: "Equilibrium and Realised Climate Change"). For the BaU "best-estimate" case in the year 2030, for example, a further 0.9°C rise would be expected, about 0.2°C of which would be realised by 2050 (in addition to changes due to further greenhouse gas increases); the rest would become apparent in decades or centuries.

Even if we were able to stabilise emissions of each of the greenhouse gases at present day levels from now on, the temperature is predicted to rise by about 0.2°C per decade for the first few decades.

The global warming will also lead to increased global average precipitation and evaporation of a few percent by 2030. Areas of sea-ice and snow are expected to diminish.

b. If emissions are subject to controls

Under the other IPCC emission scenarios which assume progressively increasing levels of controls, average rates of increase in global mean temperature over the next century are estimated to be about 0.2°C per decade (Scenario B), just above 0.1°C per decade (Scenario C) and about 0.1°C per decade (Scenario D). The results are illustrated in Figure 9, with the Business-as-Usual case shown for comparison. Only the best-estimate of the temperature rise is shown in each case.

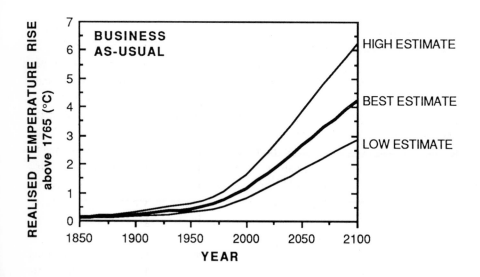

Figure 8: Simulation of the increase in global mean temperature from 1850-1990 due to observed increases in greenhouse gases, and predictions of the rise between 1990 and 2100 resulting from the Business-as-Usual emissions.

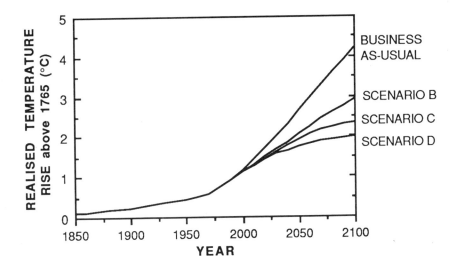

Figure 9: Simulations of the increase in global mean temperature from 1850-1990 due to observed increases in greenhouse gases, and predictions of the rise between 1990 and 2100 resulting from the IPCC Scenario B,C and D emissions, with the Business-as-Usual case for comparison.

The indicated range of uncertainty in global temperature rise given above reflects a subjective assessment of uncertainties in the calculation of climate response, but does not include those due to the transformation of emissions to concentrations, nor the effects of greenhouse gas feedbacks.

What will be the patterns of climate change by 2030?

Knowledge of the global mean warming and change in precipitation is of limited use in determining the impacts of climate change, for instance on agriculture. For this we need to know changes regionally and seasonally.

Models predict that surface air will warm faster over land than over oceans, and a minimum of warming will occur around Antarctica and in the northern North Atlantic region.

There are some continental-scale changes which are consistently predicted by the highest resolution models and for which we understand the physical reasons. The warming is predicted to be 50-100% greater than the global mean in high northern latitudes in winter, and substantially smaller than the global mean in regions of sea-ice in summer. Precipitation is predicted to increase on average in middle and high latitude continents in winter (by some 5 - 10% over 35-55°N).

Five regions, each a few million square kilometres in area and representative of different climatological regimes, were selected by IPCC for particular study (see Figure 10). In the box (over page) are given the changes in temperature, precipitation and soil moisture, which are predicted to occur by 2030 on the Business-as-Usual scenario, as an average over each of the five regions. There

may be considerable variations within the regions. In general, confidence in these regional estimates is low, especially for the changes in precipitation and soil moisture, but they are examples of our best estimates. We cannot yet give reliable regional predictions at the smaller scales demanded for impacts assessments.

How will climate extremes and extreme events change?

Changes in the variability of weather and the frequency of extremes will generally have more impact than changes in the mean climate at a particular location. With the possible exception of an increase in the number of intense showers, there is no clear evidence that weather variability will change in the future. In the case of temperatures, assuming no change in variability, but with a modest increase in the mean, the number of days with temperatures above a given value at the high end of the distribution will increase substantially. On the same assumptions, there will be a decrease in days with temperatures at the low end of the distribution. So the number of very hot days or frosty nights can be substantially changed without any change in the variability of the weather. The number of days with a minimum threshold amount of soil moisture (for viability of a certain crop, for example) would be even more sensitive to changes in average precipitation and evaporation.

If the large-scale weather regimes, for instance depression tracks or anticyclones, shift their position, this would effect the variability and extremes of weather at a particular location, and could have a major effect. However, we do not know if, or in what way, this will happen.

ESTIMATES FOR CHANGES BY 2030

(IPCC Business-as-Usual scenario; changes from **pre-industrial**)

The numbers given below are based on high resolution models, scaled to be consistent with our best estimate of global mean warming of 1.8°C by 2030. For values consistent with other estimates of global temperature rise, the numbers below should be reduced by 30% for the low estimate or increased by 50% for the high estimate. Precipitation estimates are also scaled in a similar way.

Confidence in these regional estimates is low

Central North America (35°-50°N 85°-105°W)

The warming varies from 2 to 4°C in winter and 2 to 3°C in summer. Precipitation increases range from 0 to 15% in winter whereas there are decreases of 5 to 10% in summer. Soil moisture decreases in summer by 15 to 20%.

Southern Asia (5°-30°N 70°-105°E)

The warming varies from 1 to 2°C throughout the year. Precipitation changes little in winter and generally increases throughout the region by 5 to 15% in summer. Summer soil moisture increases by 5 to 10%.

Sahel (10°-20°N 20°W-40°E)

The warming ranges from 1 to 3°C. Area mean precipitation increases and area mean soil moisture decreases marginally in summer. However, throughout the region, there are areas of both increase and decrease in both parameters throughout the region.

Southern Europe (35°-50°N 10°W- 45°E)

The warming is about 2°C in winter and varies from 2 to 3°C in summer. There is some indication of increased precipitation in winter, but summer precipitation decreases by 5 to 15%, and summer soil moisture by 15 to 25%.

Australia (12°-45°S 110°-115°E)

The warming ranges from 1 to 2°C in summer and is about 2°C in winter. Summer precipitation increases by around 10%, but the models do not produce consistent estimates of the changes in soil moisture. The area averages hide large variations at the sub-continental level.

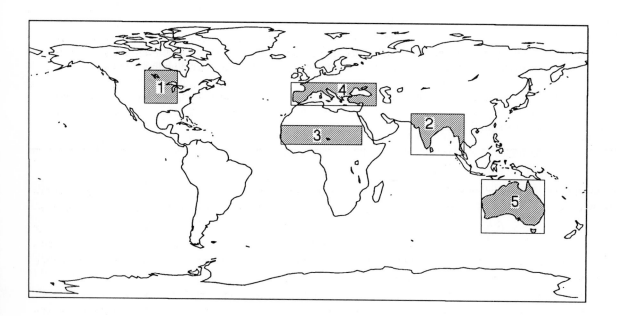

Figure 10: Map showing the locations and extents of the five areas selected by IPCC

WHAT TOOLS DO WE USE TO PREDICT FUTURE CLIMATE, AND HOW DO WE USE THEM?

The most highly developed tool which we have to predict future climate is known as a **general circulation model or GCM**. These models are based on the laws of physics and use descriptions in simplified physical terms (called parameterisations) of the smaller-scale processes such as those due to clouds and deep mixing in the ocean. In a climate model an atmospheric component, essentially the same as a weather prediction model, is coupled to a model of the ocean, which can be equally complex.

Climate forecasts are derived in a different way from weather forecasts. A weather prediction model gives a description of the atmosphere's state up to 10 days or so ahead, starting from a detailed description of an initial state of the atmosphere at a given time. Such forecasts describe the movement and development of large weather systems, though they cannot represent very small scale phenomena; for example, individual shower clouds.

To make a climate forecast, the climate model is first run for a few (simulated) decades. The statistics of the model's output is a description of the model's simulated climate which, if the model is a good one, will bear a close resemblance to the climate of the real atmosphere and ocean. The above exercise is then repeated with increasing concentrations of the greenhouse gases in the model. The differences between the statistics of the two simulations (for example in mean temperature and interannual variability) provide an estimate of the accompanying climate change.

The long term change in **surface air temperature** following a doubling of carbon dioxide (referred to as the **climate sensitivity**) is generally used as a benchmark to compare models. The range of results from model studies is 1.9 to 5.2°C. Most results are close to 4.0°C but recent studies using a more detailed but not necessarily more accurate representation of cloud processes give results in the lower half of this range. Hence the models results do not justify altering the previously accepted range of 1.5 to 4.5°C.

Although scientists are reluctant to give a single best estimate in this range, it is necessary for the presentation of climate predictions for a choice of best estimate to be made. Taking into account the model results, together with observational evidence over the last century which is suggestive of the climate sensitivity being in the lower half of the range, (see section: "Has man already begun to change global climate?") a value of climate sensitivity of 2.5°C has been chosen as the best estimate. Further details are given in Section 5 of the report.

In this Assessment, we have also used much simpler models, which simulate the behaviour of GCMs, to make predictions of the evolution with time of global temperature from a number of emission scenarios. These so-called box-diffusion models contain highly simplified physics but give similar results to GCMs when globally averaged.

A completely different, and potentially useful, way of predicting patterns of future climate is to search for periods in the past when the global mean temperatures were similar to those we expect in future, and then use the past spatial patterns as **analogues** of those which will arise in the future. For a good analogue, it is also necessary for the forcing factors (for example, greenhouse gases, orbital variations) and other conditions (for example, ice cover, topography, etc.) to be similar; direct comparisons with climate situations for which these conditions do not apply cannot be easily interpreted. Analogues of future greenhouse-gas-changed climates have not been found.

We cannot therefore advocate the use of palaeo-climates as predictions of regional climate change due to future increases in greenhouse gases. However, palaeo-climatological information can provide useful insights into climate processes, and can assist in the validation of climate models.

Will storms increase in a warmer world?

Storms can have a major impact on society. Will their frequency, intensity or location increase in a warmer world?

Tropical storms, such as typhoons and hurricanes, only develop at present over seas that are warmer than about 26°C. Although the area of sea having temperatures over this critical value will increase as the globe warms, the critical temperature itself may increase in a warmer world.

Although the theoretical maximum intensity is expected to increase with temperature, climate models give no consistent indication whether tropical storms will increase or decrease in frequency or intensity as climate changes; neither is there any evidence that this has occurred over the past few decades.

Mid-latitude storms, such as those which track across the North Atlantic and North Pacific, are driven by the equator-to-pole temperature contrast. As this contrast will

EQUILIBRIUM AND REALISED CLIMATE CHANGE

When the radiative forcing on the earth-atmosphere system is changed, for example by increasing greenhouse gas concentrations, the atmosphere will try to respond (by warming) immediately. But the atmosphere is closely coupled to the oceans, so in order for the air to be warmed by the greenhouse effect, the oceans also have to be warmed; because of their thermal capacity this takes decades or centuries. This exchange of heat between atmosphere and ocean will act to slow down the temperature rise forced by the greenhouse effect.

In a hypothetical example where the concentration of greenhouse gases in the atmosphere, following a period of constancy, rises suddenly to a new level and remains there, the radiative forcing would also rise rapidly to a new level. This increased radiative forcing would cause the atmosphere and oceans to warm, and eventually come to a new, stable, temperature. A commitment to this **equilibrium** temperature rise is incurred as soon as the greenhouse gas concentration changes. But at any time before equilibrium is reached, the actual temperature will have risen by only part of the equilibrium temperature change, known as the **realised** temperature change.

Models predict that, for the present day case of an increase in radiative forcing which is approximately steady, the realised temperature rise at any time is about 50% of the committed temperature rise if the climate sensitivity (the response to a doubling of carbon dioxide) is 4.5°C and about 80% if the climate sensitivity is 1.5°C. If the forcing were then held constant, temperatures would continue to rise slowly, but it is not certain whether it would take decades or centuries for most of the remaining rise to equilibrium to occur.

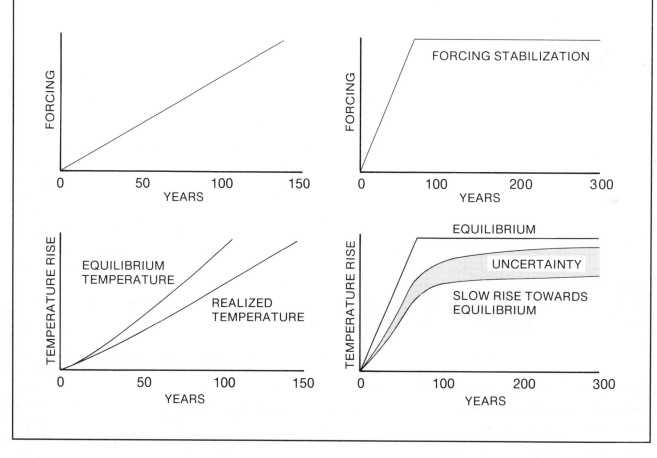

probably be weakened in a warmer world (at least in the Northern Hemisphere), it might be argued that mid-latitude storms will also weaken or change their tracks, and there is some indication of a general reduction in day-to-day variability in the mid-latitude storm tracks in winter in model simulations, though the pattern of changes vary from model to model. Present models do not resolve smaller-scale disturbances, so it will not be possible to assess

changes in storminess until results from higher resolution models become available in the next few years.

Climate change in the longer term
The foregoing calculations have focussed on the period up to the year 2100; it is clearly more difficult to make calculations for years beyond 2100. However, while the timing of a predicted increase in global temperatures has

substantial uncertainties, the prediction that an increase will eventually occur is more certain. Furthermore, some model calculations that have been extended beyond 100 years suggest that, with continued increases in greenhouse climate forcing, there could be significant changes in the ocean circulation, including a decrease in North Atlantic deep water formation.

Other factors which could influence future climate

Variations in the output of **solar energy** may also affect climate. On a decadal time-scale solar variability and changes in greenhouse gas concentration could give changes of similar magnitudes. However the variation in solar intensity changes sign so that over longer time-scales the increases in greenhouse gases are likely to be more important. **Aerosols** as a result of volcanic eruptions can lead to a cooling at the surface which may oppose the greenhouse warming for a few years following an eruption. Again, over longer periods the greenhouse warming is likely to dominate.

Human activity is leading to an increase in aerosols in the lower atmosphere, mainly from sulphur emissions. These have two effects, both of which are difficult to quantify but which may be significant particularly at the regional level. The first is the direct effect of the aerosols on the radiation scattered and absorbed by the atmosphere. The second is an indirect effect whereby the aerosols affect the microphysics of clouds leading to an increased cloud reflectivity. Both these effects might lead to a significant regional cooling; a decrease in emissions of sulphur might be expected to increase global temperatures.

Because of long-period couplings between different components of the climate system, for example between ocean and atmosphere, the Earth's climate would still vary without being perturbed by any external influences. This **natural variability** could act to add to, or subtract from, any human-made warming; on a century time-scale this would be less than changes expected from greenhouse gas increases.

How much confidence do we have in our predictions?

Uncertainties in the above climate predictions arise from our imperfect knowledge of: ·

* future rates of human-made emissions
* how these will change the atmospheric concentrations of greenhouse gases
* the response of climate to these changed concentrations

Firstly, it is obvious that the extent to which climate will change depends on the rate at which greenhouse gases (and other gases which affect their concentrations) are emitted. This in turn will be determined by various complex economic and sociological factors. Scenarios of future emissions were generated within IPCC WGIII and are described in the Annex to this Summary.

Secondly, because we do not fully understand the sources and sinks of the greenhouse gases, there are uncertainties in our calculations of future concentrations arising from a given emissions scenario. We have used a number of models to calculate concentrations and chosen a best estimate for each gas. In the case of carbon dioxide, for example, the concentration increase between 1990 and 2070 due to the Business-as-Usual emissions scenario spanned almost a factor of two between the highest and lowest model result (corresponding to a range in radiative forcing change of about 50%).

Furthermore, because natural sources and sinks of greenhouse gases are sensitive to a change in climate, they may substantially modify future concentrations (see earlier section: "Greenhouse gas feedbacks"). It appears that, as climate warms, these feedbacks will lead to an overall increase, rather than decrease, in natural greenhouse gas abundances. For this reason, climate change is likely to be greater than the estimates we have given.

Thirdly, climate models are only as good as our understanding of the processes which they describe, and this is far from perfect. The ranges in the climate predictions given above reflect the uncertainties due to model imperfections; the largest of these is cloud feedback (those factors affecting the cloud amount and distribution and the interaction of clouds with solar and terrestrial radiation), which leads to a factor of two uncertainty in the size of the warming. Others arise from the transfer of energy between the atmosphere and ocean, the atmosphere and land surfaces, and between the upper and deep layers of the ocean. The treatment of sea-ice and convection in the models is also crude. Nevertheless, for reasons given in the box overleaf, we have substantial confidence that models can predict at least the broad-scale features of climate change.

Furthermore, we must recognise that our imperfect understanding of climate processes (and corresponding ability to model them) could make us vulnerable to surprises; just as the human-made ozone hole over Antarctica was entirely unpredicted. In particular, the ocean circulation, changes in which are thought to have led to periods of comparatively rapid climate change at the end of the last ice age, is not well observed, understood or modelled.

Will the climate of the future be very different?

When considering future climate change, it is clearly essential to look at the record of climate variation in the past. From it we can learn about the range of natural climate variability, to see how it compares with what we

---

CONFIDENCE IN PREDICTIONS FROM CLIMATE MODELS

What confidence can we have that climate change due to increasing greenhouse gases will look anything like the model predictions? Weather forecasts can be compared with the actual weather the next day and their skill assessed; we cannot do that with climate predictions. However, there are several indicators that give us some confidence in the predictions from climate models.

When the latest atmospheric models are run with the present atmospheric concentrations of greenhouse gases and observed boundary conditions their simulation of present climate is generally realistic on large scales, capturing the major features such as the wet tropical convergence zones and mid-latitude depression belts, as well as the contrasts between summer and winter circulations. The models also simulate the observed variability; for example, the large day-to-day pressure variations in the middle latitude depression belts and the maxima in interannual variability responsible for the very different character of one winter from another both being represented. However, on regional scales (2,000km or less), there are significant errors in all models.

Overall confidence is increased by atmospheric models' generally satisfactory portrayal of aspects of variability of the atmosphere, for instance those associated with variations in sea surface temperature. There has been some success in simulating the general circulation of the ocean, including the patterns (though not always the intensities) of the principal currents, and the distributions of tracers added to the ocean.

Atmospheric models have been coupled with simple models of the ocean to predict the equilibrium response to greenhouse gases, under the assumption that the model errors are the same in a changed climate. The ability of such models to simulate important aspects of the climate of the last ice age generates confidence in their usefulness. Atmospheric models have also been coupled with multi-layer ocean models (to give coupled ocean-atmosphere GCMs) which predict the gradual response to increasing greenhouse gases. Although the models so far are of relatively coarse resolution, the large-scale structures of the ocean and the atmosphere can be simulated with some skill. However, the coupling of ocean and atmosphere models reveals a strong sensitivity to small-scale errors which leads to a drift away from the observed climate. As yet, these errors must be removed by adjustments to the exchange of heat between ocean and atmosphere. There are similarities between results from the coupled models using simple representations of the ocean and those using more sophisticated descriptions, and our understanding of such differences as do occur gives us some confidence in the results.

---

expect in the future, and also look for evidence of recent climate change due to man's activities.

Climate varies naturally on all time-scales from hundreds of millions of years down to the year-to-year. Prominent in the Earth's history have been the 100,000 year glacial-interglacial cycles when climate was mostly cooler than at present. Global surface temperatures have typically varied by 5-7°C through these cycles, with large changes in ice volume and sea level, and temperature changes as great as 10-15°C in some middle and high latitude regions of the Northern Hemisphere. Since the end of the last ice age, about 10,000 years ago, global surface temperatures have probably fluctuated by little more than 1°C. Some fluctuations have lasted several centuries, including the Little Ice Age which ended in the nineteenth century and which appears to have been global in extent.

The changes predicted to occur by about the middle of the next century due to increases in greenhouse gas concentrations from the Business-as-Usual emissions will make global mean temperatures higher than they have been in the last 150,000 years.

The **rate of change** of global temperatures predicted for Business-as-Usual emissions will be greater than those which have occured naturally on Earth over the last 10,000 years, and the rise in sea level will be about three to six times faster than that seen over the last 100 years or so.

Has man already begun to change the global climate?

The instrumental record of **surface temperature** is fragmentary until the mid-nineteenth century, after which it slowly improves. Because of different methods of measurement, historical records have to be harmonised with modern observations, introducing some uncertainty. Despite these problems we believe that a real warming of the globe of 0.3°C - 0.6°C has taken place over the last century; any bias due to urbanisation is likely to be less than 0.05°C.

Moreover since 1900 similar temperature increases are seen in three independent data sets: one collected over land and two over the oceans. Figure 11 shows current estimates of smoothed global-mean surface temperature over land and ocean since 1860. Confidence in the record has been increased by their similarity to recent satellite measurements of mid-tropospheric temperatures.

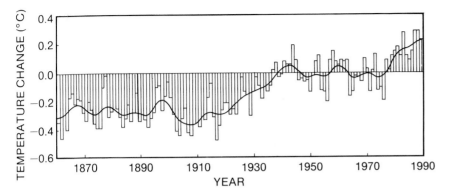

Figure 11: Global-mean combined land-air and sea-surface temperatures, 1861 - 1989, relative to the average for 1951-80.

Although the overall temperature rise has been broadly similar in both hemispheres, it has not been steady, and differences in their rates of warming have sometimes persisted for decades. Much of the warming since 1900 has been concentrated in two periods, the first between about 1910 and 1940 and the other since 1975; the five warmest years on record have all been in the 1980s. The Northern Hemisphere cooled between the 1940s and the early 1970s when Southern Hemisphere temperatures stayed nearly constant. The pattern of global warming since 1975 has been uneven with some regions, mainly in the northern hemisphere, continuing to cool until recently. This regional diversity indicates that future regional temperature changes are likely to differ considerably from a global average.

The conclusion that global temperature has been rising is strongly supported by the retreat of most **mountain glaciers** of the world since the end of the nineteenth century and the fact that global **sea level** has risen over the same period by an average of 1 to 2mm per year. Estimates of thermal expansion of the oceans, and of increased melting of mountain glaciers and the ice margin in West Greenland over the last century, show that the major part of the sea level rise appears to be related to the observed global warming. This apparent connection between observed sea level rise and global warming provides grounds for believing that future warming will lead to an acceleration in sea level rise.

The size of the warming over the last century is broadly consistent with the predictions of climate models, but is also of the same magnitude as natural climate variability. If the sole cause of the observed warming were the human-made greenhouse effect, then the implied climate sensitivity would be near the lower end of the range inferred from the models. The observed increase could be largely due to natural variability; alternatively this variability and other man-made factors could have offset a still larger man-made greenhouse warming. The unequivocal detection of the enhanced greenhouse effect from observations is not likely for a decade or more, when the committment to future climate change will then be considerably larger than it is today.

Global-mean temperature alone is an inadequate indicator of greenhouse-gas-induced climatic change. Identifying the causes of any global-mean temperature change requires examination of other aspects of the changing climate, particularly its spatial and temporal characteristics - the man-made climate change "signal". Patterns of climate change from models such as the Northern Hemisphere warming faster than the Southern Hemisphere, and surface air warming faster over land than over oceans, are not apparent in observations to date. However, we do not yet know what the detailed "signal" looks like because we have limited confidence in our predictions of climate change patterns. Furthermore, any changes to date could be masked by natural variability and other (possibly man-made) factors, and we do not have a clear picture of these.

How much will sea level rise ?

Simple models were used to calculate the rise in sea level to the year 2100; the results are illustrated below. The calculations necessarily ignore any long-term changes, unrelated to greenhouse forcing, that may be occurring but cannot be detected from the present data on land ice and the ocean. The sea level rise expected from 1990-2100 under the IPCC Business-as-Usual emissions scenario is shown in Figure 12. An average rate of global mean sea level rise of

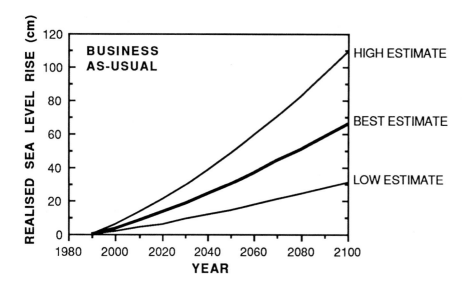

Figure 12: Sea level rise predicted to result from Business-as-Usual emissions, showing the best-estimate and range.

about 6cm per decade over the next century (with an uncertainty range of 3 - 10cm per decade). The predicted rise is about 20cm in global mean sea level by 2030, and 65cm by the end of the next century. There will be significant regional variations.

The best estimate in each case is made up mainly of positive contributions from thermal expansion of the oceans and the melting of glaciers. Although, over the next 100 years, the effect of the Antarctic and Greenland ice sheets is expected to be small, they make a major contribution to the uncertainty in predictions.

Even if greenhouse forcing increased no further, there would still be a commitment to a continuing sea level rise for many decades and even centuries, due to delays in climate, ocean and ice mass responses. As an illustration, if the increases in greenhouse gas concentrations were to suddenly stop in 2030, sea level would go on rising from 2030 to 2100, by as much again as from 1990-2030, as shown in Figure 13.

Predicted sea level rises due to the other three emissions scenarios are shown in Figure 14, with the Business-as-Usual case for comparison; only best-estimate calculations are shown.

The West Antarctic Ice Sheet is of special concern. A large portion of it, containing an amount of ice equivalent to about 5m of global sea level, is grounded far below sea level. There have been suggestions that a sudden outflow of ice might result from global warming and raise sea level quickly and substantially. Recent studies have shown that individual ice streams are changing rapidly on a decade-to-century time-scale; however this is not necessarily related to climate change. Within the next century, it is not likely

that there will be a major outflow of ice from West Antarctica due directly to global warming.

Any rise in sea level is not expected to be uniform over the globe. Thermal expansion, changes in ocean circulation, and surface air pressure will vary from region to region as the world warms, but in an as yet unknown way. Such regional details await further development of more realistic coupled ocean-atmosphere models. In addition, vertical land movements can be as large or even larger than changes in global mean sea level; these movements have to be taken into account when predicting local change in sea level relative to land.

The most severe effects of sea level rise are likely to result from extreme events (for example, storm surges) the incidence of which may be affected by climatic change.

What will be the effect of climate change on ecosystems?

Ecosystem processes such as photosynthesis and respiration are dependent on climatic factors and carbon dioxide concentration in the short term. In the longer term, climate and carbon dioxide are among the factors which control ecosystem structure, i.e., species composition, either directly by increasing mortality in poorly adapted species, or indirectly by mediating the competition between species. Ecosystems will respond to local changes in temperature (including its rate of change), precipitation, soil moisture and extreme events. Current models are unable to make reliable estimates of changes in these parameters on the required local scales.

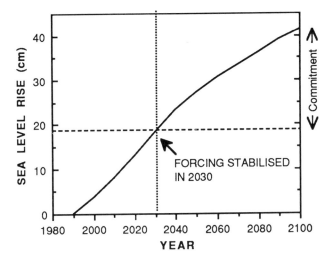

Figure 13: Commitment to sea level rise in the year 2030. The curve shows the sea level rise due to Business-as-Usual emissions to 2030, with the additional rise that would occur in the remainder of the century even if climate forcing was stabilised in 2030.

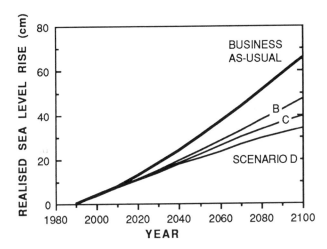

Figure 14: Model estimates of sea-level rise from 1990-2100 due to all four emissions scenarios.

Photosynthesis captures atmospheric carbon dioxide, water and solar energy and stores them in organic compounds which are then used for subsequent plant growth, the growth of animals or the growth of microbes in the soil. All of these organisms release carbon dioxide via respiration into the atmosphere. Most land plants have a system of photosynthesis which will respond positively to increased atmospheric carbon dioxide ("the carbon dioxide fertilization effect") but the response varies with species. The effect may decrease with time when restricted by other ecological limitations, for example, nutrient availability. It should be emphasized that the carbon content of the terrestrial biosphere will increase only if the forest

ecosystems in a state of maturity will be able to store more carbon in a warmer climate and at higher concentrations of carbon dioxide. We do not yet know if this is the case.

The response to increased carbon dioxide results in greater efficiencies of water, light and nitrogen use. These increased efficiencies may be particularly important during drought and in arid/semi-arid and infertile areas.

Because species respond differently to climatic change, some will increase in abundance and/or range while others will decrease. Ecosystems will therefore change in structure and composition. Some species may be displaced to higher latitudes and altitudes, and may be more prone to local, and possibly even global, extinction; other species may thrive.

As stated above, ecosystem structure and species distribution are particularly sensitive to the rate of change of climate. We can deduce something about how quickly global temperature has changed in the past from palaeo-climatological records. As an example, at the end of the last glaciation, within about a century, temperature increased by up to 5°C in the North Atlantic region, mainly in Western Europe. Although during the increase from the glacial to the current interglacial temperature simple tundra ecosystems responded positively, a similar rapid temperature increase applied to more developed ecosystems could result in their instability.

What should be done to reduce uncertainties, and how long will this take?

Although we can say that some climate change is unavoidable, much uncertainty exists in the prediction of global climate properties such as the temperature and rainfall. Even greater uncertainty exists in predictions of regional climate change, and the subsequent consequences for sea level and ecosystems. The key areas of scientific uncertainty are:

- **clouds:** primarily cloud formation, dissipation, and radiative properties, which influence the response of the atmosphere to greenhouse forcing;
- **oceans:** the exchange of energy between the ocean and the atmosphere, between the upper layers of the ocean and the deep ocean, and transport within the ocean, all of which control the rate of global climate change and the patterns of regional change;
- **greenhouse gases:** quantification of the uptake and release of the greenhouse gases, their chemical reactions in the atmosphere, and how these may be influenced by climate change;
- **polar ice sheets:** which affect predictions of sea level rise.

Studies of land surface hydrology, and of impact on ecosystems, are also important.

DEFORESTATION AND REFORESTATION

Man has been deforesting the Earth for millennia. Until the early part of the century, this was mainly in temperate regions, more recently it has been concentrated in the tropics. Deforestation has several potential impacts on climate: through the carbon and nitrogen cycles (where it can lead to changes in atmospheric carbon dioxide concentrations), through the change in reflectivity of terrain when forests are cleared, through its effect on the hydrological cycle (precipitation, evaporation and runoff) and surface roughness and thus atmospheric circulation which can produce remote effects on climate.

It is estimated that each year about 2 Gt of carbon (GtC) is released to the atmosphere due to tropical deforestation. The rate of forest clearing is difficult to estimate; probably until the mid-20th century, temperate deforestation and the loss of organic matter from soils was a more important contributor to atmospheric carbon dioxide than was the burning of fossil fuels. Since then, fossil fuels have become dominant; one estimate is that around 1980, 1.6 GtC was being released annually from the clearing of tropical forests, compared with about 5 GtC from the burning of fossil fuels. If all the tropical forests were removed, the input is variously estimated at from 150 to 240 GtC; this would increase atmospheric carbon dioxide by 35 to 60 ppmv.

To analyse the effect of reforestation we assume that 10 million hectares of forests are planted each year for a period of 40 years, i.e., 4 million km^2 would then have been planted by 2030, at which time 1 GtC would be absorbed annually until these forests reach maturity. This would happen in 40-100 years for most forests. The above scenario implies an accumulated uptake of about 20 GtC by the year 2030 and up to 80 GtC after 100 years. This accumulation of carbon in forests is equivalent to some 5-10% of the emission due to fossil fuel burning in the Business-as-Usual scenario.

Deforestation can also alter climate directly by increasing reflectivity and decreasing evapotranspiration. Experiments with climate models predict that replacing all the forests of the Amazon Basin by grassland would reduce the rainfall over the basin by about 20%, and increase mean temperature by several degrees.

To reduce the current scientific uncertainties in each of these areas will require internationally coordinated research, the goal of which is to improve our capability to observe, model and understand the global climate system. Such a program of research will reduce the scientific uncertainties and assist in the formulation of sound national and international response strategies.

Systematic long-term **observations** of the system are of vital importance for understanding the natural variability of the Earth's climate system, detecting whether man's activities are changing it, parameterising key processes for models, and verifying model simulations. Increased accuracy and coverage in many observations are required. Associated with expanded observations is the need to develop appropriate comprehensive global information bases for the rapid and efficient dissemination and utilization of data. The main observational requirements are:

i)   the maintenance and improvement of observations (such as those from satellites) provided by the World Weather Watch Programme of WMO.

ii) the maintenance and enhancement of a programme of monitoring, both from satellite-based and surface-based instruments, of key climate elements for which accurate observations on a continuous basis are required, such as the distribution of important atmospheric constituents, clouds, the Earth's radiation budget, precipitation, winds, sea surface temperatures and terrestrial ecosystem extent, type and productivity.

iii) the establishment of a global ocean observing system to measure changes in such variables as ocean surface topography, circulation, transport of heat and chemicals, and sea-ice extent and thickness.

iv) the development of major new systems to obtain data on the oceans, atmosphere and terrestrial ecosystems using both satellite-based instruments and instruments based on the surface, on automated instrumented vehicles in the ocean, on floating and deep sea buoys, and on aircraft and balloons.

v)   the use of palaeo-climatological and historical instrumental records to document natural variability and changes in the climate system, and subsequent environmental response.

The **modelling** of climate change requires the development of global models which couple together atmosphere, land, ocean and ice models and which incorporate more realistic formulations of the relevant processes and the interactions between the different components. Processes in the biosphere (both on land and in the ocean) also need to be included. Higher spatial resolution than is currently generally used is required if regional patterns are to be predicted. These models will require the largest computers which are planned to be available during the next decades.

Understanding of the climate system will be developed from analyses of observations and of the results from model simulations. In addition, detailed studies of particular processes will be required through targetted observational campaigns. Examples of such field campaigns include combined observational and small-scale modelling studies for different regions, of the formation, dissipation, radiative, dynamical and microphysical properties of clouds, and ground-based (ocean and land) and aircraft measurements of the fluxes of greenhouse gases from specific ecosystems. In particular, emphasis must be placed on field experiments that will assist in the development and improvement of sub grid-scale parametrizations for models.

The required program of research will require unprecedented international cooperation, with the World Climate Research Programme (WCRP) of the World Meteorological Organization and International Council of Scientific Unions (ICSU), and the International Geosphere-Biosphere Programme (IGBP) of ICSU both playing vital roles. These are large and complex endeavours that will require the involvement of all nations, particularly the developing countries. Implementation of existing and planned projects will require increased financial and human resources; the latter requirement has immediate implications at all levels of education, and the international community of scientists needs to be widened to include more members from developing countries.

The WCRP and IGBP have a number of ongoing or planned research programmes, that address each of the three key areas of scientific uncertainty. Examples include:

- **clouds**:
  International Satellite Cloud Climatology Project (ISCCP);
  Global Energy and Water Cycle Experiment (GEWEX).
- **oceans**:
  World Ocean Circulation Experiment (WOCE);
  Tropical Oceans and Global Atmosphere (TOGA).
- **trace gases**:
  Joint Global Ocean Flux Study (JGOFS);
  International Global Atmospheric Chemistry (IGAC);
  Past Global Changes (PAGES).

As research advances, increased understanding and improved observations will lead to progressively more reliable climate predictions. However considering the complex nature of the problem and the scale of the scientific programmes to be undertaken we know that rapid results cannot be expected. Indeed further scientific advances may expose unforeseen problems and areas of ignorance.

Time-scales for narrowing the uncertainties will be dictated by progress over the next 10-15 years in two main areas:

- Use of the fastest possible computers, to take into account coupling of the atmosphere and the oceans in models, and to provide sufficient resolution for regional predictions.
- Development of improved representation of small-scale processes within climate models, as a result of the analysis of data from observational programmes to be conducted on a continuing basis well into the next century.

Annex

EMISSIONS SCENARIOS FROM WORKING GROUP III OF THE INTERGOVERNMENTAL PANEL ON CLIMATE CHANGE

The Steering Group of the Response Strategies Working Group requested the USA and the Netherlands to develop emissions scenarios for evaluation by the IPCC Working Group I. The scenarios cover the emissions of carbon dioxide (CO_2), methane (CH_4), nitrous oxide (N_2O), chlorofluorocarbons (CFCs), carbon monoxide (CO) and nitrogen oxides (NO_x) from the present up to the year 2100. Growth of the economy and population was taken common for all scenarios. Population was assumed to approach 10.5 billion in the second half of the next century. Economic growth was assumed to be 2-3% annually in the coming decade in the OECD countries and 3-5 % in the Eastern European and developing countries. The economic growth levels were assumed to decrease thereafter. In order to reach the required targets, levels of technological development and environmental controls were varied.

In the **Business-as-Usual scenario** (Scenario A) the energy supply is coal intensive and on the demand side only modest efficiency increases are achieved. Carbon monoxide controls are modest, deforestation continues until the tropical forests are depleted and agricultural emissions of methane and nitrous oxide are uncontrolled. For CFCs the Montreal Protocol is implemented albeit with only partial participation. Note that the aggregation of national projections by IPCC Working Group III gives higher emissions (10 - 20%) of carbon dioxide and methane by 2025.

In **Scenario B** the energy supply mix shifts towards lower carbon fuels, notably natural gas. Large efficiency increases are achieved. Carbon monoxide controls are stringent, deforestation is reversed and the Montreal Protocol implemented with full participation.

In **Scenario C** a shift towards renewables and nuclear energy takes place in the second half of next century. CFCs are now phased out and agricultural emissions limited.

For **Scenario D** a shift to renewables and nuclear in the first half of the next century reduces the emissions of carbon dioxide, initially more or less stabilizing emissions in the industrialized countries. The scenario shows that stringent controls in industrialized countries combined with moderated growth of emissions in developing countries could stabilize atmospheric concentrations. Carbon dioxide emissions are reduced to 50% of 1985 levels by the middle of the next century.

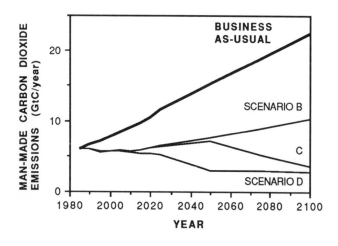 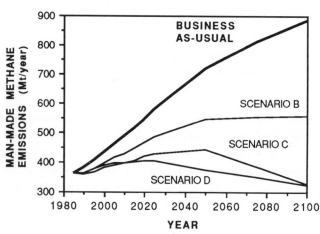

Man-made emissions of carbon dioxide and methane (as examples) to the year 2100, in the four scenarios developed by IPCC Working Group III.

Introduction

Purpose of the Report

The purpose of this report is to provide a scientific assessment of:

1. the factors which may affect climate change during the next century, especially those which are due to human activity;
2. the responses of the atmosphere-ocean-land-ice system to those factors;
3. the current ability to model global and regional climate changes and their predictability;
4. the past climate record and presently observed climate anomalies.

On the basis of this assessment, the report presents current knowledge regarding predictions of climate change (including sea-level rise and the effect on ecosystems) over the next century, the timing of changes together with an assessment of the uncertainties associated with these predictions.

This introduction provides some of the basic scientific ideas concerned with climate change, and gives an outline of the structure of the report.

The Climate System

A simple definition of climate is the average weather. A description of the climate over a period (which may typically be from a few years to a few centuries) involves the averages of appropriate components of the weather over that period, together with the statistical variations of those components.

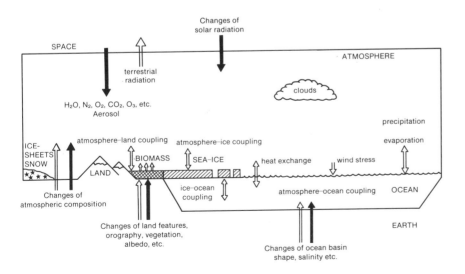

Schematic illustration of the climate system components and interactions. (from Houghton, J.T. (ed), 1984: The Global Climate; Cambridge University Press, Cambridge, UK, 233pp).

Fluctuations of climate occur on many scales as a result of natural processes; this is often referred to as natural **climate variability**. The **climate change** which we are addressing in this report is that which may occur over the next century as a result of human activities. More complete definitions of these terms can be found in WMO (1979) and WMO (1984).

The climate variables which are commonly used are concerned mainly with the atmosphere. But, in considering the climate system we cannot look at the atmosphere alone. Processes in the atmosphere are strongly coupled to the land surface, to the oceans and to those parts of the Earth covered with ice (known as the cryosphere). There is also strong coupling to the biosphere (the vegetation and other living systems on the land and in the ocean). These five components (atmosphere, land, ocean, ice and biosphere) together form the **climate system**.

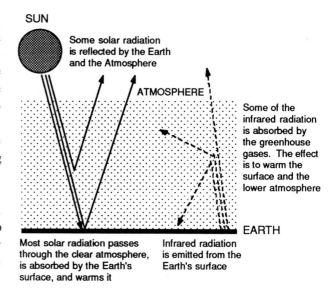

A simplified diagram illustrating the greenhouse effect

Forcing of the Climate System

The driving force for weather and climate is energy from the Sun. The atmosphere and surface of the Earth intercept solar radiation (in the short-wave, including visible, part of the spectrum); about a third of it is reflected, the rest is absorbed. The energy absorbed from solar radiation must be balanced by outgoing radiation from the Earth (terrestrial radiation); this is in the form of long-wave invisible infra-red energy. As the amount of outgoing terrestrial radiation is determined by the temperature of the Earth, this temperature will adjust until there is a balance between incoming and outgoing radiation.

There are several important factors (known as **climate forcing agents**) which can change the balance between the energy (in the form of solar radiation) absorbed by the Earth and that emitted by it in the form of long-wave infra-red radiation - the radiative forcing on climate. The most obvious of these is a change in the amount or seasonal distribution of **solar radiation** which reaches the Earth (orbital changes were probably responsible for initiating the ice ages). Any change in the albedo (reflectivity) of the land, due to **desertification or deforestation** will also affect the amount of solar energy absorbed, as will absorption of solar radiation (and outgoing long-wave radiation) by **aerosols** in the lower atmosphere (where they can be man-made) or the upper atmosphere (where they are predominantly natural, originating mainly from volcanoes).

The Greenhouse Effect

Apart from solar radiation itself, the most important radiative forcing arises from the **greenhouse effect**.

Short-wave solar radiation can pass through the clear atmosphere relatively unimpeded, but long-wave terrestrial radiation emitted by the warm surface of the Earth is partially absorbed and then re-emitted out to space by a number of trace gases in the cooler atmosphere above. This process adds to the net energy input to the lower atmosphere and the underlying surface thereby increasing their temperature. This is the basic greenhouse effect; the trace gases are often thought of as acting in a way somewhat analogous to the glass in a greenhouse. The main greenhouse gases are not the major constituents, nitrogen and oxygen, but water vapour (the biggest contributor), carbon dioxide, methane, nitrous oxide and (in recent years) chlorofluorocarbons.

How do we Know that the Greenhouse Effect is Real?

We know that the greenhouse effect works in practice, for several reasons. Firstly, the mean temperature of the Earth's surface is already 32°C warmer than it would be if the natural greenhouse gases (mainly carbon dioxide and water vapour) were not present. Satellite measurements of the radiation emitted from the Earth's surface and atmosphere demonstrate the absorption due to the greenhouse gases.

Secondly, we know that the composition of the atmospheres of Venus, Earth and Mars are very different, and their surface temperatures (shown in the table below) are in good agreement with those calculated on the basis of greenhouse effect theory.

Thirdly, measurements from ice cores, dating back 160,000 years, show that the Earth's temperature was closely related to the concentration of greenhouse gases in the atmosphere. The ice core record shows that the

	Surface Pressure (Relative to Earth)	Main Greenhouse Gases	Surface temperature in absence of Greenhouse effect	Observed Surface Temperature	Warming due to Greenhouse Effect
VENUS	90	> 90% CO_2	-46^oC	477^oC	523^oC
EARTH	1	~0.04% CO_2 ~1% H_2O	-18^oC	15^oC	33^oC
MARS	0.007	> 80% CO_2	-57^oC	-47^oC	10^oC

atmospheric levels of carbon dioxide, methane, and nitrous oxide were much lower during the ice ages than during interglacial periods. It is likely that changes in greenhouse gas concentrations contributed, in part, to the large (4 - 5°C) temperature swings between ice ages and interglacial periods.

The Enhanced Greenhouse Effect

An increase in concentrations of greenhouse gases is expected to raise the global-mean surface-air temperature which, for simplicity, is usually referred to as the "global temperature". Strictly, this is an *enhanced* greenhouse effect - above that occurring due to natural greenhouse gas concentrations. The word "enhanced" is frequently omitted, but should not be forgotten in this context.

Changes in the Abundances of the Greenhouse Gases

We know, with certainty, that the concentrations of naturally occurring greenhouse gases in the atmosphere have varied on palaeo time-scales. For a thousand years prior to the industrial revolution, the abundances of these gases were relatively constant. However, as the world's population increased, emissions of greenhouse gases such as carbon dioxide, methane, chlorofluorocarbons, nitrous oxide, and tropospheric ozone have increased substantially due to industrialisation and changes in agriculture and land-use. Carbon dioxide, methane, and nitrous oxide all have significant natural and man-made sources, while the chlorofluorocarbons (CFCs) are recent man-made gases. **Section 1** of the report summarises our knowledge of the various greenhouse gases, their sources, sinks and lifetimes, and their likely rate of increase.

Relative Importance of Greenhouse Gases

So far as radiative forcing of the climate is concerned, the increase in carbon dioxide has been the most important (contributing about 60% of the increased forcing over the last 200 years); methane is of next importance contributing about 20%; chloroflourocarbons contribute about 10% and all the other gases the remaining 10%. **Section 2** of the report reviews the contributions of the different gases to radiative forcing in more detail.

Feedbacks

If everything else in the climate system remained the same following an increase in greenhouse gases, it would be relatively easy to calculate, from a knowledge of their radiative properties, what the increase in average global temperature would be. However, as the components of the system begin to warm, other factors come into play which are called feedbacks. These factors can act to amplify the initial warming (positive feedbacks) or reduce it (negative feedbacks). Negative feedbacks can reduce the warming but cannot produce a global cooling. The simplest of these feedbacks arises because as the atmosphere warms the amount of water vapour it holds increases. Water vapour is an important greenhouse gas and will therefore amplify the warming. Other feedbacks occur through interactions with snow and sea-ice, with clouds and with the biosphere: **Section 3** explores these more fully.

The Role of the Oceans

The oceans play a central role in shaping the climate through three distinct mechanisms. Firstly, they absorb carbon dioxide and exchange it with the atmosphere (**Section 1** addresses this aspect of the carbon cycle). Secondly, they exchange heat, water vapour and momentum with the atmosphere. Wind stress at the sea surface drives the large-scale ocean circulation. Water vapour, evaporated from the ocean surface, is transported by the atmospheric circulation and provides latent heat energy to the atmosphere. The ocean circulations in their turn redistribute heat, fresh water and dissolved chemicals around the globe. Thirdly, they sequester heat, absorbed at the surface, in the deepest regions for periods of a thousand years or more through vertical circulation and convective mixing.

Therefore, any study of the climate and how it might change must include a detailed description of processes in the ocean together with the coupling between the ocean and the atmosphere. A description of ocean processes is presented in **Section 3** and the results from ocean atmosphere coupled models appear in **Section 6**.

Climate Forecasting

To carry out a climate forecast it is necessary to take into account all the complex interactions and feedbacks between the different components of the climate system. This is done through the use of a **numerical model** which as far as possible includes a description of all the processes and interactions. Such a model is a more elaborate version of the global models currently employed for weather forecasting.

Global forecasting models concentrate on the circulation of the atmosphere (for that reason they are often called **atmospheric general circulation models** (or atmospheric GCMs). They are based on equations describing the atmosphere's basic dynamics, and include descriptions in simple physical terms (called parameterizations) of the physical processes. Forecasts are made for several days ahead from an analysis derived from weather observations. Such forecasts are called deterministic weather forecasts because they describe the detailed weather to be expected at any place and time on the synoptic scale (of the order of a few hundred kilometres). They cannot of course, be deterministic so far as small-scale phenomena, such as individual shower clouds, are concerned.

The most elaborate climate model employed at the present time consists of an atmospheric GCM coupled to an **ocean GCM** which describes the structure and dynamics of the ocean. Added to this coupled model are appropriate descriptions, although necessarily somewhat crude, of the other components of the climate system (namely, the land surface and the ice) and the interactions between them. If the model is run for several years with parameters and forcing appropriate to the current climate, the model's output should bear a close resemblance to the observed climate. If parameters representing, say, increasing greenhouse gases are introduced into the model, it can be used to simulate or predict the resulting climate change.

To run models such as these requires very large computer resources indeed. However, simplified models are also employed to explore the various sensitivities of the climate system and to make simulations of the time evolution of climate change. In particular, simplifications of the ocean structure and dynamics are included; details are given in **Section 3**. **Section 4** describes how well the various models simulate current climate and also how well they have been able to make reconstructions of past climates.

Equilibrium and Time-Dependent Response

The simplest way of employing a climate model to determine the response to a change in forcing due to increases in greenhouse gases is to first run the model for several years with the current forcing; then to change the forcing (for instance by doubling the concentration of carbon dioxide in the appropriate part of the model) and run the model again. Comparing the two model climates will then provide a forecast of the change in climate to be expected under the new conditions. Such a forecast will be of the **equilibrium response**; it is the response expected to that change when the whole climate system has reached a steady state. Most climate forecasting models to date have been run in this equilibrium response mode. **Section 5** summarises the results obtained from such models.

A more complicated and difficult calculation can be carried out by changing the forcing in the model slowly on the appropriate natural time-scale. Again, comparison with the unperturbed model climate is carried out to obtain the **time-dependent response** of the model to climate change.

These time-dependent models, results from which are presented in **Section 6**, are the ones which describe the climate system most realistically. However, rather few of them have been run so far. Comparison of the magnitude and patterns of climate change as predicted by these models has been made with results from models run in the equilibrium response mode. The results of this comparison provide guidance on how to interpret some of the more detailed results from the equilibrium model runs.

Detection of Climate Change

Of central importance to the study of climate and climate change are observations of climate. From the distant past we have palaeo-climatic data which provide information on the response of the climate system to different historical forcings. **Section 4** describes how climate models can be validated in these differing climate regimes. It is only within about the last hundred years, however, that accurate observations with good global coverage exist. Even so, there have been numerous changes in instruments and observational practices during this period, and quite sophisticated numerical corrections are required to standardize the data to a self-consistent record.

Section 7 discusses these issues and provides evidence, from land and sea temperature records and glacier measurements, that a small global warming has occurred since the late nineteenth century. The temperature and precipitation records are examined regionally as well, and recent data on sea-ice and snow-cover are shown.

Within these time-series of data we can examine the natural variability of climate and search for a possible climate change signal due to increasing greenhouse gases. **Section 8** compares the expectations from model predictions with the observed change in climate. At a global level the change is consistent with predictions from models but there may be other effects producing it. Problems arise at a regional level because there are differences between the various predictions and because the changes observed so far are small and comparable to spatial and temporal noise. In this Section, however, an estimate is made of the likely time-scale for detection of the enhanced greenhouse effect.

Changes in Sea Level

An important consequence of a rise in global temperature would be an increase in sea level. **Section 9** assesses the contribution from thermal expansion of the oceans, melting of mountain glaciers and changes to the Greenland and Antarctic ice sheets under the four IPCC Scenarios of future temperature rise. Measurements of sea level from tide gauges around the world date back a hundred years and provide evidence for a small increase which appears to be fairly steady. The stability of the West Antarctic Ice Sheet, which has sometimes been invoked as a possible mechanism for large sea level rise in the future, is examined.

Climate Change and Ecosystems

Ecosystems (both land and marine based plant-life) will respond to climate change and, through feedback processes, influence it. **Section 10** looks at the direct effect of climate change on crops, forests and tundra. Plant growth and metabolism are functions of temperature and soil moisture, as well as carbon dioxide itself; changes in the activity of ecosystems will therefore modify the carbon cycle. Plant species have migrated in the past, but their ability to adapt in future may be limited by the presence of artificial barriers caused by human activities and by the speed of climate change. This Section also looks at the effects of deforestation and reforestation on the global carbon budget.

Improving our Predictions

Despite our confidence in the general predictions from numerical models, there will be uncertainties in the detailed timing and patterns of climate change due to the enhanced greenhouse effect for some time to come. **Section 11** lists the many programs which are already underway or are planned to narrow these uncertainties. These cover the full range of Earth and Space based observing systems, process studies to unravel the details of feedbacks between the many components of the climate system, and expected developments in computer models.

The Climate Implications of Emission Controls

In order that any policy decisions on emission controls are soundly based it is useful to quantify the climate benefits of different levels of controls on different time-scales. The **Annex** to this Report shows the full pathway of emissions to temperature change and sea-level rise for the four IPCC Policy Scenarios plus four other Science Scenarios. The Policy Scenarios were derived by IPCC Working Group III and assume progressively more stringent levels of emission controls. The Science Scenarios were chosen artificially to illustrate the effects of sooner, rather than later, emission controls, and to show the changes in temperature and sea level which we may be committed to as a result of past emissions of greenhouse gases.

References

WMO (1979) Proceedings of the World Climate Conference, Geneva, 12-23 February 1979. WMO 537.

WMO (1984) Scientific Plan for the World Climate Research Program. WCRP Pub series No 2, WMO/TD No 6.

1

Greenhouse Gases and Aerosols

R.T. WATSON, H. RODHE, H. OESCHGER, U. SIEGENTHALER

Contributors:
M. Andreae; R. Charlson; R. Cicerone; J. Coakley; R. Derwent; J. Elkins;
F. Fehsenfeld; P. Fraser; R. Gammon; H. Grassl; R. Harriss; M. Heimann;
R. Houghton; V. Kirchhoff; G. Kohlmaier; S. Lal; P. Liss; J. Logan; R. Luxmoore;
L. Merlivat; K. Minami; G. Pearman; S. Penkett; D. Raynaud; E. Sanhueza; P. Simon;
W. Su; B. Svensson; A. Thompson; P. Vitousek; A. Watson; M. Whitfield; P. Winkler;
S. Wofsy.

CONTENTS

EXECUTIVE SUMMARY

The Earth's climate is dependent upon the radiative balance of the atmosphere, which in turn depends upon the input of solar radiation and the atmospheric abundances of radiatively active trace gases (i.e., greenhouse gases), clouds and aerosols.

Since the industrial revolution the atmospheric concentrations of several greenhouse gases, i.e., carbon dioxide (CO_2), methane (CH_4), chlorofluorocarbons (CFCs), nitrous oxide (N_2O), and tropospheric ozone (O_3), have been increasing, primarily due to human activities. Several of these greenhouse gases have long atmospheric lifetimes, decades to centuries, which means that their atmospheric concentrations respond slowly to changes in emission rates. In addition, there is evidence that the concentrations of tropospheric aerosols have increased at least regionally.

Carbon Dioxide

The atmospheric CO_2 concentration, at 353 ppmv in 1990, is now about 25% greater than the pre-industrial (1750-1800) value of about 280 ppmv, and higher than at any time in at least the last 160,000 years. Carbon dioxide is currently rising at about 1.8 ppmv (0.5%) per year due to anthropogenic emissions. Anthropogenic emissions of CO_2 are estimated to be 5.7±0.5 Gt C (in 1987) due to fossil fuel burning, plus 0.6 - 2.5 Gt C (in 1980) due to deforestation. The atmospheric increase during the past decade corresponds to (48±8)% of the total emissions during the same period with the remainder being taken up by the oceans and land. Indirect evidence suggests that the land and oceans sequester CO_2 in roughly equal proportions, though the mechanisms are not all well understood. The time taken for atmospheric CO_2 to adjust to changes in sources or sinks is of order 50-200 years, determined mainly by the slow exchange of carbon between surface waters and deeper layers of the ocean. Consequently, CO_2 emitted into the atmosphere today will influence the atmospheric concentration of CO_2 for centuries into the future. Three models have been used to estimate that even if anthropogenic emissions of CO_2 could be kept constant at present day rates, atmospheric CO_2 would increase to 415 - 480 ppmv by the year 2050, and to 460 - 560 ppmv by the year 2100. In order to stabilize concentrations at present day levels, an immediate reduction in global anthropogenic emissions by 60-80 percent would be necessary.

Methane

Current atmospheric CH_4 concentration, at 1.72 ppmv, is now more than double the pre-industrial (1750-1800) value of about 0.8 ppmv, and is increasing at a rate of about 0.015 ppmv (0.9%) per year. The major sink for CH_4, reaction with hydroxyl (OH) radicals in the troposphere, results in a relatively short atmospheric lifetime of about 10 years. Human activities such as rice cultivation, domestic ruminant rearing, biomass burning, coal mining, and natural gas venting have increased the input of CH_4 into the atmosphere, which combined with a possible decrease in the concentration of tropospheric OH, yields the observed rise in global CH_4. However, the quantitative importance of each of the factors contributing to the observed increase is not well known at present. In order to stabilize concentrations at present day levels, an immediate reduction in global anthropogenic emissions by 15-20 percent would be necessary.

Chlorofluorocarbons

The current atmospheric concentrations of the anthropogenically produced halocarbons, CCl_3F (CFC-11), CCl_2F_2 (CFC-12), $C_2Cl_3F_3$ (CFC-113) and CCl_4 (carbon tetrachloride) are about 280 pptv, 484 pptv, 60 pptv, and 146 pptv, respectively. Over the past few decades their concentrations, except for CCl_4, have increased more rapidly (on a percentage basis) than the other greenhouse gases, currently at rates of at least 4% per year. The fully halogenated CFCs and CCl_4 are primarily removed by photolysis in the stratosphere, and have atmospheric lifetimes in excess of 50 years. Future emissions will, most likely, be eliminated or significantly lower than today's because of current international negotiations to strengthen regulations on chlorofluorocarbons. However, the atmospheric concentrations of CFCs 11, 12 and 113 will still be significant (30 - 40% of current) for at least the next century because of their long atmospheric lifetimes.

Nitrous Oxide

The current atmospheric N_2O concentration, at 310 ppbv, is now about 8% greater than in the pre-industrial era, and is increasing at a rate of about 0.8 ppbv (0.25%) per year. The major sink for N_2O, photolysis in the stratosphere, results in a relatively long atmospheric lifetime of about 150 years. It is difficult to quantitatively account for the source of the current increase in the atmospheric concentration of N_2O but it is thought to be due to human activities. Recent data suggest that the total annual flux of N_2O from combustion and biomass burning is much less than previously believed. Agricultural practices may stimulate emissions of N_2O from soils and play a major role. In order to stabilize concentrations at present day levels, an immediate reduction of 70 - 80% of the additional flux of N_2O that has occurred since the pre-industrial era would be necessary.

Ozone

Ozone is an effective greenhouse gas especially in the middle and upper troposphere and lower stratosphere. Its concentration in the troposphere is highly variable because of its short lifetime. It is photochemically produced in-situ through a series of complex reactions involving carbon monoxide (CO), CH_4, non-methane hydrocarbons (NMHC), and nitrogen oxide radicals (NO_x), and also transported downward from the stratosphere. The limited observational data support positive trends of about 1% per year for O_3 below 8 km in the northern hemisphere (consistent with positive trends in several of the precursor gases, especially NO_x, CH_4, and CO) but probably close to zero trend in the southern hemisphere. There is also evidence that O_3 has decreased by a few percent globally in the lower stratosphere (below 25 km) within the last decade. Unfortunately, there are no reliable long-term data near the tropopause.

Aerosol particles

Aerosol particles have a lifetime of at most a few weeks in the troposphere and occur in highly variable concentrations. A large proportion of the particles that influence cloud processes and the radiative balance is derived from gaseous sulphur emissions. Due to fossil fuel combustion, these emissions have more than doubled globally, causing a large increase in the concentration of aerosol sulphate especially over and around the industrialized regions of Europe and North America. Future concentrations of aerosol sulphate will vary in proportion to changes in anthropogenic emissions. Aerosol particles derived from natural (biological) emissions may contribute to climate feedback processes. During a few years following major volcanic eruptions the concentrations of natural aerosol particles in the stratosphere can be greatly enhanced.

1.1 Introduction

The Earth's climate is dependent upon the radiative balance of the atmosphere, which in turn depends upon the input of solar radiation and the atmospheric abundances of radiatively active trace gases (i.e., greenhouse gases), clouds and aerosols. Consequently, it is essential to gain an understanding of how each of these climate "forcing agents" varies naturally, and how some of them might be influenced by human activities.

The chemical composition of the Earth's atmosphere is changing, largely due to human activities (Table 1.1). Air trapped in Antarctic and Greenland ice shows that there have been major increases in the concentrations of radiatively active gases such as carbon dioxide (CO_2), methane (CH_4), and nitrous oxide (N_2O) since the beginning of the industrial revolution. In addition, industrially-produced chlorofluorocarbons (CFCs) are now present in the atmosphere in significant concentrations, and there is evidence that the concentrations of tropospheric O_3 and aerosols have increased at least regionally.

Atmospheric measurements indicate that in many cases the rates of change have increased in recent decades. Many of the greenhouse gases have long atmospheric life-times, decades to centuries, which implies that their atmospheric concentrations respond slowly to changes in emission rates.

The effectiveness of a greenhouse gas in influencing the Earth's radiative budget is dependent upon its atmospheric concentration and its ability to absorb outgoing long-wave terrestrial radiation. Tropospheric water vapour is the single most important greenhouse gas, but its atmospheric concentration is not significantly influenced by direct anthropogenic emissions. Of the greenhouse gases that are directly affected by human activities, CO_2 has the largest radiative effect, followed by the CFCs, CH_4, tropospheric O_3, and N_2O. Although the present rate of increase in the atmospheric concentration of CO_2 is about a factor of 70,000 times greater than that of CCl_3F (CFC-11) and CCl_2F_2 (CFC-12) combined, and a factor of about 120 times greater than that of CH_4, its contribution to changes in the radiative forcing during the decade of the 1980s was

Table 1.1 Summary of Key Greenhouse Gases Influenced by Human Activities [1]

Parameter	CO_2	CH_4	CFC-11	CFC-12	N_2O
Pre-industrial atmospheric concentration (1750-1800)	280 ppmv[2]	0.8 ppmv	0	0	288 ppbv[2]
Current atmospheric concentration (1990)[3]	353 ppmv	1.72 ppmv	280 pptv[2]	484 pptv	310 ppbv
Current rate of annual atmospheric accumulation	1.8 ppmv (0.5%)	0.015 ppmv (0.9%)	9.5 pptv (4%)	17 pptv (4%)	0.8 ppbv (0.25%)
Atmospheric lifetime[4] (years)	(50-200)	10	65	130	150

1   Ozone has not been included in the table because of lack of precise data.
2   ppmv = parts per million by volume;   ppbv = parts per billion by volume; pptv = parts per trillion by volume.
3   The current (1990) concentrations have been estimated based upon an extrapolation of measurements reported for earlier years, assuming that the recent trends remained approximately constant.
4   For each gas in the table, except CO_2, the "lifetime" is defined here as the ratio of the atmospheric content to the total rate of removal. This time scale also characterizes the rate of adjustment of the atmospheric concentrations if the emission rates are changed abruptly. CO_2 is a special case since it has no real sinks, but is merely circulated between various reservoirs (atmosphere, ocean, biota). The "lifetime" of CO_2 given in the table is a rough indication of the time it would take for the CO_2 concentration to adjust to changes in the emissions (see section 1.2.1 for further details).

about 55%, compared to 17% for CFCs (11 and 12), and 15% for CH_4 (see Section 2). Other CFCs and N_2O accounted for about 8%, and 5%, respectively, of the changes in the radiative forcing. While the contribution from tropospheric O_3 may be important, it has not been quantified because the observational data is inadequate to determine its trend. This pattern arises because of differences in the efficiencies of the gases to absorb terrestrial radiation.

Aerosol particles play an important role in the climate system because of their direct interaction (absorption and scattering) with solar and terrestrial radiation, as well as through their influence on cloud processes and thereby, indirectly, on radiative fluxes.

There is a clear need to document the historical record of the atmospheric concentrations of greenhouse gases and aerosols, as well as to understand the physical, chemical, geological, biological and social processes responsible for the observed changes. A quantitative understanding of the atmospheric concentrations of these gases requires knowledge of: the cycling and distribution of carbon, nitrogen and other key nutrients within and between the atmosphere, terrestrial ecosystems, oceans and sediments; and the influence of human actions on these cycles. Without knowledge of the processes responsible for the observed past and present changes in the atmospheric concentrations of greenhouse gases and aerosols it will not be possible to predict with confidence future changes in atmospheric composition, nor therefore the resulting changes in the radiative forcing of the atmosphere.

1.2 Carbon Dioxide

1.2.1 The Cycle of Carbon in Nature

Carbon in the form of CO_2, carbonates, organic compounds, etc. is cycled between various reservoirs, atmosphere, oceans, land biota and marine biota, and, on geological time scales, also sediments and rocks (Figure 1.1; for more detailed reviews see Sundquist, 1985: Bolin, 1981, 1986; Trabalka, 1985; Siegenthaler, 1986). The largest natural exchange fluxes occur between the atmosphere and the terrestrial biota and between the atmosphere and the surface water of the oceans. By comparison, the net inputs into the atmosphere from fossil fuel combustion and deforestation are much smaller, but are large enough to modify the natural balance.

The turnover time of CO_2 in the atmosphere, measured as the ratio of the content to the fluxes through it, is about 4 years. This means that on average it takes only a few years before a CO_2 molecule in the atmosphere is taken up by plants or dissolved in the ocean. This short time scale must not be confused with the time it takes for the atmospheric CO_2 level to adjust to a new equilibrium if sources or sinks change. This adjustment time, corresponding to the lifetime

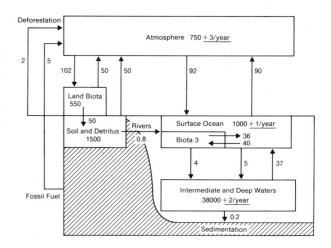

Figure 1.1: Global carbon reservoirs and fluxes. The numbers apply for the present-day situation and represent typical literature values. Fluxes, e.g. between atmosphere and surface ocean, are gross annual exchanges. Numbers underlined indicate net annual CO_2 accumulation due to human action. Units are gigatons of carbon (GtC; $1Gt = 10^9$ metric tons = 10^{12}kg) for reservoir sizes and GtC yr^{-1} for fluxes. More details and discussions are found in several reviews (Sundquist, 1985; Trabalka, 1985; Bolin, 1986; Siegenthaler, 1986).

in Table 1.1, is of the order of 50 - 200 years, determined mainly by the slow exchange of carbon between surface waters and the deep ocean. The adjustment time is important for the discussions on global warming potential, cf. Section 2.2.7.

Because of its complex cycle, the decay of excess CO_2 in the atmosphere does not follow a simple exponential curve, and therefore a single time scale cannot be given to characterize the whole adjustment process toward a new equilibrium. The two curves in Figure 1.2, which represent simulations of a pulse input of CO_2 into the atmosphere using atmosphere-ocean models (a box model and a General Circulation Model (GCM)), clearly show that the initial response (governed mainly by the uptake of CO_2 by ocean surface waters) is much more rapid than the later response (influenced by the slow exchange between surface waters and deeper layers of the oceans). For example, the first reduction by 50 percent occurs within some 50 years, whereas the reduction by another 50 percent (to 25 percent of the initial value) requires approximately another 250 years. The concentration will actually never return to its original value, but reach a new equilibrium level; about 15 percent of the total amount of CO_2 emitted will remain in the atmosphere.

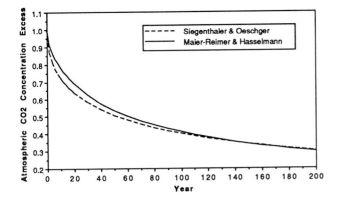

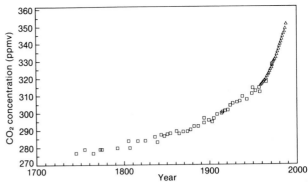

Figure 1.2: Atmospheric CO_2 concentration excess after a pulse input at time 0 (initially doubling the atmospheric CO_2 concentration), as calculated with two ocean-atmosphere models. Solid line: 3-dimensional ocean-circulation model of Maier-Reimer and Hasselmann (1987), dashed line: 1-dimensional box-diffusion model of Siegenthaler and Oeschger (1987). The adjustment towards a new equilibrium does not follow an exponential curve; it is very fast during the first decade, then slows down more and more. The concentration excess does not go to zero; after a long time, a new equilibrium partitioning between atmosphere and ocean will be reached, with about 15 percent of the input residing in the atmosphere.

Figure 1.3: Atmospheric CO_2 increase in the past 250 years, as indicated by measurements on air trapped in ice from Siple Station, Antarctica (squares; Neftel et al., 1985a; Friedli et al., 1986) and by direct atmospheric measurements at Mauna Loa, Hawaii (triangles; Keeling et al., 1989a).

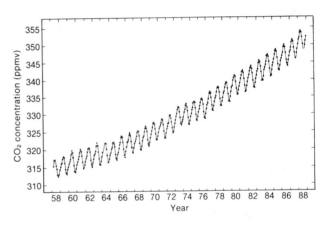

Figure 1.4: Monthly average CO_2 concentration in parts per million of dry air, observed continuously at Mauna Loa, Hawaii (Keeling et al., 1989a). The seasonal variations are due primarily to the withdrawal and production of CO_2 by the terrestrial biota.

1.2.1.1 The role of the atmosphere

The mean annual concentration of CO_2 is relatively homogeneous throughout the troposphere because the troposphere is mixed on a time scale of about 1 year. The pre-industrial atmospheric CO_2 concentration was about 280 ppmv, as reconstructed from ice core analyses (c.f. Section 1.2.4.1), corresponding to an atmospheric amount of 594 Gigatonnes of carbon (GtC: $1 Gt = 10^9 t = 10^{15} g$; 1 ppmv CO_2 of the global atmosphere equals 2.12 GtC and 7.8 Gt CO_2); today, the level is about 353 ppmv (Figures 1.3 and 1.4). The atmospheric increase has been monitored since 1958 at a growing number of stations (Keeling and Heimann, 1986; Keeling et al., 1989a; Beardsmore and Pearman, 1987; Conway et al., 1988).

1.2.1.2 The role of the ocean

On time scales of decades or more, the CO_2 concentration of the unperturbed atmosphere is mainly controlled by the exchange with the oceans, since this is the largest of the carbon reservoirs. There is a continuous exchange of CO_2 in both directions between the atmosphere and oceans. The net flux into (or out of) the ocean is driven by the difference between the atmospheric partial pressure of CO_2 and the equilibrium partial pressure of CO_2 (pCO_2) in surface waters.

The exchange of carbon between the surface and deeper layers is accomplished mainly through transport by water

motions. Ventilation of the thermocline (approximately the uppermost km of the ocean) is particularly important for the downward transport of anthropogenic CO_2. The deep circulation is effective on time scales of 100-1000 years.

The natural carbon cycle in the ocean and in particular pCO_2 in surface ocean water are strongly influenced also by biological processes. The marine biota serve as a "biological pump", transporting organic carbon from surface waters to deeper layers as a rain of detritus at a rate of about 4 GtC per year (Eppley and Peterson, 1979), which is balanced by an equal upward transport of carbon by deeper water richer in CO_2 than surface water. This "biological pump" has the effect of reducing surface pCO_2

very substantially: without the biological pump ("dead ocean") the pre-industrial CO_2 level would have been higher than the observed value of 280 ppmv, at perhaps 450 ppmv (Wenk, 1985; Bacastow and Maier-Reimer, 1990). Alterations in the marine biota due to climatic change could therefore have a substantial effect on CO_2 levels in the future. Note, however, that the "biological pump" does not help to sequester anthropogenic CO_2 (see Section 1.2.4.2).

1.2.1.3 The role of terrestrial vegetation and soils

The most important processes in the exchange of carbon are those of photosynthesis, autotrophic respiration (i.e., CO_2 production by the plants) and heterotrophic (i.e., essentially microbial) respiration converting the organic material back into CO_2 mainly in soils (c.f. Section 10 for a detailed discussion). Net primary production (NPP) is the net annual uptake of CO_2 by the vegetation; NPP is equal to the gross uptake (gross primary production, GPP) minus autotrophic respiration. In an unperturbed world, NPP and decomposition by heterotrophic respiration are approximately balanced on an annual basis; formation of soils and peat corresponds to a (relatively small) excess of NPP.

The carbon balance can be changed considerably by the direct impact of human activities (land use changes, particularly deforestation), by climate changes, and by other changes in the environment, e.g., atmospheric composition. Since the pools and fluxes are large (NPP 50-60 GtC per year, GPP 90-120 GtC per year; Houghton et al., 1985b), any perturbations can have a significant effect on the atmospheric concentration of CO_2.

1.2.2 Anthropogenic Perturbations

The concentrations of CO_2 in the atmosphere are primarily affected by two anthropogenic processes: release of CO_2 from fossil fuel combustion, and changes in land use such as deforestation.

1.2.2.1 Historical fossil fuel input

The global input of CO_2 to the atmosphere from fossil fuel combustion, plus minor industrial sources like cement production, has shown an exponential increase since 1860 (about 4% per year), with major interruptions during the two world wars and the economic crisis in the thirties (Figure 1.5). Following the "oil crisis" of 1973, the rate of increase of the CO_2 emissions first decreased to approximately 2% per year, and after 1979 the global emissions remained almost constant at a level of 5.3 GtC per year until 1985, when they started to rise again, reaching 5.7 GtC per year in 1987 (Figure 1.5). The cumulative release of CO_2 from fossil fuel use and cement manufacturing from 1850 to 1987 is estimated at 200 GtC ± 10% (Marland, 1989).

Ninety five percent of the industrial CO_2 emissions are from the Northern Hemisphere, dominated by industrial

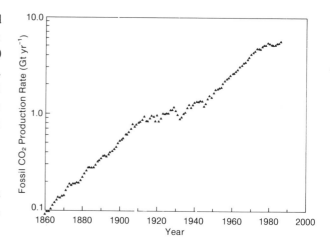

Figure 1.5: Global annual emissions of CO_2 from fossil fuel combustion and cement manufacturing, expressed in GtC yr^{-1} (Rotty and Marland, 1986; Marland, 1989). The average rate of increase in emissions between 1860 and 1910 and between 1950 and 1970 is about 4% per year.

countries, where annual releases reach up to about 5 tC per capita (Rotty and Marland, 1986). In contrast, CO_2 emission rates in most developing countries lie between 0.2 and 0.6 tC per capita per year. However, the relative rate of increase of the CO_2 emissions is much larger in the developing countries (~ 6% per year), showing almost no slowing down after 1973 in contrast to Western Europe and North America where the rate of increase decreased from about 3% per year (1945-72) to less than 1% per year (1973-84).

1.2.2.2 Historical land use changes

The vegetation and soils of unmanaged forests hold 20 to 100 times more carbon per unit area than agricultural systems. The amount of carbon released to the atmosphere compared to that accumulated on land as a result of land use change depends on the amounts of carbon held in biomass and soils, rates of oxidation of wood products (either rapidly through burning or more slowly through decay), rates of decay of organic matter in soils, and rates of regrowth of forests following harvest or abandonment of agricultural land. The heterogeneity of terrestrial ecosystems makes estimation of global inventories and fluxes difficult.

The total release of carbon to the atmosphere from changes in land use, primarily deforestation, between 1850 and 1985 has been estimated to be about 115 GtC (Houghton and Skole, 1990), with an error limit of about ±35 GtC. The components of the flux to the atmosphere are: (1) burning associated with land use change; (2) decay of biomass on site (roots, stumps, slash, twigs etc.); (3) oxidation of wood products removed from site (paper, lumber, waste etc.); (4) oxidation of soil carbon; minus (5) regrowth of trees and redevelopment of soil organic matter following harvest. Although the greatest releases of carbon in the nineteenth and early twentieth centuries were from lands in the temperate zone (maximum 0.5 GtC per year), the major source of carbon during the past several decades has been from deforestation in the tropics, with a significant increase occurring since 1950. Over the entire 135 yr period, the release from tropical regions is estimated to have been 2-3 times greater than the release from middle and high latitudes. Estimates of the flux in 1980 range from 0.6 to 2.5 GtC (Houghton et al., 1985a, 1987, 1988; Detwiler and Hall, 1988): virtually all of this flux is from the tropics. The few regions for which data exist suggest that the annual flux is higher now than it was in 1980.

1.2.3 Long-Term Atmospheric Carbon Dioxide Variations

The most reliable information on past atmospheric CO_2 concentrations is obtained by the analysis of polar ice cores. The process of air occlusion lasts from about 10 up to 1000 years, depending on local conditions (e.g., precipitation rate), so that an air sample in old ice reflects the atmospheric composition averaged over a corresponding time interval.

Measurements on samples representing the last glacial maximum (18,000 yr before present) from ice cores from Greenland and Antarctica (Neftel et al., 1982, 1988; Delmas et al., 1980) showed CO_2 concentrations of 180-200 ppmv, i.e., about 70 percent of the pre-industrial value. Analyses on the ice cores from Vostok, Antarctica, have provided new data on natural variations of CO_2, covering a full glacial-interglacial cycle (Figure 1.6; Barnola et al., 1987). Over the whole period there is a remarkable correlation between polar temperature, as deduced from deuterium data, and the CO_2 profile. The glacial-interglacial shifts of CO_2 concentrations must have been linked to large-scale changes in the circulation of the ocean, and in the whole interplay of biological, chemical and physical processes, but the detailed mechanisms are not yet very clear. The CO_2 variations were large enough to potentially contribute, via the greenhouse effect, to a substantial (although not the major) part of the glacial-interglacial climate change (Hansen et al., 1984; Broccoli and Manabe, 1987).

Ice core studies on Greenland ice indicate that during the last glaciation, CO_2 concentration shifts of the order of 50 ppmv may have occurred within less than 100 years (Stauffer et al., 1984), parallel to abrupt, drastic climatic events (temperature changes of the order of 5°C). These rapid CO_2 changes have not yet been identified in ice cores from Antarctica (possibly due to long occlusion times,

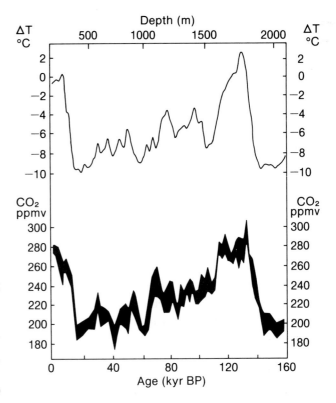

Figure 1.6: CO_2 concentrations (bottom) and estimated temperature changes (top) during the past 160,000 years, as determined on the ice core from Vostok, Antarctica (Barnola et al., 1987). Temperature changes were estimated based on the measured deuterium concentrations.

Neftel et al., 1988), therefore, it is not yet clear if they are real or represent artefacts in the ice record.

1.2.4 The Contemporary Record of Carbon Dioxide - Observations and Interpretation

1.2.4.1 The carbon dioxide increase from pre-industrial period

Relatively detailed CO_2 data have been obtained for the last millennium from Antarctic ice cores (Neftel et al., 1985a; Friedli et al., 1986; Siegenthaler et al., 1988; Raynaud and Barnola, 1985; Pearman et al., 1986). They indicate that during the period 1000 to 1800, the atmospheric concentration was between 270 and 290 ppmv. The relative constancy seems surprising in view of the fact that the atmosphere exchanges about 30 percent of its CO_2 with the oceans and biota each year. This indicates that the sensitivity of atmospheric CO_2 levels to minor climatic changes such as the Little Ice Age (lasting from the end of the 16th to the middle of the 19th century), when global mean temperatures probably decreased by about 1°C, is small.

A precise reconstruction of the CO_2 increase during the past two centuries has been obtained from an ice core from Siple Station, Antarctica (Figure 1.3; Neftel et al., 1985a; Friedli et al., 1986). These results indicate that CO_2 started to rise around 1800 and had already increased by about 15 ppmv by 1900. Precise direct atmospheric measurements started in 1958, when the level was about 315 ppmv and the rate of increase 0.6 ppmv per year. The present atmospheric CO_2 level has reached 353 ppmv, and the mean growth rate has now reached about 1.8 ppmv per year (Figure 1.4; Keeling et al., 1989a).

1.2.4.2 Uptake by the ocean

The ocean is an important reservoir for taking up anthropogenic CO_2. The relative increase of dissolved inorganic carbon (total CO_2) in ocean water is smaller than in the atmosphere (only 2-3 percent until now, see below). Precise measurements of dissolved inorganic carbon can be made with present analytical tools. However, an accurate determination of the trend in dissolved inorganic carbon is difficult because of its variability in time and space. Hence, repeated transects and time series will be required to assess the total oceanic CO_2 uptake with good precision.

The net flux of CO_2 into (or out of) the ocean is given by the product of a gas transfer coefficient and ΔpCO_2 (the CO_2 partial pressure difference between ocean and atmosphere). The gas transfer coefficient increases with increasing wind speed and also depends on water temperature. Therefore, the net flux into the ocean can be estimated from a knowledge of the atmospheric CO_2 concentration, pCO_2 in surface water (for which the data are still sparse), the global distribution of wind speeds over the ocean as well as the relation between wind speed and gas transfer coefficient (which is known to $\pm 30\%$ only). There have been several estimates of the global net uptake of CO_2 by the oceans using observations (e.g., Enting and Pearman, 1982, 1987). The most recent estimate yields 1.6 GtC per year (Tans et al., 1990); the error of this estimate is, according to the authors, not easy to estimate.

Estimates of oceanic CO_2 uptake, in the past and in the future, require models of the global carbon cycle that take into account air-sea gas exchange, aqueous carbonate chemistry and the transport from the surface to deep ocean layers. The aqueous carbonate chemistry in sea water operates in a mode that if the atmospheric CO_2 concentration increases by e.g., 10%, then the concentration of dissolved inorganic carbon in sea water increases by only about 1% at equilibrium. Therefore, the ocean is not such a powerful sink for anthropogenic CO_2 as might seem at first when comparing the relative sizes of the reservoirs (Figure 1.1).

The rate at which anthropogenic CO_2 is transported from the surface to deeper ocean layers is determined by the rate of water exchange in the vertical. It is known from measurements of the radioactive isotope ^{14}C that on average it takes hundreds to about one thousand years for water at the surface to penetrate to well below the mixed layer of the major oceans (e.g., Broecker and Peng, 1982). Thus, in most oceanic regions only the top several hundred metres of the oceans have at present taken up significant amounts of anthropogenic CO_2. An exception is the North Atlantic Ocean where bomb-produced tritium has been observed even near the bottom of the sea, indicating the active formation of new deep water.

The rain of biogenic detrital particles, which is important for the natural carbon cycle, does not significantly contribute to a sequestering of excess CO_2, since the marine biota do not directly respond to the CO_2 increase. Their activity is controlled by other factors, such as light, temperature and limiting nutrients (e.g., nitrogen, phosphorus, silicon). Thus, only the input of fertilizers (phosphate, nitrate) into the ocean through human activities may lead to an additional sedimentation of organic carbon in the ocean; different authors have estimated the size of this additional sink at between 0.04 and 0.3 GtC per year (see Baes et al., 1985). It seems thus justified to estimate the fossil fuel CO_2 uptake to date considering the biological flux to be constant, as long as climatic changes due to increasing greenhouse gases, or natural causes, do not modify the marine biotic processes. Although this appears a reasonable assumption for the past and present situation, it may well not be so in the future.

The carbon cycle models used to date to simulate the atmosphere-ocean system have often been highly simplified, consisting of a few well-mixed or diffusive reservoirs (boxes) (e.g., Oeschger et al., 1975; Broecker et al., 1980; Bolin, 1981; Enting and Pearman, 1987; Siegenthaler, 1983). Even though these box models are highly simplified they are a powerful means for identifying the importance of the different processes that determine the flux of CO_2 into the ocean (e.g., Broecker and Peng, 1982; Peng and Broecker, 1985). The results of these models are considered to be reasonable because, as long as the ocean circulation is not changing, the models need only simulate the transport of excess CO_2 from the atmosphere into the ocean, but not the actual dynamics of the ocean. In the simple models, the oceanic transport mechanisms, e.g., formation of deep water, are parameterized. The transport parameters (e.g., eddy diffusivity) are determined from observations of transient tracers that are analogues to the flux of anthropogenic CO_2 into the ocean. If a model reproduces correctly the observed distribution of, e.g., bomb-produced ^{14}C, then it might be expected to simulate reasonably the flux of CO_2 into the ocean. A 1-D box-diffusion model yields an oceanic uptake of 2.4 GtC per year on average for the decade 1980 - 1989, and an outcrop-diffusion model (both described by Siegenthaler, 1983) 3.6 GtC per year. The latter model most probably overpredicts the flux into the ocean, because it includes an infinitely fast exchange between high-latitude surface waters and the deep ocean.

However, it is obviously desirable to use 3-dimensional (3-D) general circulation models of the oceans for this purpose. At this time, only a few modelling groups have started to do this. One 3-D model (Maier-Reimer and Hasselmann, 1987) gives a similar CO_2 uptake as a 1-D box-diffusion model of Siegenthaler (1983), as illustrated by the model response to a pulse input of CO_2 (Figure 1.2). In a recent revised version of this model (Maier-Reimer et al., personal communication) the ocean takes up less CO_2, about 1.2 GtC per year on average for the decade 1980 - 1989. The GFDL 3-D ocean model (Sarmiento et al., 1990) has an oceanic uptake of 1.9 GtC per year for the same period. 3-D ocean models and especially coupled atmosphere-ocean models are the only means to study in a realistic way the feedback effects that climate change may have on atmospheric CO_2 via alteration of the ocean circulation (cf. Section 1.2.7.1). However, models need to be constrained by more data than are presently available.

The oceanic uptake of CO_2 for the decade 1980 - 1989, as estimated based on carbon models (e.g., Siegenthaler and Oeschger, 1987; Maier-Reimer et al., personal communication, 1990; Goudriaan, 1989; Sarmiento et al., 1990) is in the range 2.0±0.8 GtC per year.

1.2.4.3 Redistribution of anthropogenic carbon dioxide

During the period 1850 to 1986, 195±20 GtC were released by fossil fuel burning and 117±35 GtC by deforestation and changes in land use, adding up to a cumulative input of 312±40 GtC.

Atmospheric CO_2 increased from about 288 ppmv to 348 ppmv during this period, corresponding to (41±6)% of the cumulative input. This percentage is sometimes called the "airborne fraction", but that term should not be misunderstood: all CO_2, anthropogenic and non-anthropogenic, is continuously being exchanged between atmosphere, ocean and biosphere. Conventionally, an "airborne fraction" referring to the fossil fuel input only has often been quoted, because only the emissions due to fossil fuel burning are known with good precision. However, this may be misleading, since the atmospheric increase is a response to the total emissions. We therefore prefer the definition based on the latter. The airborne fraction for the period 1980 - 1989 (see calculation below) corresponds to (48±8)% of the cumulative input.

In model simulations of the past CO_2 increase, using estimated emissions from fossil fuels and deforestation, it has generally been found that the simulated increase is larger than that actually observed. An estimate for the decade 1980-1989 is:

	GtC/yr
Emissions from fossil fuels into the atmosphere (Figure 1.5)	5.4±0.5
Emissions from deforestation and land use	1.6±1.0
Accumulation in the atmosphere	3.4±0.2
Uptake by the ocean	2.0±0.8
Net imbalance	1.6±1.4

The result from this budget and from other studies is that the estimated emissions exceed the sum of atmospheric increase plus model-calculated oceanic uptake by a significant amount. The question therefore arises whether an important mechanism has been overlooked. All attempts to identify such a "missing sink" in the ocean have, however, failed so far. A possible exception is that a natural fluctuation in the oceanic carbon system could have caused a decreasing atmospheric baseline concentration in the past few decades; this does not appear likely in view of the relative constancy of the pre-industrial CO_2 concentration. There are possible processes on land, which could account for the missing CO_2 (but it has not been possible to verify them). They include the stimulation of vegetative growth by increasing CO_2 levels (the CO_2 fertilization effect), the possible enhanced productivity of vegetation under warmer conditions, and the direct effect of fertilization from agricultural fertilizers and from nitrogenous releases into the atmosphere. It has been estimated that increased fertilization by nitrogenous releases could account for a sequestering of up to a maximum of 1 GtC per year in terrestrial ecosystems (Melillo, private communication, 1990). In addition, changed forest management practices may also result in an increase in the amount of carbon stored in northern mid-latitude forests. The extent to which mid-latitude terrestrial systems can sequester carbon before becoming saturated and ineffective is unknown. As mid-latitude terrestrial systems become close to saturation, and hence ineffective in sequestering carbon, this would allow more of the CO_2 to remain in the atmosphere.

A technique for establishing the global distribution of surface sources and sinks has been to take global observations of atmospheric CO_2 concentration and isotopic composition and to invert these by means of atmospheric transport models to deduce spatial and temporal patterns of surface fluxes (Pearman et al., 1983; Pearman and Hyson, 1986; Keeling and Heimann, 1986). The observed inter-hemispheric CO_2 concentration difference (currently about 3 ppmv) is smaller than one would expect given that nearly all fossil releases occur in the Northern Hemisphere. The results of this approach suggest that there is an unexpectedly large sink in the Northern Hemisphere, equivalent to more than half of the fossil fuel CO_2 release (Enting and Mansbridge, 1989; Tans et al., 1990; Keeling et al., 1989b). Furthermore, it has been concluded that the oceanic uptake compatible with oceanic and atmospheric CO_2 data and with a 3-dimensional atmospheric transport model is at most 1 GtC per year (Tans et al., 1990). Thus, a significant terrestrial sink, possibly larger than the oceanic uptake, is suggested by these model analyses.

1.2.4.4 Seasonal variations

Atmospheric CO_2 exhibits a seasonal cycle, dominated by the seasonal uptake and release of atmospheric CO_2 by land plants. Its amplitude is small (1.2 ppmv peak-to-peak) in the Southern Hemisphere and increases northward to a maximum of order 15 ppmv peak-to-peak in the boreal forest zone (55-65° N).

The amplitude of the seasonal cycle has been observed to be increasing (e.g., Pearman and Hyson, 1981: Bacastow et al., 1985; Thompson et al., 1986). For example, at Mauna Loa, Hawaii, the seasonal amplitude has increased by nearly 20% since 1958. The increase has however, not been monotonic, and different evaluation methods yield somewhat different values; still, it is statistically significant. This increasing amplitude could point to a growing productivity (NPP) of the terrestrial ecosystems, and to a sequestering of carbon by a growing biomass, provided the increase in biomass is not fully compensated by respiration. It is important to note that such a change does not necessarily indicate increased productivity or increased storage of carbon (Pearman and Hyson, 1981; Kohlmaier et al., 1989; Houghton, 1987); it could also be due to, e.g., accelerated soil respiration in winter.

1.2.4.5 Interannual variations

Small imbalances in natural exchange fluxes are reflected in interannual CO_2 concentration fluctuations (±1 ppmv over 1-2 years). They are correlated with the El Niño-Southern Oscillation (ENSO) phenomenon (Thompson et al., 1986; Keeling et al., 1989a), which suggests a relation to changes in the equatorial Pacific Ocean, where normally the upwelling causes a high pCO_2 peak and outgassing of CO_2 into the atmosphere. However, a closer inspection shows that this cannot be the dominating mechanism, since during El Niño, the equatorial pCO_2 peak disappears (Feely et al., 1987), while atmospheric CO_2 grows more strongly than normally. Alternatively, processes in the land biosphere, perhaps in response to climatic events connected with ENSO events, may be responsible. This explanation is supported by one set of stable carbon isotope data on atmospheric CO_2 (Keeling et al., 1989a); but not supported by a second set (Goodman and Francey, 1988).

1.2.4.6 Temporal variations of carbon isotopes

The release of CO_2 from biospheric carbon and fossil fuels, both having lower $^{13}C/^{12}C$ ratios than atmospheric CO_2, has led to a decrease of the isotope ratio $^{13}C/^{12}C$ in the atmosphere by about $1^O/_{OO}$. The man-made emissions of ^{14}C-free fossil fuel CO_2 have likewise caused a decrease of the atmospheric ^{14}C concentration (measured on tree-rings)

of the order of 2% from 1800 to 1950. Both isotopic perturbations can be used to constrain the history of the anthropogenic release of CO_2. The observed decrease of ^{13}C, as observed in air trapped in ice cores (Friedli et al., 1986) and ^{14}C, observed in tree rings, agree, within experimental uncertainty, with those expected from model calculations with the same carbon cycle models as used for studying the CO_2 increase (Stuiver and Quay, 1981; Siegenthaler and Oeschger, 1987). The interpretation of ^{13}C trends in tree rings has proven to be difficult because of plant physiological effects on isotope fractionation (Francey and Farquhar, 1982).

1.2.5 Evidence that the Contemporary Carbon Dioxide Increase is Anthropogenic

How do we know that in fact human activity has been responsible for the well documented 25% increase in atmospheric CO_2 since the early 19th century? Couldn't this rise instead be the result of some long-term natural fluctuation in the natural carbon cycle? Simple arguments allow us to dismiss this possibility.

First, the observational CO_2 records from ice cores with good time resolution clearly show that the maximum range of natural variability about the mean of 280 ppmv during the past 1000 years was small (10 ppmv over a 100 year time-scale), that is an order of magnitude less than the observed rise over the last 150 years. A value as high as the current level of 353 ppmv is not observed anywhere in the measured ice core record for the atmospheric history during the past 160,000 years; the maximum value is 300 ppmv during the previous interglacial, 120,000 years ago.

Second, the observed rate of CO_2 increase closely parallels the accumulated emission trends from fossil fuel combustion and from land use changes (c.f. Section 1.2.2). Since the start of atmospheric monitoring in 1958, the annual atmospheric increase has been smaller each year than the fossil CO_2 input. Thus, oceans and biota together must have been a global sink rather than a source during all these years. Further evidence is provided by the fact that the north-to-south CO_2 concentration difference has been observed to increase from 1 ppmv in 1960 to 3 ppmv in 1985, parallel to the growth of the (Northern Hemisphere) fossil fuel combustion sources (Keeling et al., 1989a).

Third, the observed isotopic trends of ^{13}C and ^{14}C agree qualitatively with those expected due to the CO_2 emissions from fossil fuels and the biosphere, and they are quantitatively consistent with results from carbon cycle modelling.

1.2.6 Sensitivity Analyses for Future Carbon Dioxide Concentrations

Future atmospheric CO_2 concentrations depend primarily on emission rates from energy use and deforestation, and on the effectiveness of the ocean and land biota as CO_2

sinks. For the sake of illustration, several schematic scenarios are shown in Figures 1.7 and 1.8. Those of Figure 1.7 are based on prescribed total CO_2 emission rates after 1990; for those in Figure 1.8 atmospheric concentrations after 1990 were prescribed and the corresponding emission rates were calculated to fit these concentrations. A box-diffusion model of the global cycle was used for these simulations (Enting and Pearman, 1982, 1987), with an oceanic eddy diffusivity of 5350 m^2year^{-1} and an air-sea gas exchange rate corresponding to an exchange coefficient of 0.12 $year^{-1}$. The calculations assume no biospheric-climate feedbacks, and also assume that after 1990 the net biospheric input of CO_2 is zero, i.e., the input of CO_2 from tropical deforestation is balanced by uptake of CO_2 by terrestrial ecosystems.

In case a (all emissions stopped; Figure 1.7), the atmospheric concentration declines, but only slowly (from 351 ppmv in 1990 to 331 ppmv in 2050 and 324 ppmv in 2100), because the penetration of man-made CO_2 to deeper ocean layers takes a long time. Even if the emissions were reduced by 2% per year from 1990 on (case b), atmospheric CO_2 would continue to increase for several decades. Case c (constant emission rate after 1990) gives CO_2 levels of about 450 ppmv in 2050 and 520 ppmv in 2100. A constant relative growth rate of 2% per year (case d) would yield 575 ppmv in 2050 and 1330 ppmv in 2100. Comparison of cases b, c and d clearly shows that measures to reduce emissions will result in slowing down the rate of atmospheric CO_2 growth.

Cases b' and c', in comparison to b and c, schematically illustrate the effect of reducing emissions in 2010 instead of in 1990.

If an (arbitrary) threshold of 420 ppmv. i.e., 50% above pre-industrial, is not to be exceeded (case e, Figure 1.8), then CO_2 production rates should slowly decline, reaching about 50% of their present value by 2050 and 30% by 2100. In order to keep the concentration at the present level (case f), emissions would have to be reduced drastically to 30% of present immediately and to less than 20% by 2050.

The results of scenario calculations with a 3-D ocean-atmosphere model (Maier-Reimer and Hasselmann, 1987; Maier-Reimer et al., personal communication, 1990 - revised model) give higher concentrations than those shown in Figure 1.7 obtained with a box-diffusion model; for instance, about 480 ppmv in the year 2050 and about 560 ppmv in the year 2100 for Scenario C, compared to about 450 ppmv and 520 ppmv. On the other hand, calculations with a box model that includes a biospheric CO_2 sink (Goudriaan, 1989) yields somewhat lower concentrations than shown in Figure 1.7; for instance about 415 ppmv in the year 2050 and 460 ppmv in the year 2100 for Scenario C.

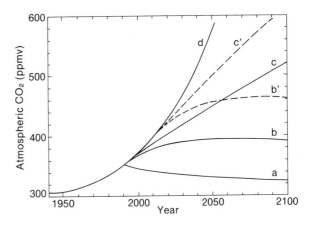

Figure 1.7: Future atmospheric CO_2 concentrations as simulated by means of a box-diffusion carbon cycle model (Enting and Pearman, 1982, 1987) for the following scenarios: (a) - (d): anthropogenic CO_2 production rate p prescribed after 1990 as follows: (a) p = 0; (b) p decreasing by 2% per year; (c) p = constant; (d) p increasing at 2% per year. Scenarios (b') and (c'): p grows by 2% per year from 1990-2010, then decreases by 2% per year (b') or is constant (c'). Before 1990, the concentrations are those observed (cf. Figure 1.3), and the production rate was calculated to fit the observed concentrations.

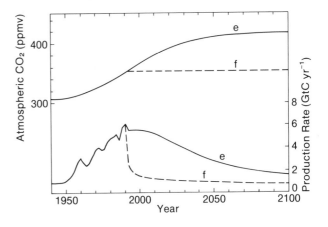

Figure 1.8: Future CO_2 production rates calculated by means of a box-diffusion carbon cycle model (Enting and Pearman, 1982, 1987) so as to yield the prescribed atmospheric CO_2 concentrations after 1990. (e) concentration increasing steadily (logistic function of time) to 420 ppmv. (f) concentration constant after 1990.

1.2.7 Feedbacks from Climate Change into the Carbon Dioxide Cycle

As increasing greenhouse gas concentrations alter the Earth's climate, changing climate and environmental conditions in their turn act back on the carbon cycle and atmospheric CO_2. The climate change Earth has experienced in the recent past is still within the range of natural short-term variability, and so are probably, therefore, the feedback effects of anthropogenic climate change. However, as the changes in the climate become larger than natural climatic variation the magnitude of the feedback effects should begin to have a significant effect. These feedbacks could in general be either positive (amplifying the initial changes) or negative (attenuating them).

1.2.7.1 Oceanic feedback effects
The following are possible feedback effects on the ocean-atmosphere carbon system:

1.2.7.1.1 Ocean temperature: Ocean temperature changes can affect sea water CO_2 chemistry. Surface-water pCO_2 will increase with increasing temperature, tending to decrease the net uptake by the oceans. The future atmospheric CO_2 increase may be amplified by something like 5 percent due to this effect (Lashof, 1989).

1.2.7.1.2 Ocean circulation: The ocean circulation may change in response to climatic change. As a consequence of increasing surface water temperatures, the thermocline may become more resistant to vertical mixing and slow down the uptake of anthropogenic CO_2. Modified wind stress may affect the ocean circulation. However, the overall change in ocean dynamics, and consequently in CO_2 uptake, due to a climatic change cannot be estimated from simple considerations; a proper evaluation of such an effect can only be done using dynamical ocean models. Studies on Greenland ice cores indicates that during the last glaciation, significant CO_2 concentration shifts may have occurred within less than 100 years (c.f. Section 1.2.3), probably caused by strong changes of large-scale ocean circulation. Therefore, the possibility that, due to climatic changes, unexpected abrupt events may take place in the natural carbon system cannot be excluded.

1.2.7.1.3 Gas exchange rates: A change in the global wind pattern could influence the gas transfer from the atmosphere to the sea surface. Carbon cycle models show that the net CO_2 uptake by the global ocean is not sensitive to the gas transfer coefficients (because it is controlled mainly by vertical mixing, not by gas exchange; Oeschger et al., 1975; Broecker et al., 1980; Sarmiento et al., 1990), so this effect would probably be of minor influence.

1.2.7.1.4 Modification of oceanic biogeochemical cycling: The rain of dead organic particles corresponds to a continuous export flux of carbon (and nutrients) out of the ocean surface, which under non-perturbed conditions is balanced by an equal upward transport of dissolved carbon

(and dissolved nutrients) by water motion. In polar regions and strong upwelling zones, where productivity is not limited by nitrogen or phosphorus, the balance could become disturbed consequent on variations in ocean dynamics (c.f. Section 1.2.7.1.2), so as to influence atmospheric CO_2. As a result of climate change, the distribution of marine ecosystems and species composition could change, which could affect pCO_2 in surface waters. It is not possible at present to predict the direction and magnitude of such effects.

Warming of the oceans might lead to accelerated decomposition of dissolved organic carbon, converting it into CO_2, and thus amplify the atmospheric increase (Brewer, personal communication, 1990).

1.2.7.1.5 UV-B radiation: A reduction in stratospheric O_3 would increase the intensity of UV-B radiation at the Earth's surface. This might have negative effects on the marine biota due to a decrease of marine productivity and thus on the biological carbon pump. This could lead to an increase in the concentration of CO_2 in surface waters and consequently in the atmosphere.

1.2.7.2 Terrestrial biospheric feedbacks
The following are probable feedback effects on the terrestrial biosphere-atmospheric carbon system:

1.2.7.2.1 Carbon dioxide fertilization: Short-term experiments under controlled conditions with crops and other annuals, as well as with a few perennials, show an increase in the rates of photosynthesis and growth in most plants under elevated levels of CO_2 (Strain and Cure, 1985). If elevated levels of CO_2 increase the productivity of natural ecosystems, more carbon may be stored in woody tissue or soil organic matter. Such a storage of carbon will withdraw carbon from the atmosphere and serve as a negative feedback on the CO_2 increase. Of particular importance is the response of forests (Luxmoore et al., 1986), given that forests conduct about 2/3 of global photosynthesis (50% of this cycles annually through leaves, while 50% is stored in woody tissue). However, it is not clear whether the increases in photosynthesis and growth will persist for more than a few growing seasons, whether they will occur at all in natural ecosystems and to what degree they will result in an increased storage of carbon in terrestrial ecosystems.

1.2.7.2.2 Eutrophication and toxification: The increased availability of nutrients such as nitrate and phosphate from agricultural fertilizers and from combustion of fossil fuels may stimulate the growth of plants. It has been estimated that the effect of eutrophication, both on land and in the oceans, could be as large as 1 GtC per year (Melillo, private communication, 1990). However, it should be noted that the greater availability of nutrients has often been associated with increasing levels of acid precipitation and air pollution, which have been associated with a reduction in the growth of terrestrial biota.

1.2.7.2.3 Temperature: Under non-tropical conditions, photosynthesis and respiration by plants and by microbes both tend to increase with increasing temperature; but respiration is the more sensitive process, so that a warming of global air temperature is likely to result in an initially increased release of carbon to the atmosphere. Estimates indicate that the additional flux might be significant, perhaps as large as one or a few GtC per year (Woodwell, 1983; Kohlmaier, 1988; Lashof, 1989; Houghton and Woodwell, 1989). This temperature-enhanced respiration would be a positive feedback on global warming.

1.2.7.2.4 Water: Changes in soil water may affect carbon fixation and storage. Increased moisture can be expected to stimulate plant growth in dry ecosystems and to increase the storage of carbon in tundra peat. There is a possibility that stresses brought about by climatic change may be alleviated by increased levels of atmospheric CO_2. At present however, it is not possible to predict reliably either the geographical distribution of changes in soil water or the net effect of these changes on carbon fluxes and storage in different ecosystems. Changes in climate are generally believed to be more important than changes in the atmospheric concentration of CO_2 in affecting ecosystem processes (c.f. Section 10.)

1.2.7.2.5 Change in geographical distribution of vegetation types: In response to environmental change, the structure and location of vegetation types may change. If the rate of change is slow, plant distributions may adjust. If, however, the rate of change is fast, large areas of forests might not be able to adapt rapidly enough, and hence be negatively affected with a subsequent release of CO_2 to the atmosphere.

1.2.7.2.6 UV-B radiation: A reduction in stratospheric O_3 would increase the intensity of UV-B radiation at the Earth's surface. Increased UV-B may have a detrimental effect on many land biota, including crops (Teramura, 1983), thus affecting the strength of the biospheric sink of CO_2 over land.

1.2.8 Conclusions

The atmospheric CO_2 concentration is now about 353ppmv, 25% higher than the pre-industrial (1750-1800) value and higher than at any time in at least the last 160,000 years. This rise, currently amounting to about 1.8 ppmv per year, is beyond any doubt due to human activities. Anthropogenic emissions of CO_2 were 5.7 ± 0.5

GtC due to fossil fuel burning in 1987, plus 0.6 to 2.5 GtC due to deforestation (estimate for 1980). During the last decade (1980 - 1989) about 48% of the anthropogenic emissions have stayed in the atmosphere, the remainder has been taken up by the oceans and possibly by land ecosystems. Our qualitative knowledge of the global carbon cycle is, in view of the complexity of this cycle, relatively good. However, the current quantitative estimates of sources and of sinks of CO_2 do not balance; the atmospheric increase is less rapid than expected from carbon cycle models (in which CO_2 fertilization or environmental responses of the biosphere are not included). This, and model analyses of the inter-hemispheric CO_2 gradient, indicate that the Northern Hemisphere terrestrial ecosystems may act as a significant sink of carbon. Such a sink has, however, not been directly identified. To summarize: the total annual input of anthropogenic CO_2 is currently (1980-1989) about 7.0 ± 1.1 GtC, assuming a central value for the input of CO_2 from tropical deforestation; the annual uptake by the oceans is estimated (based on the box models, GCMs and Tans et al., 1990) to be about 2.0 ± 1.0 GtC; and the annual atmospheric accumulation is about 3.4 ± 0.2 GtC. Thus, the annual sequestering by the terrestrial biosphere should be about 1.6 ± 1.5 GtC. While several mechanisms have been suggested that could sequester carbon in terrestrial ecosystems, it is difficult to account for the total required sink. Therefore, it appears likely that, (i) the uptake of CO_2 by the oceans is underestimated, (ii) there are important unidentified processes in terrestrial ecosystems that can sequester CO_2, and/or (iii) the amount of CO_2 released from tropical deforestation is at the low end of current estimates.

If the land biota presently act as a sink of carbon due to a fertilization effect, then they might become saturated with respect to this fertilization at some time in the future. This means that we cannot assume that the terrestrial sink, which may be active currently, will continue to exist unchanged through the next century.

In order to avoid a continued rapid growth of CO_2 in the atmosphere, severe reductions on emissions will be necessary. The time taken for atmospheric CO_2 to adjust to changes in sources or sinks is of the order of 50-200 years, determined mainly by the slow exchange of carbon between surface waters and deeper layers of the ocean. Even if all anthropogenic emissions of CO_2 were halted, the atmospheric concentration would decline only slowly, and it would not approach its pre-industrial level for many hundreds of years. Thus, any reductions in emissions will only become fully effective after a time of the order of a century or more. Based on some model estimates, which neglect the feedbacks discussed earlier, the atmospheric concentration in the year 2050 would be between 530 - 600 ppmv for a constant relative growth of the annual anthropogenic emissions by 2% per year, and between 415 - 480 ppmv (increasing to 460 - 560 ppmv by the year 2100) for a constant anthropogenic emission rate at the 1990 level. In order not to exceed 420 ppmv (50% above pre-industrial), annual anthropogenic emissions would have to be reduced continuously to about 50% of their present value by the year 2050. In order to stabilize concentrations at present day concentrations (353 ppmv), an immediate reduction in global anthropogenic emissions by 60-80 percent would be necessary. The size of the estimated reduction depends on the carbon cycle model used.

During the millennium preceding the anthropogenic CO_2 growth, the concentration was relatively constant near 280 ppmv, with a variability of less than ± 10 ppmv. This indicates that the sensitivity of atmospheric CO_2 levels to minor climatic changes such as the Little Ice Age, where global mean temperatures probably decreased by about 1°C, is within this range. However, the anticipated climatic and environmental changes may soon become large enough to act back on the oceanic and terrestrial carbon cycle in a more substantial way. A close interaction between climate variations and the carbon cycle is indicated by the glacial-interglacial CO_2 variations. The ice-core record shows that CO_2 concentrations during the coldest part of the last glaciation were about 30% lower than during the past 10,000 years. The glacial-interglacial CO_2 variations were probably due to changes in ocean circulation and marine biological activity, and were correlated to variations in global climate. There is some (not fully clear) evidence from ice cores that rapid changes of CO_2, ca. 50 ppmv within about a century, occurred during and at the end of the ice age.

If global temperatures increase, this could change the natural fluxes of carbon, thus having feedback effects on atmospheric CO_2. Some of the identified feedbacks are potentially large and could significantly influence future CO_2 levels. They are difficult to quantify, but it seems likely that there would be a net positive feedback, i.e., they will enhance the man-made increase. On the longer term, the possibility of unexpected large changes in the mechanisms of the carbon cycle due to a human-induced change in climate cannot be excluded.

1.3 Methane

Methane is a chemically and radiatively active trace gas that is produced from a wide variety of anaerobic (i.e., oxygen deficient) processes and is primarily removed by reaction with hydroxyl radicals (OH) in the troposphere. Oxidation of CH_4 by OH in the stratosphere is a significant source of stratospheric water (H_2O) where it is an important greenhouse gas.

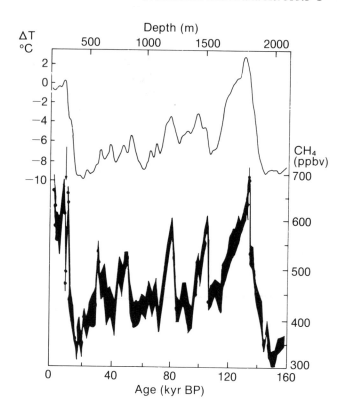

Figure 1.9: Methane concentrations (bottom) and estimated temperature changes (top) during the past 160,000 years as determined on the ice core from Vostok, Antarctica (Chappelaz et al., 1990). Temperature changes were estimated based on the measured deuterium concentrations.

1.3.1 Atmospheric Distribution of Methane

1.3.1.1 Palaeo-atmospheric record of methane

There are good data on the atmospheric concentration of CH_4 (Figure 1.9) from Antarctic and Greenland ice cores for the period between 10,000 and 160,000 years ago (Raynaud et al., 1988; Stauffer et al., 1988; Craig and Chou, 1982; Chappellaz et al., 1990). The minimum concentration during the last glacial periods (about 20,000 and 150,000 years ago) was around 0.35 ppmv, and rose rapidly, in phase with the observed temperature increases, to about 0.65 ppmv during the glacial-interglacial transitions (about 15,000 and 130,000 years ago). The atmospheric concentrations of CH_4 decreased rapidly, prior to, and during the last deglaciation period about 10,000 - 11,000 years ago (the Younger Dryas period when there were abrupt temperature decreases in Greenland and northern Europe), and increased rapidly thereafter.

Because of the brittle nature of the ice cores, data on the atmospheric concentrations of CH_4 are reliable only during the last 2,000 years of the Holocene period (last 10,000 years).

1.3.1.2 Contemporary record of methane

Ice core data (Figure 1.10) indicate that the atmospheric concentrations of CH_4 averaged around 0.8 ppmv between two hundred and two thousand years ago, increasing to 0.9 ppmv one hundred years ago (Craig and Chou, 1982; Rasmussen and Khalil, 1984; Stauffer et al., 1985; Pearman and Fraser, 1988; Pearman et al., 1986; Etheridge et al., 1988). Since then, the atmospheric concentration of CH_4 has increased smoothly to present levels, highly correlated with global human population. Analysis of infrared solar spectra has shown that the atmospheric concentration of CH_4 has increased by about 30% over the last 40 years (Rinsland et al., 1985; Zander et al., 1990).

Atmospheric concentrations of CH_4 have been measured directly since 1978 when the globally averaged value was 1.51 ppmv (e.g., Rasmussen and Khalil, 1981; Blake and Rowland, 1988). Currently the value is 1.72 ppmv, corresponding to an atmospheric reservoir of about 4900 Tg (1 Tg = 10^{12} g) and it is increasing at a rate of 14 to 17 ppbv per year (40 to 48 Tg per year), i.e., 0.8 to 1.0% per year (Blake and Rowland, 1988; Steele et al., 1987). The atmospheric concentration of CH_4 in the Northern Hemisphere is 1.76 ppmv, compared to 1.68 ppmv in the Southern Hemisphere (Figure 1.11). The magnitude of the seasonal variability varies with latitude (Steele et al. 1987; Fraser et al. 1984), being controlled by the temporal variability in source strengths and atmospheric concentration of OH radicals.

1.3.1.3 Isotopic composition of methane

Methane is produced from different sources with distinctive proportions of carbon ^{12}C, ^{13}C, and ^{14}C, and hydrogen isotopes H, D (^{2}H), and T (^{3}H). Similarly, the rates of processes that destroy CH_4 depend upon its isotopic composition. Consequently, the CH_4 budget can be constrained by knowledge of the isotopic composition of atmospheric CH_4, the extent of isotopic fractionation during removal, and the isotopic signatures of CH_4 from different sources. Recent work to elucidate the sources of CH_4 has proceeded through an analysis of carbon isotopic signatures (Cicerone and Oremland, 1988; Wahlen et al., 1989; Lowe et al., 1988; and references therein). One example of this is an analysis of ^{14}C data which suggests that about 100 Tg CH_4 per year may arise from fossil sources (Cicerone and Oremland, 1988; Wahlen et al., 1989). Such a distinction is possible because CH_4 from fossil sources is ^{14}C-free, while that from other sources has essentially the ^{14}C concentration of modern carbon.

1.3.2 Sinks of Methane

The major sink for atmospheric CH_4 is reaction with OH in the troposphere, the OH concentration being controlled by a complex set of reactions involving CH_4, CO, NMHC, NO_x, and tropospheric O_3 (discussed in Section 1.7; Sze,

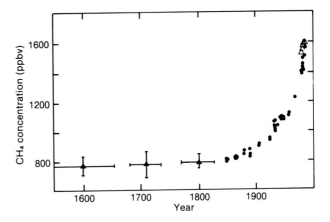

Figure 1.10: Atmospheric methane variations in the past few centuries measured from air in dated ice cores (Etheridge et al., 1988; Pearman and Fraser, 1990).

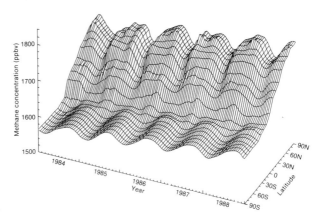

Figure 1.11: The global distribution, seasonality, and trend of methane from the GMCC network (Steele et al., 1987, and unpublished data)

1977; Crutzen, 1987). Based on the reaction rate coefficient between CH_4 and OH, and the estimated tropospheric distribution of OH, an atmospheric lifetime for CH_4 of between 8 and 11.8 years has been estimated (Prinn et al. 1987). This estimate is supported by the fact that models of global OH are tested by analyses of the budgets for CH_3CCl_3 (Logan et al., 1981; Fraser et al., 1986a; Prinn et al., 1987) and ^{14}CO (Appendix to WMO, 1989b). The reaction between CH_4 and OH currently represents a sink of 400 to 600 Tg of CH_4 per year. The efficiency of this sink may, however, have decreased during the last century because the atmospheric concentration of OH in the troposphere may have decreased, hence the lifetime of CH_4 would have increased, in response to increasing concentrations of CO, NMHC, and CH_4 (Sze, 1977).

Table 1.2 Estimated Sources and Sinks of Methane

	Annual Release (Tg CH$_4$)	Range (Tg CH$_4$)
Source		
Natural Wetlands (bogs, swamps, tundra, etc)	115	100 - 200
Rice Paddies	110	25 - 170
Enteric Fermentation (animals)	80	65 - 100
Gas Drilling, venting, transmission	45	25 - 50
Biomass Burning	40	20 - 80
Termites	40	10 - 100
Landfills	40	20 - 70
Coal Mining	35	19 - 50
Oceans	10	5 - 20
Freshwaters	5	1 - 25
CH$_4$ Hydrate Destabilization	5	0 - 100
Sink		
Removal by soils	30	15 - 45
Reaction with OH in the atmosphere	500	400 - 600
Atmospheric Increase	44	40 - 48

Soils may represent a removal mechanism for CH$_4$. The magnitude of this sink has been estimated (this assessment) to be 30±15 Tg CH$_4$ per year from the work of Harriss et al., 1982 and Seiler and Conrad, 1987.

1.3.3 Sources of Methane

Methane is produced from a wide variety of anaerobic sources (Cicerone and Oremland, 1988). Two main pathways for CH$_4$ production have been identified: (i) reduction of CO$_2$ with hydrogen, fatty acids or alcohols as hydrogen donors, or (ii) transmethylation of acetic acid or methyl alcohol by CH$_4$-producing bacteria. Table 1.2 summarizes identified sources of CH$_4$ with ranges of likely annual emissions. The total annual CH$_4$ source must equal the atmospheric sink of about 500 (400 to 600) Tg CH$_4$ per year, the possible soil sink of about 30 (15 to 45) Tg CH$_4$ per year, and the annual growth of 40 to 48 Tg CH$_4$ in the atmosphere. The sum of the present best estimates of the sizes of the individual sources identified in Table 1.2 equal 525 Tg CH$_4$ per year. It should be noted that the newest data for rice paddies, biomass burning, and coal mining sources suggest that the values may be even less than those of Table 1.2, possibly indicating a missing source of CH$_4$, or an overestimate of the sink for CH$_4$.

1.3.3.1 Natural wetlands

Significant progress has been made in quantifying the magnitude of the source of CH$_4$ from natural wetlands (Svensson and Rosswall, 1984; Sebacher et al., 1986; Whalen and Reeburgh, 1988; Moore and Knowles, 1987; Mathews and Fung, 1987; Harriss et al., 1985; Crill et al., 1988; Andronova, 1990; Harriss and Sebacher, 1981; Burke et al., 1988; Harriss et al., 1988; Aselmann and Crutzen, 1989). Recent data support earlier estimates of a global flux of 110 - 115 Tg CH$_4$ per year, but reverses the relative importance of tropical and high latitude systems (Bartlett et al., 1990). The data base, which is still quite limited (no data from Asia), suggests 55 Tg CH$_4$ per year (previously 32 Tg CH$_4$ per year) from tropical wetlands, and 39 Tg CH$_4$ per year (previously 63 Tg CH$_4$ per year) from high latitude wetlands. Since CH$_4$ is produced through biological processes under anaerobic conditions, any factors affecting the physical, chemical or biological characteristics of soils could affect CH$_4$ emission rates.

1.3.3.2 Rice paddies

Rice paddies are an important source of CH$_4$ with estimates of the globally averaged flux ranging from 25 - 170 Tg CH$_4$ per year (Neue and Scharpenseel, 1984; Yagi and Minami, 1990; Holzapfel-Pschorn and Seiler, 1986; Cicerone and Shetter, 1981; Cicerone et al., 1983). The flux of CH$_4$ from rice paddies is critically dependent upon several factors including: (i) agricultural practices (e.g., fertilization, water management, density of rice plants, double cropping systems, application of manure or rice straw), (ii) soil / paddy characteristics (soil type, acidity, redox potential, temperature, nutrient availability, substrate, profile of anaerobic environment), and (iii) time of season. One difficulty in obtaining accurate estimates is that almost 90% of the world's harvested area of rice paddies is in Asia, and of this about 60% are in China and India from which no detailed data are available. The annual production of rice since 1940 has approximately doubled as a result of double cropping practices and an increased area of cultivation. It is likely that CH_4 emissions have increased proportionally as well.

1.3.3.3 Biomass burning

Biomass burning in tropical and sub-tropical regions is thought to be a significant source of atmospheric CH_4, with estimates of global emission rates ranging from 20 to 80 Tg CH_4 per year (Andreae et al., 1988; Bingemer and Crutzen, 1987; Crutzen et al., 1979; Crutzen et al., 1985; Crutzen, 1989; Greenberg et al., 1984; Stevens et al., 1990; Quay et al., 1990). Improved estimates require an enhanced understanding of: (i) CH_4 emission factors, (ii) the amount, by type, of vegetation burnt each year on an area basis, and (iii) type of burning (smouldering vs flaming). Current estimates indicate that over the last century the rate of forest clearing by burning has increased (c.f. Section 1.2.2.2).

1.3.3.4 Enteric fermentation (animals)

Methane emissions from enteric fermentation in ruminant animals, including all cattle, sheep and wild animals, is estimated to provide an atmospheric source of 65 - 100 Tg CH_4 per year (Crutzen et al., 1986; Lerner et al., 1988). Methane emissions depend upon animal populations, as well as the amount and type of food. It is difficult to estimate the change in this source over the last century accurately because the significant increase in the number of cattle and sheep has been partially offset by decreases in the populations of elephants and North American bison. One estimate suggests that the magnitude of this source has increased from 21 Tg CH_4 per year in 1890 to 78 Tg CH_4 per year in 1983 (Crutzen et al., 1986).

1.3.3.5 Termites

There is a large range in the magnitude of the estimated fluxes of CH_4 from termites; 10 - 100 Tg CH_4 per year (Cicerone and Oremland, 1988; Zimmerman et al., 1982; Rasmussen and Khalil, 1983; Seiler et al., 1984; Fraser et al., 1986b). The values are based on the results of laboratory experiments, applied to estimates of global termite populations and the amount of biomass consumed by termites, both of which are uncertain, and field

experiments. It is important to determine whether the global termite population is currently increasing, and whether it is likely to respond to changes in climate.

1.3.3.6 Landfills

The anaerobic decay of organic wastes in landfills may be a significant anthropogenic source of atmospheric CH_4, 20 - 70 Tg CH_4 per year. However, several factors need to be studied in order to quantify the magnitude of this source more precisely, including amounts, trends, and types of waste materials, and landfill practices (Bingemer and Crutzen, 1987).

1.3.3.7 Oceans and freshwaters

Oceans and freshwaters are thought to be a minor source of atmospheric CH_4. The estimated flux of CH_4 from the oceans is based on a limited data set, taken in the late 1960's / early 1970's when the atmospheric concentration of CH_4 was about 20% lower. They showed that the open oceans were only slightly supersaturated in CH_4 with respect to its partial pressure in the atmosphere. There are inadequate recent data from either the open oceans or coastal waters to reduce the uncertainty in these estimates (Cicerone and Oremland, 1988).

1.3.3.8 Coal mining

Methane is released to the atmosphere from coal mine ventilation, and degassing from coal during transport to an end-use site. A recent unpublished study estimated the flux of CH_4 from coal mining, on a country basis, for the top twenty coal producing countries, and deduced a global minimum emission of 19 Tg CH_4 per year. Global CH_4 fluxes from coal mining have been estimated to range from 10 - 50 Tg CH_4 per year (Cicerone and Oremland, 1988; ICF, 1990, and recent unpublished studies by others).

1.3.3.9 Gas drilling, venting and transmission

Methane is the major component of natural gas, hence leakage from pipelines and venting from oil and gas wells could represent a significant source of atmospheric CH_4 (Cicerone and Oremland, 1988). The global flux from these sources is estimated, based on limited data of questionable reliability, to range from 25 - 50 Tg CH_4 per year.

1.3.4 Feedbacks from Climate Change into the Methane Cycle

Future atmospheric concentrations of CH_4 will depend on changes in the strengths of either the sources or sinks, which are dependent upon social, economic, and political, and also environmental factors and in particular changes in climate. Methane emissions from wetlands are particularly sensitive to temperature and soil moisture, and hence future climatic changes could significantly change the fluxes of CH_4 from both natural wetlands and rice paddies.

Tropospheric OH, which provides the atmospheric sink for CH_4, is dependent upon a number of factors, including the intensity of UV-B radiation, and the ambient concentrations of H_2O, CO, CH_4, reactive nitrogen oxides, and tropospheric O_3 (See Section 1.7) (Crutzen, 1987; Isaksen and Hov, 1987; Thompson and Cicerone, 1986).

1.3.4.1 Tropical methane sources

The major sources of CH_4 in tropical regions (natural wetlands and rice paddies) are quite sensitive to variations in soil moisture. Consequently, changes in soil moisture, which would result from changes in temperature and precipitation, could significantly alter the magnitude of these large sources of atmospheric CH_4. Increased soil moisture would result in larger fluxes, whereas a decrease in soil moisture would result in smaller fluxes.

1.3.4.2 High latitude methane sources

Methane fluxes from the relatively flat tundra regions would be sensitive to changes of only a few centimetres in the level of the water table, with flooded soils producing a factor of 100 more CH_4 than dry soils. Similarly, emissions of CH_4 are significantly larger at warmer temperatures, due to accelerated microbiological decomposition of organic material in the near-surface soils (Whalen and Reeburgh, 1988; Crill et al., 1988). Consequently, an increase in soil moisture and temperatures in high latitude wetlands would result in enhanced CH_4 emissions, whereas warmer dryer

soils might have decreased CH_4 emissions.

Higher temperatures could also increase the fluxes of CH_4 at high northern latitudes from; (i) CH_4 trapped in permafrost, (ii) decomposable organic matter frozen in the permafrost, and (iii) decomposition of CH_4 hydrates (Cicerone and Oremland, 1988; Kvenvolden, 1988; Nisbet, 1989). Quantifying the magnitudes of these positive feedbacks is difficult. Time-scales for thawing the permafrost, located between a few centimetres to metres below the surface, could be decades to centuries, while the time for warming the CH_4 hydrates could be even longer, although one study (Kvenvolden, 1988) estimated that the flux of CH_4 from hydrate decomposition could reach 100 Tg CH_4 per year within a century.

1.3.5 Conclusions

Current atmospheric CH_4 concentrations, at 1.72 ppmv, are now more than double the pre-industrial value (1750-1800) of about 0.8 ppmv, and are increasing at a rate of 0.9% per year. The ice core record shows that CH_4 concentrations were about 0.35 ppmv during glacial periods, and increased in phase with temperature during glacial-interglacial transitions. The current atmospheric concentration of CH_4 is greater than at any time during the last 160,000 years.

Reaction with OH in the troposphere, the major sink for CH_4, results in a relatively short atmospheric lifetime of 10 ± 2 years. The short lifetime of CH_4 implies that atmospheric concentrations will respond quite rapidly, in

Table 1.3 Halocarbon Concentrations and Trends (1990) †

Halocarbon		Mixing Ratio pptv	Annual Rate of Increase pptv	%	Lifetime Years
CCl_3F	(CFC-11)	280	9.5	4	65
CCl_2F_2	(CFC-12)	484	16.5	4	130
$CClF_3$	(CFC-13)	5			400
$C_2Cl_3F_3$	(CFC-113)	60	4-5	10	90
$C_2Cl_2F_4$	(CFC-114)	15			200
C_2ClF_5	(CFC-115)	5			400
CCl_4		146	2.0	1.5	50
$CHClF_2$	(HCFC-22)	122	7	7	15
CH_3Cl		600			1.5
CH_3CCl_3		158	6.0	4	7
$CBrClF_2$	(halon 1211)	1.7	0.2	12	25
$CBrF_3$	(halon 1301)	2.0	0.3	15	110
CH_3Br		10-15			1.5

†   There are a few minor differences between the lifetimes reported in this table and the equivalent table in WMO 1989b. These differences are well within the uncertainty limits. The 1990 mixing ratios have been estimated based upon an extrapolation of measurements reported in 1987 or 1988, assuming that the recent trends remained approximately constant.

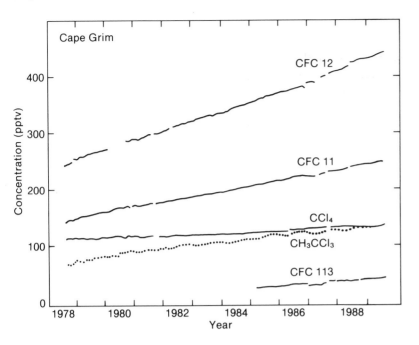

Figure 1.12: Halocarbon concentrations measured at Cape Grim, Tasmania during the period 1978-1989 (Fraser and Derek, 1989, and unpublished data).

comparison to the longer lived gases such as CO_2, N_2O, and CFCs, to changes in emissions. In order to stabilize concentrations at present day levels, an immediate reduction in global man-made emissions by 15-20 percent would be necessary (this and other scientific sensitivity analyses are discussed in the Annex). Global concentrations of OH are dependent upon the intensity of UV-B radiation, and the concentrations of gases such as H_2O, CO, CH_4, NO_x, NMHC, and O_3, and may have declined during the twentieth century due to changes in the atmospheric concentrations of these gases.

The individual sources of atmospheric CH_4 have been qualitatively identified, but there are significant uncertainties in the magnitude of their strengths. Human activities such as rice cultivation, rearing of domestic ruminants, biomass burning, coal mining, and natural gas venting have increased the input of CH_4 into the atmosphere, and these combined with an apparent decrease in the concentration of tropospheric OH, yields the observed rise in global CH_4. However, the quantitative importance of each of the factors contributing to the observed increase is not well known at present.

Several potential feedbacks exist between climate change and CH_4 emissions, in both tropical and high latitude wetland sources. In particular, an increase in high latitude temperatures could result in a significant release of CH_4 from the melting of permafrost and decomposition of CH_4 hydrates.

1.4 Halocarbons

Halocarbons containing chlorine and bromine have been shown to deplete O_3 in the stratosphere. In addition, it has been recognized that they are important greenhouse gases. Their sources, sinks, atmospheric distributions, and role in perturbing stratospheric O_3 and the Earth's radiative balance have been reviewed in detail (WMO 1985, 1989a, 1989b). Many governments, recognizing the harmful effects of halocarbons on the environment, signed the "Montreal Protocol on Substances that Deplete the Ozone Layer" (UNEP, 1987) in 1987 to limit the production and consumption of a number of fully halogenated CFCs and halons. The control measures of the Montreal Protocol freeze the production and consumption of CFCs 11, 12, 113, 114, and 115 in developed countries at their 1986 levels from the year 1990, a reduction to 80% of their 1986 levels from the year 1993, with a further reduction to 50% of their 1986 levels from the year 1998. Developing countries, with a per capita use of CFCs of less than 0.3 kg per capita, are allowed to increase their per capita use up to this limit and can delay compliance with the control measures by 10 years. All major producing and consuming developed countries, and many developing countries, have signed and ratified the Montreal Protocol.

1.4.1 Atmospheric Distribution of Halocarbons

The mean atmospheric concentrations of the most abundant radiatively active halocarbons are shown in Table 1.3. The atmospheric concentrations of the halocarbons are currently increasing more rapidly on a global scale (on a percentage basis) than the other greenhouse gases (Figure 1.12). The concentrations of the fully halogenated chloro-fluorocarbons (CFCs), slightly greater in the northern hemisphere than in the southern hemisphere, are consistent with the geographical distribution of releases (>90% from the industrialized nations), a 45°N - 45°S mixing time of about 1 year, and their very long atmospheric lifetimes.

1.4.2 Sinks for Halocarbons

There is no significant tropospheric removal mechanism for the fully halogenated halocarbons such as CCl_3F (CFC-11), CCl_2F_2 (CFC-12), $C_2Cl_3F_3$ (CFC-113), $C_2Cl_2F_4$ (CFC-114), C_2ClF_5 (CFC-115), carbon tetrachloride (CCl_4), and halon 1301 ($CBrF_3$). They have long atmospheric lifetimes, decades to centuries, and are primarily removed by photodissociation in the mid - upper stratosphere. There is currently a significant imbalance between the sources and sinks giving rise to a rapid growth in atmospheric concentrations. To stabilize the atmospheric concentrations of CFCs 11, 12 and 113 at current levels would require reductions in emissions of approximately 70-75%, 75-85%, and 85-95%, respectively (see Annex).

Non-fully halogenated halocarbons containing a hydrogen atom such as methyl chloride (CH_3Cl), methylchloroform (CH_3CCl_3), $CHClF_2$ (HCFC-22), and a number of other HCFCs and HFCs being considered as substitutes for the current CFCs (c.f. Section 1.4.4) are primarily removed in the troposphere by reaction with OH. These hydrogen containing species have atmospheric lifetimes ranging from about one to forty years, much shorter on average than the fully halogenated CFCs. To stabilize the atmospheric concentrations of HCFC-22 at current levels would require reductions in emissions of approximately 40-50%.

1.4.3 Sources of Halocarbons

Most halocarbons, with the notable exception of CH_3Cl, are exclusively of industrial origin. Halocarbons are used as aerosol propellants (CFCs 11, 12, and 114), refrigerants (CFCs 12 and 114, and HCFC-22), foam blowing agents (CFCs 11 and 12), solvents (CFC-113, CH_3CCl_3, and CCl_4), and fire retardants (halons 1211 and 1301). Current emission fluxes are approximately CFC-11: 350Gg/y; CFC-12: 450 Gg/y; CFC-113: 150 Gg/y; HCFC-22: 140 Gg/y; others are significantly smaller. The atmospheric concentration of methyl chloride is about 0.6 ppbv, and is primarily released from the oceans and during biomass burning. There is no evidence that the atmospheric concentration of CH_3Cl is increasing. Methyl bromide (CH_3Br) is produced by oceanic algae, and there is evidence that its atmospheric concentration has been increasing in recent times due to a significant anthropogenic source (Penkett et al., 1985; Wofsy et al., 1975).

1.4.4 Future Atmospheric Concentration of Halocarbons

Future emissions of CFCs 11, 12, 113, 114, and 115 will be governed by the Montreal Protocol on "Substances that Deplete the Ozone Layer " as discussed in Section 1.4. In addition, international negotiations are currently in progress that will likely (i) result in a complete global phase-out of production of these chemicals by the year 2000, and (ii) enact limitations on the emissions (via production and consumption controls) of CCl_4, and CH_3CCl_3. However, even with a complete cessation of production of CFCs 11, 12 and 113 in the year 2000 their atmospheric concentrations will still be significant for at least the next century because of their long atmospheric lifetimes. It should be noted that emissions of these gases into the atmosphere will continue for a period of time after production has ceased because of their uses as refrigerants, foam blowing agents, fire retardants, etc.

A number of hydrofluorocarbons (HFCs) and hydrochlorofluorocarbons (HCFCs) are being considered as potential replacements for the long-lived CFCs (11, 12, 113, 114, and 115) that are regulated under the terms of the Montreal Protocol. The HFCs and HCFCs primarily being considered include: HCFC-22, HCFC-123 ($CHCl_2CF_3$), HCFC-124 ($CHClFCF_3$), HFC-125 (CHF_2CF_3), HFC-134a (CH_2FCF_3), HCFC-141b (CH_3CCl_2F), HCFC-142b (CH_3CClF_2), HFC-143a (CH_3CF_3), and HFC-152a (CH_3CHF_2). The calculated atmospheric lifetimes of these chemicals are controlled primarily by reaction with tropospheric OH and range between about 1 and 40 years. It has been estimated (UNEP, 1989) that a mix of HFCs and HCFCs will replace the CFCs currently in use at a rate of about 0.4 kg of substitute for every kg of CFCs currently produced, with an annual growth rate of about 3%. Because of their shorter lifetimes, and expected rates of substitution and emissions growth rates, the atmospheric concentrations of HFCs and HCFCs will be much lower for the next several decades than if CFCs had continued to be used, even at current rates. However, continued use, accompanied by growth in the emission rates of HFCs and HCFCs for more than several decades would result in atmospheric concentrations that would be radiatively important.

1.4.5 Conclusions

The atmospheric concentrations of the industrially-produced halocarbons, primarily CCl_3F, CCl_2F_2, $C_2Cl_3F_3$, and CCl_4 are about 280 pptv, 484 pptv, 60 pptv, and 146 pptv, respectively. Over the past few decades their concentrations (except CCl_4) have increased more rapidly (on a percentage basis) than the other greenhouse gases, currently at rates of at least 4% per year. The fully halogenated CFCs and CCl_4 are primarily removed by photolysis in the stratosphere, and have atmospheric lifetimes in excess of 50 years.

Most halocarbons, with the notable exception of methyl chloride, are exclusively anthropogenic and their sources (solvents, refrigerants, foam blowing agents, and aerosol propellants) are well understood.

To stabilize, and then reduce, the current atmospheric concentrations of the fully halogenated CFCs (e.g., 11, 12 and 113) would require approximate reductions in emissions of 70-75%, 75-85%, and 85-95%, respectively. Future emissions of CFCs and CCl4 will, most likely, be eliminated or be significantly lower than today's because the stringency, scope, and timing of international regulations on chlorine and bromine containing chemicals, (i.e., the Montreal Protocol on Substances that Deplete the Ozone Layer) are currently being renegotiated. However, the atmospheric concentrations of CFCs 11, 12 and 113 will still be significant (30 - 40% of current) for at least the next century because of their long atmospheric lifetimes.

1.5 Nitrous Oxide

Nitrous oxide is a chemically and radiatively active trace gas that is produced from a wide variety of biological sources in soils and water and is primarily removed in the stratosphere by photolysis and reaction with electronically excited oxygen atoms.

1.5.1 Atmospheric Distribution of Nitrous Oxide

The mean atmospheric concentration of N_2O in 1990 is about 310 ppbv, corresponding to a reservoir of about 1500 TgN, and increasing at a rate of 0.2 - 0.3% per year (Figure 1.13; Weiss, 1981; Prinn et al., 1990; Robinson et al., 1988; Elkins and Rossen, 1989; Rasmussen and Khalil, 1986). This observed rate of increase represents an atmospheric growth rate of about 3 to 4.5 TgN per year. The atmospheric concentration of N_2O is higher in the Northern Hemisphere than in the Southern Hemisphere by about 1 ppbv. Ice core measurements show that the pre-industrial value of N_2O was relatively stable at about 285 ppbv for most of the past 2000 years, and started to increase around the year 1700 (Figure 1.14; Pearman et al., 1986; Khalil and Rasmussen, 1988b; Etheridge et al., 1988; Zardini et al., 1989). Figure 1.14 shows that the atmospheric concentrations of N_2O may have decreased by a few ppbv during the period of the "Little Ice Age".

1.5.2 Sinks for Nitrous Oxide

The major atmospheric loss process for N_2O is photochemical decomposition in the stratosphere, and is calculated to be 10 ± 3 Tg N per year (Table 1.4). Nitrous oxide has an atmospheric lifetime of about 150 years. The observed rate of growth represents a 30% imbalance between the sources and sinks (Hao et al., 1987). Tropospheric sinks such as surface loss in aquatic and soil

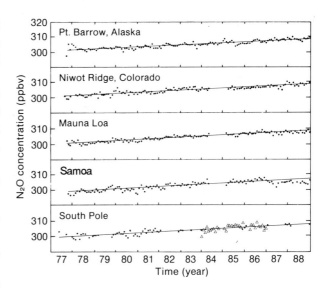

Figure 1.13: Atmospheric measurements of nitrous oxide from the NOAA/GMCC network (Elkins and Rossen, 1989).

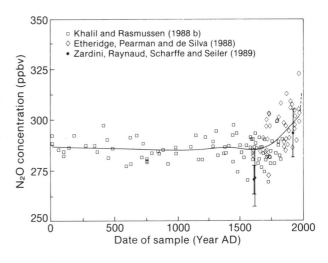

Figure 1.14: Nitrous oxide measurements from ice-core samples.

systems are considered to be small (Elkins et al., 1978, Blackmer and Bremner, 1976).

1.5.3 Sources of Nitrous Oxide

1.5.3.1 Oceans

The oceans are a significant, but not dominant source of N_2O (McElroy and Wofsy, 1986). Based on measurements of the concentration gradients between the atmosphere and surface waters (Butler et al., 1990, and NOAA GMCC unpublished data), and on estimates of the gas exchange coefficient, the current estimate of the magnitude of the

Table 1.4 Estimated Sources and Sinks of Nitrous Oxide

	Range (TgN per year)
Source	
Oceans	1.4 - 2.6
Soils (tropical forests)	2.2 - 3.7
(temperate forests)	0.7 - 1.5
Combustion	0.1 - 0.3
Biomass burning	0.02 - 0.2
Fertilizer (including ground-water)	0.01 - 2.2
TOTAL:	4.4 - 10.5
Sink	
Removal by soils	?
Photolysis in the stratosphere	7 - 13
Atmospheric Increase	3 - 4.5

ocean source ranges from 1.4 - 2.6 Tg N per year, significantly lower than earlier estimates (Elkins et al., 1978; Cohen and Gordon, 1979; Cline et al., 1987). An accurate determination of the global annual ocean flux is difficult because of uncertainties associated with quantifying the gas exchange coefficient and because the partial pressure of N_2O in the surface waters is highly variable, both spatially and temporally. The partial pressure of N_2O in surface waters varies considerably, ranging from being supersaturated by up to 40% in upwelling regions to being undersaturated by a few percent in areas around Antarctica and within gyres. Data suggest that during El Niño events, when upwelling in the Pacific ocean is suppressed, the ocean fluxes of N_2O are significantly lower (Cline et al., 1987; Butler et al., 1990). It is still unclear whether N_2O is primarily produced from nitrification in near surface waters, or denitrification in oxygen deficient deep waters. Based on vertical profile measurements of oceanic N_2O (NOAA GMCC, unpublished) the oceanic reservoir has been estimated to be between 900 and 1100 Tg N, comparable to the atmosphere. Consequently, changes in the exchange fluxes of N_2O between the ocean and the atmosphere could have a significant impact on its atmospheric concentration.

1.5.3.2 Soils

Denitrification in aerobic soils is thought to be a dominant source of atmospheric N_2O (Keller et al., 1986; Matson and Vitousek, 1987; Matson and Vitousek, 1989; Slemr et al., 1984). Nitrification under anaerobic conditions could,

however, produce higher yields of N_2O per unit of transformed nitrogen. Quantification of global N_2O emissions from soils is difficult because of the heterogeneity of terrestrial ecosystems and the variability in environmental conditions that control the fluxes of N_2O.

Estimates of global fluxes of N_2O from tropical forests range from 2.2 - 3.7 Tg N per year. The impact of deforestation on the emissions of N_2O from tropical soils is unclear, with some studies suggesting that the emissions of N_2O from deforested land are enhanced by as much as a factor of three (Luizao et al., 1990), whereas other studies concluded that N_2O fluxes decreased if vegetation did not return (Robertson and Tiedje, 1988).

Quantifying the roles of temperate forest soils and grasslands in the N_2O budget is difficult because of the paucity of data, and conflicting results. Estimates of N_2O fluxes from temperate forest soils range from 0.7 - 1.5 Tg N per year in one study (Schmidt et al., 1988), to almost none in another study (Bowden et al., 1990). One study also reported that deforestation in temperate forests would lead to enhanced emissions of N_2O (Bowden and Bormann, 1986). Reliable global N_2O fluxes from grasslands are impossible to derive from the fragmented data available. One study (Ryden, 1981) concluded that English grassland soils, with no fertilization, are a sink for N_2O, whereas limited studies of tropical grasslands and pastures suggest that they may be a moderate to significant source of N_2O (Luizao et al., 1990; Robertson and Tiedje, 1988).

1.5.3.3 Combustion

Until recently, the combustion of fossil fuels was thought to be an important source of atmospheric N_2O (Pierotti and Rasmussen, 1976; Weiss and Craig, 1976; Hao et al., 1987). However, a recent study has shown that the earlier results are incorrect because N_2O was being artificially produced in the flasks being used to collect N_2O from combustion sources (Muzio and Kramlich, 1988). The latest estimate of the global flux of N_2O from combustion sources is between 0.1 and 0.3 Tg N per year, compared to earlier values which were as high as 3.2 Tg N per year.

1.5.3.4 Biomass burning

Biomass burning is now thought to be a minor source of atmospheric N_2O with a global flux of less than 0.2 Tg N per year (Muzio and Kramlich, 1988; Crutzen 1989; Elkins et al., 1990; Winstead et al., 1990; Griffith et al., 1990). This value is 1-2 orders of magnitude less than previous estimates (Crutzen et al., 1979, 1985) which were influenced by artifacts involving N_2O analysis (Crutzen et al., 1985) and N_2O production in sampling flasks (Muzio and Kramlich, 1988).

1.5.3.5 *Fertilizer / Ground-Water*

Nitrous oxide production from the use of nitrate and ammonium fertilizers is difficult to quantify because the N_2O fluxes are dependent upon numerous factors including type of fertilizer, soil type, soil temperature, weather, and farming practices (e.g., ploughing, sowing, irrigating). Conversion of fertilizer N to N_2O ranges from 0.01 - 2.0% (Conrad et al., 1983; Bremner et al., 1981). This range, coupled with a global fertilizer production of 55 Tg N per year in 1980, results in a total N_2O emission of between 0.01 - 1.1 Tg N per year (Conrad et al., 1983). Leaching of nitrogen fertilizers from soils into groundwater may result in additional fluxes of N_2O up to 1.1 Tg N per year (Conrad et al., 1983; Ronen et al., 1988). Consequently, a range of 0.01 - 2.2 Tg N per year can be derived for the flux of N_2O from fertilizer use.

1.5.4 *Conclusions*

Nitrous oxide is a greenhouse gas whose atmospheric concentration, at 310 ppbv, is now about 8% greater than in the pre-industrial era, and is increasing at a rate of about 0.2 - 0.3% per year, corresponding to about 3 - 4.5 Tg N per year. This represents an excess of 30% of current global emissions over current sinks. The major sink for N_2O is photolysis in the stratosphere, resulting in a relatively long atmospheric lifetime of about 150 years. The magnitude of the sink for N_2O is relatively well known ($\pm$ 30%). In order to stabilize concentrations at present day levels, an immediate reduction of 70 - 80% of the additional flux of N_2O that has occurred since the pre-industrial era would be necessary.

Quantification of the various natural and anthropogenic sources is uncertain. Since the latest studies indicate that the total combined flux of N_2O from combustion and biomass burning is between 0.1 to 0.5 Tg N per year, in contrast to earlier estimates of about 5 Tg N per year, and production of N_2O from fertilizer (including groundwater) is believed to be less than or equal to 2.2 Tg N per year, it is difficult to account for the annual increase based on known sources. Stimulation of biological production due to agricultural development may account for the missing anthropogenic emissions. Estimates of the removal rate of N_2O by photodissociation in the stratosphere range from 7 - 13 Tg N per year. Therefore, the total source needed to account for the observed annual atmospheric growth is 10 - 17.5 Tg N per year against a flux of N_2O from known sources of 4.4 - 10.5 Tg N per year. These data suggest that there are missing sources of N_2O, or the strengths of some of the identified sources have been underestimated. Despite these uncertainties, it is believed that the observed increase in N_2O concentrations is caused by human activities.

1.6 Stratospheric Ozone

Stratospheric O_3 is an important constituent of the Earth's atmosphere. It protects the Earth's surface from harmful solar ultraviolet radiation and it plays an important role in controlling the temperature structure of the stratosphere by absorbing both incoming solar ultraviolet radiation and outgoing terrestrial (longwave) radiation. Part of the absorbed outgoing longwave radiation is then re-radiated back to the surface-troposphere system. Reductions in stratospheric O_3 can modify the surface temperature via two competing processes: more solar radiation is transmitted to the surface-troposphere system, thereby contributing to a surface warming; on the other hand, the cooler stratosphere (due to decreased solar and long-wave absorption) emits less to the troposphere which would tend to cool the surface. The solar warming (a function of total column amount of O_3) and longwave cooling (a function of the vertical distribution of O_3) are similar in magnitude. Therefore, the magnitude as well as the sign of the change in surface temperature depends critically on the magnitude of the O_3 change, which in turn is depends strongly on altitude, latitude and season.

The concentration and distribution of stratospheric O_3 is controlled by dynamical, radiative and photochemical processes. Stratospheric O_3 is photochemically controlled by chemically active species in the (i) oxygen, (ii) hydrogen, (iii) nitrogen, (iv) chlorine, and (v) bromine families. The precursors for the photochemically active species are (i) O_2, (ii) H_2O and CH_4; (iii) N_2O; (iv) CFCs, CCl_4, CH_3CCl_3, CH_3Cl, and (v) halons and CH_3Br, respectively.

1.6.1 *Stratospheric Ozone Trends*

1.6.1.1 *Total column ozone trends*

The Antarctic ozone hole, which formed during the mid to late 1970s, recurs every springtime. To determine O_3 trends more widely, data from the ground-based Dobson network have been re-evaluated, station by station, and used to determine changes in total column O_3 over the past two decades. Unfortunately, the network and data are adequate for only a limited geographical region, i.e., 30 - 64°N. They are inadequate to determine total column O_3 changes in the Arctic, tropics, subtropics, or southern hemisphere apart from Antarctica. Satellite data can provide the desired global coverage, but the current record is too short (about one solar cycle, 1978 to present), to differentiate between the effects of natural and human-influenced processes on O_3. The re-evaluated data was analysed for the effects of known natural geophysical processes (seasonal variation, the approximately 26-month quasi-biennial-oscillation, and the 11-year solar cycle) and possible human perturbations. After allowing for natural variability, the analyses, using a variety of statistical models and assumptions, showed measurable zonal mean

O_3 decreases in the range 3.4% to 5.1% between 30 and 64°N latitude for the winter months (December - March) between 1969 and 1988, with the larger decreases at the higher latitudes (WMO, 1989a,b). No statistically significant zonal trends were found for the summer period (May - August). Lastly, within longitudinal sectors, regional differences in the O_3 trends were indicated, with the largest values over North America and Europe and the smallest over Japan.

1.6.1.2 Changes in the vertical distribution of ozone

Substantial uncertainties remain in defining changes in the vertical distribution of O_3. Analysis of SAGE I and II satellite data, averaged over 20 to 50°N and S latitudes, indicates that near 40 km O_3 decreased by (3 ± 2)% between February 1979 - November 1981 and October 1984 - December 1988 (WMO, 1989b). Because the SAGE record is so short (less than one solar cycle), no attempt has been made to distinguish between solar-induced and human-influenced contributions to these changes. A thorough analysis of data from 10 ground-based Umkehr stations in the Northern Hemisphere for the period 1977 to 1987 indicates a statistically significant decrease in O_3 between 30 and 43 km. The decrease near 40 km of (4.8 ± 3.1)%, after allowing for seasonal and solar-cycle effects and correcting the data for aerosol interferences, is broadly consistent with theoretical predictions. Based on satellite, ground-based, and ozonesonde data, there are indications of a continuing stratospheric O_3 decrease since the late 1970s of a few percent at 25 km and below. Photochemical models (which do not take into account heterogeneous processes) do not predict these changes, but the measurements are qualitatively consistent with those required for compatibility with the total column measurements.

1.6.2 Future Changes

Future changes in stratospheric O_3 are critically dependent upon future emissions of CFCs, other halocarbons, CH_4, N_2O, and CO_2. Assuming that the current regulatory measures agreed under the Montreal Protocol are not strengthened, then the chlorine loading of the atmosphere is predicted to reach about 9 ppbv by the year 2060, about three times today's level, and a bromine loading of about 30 pptv, about twice today's level . Models predict column O_3 reductions of 0 to 4% in the tropics, and from 4 to 12% at high latitudes in late winter. These predictions do not include the effects of heterogeneous processes, which play a critical role in the formation of the Antarctic ozone hole. Consequently, models that include the effects of heterogeneous processes would predict larger O_3 depletions, at least in polar regions. Ozone is predicted to decrease by 25 - 50% at 40 km and result in stratospheric temperature decreases of 10 to 20 K. If, as expected, the

Montreal Protocol is modified to eliminate the emissions of CFCs 11, 12, 113, 114, 115, halons 1211 and 1301, and restrict the emissions of CCl_4 and CH_3CCl_3, by the year 2000, then the chlorine loading of the atmosphere by the year 2060 will probably lie between 2.5 and 4 ppbv (depending upon the emissions of CCl_4 and CH_3CCl_3, and HCFCs). Models that do not include heterogeneous processes predict that global O_3 levels would be similar to today. However, if the atmospheric chlorine loading approaches 4 ppbv the implications for polar O_3, and its subsequent impacts on O_3 at mid-latitudes, are unknown.

1.7 Tropospheric Ozone and Related Trace Gases (Carbon Monoxide, Non-Methane Hydrocarbons, and Reactive Nitrogen Oxides)

1.7.1 Tropospheric Ozone

Tropospheric O_3 is a greenhouse gas, of particular importance in the upper troposphere in the tropics and sub-tropics.Its distribution is controlled by a complex interplay between chemical, radiative, and dynamical processes. Ozone is: (i) transported down into the troposphere from the stratosphere; (ii) destroyed by vegetative surfaces; (iii) produced by the photo-oxidation of CO, CH_4, and NMHC in the presence of reactive nitrogen oxides (NO_x), and (iv) destroyed by uv-photolysis and by reaction with hydrogen oxide radicals (HO_2) (Danielsen, 1968; Mahlman and Moxim, 1978; Galbally and Roy, 1980; Crutzen, 1974; Isaksen et al., 1978). Chemical processes in clouds could have a strong influence on O_3 production and destruction rates (Lelieveld and Crutzen, 1990).

Consequently, while CO, NMHC, and NO_x are not important greenhouse gases in themselves, they are important precursors of tropospheric O_3 and they are therefore treated in some detail in this sub-section.

1.7.1.1 Atmospheric distribution

Ozone in the troposphere has a lifetime of at most several weeks, hence its concentration varies with latitude, longitude, altitude and season (Chatfield and Harrison, 1977; Logan, 1985). Near the surface, monthly mean concentrations (30 - 50 ppbv) are highest in spring and summer at northern mid-latitudes (Figure 1.15). In the middle troposphere at northern mid-latitudes values are highest also in spring and summer, 60 - 65 ppbv. The summer maximum results from photo-oxidation of O_3 precursors from fossil fuel combustion and industrial activity (Isaksen et al., 1978; Fishman et al., 1985; Logan 1985). Ozone values are highest in winter and spring at other latitudes, in part because the stratospheric source is largest then (Levy et al., 1985). There is 35% more O_3 at 40°N than at 40°S, in the middle troposphere (Logan, 1985).

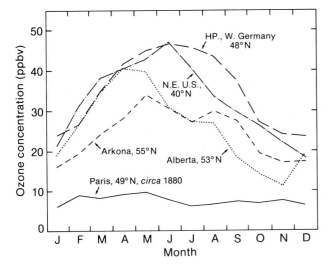

Figure 1.15: The seasonal variation of surface ozone. The solid line shows data from Montsouris, France, for 1876-86 (Volz and Kley, 1988). All other data are from the 1970s and 1980s: dashed line, Arkona, GDR (Feister and Warmbt, 1987); dotted line, Ellerslie, Alberta, Canada (Angle and Sandhu, 1986); dot-dash line, average of eight rural sites in the northeastern U.S., the SURE sites (Logan, 1988); long dashed line, Hohenpeissenberg, FRG (Logan, 1985). All the recent data are shown as monthly means of daily average values.

Concentrations of O_3 tend to be smaller in the tropics than in mid-latitudes, except in the dry season, when emissions of O_3 precursors from biomass burning provide a photochemical source (Delany et al., 1985; Crutzen et al., 1985; Logan and Kirchhoff, 1986; Fishman et al., 1990). Ozone values during the southern spring over South America can reach almost as high values as found over the industrialized mid-latitudes in summer. Large regions of the tropical troposphere appear to be influenced by sources of O_3 from biomass burning (Fishman et al., 1990). Remote marine air and continental air during the wet season may provide a photochemical sink for O_3 in the tropics; mean surface concentrations as low as 4 - 12 ppbv have been measured (Liu et al., 1980; Oltsmans and Komhyr, 1986; Kirchhoff, 1990).

1.7.1.2 Trends
Most long-term measurements of O_3 have been made at northern mid-latitudes, from surface sites and from balloons. Only sporadic data are available before the 1970s. A comparison of data obtained in Paris from 1876-1910 (Volz and Kley, 1988) with rural data from the present day from Europe and North America (Logan, 1985, 1989) suggests that surface O_3 has increased by a factor of 2 -3 on average; the increase is largest in summer, the factor then being 4 - 6 (Figure 1.15). Ozone values in Europe in

the 1970s appear to be about twice those found between 1930 and 1950 (Crutzen, 1988). Data from Europe suggest an increase of 1 - 2% per year from the mid-1950s to the early 1980s, with increases in winter and summer (Feister and Warmbt, 1987; Bojkov, 1988). Since the mid-1970s O_3 has increased by 0.8% per year at remote sites in Alaska and Hawaii; shown no annual trend at Samoa, but has decreased by 0.5% per year at the South Pole (Oltmans et al., 1988). Decreases of 1.8% per year are found at both Samoa and South Pole in summer. Trend data are lacking for tropical continental sites.

Ozonesonde data for northern mid-latitudes between 1965 and 1986 suggest that O_3 has increased by about 1% per year below 8 km, primarily over North Europe and Japan (Angell and Korshover, 1983; Logan 1985; Tiao et al., 1986; WMO, 1989a,b), but there are no clear trends in the upper troposphere. By contrast O_3 has decreased in the lower stratosphere (below 25 km), the crossover in the trend being near the tropopause. There is no trend in O_3 at the single sonde station at southern mid-latitudes, and long term sonde data are lacking in the tropics.

1.7.1.3 Relationships between ozone and its precursors
The concentration of tropospheric O_3 is dependent in a very non-linear manner on the atmospheric concentrations of its precursor gases, i.e., CO, CH_4, NMHC, and, in particular NO_x (NO_x = NO + NO_2). Nitrogen oxide concentrations and trends control changes in the concentration of O_3 (Dignon and Hameed, 1985). At low NO_x concentrations (where NO_x is less than 5 - 30 pptv; this threshold depends on the concentrations of O_3 and hydrocarbons) increases in CO, CH_4, and NMHC lead to a decrease in O_3, whereas at high NO_x concentrations increases in CO, CH_4, and NMHC lead to significant enhancements in O_3. Therefore, no simple relationship exists between increases in the precursor gases and changes in tropospheric O_3. Several model calculations have been performed to investigate the sensitivity of O_3 changes to changes in the precursor gases, both individually and collectively. All models that have attempted to simulate changes in O_3 during the past century have calculated increases in Northern Hemisphere O_3 by up to a factor of two, broadly consistent with observations, depending upon the assumptions made regarding the initial concentration, distribution, and changes in precursor gas concentrations, particularly NO_x.

Understanding the feedbacks among O_3 and its precursor gases is essential to understand tropospheric OH, which controls the atmospheric lifetimes of CH_4 and the NMHCs. The global concentration of OH, which determines the oxidizing capacity of the troposphere, can be either enhanced because of elevated levels of tropospheric O_3, NO_x or water vapour (associated with a global warming) or suppressed because of increases in

CH_4, CO, and NMHC (Crutzen, 1987; Thompson et al., 1989). Prediction of regional and global trends in OH concentrations requires an understanding of regional emissions of CH_4, CO, NMHC and NO_x, as well as transport of O_3 between its source regions and the remote troposphere. One key point is that a continued increase in levels of CO would reduce the global concentration of OH because NO_x is too short-lived to counteract that effect over much of the globe. This would increase the atmospheric lifetime of CH_4.

1.7.2 Carbon Monoxide

1.7.2.1 Atmospheric distribution of carbon monoxide
The atmospheric concentration of CO exhibits significant spatial and temporal variability because of its short atmospheric lifetime (2 - 3 months). The short atmospheric lifetime, coupled with an inadequate monitoring network, means that the global spatial variability and long-term trends in CO are not well documented. The limited observational data base (Heidt et al., 1980; Dianov-Klokov and Yurganov, 1981; Seiler and Fishman, 1981; Seiler et al., 1984; Khalil and Rasmussen, 1984, 1988a; Fraser et al., 1986a, c; Newell et al., 1989; Zander et al., 1989; Kirchhoff and Marinho, 1989; Kirchhoff et al., 1989) has demonstrated that the concentration of CO, (i) is about a factor of two greater in the Northern than in the Southern Hemisphere where the annual average is about 50 - 60 ppbv, (ii) increases with latitude in the Northern Hemisphere, (iii) exhibits strong seasonal variations in both hemispheres at mid to high latitudes, and (iv) decreases with altitude. CO appears to be increasing at about 1% per year in the Northern Hemisphere, but the evidence for increases in the Southern Hemisphere is ambiguous.

1.7.2.2 Sources and sinks of carbon monoxide
The total annual source of CO is about 2400 Tg CO, being about equally divided between direct anthropogenic (incomplete combustion of fossil fuels and biomass) and atmospheric (oxidation of natural and anthropogenic CH_4 and NMHC) sources (Logan et al., 1981; Cicerone, 1988). Atmospheric concentrations of CO may have increased in the Northern Hemisphere because of the fossil fuel source, and because of changes in the rate of oxidation of CH_4, whose atmospheric concentration has increased since pre-industrial times. Fossil fuel sources of CO are at present decreasing in North America (EPA, 1989) and possibly in Europe, but may be increasing elsewhere.

The major removal process for atmospheric CO is reaction with OH (Logan et al., 1981). The observed seasonal variability in the Southern Hemisphere, distant from seasonally varying sources, can be explained by the seasonal variability in the concentration of tropospheric OH. Soils may provide a minor sink for CO (Conrad and Seiler, 1985).

1.7.3 Reactive Nitrogen Oxides
The key constituents of tropospheric NO_y, defined as the sum of all nitrogen oxide species except for N_2O, are NO_x, nitric acid (HNO_3), peroxyacetylnitrate (PAN: $CH_3CO_3NO_2$), and organic nitrates. Most primary sources of nitrogen oxides release NO_x (mainly NO); the other species are produced by photochemical reactions in the atmosphere. While the atmospheric lifetime of NO_x is short (about 1 day), the atmospheric lifetime of NO_y can range up to several weeks. Thus NO_y can transport nitrogen compounds away from source regions to more remote locations, where photolysis of HNO_3 and PAN, and thermal decomposition of PAN, can regenerate NO_x.

1.7.3.1 Atmospheric distribution of nitrogen oxides
The atmospheric concentrations of NO_x exhibit significant spatial and temporal variability, reflecting the complex distribution of sources and the short atmospheric lifetime. The near surface and free tropospheric concentrations of NO_x each vary by several orders of magnitude, highly influenced by the proximity of source regions. Near surface concentrations of NO_x range from as low as 0.001 ppbv in remote maritime air to as high as 10 ppbv in Europe and Eastern North America (excluding urban areas), while free tropospheric concentrations range from 0.02 ppbv in remote regions to more than 5 ppbv over populated areas (Fehsenfeld et al., 1988).

The spatial inhomogeneity, coupled with a sparsity of measurements, means that the spatial and temporal distribution and long-term trends in NO_x and NO_y are not adequately documented, although reconstructed emissions inventories of NO_x suggest large increases throughout this century (Dignon and Hameed, 1989). Data from a Greenland ice core have shown that the concentration of nitrate ions (dissolved nitrate from HNO_3) remained constant from 10,000 years ago to about 1950, then doubled by the late 1970's, consistent with the increase in industrial emissions (Neftel et al. 1985b). Data from glacier ice in Switzerland indicates that nitrate ions increased by a factor of 4 -5 between 1900 and the 1970's in Western Europe (Wagenbach et al. 1988).

1.7.3.2 Sources and sinks of nitrogen oxides
The sources of atmospheric NO_x are about equally divided between anthropogenic (combustion of fossil fuels: 21 Tg N per year, and biomass burning: 2 - 5 Tg N per year), and natural (microbial processes in soils: 20 Tg N per year; lightning: 2 - 8 Tg N per year, and transport from the stratosphere: 1 Tg N per year) (Galbally, 1989). Emissions of NO_x (6.3 Tg N per year) from the combustion of fossil fuels have not increased in North America since 1970 (EPA, 1989). Soil emissions of NO are stimulated by agricultural activity (e.g., addition of fertilizer, manure,

etc.), hence, agricultural soil emissions may provide significant sources of NO_x in many areas.

The dominant removal processes for NO_x are (i) conversion to HNO_3, PAN, and organic nitrates by photochemical mechanisms, (ii) reactions involving NO_3 radicals, and possibly (iii) deposition of NO_2 on vegetation. The resulting NO_y species are then removed from the atmosphere by wet and dry deposition, or by conversion back to NO_x.

1.7.4 Non-Methane Hydrocarbons

1.7.4.1 Atmospheric distribution of non-methane hydrocarbons

The NMHC can be classified by atmospheric lifetime: (i) relatively long-lived (lifetimes > week) where the highest concentrations (up to 3 ppbv for ethane) are observed at middle to high northern latitudes; (ii) more reactive (lifetimes between half a day and one week) such as C_2-C_5 alkenes whose concentrations exhibit significant temporal and latitudinal variability from <0.1ppbv in remote areas to a few ppbv close to source regions, and (iii) extremely short lived (lifetimes of hours) such as terpenes or isoprene whose local concentrations may reach about 10 ppbv very close to their sources. Trends in the atmospheric concentrations of NMHC have not been established due to a lack of measurements.

1.7.4.2 Sources and sinks for non-methane hydrocarbons

The oceans are a major source of NMHC, mainly alkenes. Estimates of the source strength of ethene and propene range from 26 Tg C per year (Bonsang et al., 1988) to as high as 100 Tg C per year (Penkett, 1982). Emissions of NMHC from terrestrial vegetation are dependent upon environmental factors as well as the type of vegetation. Isoprene is primarily emitted from deciduous plants, whereas conifer trees are primarily a source of terpenes. Isoprene and terpene emission rates are very large, about 500 Tg per year for each (Rasmussen and Khalil, 1988). The source strength of NMHC from anthropogenic activities such as biomass burning, solvents and fossil fuel combustion has been estimated to be about 100 Tg per year.

The dominant loss mechanism for most NMHC is rapid (much faster than CH_4) reaction with OH. The products of these reactions are capable of forming O_3 in the presence of NO_x.

1.7.5 Feedbacks Between Climate and the Methane/Non-Methane Hydrocarbon/ Carbon Monoxide/Oxides of Nitrogen/ Tropospheric Ozone System

There are numerous potentially important feedbacks between climate change and tropospheric O_3 and OH. Changes in cloud cover, precipitation, and circulation patterns, as well as changes in the biospheric source strengths of CH_4, CO, NMHC and NO_x, will induce changes in homogeneous and heterogeneous reactions controlling O_3 and OH. In addition, changes in stratospheric O_3 may induce changes in tropospheric processes, through changes in ultraviolet radiation. Stratospheric O_3 depletion is likely to increase tropospheric O_3 when the levels of CO, NO_x, and NMHC are high, but reduce it in regions of very low NO_x. The importance of these feedback processes remains to be determined.

1.7.6 Conclusions

Tropospheric O_3 is a greenhouse gas that is produced photochemically through a series of complex reactions involving CO, CH_4, NMHC and NO_x. Hence, the distribution and trends of tropospheric O_3 depend upon the distribution and trends of these gases whose atmospheric concentrations are changing.

The short atmospheric lifetimes of O_3 (several weeks), and many of its precursor gases, coupled with inadequate observational networks, leave their distributions and trends inadequately documented. Most data support positive trends of about 1% per year for O_3 below 8 km altitude in the Northern Hemisphere (consistent with positive trends in several of the precursor gases, especially NO_x, CH_4, and CO), and a similar trend for CO in the Northern Hemisphere, but not in the Southern Hemisphere. While there is no systematic series of data that allow quantitative estimates of trends in NMHC and NO_x to be made, their atmospheric concentrations are likely to have increased during the past few decades because of increased anthropogenic sources. The ice core records of nitrate levels provide indirect evidence for a Northern Hemisphere increase in atmospheric NO_x.

1.8 Aerosol Particles

1.8.1 Concentrations and Trends of Aerosol Particles in the Troposphere

Aerosol particles play an important role in the climate system because of their direct interaction (absorption and scattering) with solar and terrestrial radiation, as well as through their influence on cloud processes and thereby, indirectly, on radiative fluxes. These processes are discussed in more detail in Sections 2.3.2 and 2.3.3. Two separate issues should be identified. The first is the effect of increasing or decreasing anthropogenic emissions of aerosol particles and their precursors in regions impacted by these emissions. The second is the role of feedback processes linking climate change and natural (biological) production of particles in unpolluted regions, especially over the oceans (cf. Section 10.8.3).

Total suspended particulate matter in air varies from less than 1 µg m^{-3} over polar ice caps or in the free mid-ocean troposphere to 1 mg m^{-3} in desert dust outbreaks or in dense plumes from for example, forest fires. In a typical sample of continental air, mineral dust, sulphuric acid, ammonium sulphate as well as organic material and elemental carbon (soot) may be found both as pure or mixed particles. Most of the soluble particles become solution droplets at relative humidities above 80%; thus the radiative properties of aerosol particles even vary with relative humidity at constant dry aerosol mass.

A large part of the aerosol mass in submicron size particles is derived from gas-to-particle conversion through photochemical processes involving gaseous sulphur and hydrocarbon compounds. Such conversion may take place through photochemical processes involving the oxidation of sulphur dioxide (SO_2) and other sulphur gases to sulphuric acid (H_2SO_4) by reaction with OH. The H_2SO_4 so formed, having a low equilibrium vapour pressure, immediately condenses onto existing aerosol particles or forms new ones. Transformation to sulphuric acid and sulphate also takes place in cloud droplets, the majority of which eventually evaporate leaving the sulphate in the aerosol phase. Trends in the emission of these gaseous precursors, especially the sulphur gases, are therefore of great importance for the regional aerosol burden and thereby potentially for climate.

Large quantities of aerosol particles are also emitted from the burning of savannas and forests in tropical regions. The directly emitted particles consist largely of carbonaceous materials including black carbon (soot) (Andreae et al., 1988). In addition, particles are formed from precursor gases like SO_2 and hydrocarbons emitted by fires.

The average tropospheric lifetime of aerosol particles, and of their precursor gases, is of the order of only days or weeks. This is much shorter than the lifetime of most greenhouse gases. It implies that the atmospheric loading at any one time reflects the emissions that have taken place during the past few weeks only. No long-term accumulation in the troposphere is thus possible and any reduction in anthropogenic emissions will immediately result in a corresponding reduction in tropospheric concentrations. The short lifetime also implies large spatial and temporal variability in the concentrations of aerosol particles.

It has been established from analyses of Greenland ice cores that the amounts of sulphate, nitrate and trace metals, derived mainly from atmospheric aerosols, have been increasing since industrialisation began (Neftel et al., 1985b; Mayewsky et al., 1986). However, there are almost no long-term, continuous direct observations of aerosol parameters in the atmosphere outside urban and industrial areas (Charlson, 1988). Indirect evidence from visibility observations indicates that the concentration of submicron

aerosols over much of the eastern part of the U.S. has increased during the period 1948-1978 (Husar et al., 1981).

Another example of a trend analysis of atmospheric aerosols is due to Winkler and Kaminski (1988), who concluded that submicrometer aerosol mass outside Hamburg has increased by a factor of nearly two between 1976 and 1988 due to long range transport from industrialized centres in the region.

The hypothesis by Charlson et al. (1987) of a connection between climate and phytoplankton activity in ocean surface waters is based on the role played by soluble aerosol particles in determining the microphysical properties of clouds. The proposed climate-phytoplankton feedback rests on the facts that cloud condensation nucleus (CCN) concentrations in air are low over oceans far from land, that the CCN available in clean maritime air are composed almost totally of sulphate particles, and that this sulphur originates almost entirely from emissions of reduced sulphur gases (principally dimethylsulphide (DMS)) from the ocean surface. There is a significant non-linearity in the effect on cloud microphysics of given changes in CCN concentration, depending on the starting CCN concentration characteristics of clean oceanic air.

There is abundant evidence in the literature to confirm the role played by CCN concentration in determining cloud droplet size distribution. However, at this stage neither the sign nor magnitude of the proposed climate feedback can be quantitatively estimated, though preliminary calculations based on plausible scenarios indicate that this hypothesis merits careful consideration. Preliminary attempts to test this hypothesis using existing historical data of various types have been inadequate and have yielded only equivocal conclusions.

1.8.2 The Atmospheric Sulphur Budget

Current estimates of the global sulphur cycle show that anthropogenic emissions of SO_2 are likely to be at least as large as natural emissions of volatile sulphur species, cf. Table 1.5 (based essentially on Andreae, 1989). Within the industrialized regions of Europe and North America, anthropogenic emissions dominate over natural emissions by about a factor of ten or even more (Galloway et al., 1984; Rodhe, 1976). The anthropogenic SO_2 emissions have increased from less than 3 TgS per year globally in 1860, 15 in 1900, 40 in 1940 and about 80 in 1980 (Ryaboshapko, 1983). It is evident from these numbers that the sulphur fluxes through the atmosphere have increased very substantially during the last century, especially in the Northern Hemisphere. During the past decade the anthropogenic sulphur emissions in North America and parts of Europe have started to decline.

Small amounts of carbonyl sulphide (COS) are also emitted into the atmosphere. They do not significantly affect the sulphur balance of the troposphere but they are

Table 1.5 Estimates of Global Emission to the Atmosphere of Gaseous Sulphur Compounds †

Source	Annual Flux (TgS)
Anthropogenic (mainly SO_2 from fossil fuel combustion)	80
Biomass burning (SO_2)	7
Oceans (DMS)	40
Soils and plants (H_2S, DMS)	10
Volcanoes (H_2S, SO_2)	10
TOTAL	147

†   The uncertainty ranges are estimated to be about 30% for the anthropogenic flux and a factor of two for the natural fluxes.

important in maintaining an aerosol layer in the stratosphere.

Because of the limited atmospheric lifetime of most sulphur compounds, the augmentation of the sulphur concentrations brought about by industrialization is not evenly distributed around the globe. This is illustrated by Figure 1.16, which shows an estimate of how much more aerosol sulphate there is at present in the lower atmosphere (900 hPa level) than in the pre-industrial situation (Langner and Rodhe, 1990). Over the most polluted regions of Europe and North America the sulphate levels have gone up by more than a factor of 10. Smaller increases have occurred over large parts of the oceans.

1.8.3 Aerosol Particles in the Stratosphere

The vertical profile of aerosol particle concentration normally exhibits a marked decline up through the troposphere followed by a secondary maximum in the lower stratosphere at around 20 km. The stratospheric aerosol layer is maintained by an upward flux of gaseous precursors, mainly carbonyl sulphide (COS). Concentrations may be greatly enhanced over large areas for a few years following large volcanic eruptions, such as El Chichon in 1982. No significant trends have been detected in the global background aerosol layer in the stratosphere during periods of low volcanic activity (WMO, 1989a). The potential impact on climate of stratospheric aerosols is discussed in Section 2.3.2.

1.8.4 Conclusions

Aerosol particles have a lifetime of at most a few weeks in the troposphere and occur in highly variable concentrations. A large proportion of the particles which influence cloud processes and for radiative balance, are derived from gaseous sulphur emissions. These emissions have more than doubled globally, causing a large increase in the concentration of aerosol sulphate especially over and around the industrialized regions in Europe and North America. If anthropogenic sulphur emissions are indeed a major contributor to cloud condensation nuclei concentrations on a global scale, then any climate prediction must take account of future trends in regional and global anthropogenic sulphur emission, which may be quite different from those of the greenhouse gases.

Aerosol particles derived from natural (biological) emissions may contribute in important ways to climate feedback processes. During a few years following major volcanic eruptions the concentration of aerosol particles can be greatly enhanced.

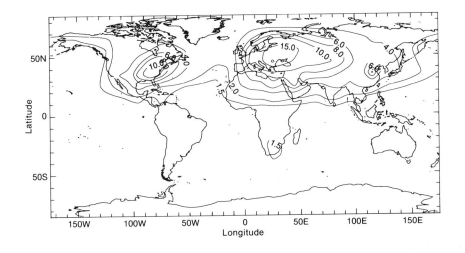

Figure 1.16: Simulated concentration of sulphate at 900 hPa: Ratio of concentrations based on total emissions (natural plus anthropogenic) divided by concentrations based on natural emissions, in July (Langner and Rodhe, 1990).

References

Andreae, M. O., E. V. Browell, M. Garstang, G. L. Gregory, R. C. Harriss, G. F. Hill, D. J. Jacob, M. C. Pereira, G. W. Sachse, A. W. Setzer, P. L. Silva Dias, R. W. Talbot, A. L. Torres, and S. C. Wofsy, 1988: Biomass burning emissions and associated haze layers over Amazonia, *J. Geophys. Res.*, **93**, 1509-1527.

Andreae, M. O., The global biogeochemical sulphur cycle, 1989: A review, *Trace Gases and the Biosphere*, B. Moore (ed.), University of Arizona Press.

Andronova, N., 1990: unpublished manuscript (USSR).

Angell, J.K. and J.Korshover, 1983: Global variation in total ozone and layer mean ozone: An update through 1981, *J. Climate Appl. Meteorol.*, **22**, 1611-1626.

Angle, R.P. and H.F. Sandhu, 1986: Rural ozone concentration in Alberta, Canada, *Atmospheric Environment*, **20**, 1221-1228.

Aselmann, I. and P.J. Crutzen, 1989: The global distribution of natural freshwater wetlands and rice paddies, their net primary productivity, seasonality and possible methane emission, *J. Atm. Chem.*, **8**, 307-358.

Bacastow, R. B., C. D. Keeling, and T. P. Whorf, 1985: Seasonal amplitude increase in atmospheric CO_2 concentration at Mauna Loa, Hawaii, 1959-1982, *J.Geophys. Res.*, **90**, 10,529-10,540.

Bacastow, R.B. and E. Maier-Reimer, 1990: Climate Dynamics, in press.

Baes, C.F., A. Björkström and P. Mulholland, 1985: Uptake of carbon dioxide by the oceans. In J.R. Trabalka (ed.): Atmospheric carbon dioxide and the global carbon cycle. U.S. Dept. of Energy DOE/ER-0239, 81-111, Washington DC, USA.

Barnola, J. M., D. Raynaud, Y. S. Korotkevitch, and C. Lorius, 1987: Vostok ice core: A 160,000 year record of atmospheric CO_2, *Nature*, **329**, 408-414.

Bartlet, K.B., P.M. Crill, J.A. Bonassi, J.E. Richey and R.C. Harriss, 1990: Methane flux form the Amazon River floodplain: emissions during rising water. *J. Geophys. Res.* In press.

Beardsmore, D.J. and G.I. Pearman, 1987: Atmospheric carbon dioxide measurements in the Australian region: Data from surface observatories. *Tellus* **39B**, 459-476.

Bingemer, H. G., and P. J. Crutzen, 1987: The production of methane from solid wastes, *J. Geophys. Res.*, **92**, 2181-2187.

Blackmer, A. M., and J. M. Bremner, 1976: Potential of soil as a sink for atmospheric nitrous oxide, *Geophys. Res. Lett.*, **3**, 739-742.

Blake, D. R., and F. S. Rowland, 1988: Continuing worldwide increase in tropospheric methane, 1978 to 1987, *Science*, **239**, 1129-1131.

Bojkov, R.D., 1988: Ozone changes at the surface and in the free troposphere, Proceedings of the NATO Advanced Research Workshop on Regional and Global Ozone and its Environmental Consequences, ed. by I.S.A. Isaksen, *NATO ASI Series C*, **227**, 83-96, Reidel, Dordrecht, Holland.

Bolin, B, (ed.) 1981: Carbon Cycle Modelling, *SCOPE 16*, John Wiley.

Bolin, B., 1986: How much CO_2 will remain in the atmosphere? In: The Greenhouse Effect, Climatic Change and Ecosystems, *SCOPE* 29, B. Bolin et al., eds., 93-155, John Wiley.

Bonsang, B. M. Kanakidou, G. Lambert, P. Monfray, 1988: The marine source of C_2-C_6 aliphatic hydrocarbons, *J. Atmos. Chem.*, **6**, 3-20.

Bowden, W. B., and F. H. Bormann, 1986: Transport and loss of nitrous oxide in soil water after forest clear-cutting, *Science*, **233**, 867-869.

Bowden, R. D., P. A Steudler, J. M. Melillo, and J. D. Aber, 1990: Annual nitrous oxide fluxes from temperate forest soils in the northeastern United States, *submitted to J. Geophys. Res.*

Bremner, J. M., G. A. Breitenbeck, and A M. Blackmer, 1981: Effect of nitropyrin on emission of nitrous oxide from soil fertilized with anhydrous ammonia, *Geophys. Res. Lett.*, **8(4)**, 353-356.

Broccoli, A. J., and S. Manabe, 1987: The influence of continental ice, atmospheric CO_2, and land albedo on the climate of the last glacial maximum, *Climate Dynamics*, **1**, 87-99.

Broecker, W. S., T. H. Peng, and R. Engh, 1980: Modeling the carbon system, *Radiocarbon*, **22**, 565-598.

Broecker, W.S., and T.H. Peng, 1982: Tracers in the Sea, *Eldigio Press*, Lamont-Doherty Geological Observatory, Palisades, N.Y.

Burke, R. A., T. R. Barber, and W. M. Sackett, 1988: Methane flux and stable hydrogen and carbon isotope composition of sedimentary methane from the Florida Everglades, *Global Biogeochem. Cycles*, **2**, 329-34.

Butler, J. H., J. W. Elkins, T. M. Thompson, and K. B. Egan, 1990: Tropospheric and dissolved N_2O of the West Pacific and East Indian oceans during the El Niño-southern oscillation event of 1987, *J. Geophys. Res.*, in press.

Chappellaz, J., J. M. Barnola, D. Raynaud, Y. S. Korotkevich, and C. Lorius, 1990: Ice-core record of atmospheric methane over the past 160,000 years., *Nature*, **345**, 127-131..

Charlson, R. J., 1988: Have concentrations of tropospheric aerosols changed? in *The Changing Atmosphere*, F. S. Rowland and I. S. A. Isaksen (eds.), Dahlem Workshop Reports. J. Wiley and Sons, Chichester.

Charlson, R.J., J.E. Lovelock, M.O. Andreae and S.,G. Warren, 1987: Oceanic phytoplankton, atmospheric sulphur, cloud albedo and climate, *Nature*, **326**, 655-661.

Chatfield, R. and H.Harrison, 1977: Tropospheric ozone, 2, Variations along a meridional band, *J. Geophys. Res.*, **82**, 5969-5976.

Cicerone, R. J., 1988: How has the atmospheric concentration of CO changed?, *The Changing Atmosphere*, F. S. Rowland and I. S. A. Isaksen (Eds.), Wiley-Interscience, 49-61.

Cicerone, R. J., and J. D. Shetter, 1981: Sources of atmospheric methane; Measurements in rice paddies and a discussion, *J. Geophys. Res.*, **86**, 7203-7209.

Cicerone, R. J., J. D. Shetter, and C. C. Delwiche, 1983: Seasonal variation of methane flux from a California rice paddy, *J. Geophys. Res.*, **88**, 11,022-11,024.

Cicerone, R., and R. Oremland, 1988: Biogeochemical aspects of atmospheric methane, *Global Biogeochem. Cycles*, **2**, 299-327.

Cline, J. D., D. P. Wisegarver, and K. Kelly-Hansen, 1987: Nitrous oxide and vertical mixing in the equatorial Pacific during the 1982-1983 El Niño, *Deep-Sea Res.*, **34**, 857-873.

Cohen, Y., and L. I. Gordon, 1979: Nitrous oxide production in the ocean, *J. Geophys. Res*, **84(C1)**, 347-353.

Conrad, R. and W. Seiler, 1985: Influence of temperature, moisture, and organic carbon on the flux of H_2 and CO between soil and atmosphere: field studies in sub-tropical regions, *J. Geophys. Res.*, **90**, 5633-5709.

Conrad, R., W. Seiler, and G. Bunse, 1983: Factors influencing the loss of fertilizer nitrogen into the atmosphere as N_2O, *J. Geophys. Res.*, **88**, 6709-6718.

Conway, T.J., P. Tans, L.S. Waterman, K.W. Thoning, K.A. Masarie, and R.H. Gammon, 1988: Atmospheric carbon dioxide measurements in the remote global troposphere, 1981-1984, *Tellus*, **40B**, 81-115.

Craig, J., and C. C. Chou, Methane: 1982: The record in polar ice cores, *Geophys. Res. Lett.*, **9**, 1221-1224.

Crill, P. M., K. B. Bartlett, R. C. Harriss, E. Gorham, E. S. Verry, D. I. Sebacher, L. Madzar, and W. Sanner, 1988: Methane flux from Minnesota peatlands, *Global Biogeochem. Cycles*, **2**, 371-384.

Crutzen, P. J., 1974: Photochemical reactions initiated by and influencing ozone in unpolluted tropospheric air, *Tellus*, **26**, 47-57.

Crutzen, P. J., L. E. Heidt, J. P. Krasnec, W. H. Pollock, and W. Seiler, 1979: Biomass burning as a source of atmospheric gases CO, H_2, N_2O, NO, CH_3Cl, and COS, *Nature*, **282**, 253-256.

Crutzen, P. J., A. C. Delany, J. Greenberg, P. Haagenson, L. Heidt, R. Lueb, W. Pollock, W. Seiler, A. Wartburg, and P. Zimmerman, 1985: Tropospheric chemical composition measurements in Brazil during the dry season, *J. Atmos. Chem.*, **2**, 233-256.

Crutzen, P. J., I. Aselmann, and W. S. Seiler, 1986: Methane production by domestic animals, wild ruminants, other herbivorous fauna, and humans, *Tellus*, **38B**, 271-284.

Crutzen, P. J., 1987: Role of the tropics in atmospheric chemistry, in *The Geophysiology of Amazonia*, edited by R. E. Dickinson, pp. 107-130, John Wiley, New York.

Crutzen, P. J., 1989: Emissions of CO_2 and other trace gases to the atmosphere from fires in the tropics, *28th Liége International Astrophysical Colloquium, University de Liége*, Belgium, June 26-30.

Crutzen, P.J., 1988: Tropospheric ozone, an overview, Proceedings of the NATO Advanced Research Workshop on Regional and Global Ozone and its Environmental Consequences, ed. By I.S.A. Isaksen, *NATO ASI Series C*, **227**, 3-23, Reidel Dordrecht, Holland.

Danielsen, E.F., 1968: Stratosphere-troposphere exchange based on radioactivity, ozone and potential vorticity, *J. Atmos. Sci.*, **25**, 502-518.

Delany, A.C., P.J. Crutzen, P. Haagenson, S. Walters, and A.F. Wartburg, 1985: Photochemically produced ozone in the emissions from large-scale tropical vegetation fires, *J. Geophys. Res.*, **90**, 2425-2429.

Delmas, R.J., J.M. Ascensio, and M. Legrand, 1980: Polar ice evidence that atmospheric CO_2 20,000 years B.P. was 50% of present, *Nature*, **284**, 155-157.

Detwiler, R. P., and C. A. S. Hall, 1988: Tropical forests and the global carbon cycle, *Science*, **239**, 43-47.

Dianov-Klokov, V.I., and L.N. Yurganov, 1981: A spectroscopic study of the global space-time distribution of atmospheric CO. *Tellus*, **33**, 262-273.

Dignon, J. and S. Hameed, 1985: A model investigation of the impact of increases in anthropogenic NOx emissions between 1967 and 1980 on tropospheric ozone, *J. Atmos. Chem.*, **3**, 491-506.

Dignon, J. and S. Hameed, 1989: Global emissions of nitrogen and suphur oxides from 1860 to 1980, *J. Air Pollut. Contr. Assoc.*, **39**, 180-186.

Elkins, J. W., S. C. Wofsy, M. B. McElroy, C. E. Kolb, and W. A. Kaplan, 1978: Aquatic sources and sinks for nitrous oxide, *Nature*, **275**, 602-606.

Elkins, J. W., and R. Rossen, 1989: *Summary Report 1988: Geophysical Monitoring for Climatic Change*, NOAA ERL, Boulder, CO.

Elkins, J. W., B. D. Hall, and J. H. Butler, 1990: Laboratory and field investigations of the emissions of nitrous oxide from biomass burning, *Chapman Conference on Global Biomass Burning: Atmospheric, Climatic, and Biospheric Implications*, Williamsburgh, Virginia, USA, March 19-23, J. S. Levine, Editor.

Enting, I.G. and G.I. Pearman, 1982: *Description of a one-dimensional global carbon cycle model*. CSIRO, Division of Atmospheric Physics, Technical Paper No. 42. 95 pp, Australia.

Enting, I.G., and G.I. Pearman, 1987: Description of a one-dimensional carbon cycle model calibrated using techniques of constrained inversion, *Tellus*, **39B**, 459-476.

Enting, I. G., and J. V. Mansbridge, 1989: Seasonal sources and sinks of atmosphere CO_2 direct inversion of filtered data, *Tellus*, **41B**, 111-126.

EPA, 1989: *National air pollutant emission estimates 1940 - 1987*. US Environment Protection Board Report EPA/450/4/88/022, Research Triangle Park, North Carolina, USA.

Eppley, R.W., and B.J. Peterson, 1979: Particulate organic matter flux and planktonic new production in the deep ocean, *Nature*, **282**, 677-680.

Etheridge, D. M., G. I. Pearman, and F. de Silva, 1988: Atmospheric trace-gas variations as revealed by air trapped in an ice core from Law Dome, Antarctica, *Annal. Glaciology*, **10**, 28-33.

Feely, R.A., R.H. Gammon, B.A. Taft, P.E. Pullen, L.E. Waterman, T.J. Conway, J.F. Gendron, and D.P. Wisegarver, 1987: Distribution of chemical tracers in the Eastern Equatorial Pacific during and after the 1982-1983 El-Nino/Southern Oscillation event, *J. Geophysical Res.*, **92**, 6545-6558.

Fehsenfeld, F.C., D.W. Parrish, and D.W. Fahey, 1988: The measurements of NO_x in the non-urban troposphere, *Tropospheric Ozone: Regional and Global Scale Interactions*, I. S. A. Isaksen ed., D. Reidel Publishing Co., Boston, MA, 185-217.

Feister, U., and W. Warmbt, 1987: Long-term measurements of surface ozone in the German Democratic Republic, *J. Atmos. Chem.*, **5**, 1-21.

Fishman, J., F.M. Vukovich, and E.V. Browell, 1985: The photochemistry of synoptic-scale ozone synthesis: Implications for the global tropospheric ozone budget, *J. Atmos. Chem.*, **3**, 299-320.

Fishman, J., C.E. Watson, J.C. Larsen, and J.A. Logan, 1990: The distribution of tropospheric ozone obtained from satellite data, *J. Geophys. Res.*, in press.

Francey, R.J., and G. Farquhar, 1982: An explanation for $^{13}C/^{12}C$ variations in tree rings, *Nature*, **291**, 28-31.

Fraser, P.J., M.A.K.Khalil, R.A. Rasmussen, and L.P. Steele, 1984: Tropospheric methane in the mid-latitudes of the southern hemisphere, *J. Atmos. Chem.*, **1**, 125-135.

Fraser, P.J., and N. Derek, 1989: Atmospheric halocarbons, nitrous oxide, and methane-the GAGE program, *Baseline 87*, 34-38, Eds. B. Forgan and G. Ayers, Bureau of Meteorology-CSIRO, Australia.

Fraser, P.J., P.Hyson, R.A. Rasmussen, A.J. Crawford, and M.A. Khahil, 1986a: Methane, and methylchloroform in the Southern Hemisphere, *J. Atmos. Chem.*, **4**, 3-42.

Fraser, P.J., R. Rasmussen, J.W. Creffield, J.R. French and M.A.K. Khalil, 1986b: Termites and global methane: another assessment, *J. Atmospheric Chemistry*, **4**, 295-310.

Fraser, P.J., P. Hyson, S. Coram, R.A. Rasmussen, A. J. Crawford, and M.A. Khahil, 1986c: Carbon monoxide in the Southern Hemisphere, *Proceedings of the Seventh World Clean Air Congress*, **2**, 341-352.

Friedli, H., H. Loetscher, H. Oeschger, U. Siegenthaler, and B. Stauffer, 1986: Ice core record of the $^{13}C/^{12}C$ record of atmospheric CO_2 in the past two centuries, *Nature*, **324**, 237-238.

Galbally, I.E., 1989: Factors controlling NO_x emissions from soils, *Exchange of Trace Gases between Terrestrial Ecosystems and the Atmosphere*, Dahlem publications, John Wiley and Sons, 23-27.

Galbally, I.E., and C.R. Roy, 1980: Destruction of ozone at the Earth's surface, *Q. J. Roy. Meteorol. Soc.*, **106**, 599-620.

Galloway, J.N., Likens, G.E., and M.E. Hawley, 1984: Acid deposition: Natural versus anthropogenic sources, *Science*, **226**, 829- 831.

Goodman, and R.J. Francey, 1988: $^{13}C/^{12}C$ and $^{18}O/^{16}O$ in Baseline CO_2, in: *Atmospheric Programme (Australia) 1986*, Eds. B W Forgan and P J Fraser, 54-58, CSIRO Division of Atmospheric Research, Australia.

Goudriaan, J., 1989: Modelling biospheric control of carbon fluxes between atmosphere, ocean and land in view of climatic change.In A. Berger et al. (eds.) Climate and Geo-sciences. NATO-ASI Series C:285, 481-499, Kluwer.

Griffith, D.W.T., W.G. Mankin, M.T. Coffey, D.E. Ward, and A. Riebau, 1990: FTIR remote sensing of biomass burning emissions of CO_2, CO, CH_4, CH_2O, NO, NO_2, NH_3, and N_2O, *Chapman Conference on Global Biomass Burning: Atmospheric, Climatic, and Biospheric Implications*, Williamsburgh, Virginia, USA, March 19-23, J. S. Levine, Editor.

Greenberg, J.P., P.R. Zimmerman, L. Heidt, and W. Pollock, 1984: Hydrocarbon and carbon monoxide emissions from biomass burning in Brazil, *J. Geophys. Res.*, **89**, 1350-1354.

Hao, W.M., S.C. Wofsy, M.B. McElroy, J.M. Beer, and M.A. Togan, 1987: Sources of atmospheric nitrous oxide from combustion, *J. Geophys. Res.*, **92**, 3098-3104.

Hansen, J., A. Lacis, D. Rind, G. Russell, P. Stone, I Fung, R. Ruedy, and J. Lerner, 1984: Climate Sensitivity: Analysis of feedback effects. In Climate Processes and Climate Sensitivity, *Geophysical Monograph* , 29, 130-163, AGU.

Harriss, R.C., and D.I. Sebacher, 1981: Methane flux in forested freshwater swamps of the southeastern United Statues, *Geophys. Res. Lett.*, **8**, 1002-1004.

Harriss, R.C., D.I. Sebacher, and F. Day, 1982: Methane flux in the Great Dismal Swamp, *Nature,* **297**, 673-674.

Harriss, R.C., E. Gorham, D.I. Sebacher, K.B. Bartlett, and P.A. Flebbe, 1985: Methane flux from northern peat-lands, *Nature,* **315**, 652-654.

Harriss, R.C., D.I. Sebacher, K.B. Bartlett, D.S. Bartlett, and P.M. Crill, 1988: Sources of atmospheric methane in the South Florida environment, *Global Biogeochem. Cycles,* **2**, 231-244.

Heidt, L.E., J.P. Krasnec, R.A. Lueb, W.H. Pollack, B.E. Henry, and P.J. Crutzen, 1980: Latitudinal distribution of CO and CH_4 over the Pacific, *J. Geophys. Res.*, **85**, 7329-7336.

Holzapfel-Pschorn, A., and W. Seiler, 1986: Methane emission during a cultivation period from an Italian rice paddy. *J. Geophys. Res.*, **91**, 11,803-11,814.

Houghton, R.A., 1987: Biotic changes consistent with the increased seasonal amplitude of atmospheric CO_2 concentrations, *J.Geophys. Res.*, **92**, 4223-4230.

Houghton, R.A., R.D. Boone, J.M. Melillo, C.A. Palm, G.M. Woodwell, N. Myers, B. Moore, and D.L. Skole, 1985a: Net flux of CO_2 from tropical forests in 1980, *Nature,* **316**, 617-620.

Houghton, R. A., W. H. Schlesinger, S. Brown, and J. F. Richards, 1985b: Carbon dioxide exchange between the atmosphere and terrestrial ecosystems, in *Atmospheric Carbon Dioxide and the Global Carbon Cycle*, J. R. Trabalka, Ed., U.S. Department of Energy, DOE/ER-0239, Washington, DC.

Houghton, R. A., R. D. Boone, J. R. Fruci, J. E. Hobbie, J. M. Melillo, C. A. Palm, B. J. Peterson, G. R. Shaver, G. M. Woodwell, B. Moore, D. L. Skole, and N. Myers, 1987: The flux of carbon from terrestrial ecosystems to the atmosphere in 1980 due to changes in land use: Geographic distribution of the global flux, *Tellus*, **39B**, 122-139.

Houghton, R. A., G. M. Woodwell, R. A. Sedjo, R. P. Detwiler, C. A. S. Hall, and S. Brown, 1988: The global carbon cycle, *Science*, **241**, 1736-1739.

Houghton, R. A., and D. L. Skole, 1990: Changes in the global carbon cycle between 1700 and 1985, in *The Earth Transformed by Human Action*, B. L. Turner, ed., Cambridge University Press, in press.

Houghton, R. A., and G. M. Woodwell, 1989: Global Climatic Change, *Scientific American*, **260** No. 4, 36-44

Husar, R. B., J. M. Holloway, and D. E. Patterson, 1981: Spatial and temporal pattern of eastern U.S. haziness: A summary, *Atmos. Environ.*, **15**, 1919-1928.

ICF, 1990: Conference organised as part of IPCC Working Group 3, *International Workshop on methane emissions from natural gas systems, coal mining and waste management systems*, Washington DC, USA. 9-13 April.

Isaksen, I. S. A., and Ö. Hov, 1987: Calculation of trends in the tropospheric concentration of O_3, OH, CH_4, and NO_x, *Tellus*, **39B**, 271-285.

Isaksen, I.S.A., Ö. Hov, and E. Hesstvedt, 1978: Ozone generation over rural areas, *Environ. Sci. Technol.,* **12**, 1279-1284.

Jacob, D.J., and S.C. Wofsy, 1990: Budgets of reactive nitrogen, hydrocarbons and ozone over the Amazon forest during the wet season, J. Geophys. Res., submitted.

Keeling, C. D., and M. Heimann, 1986: Meridional eddy diffusion model of the transport of atmospheric carbon dioxide. 2. Mean annual carbon cycle, *J. Geophys. Res.,* **91**, 7782-7798.

Keeling , C.D., R.B. Bacastow, A.F. Carter, S.C. Piper, T.P. Whorf, M. Heimann, W.G. Mook, and H. Roeloffzen, 1989a: A three dimensional model of atmospheric CO_2 transport based on observed winds: 1. Analysis of observational data in: Aspects of climate variability in the Pacific and the Western Americas, D.H. Peterson (ed.), *Geophysical Monograph* , **55**, AGU, Washington (USA), 165-236.

Keeling , C.D., S.C. Piper, and M. Heimann, 1989b: A three dimensional model of atmospheric CO_2 transport based on observed winds: 4. Mean annual gradients and interannual variations, in: Aspects of climate variability in the Pacific and the Western Americas, D.H. Peterson (ed.), *Geophysical Monograph* , **55**, AGU, Washington (USA), 305-363.

Keller, M., W. A. Kaplan, and S. C. Wofsy, 1986: Emissions of N_2O, CH_4, and CO_2 from tropical soils, *J. Geophys. Res.,* **91**, 11,791-11,802.

Khalil, M. A. K., and R. A Rasmussen, 1984: Carbon monoxide in the Earth's atmosphere: Increasing trend, *Science,* **224**, 54-56.

Khalil, M. A. K., and R. A Rasmussen, 1988a: Carbon monoxide in the Earth's atmosphere: Indications of a global increase, *Nature,* **332**, 242-245.

Khalil, M. A. K., and R. A Rasmussen, 1988b: Nitrous oxide: Trends and global mass balance over the last 3000 years, *Annal. Glaciology,* **10**, 73-79.

Kirchhoff, V.W.J.H., A.W. Setzer, and M.C. Pereira, 1989: Biomass burning in Amazonia: Seasonal effects on atmospheric O_3 and CO, *Geophys. Res. Lett.,* **16**, 469-472.

Kirchhoff, V. W. J. H., 1990: Ozone measurements in Amazonia: Dry versus wet season, accepted for publication, *J. Geophys. Res.*

Kirchhoff, V. W. J. H. and E. V. A. Marinho, 1989: A survey of continental concentrations of atmospheric CO in the Southern Hemisphere, *Atmospheric Environment,* **23**, 461-466.

Kohlmaier, G.H., E.O. Sire, A. Janecek, C.D. Keeling, S.C. Piper, and R.Revelle, 1989: Modelling the seasonal contribution of a CO_2 fertilization effect of the terrestrial vegetation to the amplitude increase in atmospheric CO_2 at Mauna Loa Observatory, *Tellus,* **41B**, 487-510.

Kohlmaier, G. H., 1988: The CO_2 response of the biota-soil system to a future global temperature change induced by CO_2 and other greenhouse gases. Roger Chapman Conference, San Diego. in press.

Kvenvolden, K. A., 1988: Methane hydrates and global climate, *Global Biogeochem. Cycles,* **2**, 221-230.

Langner, J. and H. Rodhe, 1990: Anthropogenic impact on the global distribution of atmospheric sulphate. Proceedings of First International Conference on Global and Regional Atmospheric Chemistry, Beijing, China, 3-10 May 1989.L.

Newman , L, W, Wang and C.S. Kiang (eds.), US Dept of Energy Report, Washington DC, *In Press.*

Lashof, D. A., 1989: The dynamic greenhouse: Feedback processes that may influence future concentrations of atmospheric trace gases and climate change, *Climatic Change,* **14**, 213-242.

Lelieveld, J. and P.J. Crutzen, 1990: Influence of cloud photochemical processes on tropospheric ozone.*Nature,* **343**, 227-233.

Lerner, J., G. Mathews and I.Fung, 1988: Methane emissions for animals: A global high-resolution data base, *Global Biogeochem. Cycles,* **2**, 139-156.

Levy, H., II, J.D. Mahlman, W.J. Moxim, and S.C. Liu, 1985: Tropospheric ozone: The role of transport, *J. Geophys. Res.,* **90**, 3753-3772.

Liu, S.C., D. Kley, M. McFarland, J.D. Mahlman, and H. Levy, 1980: On the origin of tropospheric ozone, *J. Geophys. Res.,* **85**, 7546-7552.

Logan, J. A., 1988: The ozone problem in rural areas of the United States, Proceedings of the NATO Advanced Research Workshop on regional and global ozone and its environmental consequences, ed. by I. S. A. Isaksen, NATO ASI Series C, 227, 327-344, Reidel, Dordrecht, Holland.

Logan, J. A., 1985: Tropospheric Ozone: Seasonal behavior, trends, and anthropogenic influence, *J. Geophys. Res.,* **90**, 10,463-10,482.

Logan, J. A., M. J. Prather, S. C. Wofsy, and M. B. McElroy, 1981: Tropospheric chemistry: A global perspective, *J. Geophys. Res.,* **86**, 7210-7254.

Logan, J.A., 1989: Ozone in rural areas of the United States, *J. Geophys. Res.,* **94**, 8511-8532.

Logan, J.A. and V.W.J.H. Kirchhoff, 1986: Seasonal variations of tropospheric ozone at Natal Brazil. *J. Geophys. Res.,* **91**, 7875-7881.

Lowe, D. C., C. A. M. Brenninkmeijer, M. R. Manning, R. Sparkes, and G. Wallace, 1988: Radiocarbon determination of atmospheric methane at Baring Head, New Zealand, *Nature,* **332**, 522-525.

Luizão, F., P. Matson, G. Livingston, R. Luizão, and P. Vitousek, 1990: Nitrous oxide flux following tropical land clearing, *Global Biogeochem. Cycles,* in press.

Luxmoore, R. J., E. G. O'Neill, J. M. Ells, H. H. Rogers, 1986: Nutrient uptake and growth responses of Virginia pine to elevated atmospheric carbon dioxide, *J. Environ. Qual.,* **15**, 244-251.

Mahlman, J.D., and W.J. Moxim, 1978: Tracer simulation using a global general circulation model: Results from a mid-latitude instantaneous source experiment, *J. Atmos. Sci.,* **35**, 1340-1374.

Maier-Reimer, E. and K.Hasselmann, 1987: Transport and storage of carbon dioxide in the ocean, and in organic ocean-circulation carbon cycle model, *Climate Dynamics,* **2**, 63-90.

Marland, G., 1989: Fossil fuels CO_2 emissions: Three countries account for 50% in 1988. CDIAC Communications, Winter 1989, 1-4, Carbon Dioxide Information Analysis Center, Oak Ridge National Laboratory, USA.

Mathews, E. and I. Fung, 1987: Methane emissions from natural wetlands: Global distribution, area and environment of characteristics of sources, *Global Biogeochem Cycles,* **1**, 61-86.

Matson, P. A., and P. M. Vitousek, 1987: Cross-system comparisons of soil nitrogen transformations and nitrous oxide flux in tropical forest ecosystems, *Global Biogeochem. Cycles,* **1**, 163-170.

Matson, P. A., and P. M. Vitousek, 1989: Ecosystem approaches for the development of a global nitrous oxide budget, *Bioscience,* in press.

Mayewsky, P.A., W. B. Lyons, M. Twickler, W. Dansgaard, B. Koci, C. I. Davidson, and R. E. Honrath, 1986: Sulfate and nitrate concentrations from a South Greenland ice core. *Science* **232**, 975-977.

McElroy, M. B., and S. C. Wofsy, 1986: Tropical forests: Interactions with the atmosphere, *Tropical Rain Forests and the World Atmosphere,* AAAS Selected symposium 101, edited by Prance, G. T., pp. 33-60, Westview Press, Inc., Boulder, CO.

Moore, T. R., and R. Knowles, 1987: Methane and carbon dioxide evolution from subarctic fens, *Can. J. Soil Sci.,* **67**, 77-81.

Muzio, L. J., and J. C. Kramlich, 1988: An artifact in the measurement of N_2O from combustion sources, *Geophys. Res. Lett.,* **15(12)**, 1369-1372.

Neftel, A. H. Oeschger, J. Schwander, B. Stauffer, and R. Zumbrunn, 1982: Ice core measurements give atmospheric CO_2 content during the past 40,000 years, *Nature,* **295**, 222-223.

Neftel, A., E. Moor, H. Oeschger, and B. Stauffer, 1985a: Evidence from polar ice cores for the increase in atmospheric CO_2 in the past two centuries, *Nature,* **315**, 45-47.

Neftel, A. J. Beer, H. Oeschger, F. Zurcher, and R. C. Finkel, 1985b: Sulfate and nitrate concentrations in snow from South Greenland 1895-1978, *Nature,* **314**, 611-613.

Neftel, A., H. Oeschger, T. Staffelbach and B. Stauffer, 1988: CO_2 record in the Byrd ice core 50,000-5,000 years BP, *Nature,* **331**, 609-611.

Newell, R. E., H. G. Reichle, and W. Seiler, 1989: Carbon monoxide and the burning Earth, *Scientific American, October 1989,* 82-88.

Neue, H. H., and H. W. Scharpenseel, 1984: Gaseous products of decomposition of organic matter in submerged soils, *Int. Rice Res. Inst., Organic Matter and Rice,* 311-328, Los Banos, Phillipines.

Nisbet, E.G., 1989: Some Northern sources of atmospheric methane: production, history and future implications. *Can. J. Earth Sci.,* **26**, 1603-1611.

Oeschger, H., U. Siegenthaler, U. Schotterer, and A. Gugelmann, 1975: A box diffusion model to study the carbon dioxide exchange in nature, *Tellus,* **27**, 168-192.

Oltmans, S.J., and W.D. Kohmyr, 1986: surface ozone distributions and variations from 1973-1984 measurements at the NOAA Geophysical Monitoring for Climate Change baselines observatories, *J. Geophys. Res.,* **91**, 5229-5236.

Oltmans, S.J., W.D. Kohmyr, P.R. Franchois, and W.A. Matthews, 1988: Tropospheric ozone: Variations from surface and ECC ozonesonde observations, in Proc. 1988 Quadrennial Ozone Symposium, ed. R.D. Bojkov and P. Fabian, A. Deepak Publ., Hampton, VA, USA.

Pearman, G. I., and P. Hyson, 1981: The annual variation of atmospheric CO_2 concentration observed in the northern hemisphere, *J. Geophys. Res.,* **86**, 9839-9843.

Pearman, G. I., P. Hyson and P J Fraser, 1983: The global distribution of atmospheric.carbon dioxide, I, Aspects of Observation and Modelling, *J. Geophys Res,* **88**, 3581-3590.

Pearman, G. I., and P. Hyson, 1986: Global transport and inter-reservoir exchange of carbon dioxide with particular reference to stable isotopic distributions, *J. Atmos. Chem.,* **4**, 81-124.

Pearman, G. I., and P. J. Fraser, 1988: Sources of increased methane, *Nature,* **332**, 489-490.

Pearman, G.I., D. Etheridge, F. de Silva, P.J. Fraser, 1986: Evidence of changing concentrations of atmospheric CO_2, N_2O, and CH_4 from air bubbles in Antarctic ice, *Nature,* **320**, 248-250.

Peng, T.-H., and W. S. Broecker, 1985: The utility of multiple tracer distributions in calibrating models for uptake of anthropogenic CO_2 by the ocean thermocline, *J. Geophys. Res.,* **90**, 7023-7035.

Penkett, S.A., 1982: Non-methane organics in the remote troposphere. In E.D. Goldberg (ed.) Atmospheric Chemistry ,329-355. Dahlem publications, Springer Verlag, Berlin.

Penkett, S. A., B. M. R. Jones, M. J. Rycroft, and D. A. Simmons, 1985: An interhemispheric comparison of the concentration of bromine compounds in the atmosphere, *Nature,* **318**, 550-553.

Pierotti, D., and R. A. Rasmussen, 1976: Combustion as a source of nitrous oxide in the atmosphere, *Geophys. Res. Lett.,* **3**, 265-267.

Prinn, R., D. Cunnold, R. Rasmussen, P. Simmonds, F. Alyea, A. Crawford, P. Fraser, and R. Rosen, 1987: Atmospheric trends in methylchloroform and the global average for the hydroxyl radical, *Science,* **238**, 945-950.

Prinn, R., D. Cunnold, R. Rasmussen, P. Simmonds, F. Alyea, A. Crawford, P. Fraser, and R. Rosen, 1990: Atmospheric trends and emissions of nitrous oxide deducted from ten years of ALE-GAGE data, *submitted to J. Geophys. Res.*

Quay, P. D., S. L. King, J. M. Landsdown, and D. O. Wilbur, 1990: Methane release from biomass burning: estimates derived from ^{13}C composition of atmospheric methane, *Chapman Conference on Global Biomass Burning: Atmospheric, Climatic, and Biospheric Implications,* Williamsburgh, Virginia, USA, March 19-23, J. S. Levine, Editor.

Rasmussen, R. A., and M. A. K. Khalil, 1981: Atmospheric methane (CH_4): Trends and seasonal cycles, *J. Geophys. Res.,* **86**, 9826-9832.

Rasmussen, R. A., and M. A. K. Khalil, 1983: Global production of methane by termites, *Nature, 301,* 700-702.

Rasmussen, R. A., and M. A. K. Khalil, 1984: Atmospheric methane in the recent and ancient atmospheres: Concentrations, trends, and inter-hemispheric gradient, *J. Geophys. Res., 89,* 11,599-11,605.

Rasmussen, R. A., and M. A. K. Khalil, 1986: Atmospheric trace gases: Trends and distributions over the last decade, *Science,* **232**, 1623-1624.

Rasmussen, R. A., and M. A. K. Khalil, 1988: Isoprene over the Amazon basin, *J. Geophys. Res., 93,* 1417-1421.

Raynaud, D., and J. M. Barnola, 1985: An Antarctic ice core reveals atmospheric CO_2 variations over the past few centuries, *Nature*, **315**, 309-311.

Raynaud, D., J. Chappellaz, J. M. Barnola, Y. S. Korotkevich, and C. Lorius,1988: Climatic and CH_4 cycle implications of glacial-interglacial CH_4 change in the Vostok ice core, *Nature*, **333**, 655-657.

Rinsland, C. P., J. S. Levine, and T. Miles, 1985: Concentration of methane in the troposphere deduced from 1951 infrared solar spectra, *Nature*, **330**, 245-249.

Robertson, G. P., and J. M. Tiedje, 1988: Deforestation alters denitrification in a lowland tropical rain forest, *Nature*, **336**, 756-759.

Robinson, E., B. A. Bodhaine, W. D. Komhyr, S. J. Oltmans, L. P. Steele, P. Tans, and T. M. Thompson, 1988: Long-term air quality monitoring at the South Pole by the NOAA program Geophysical Monitoring for Climatic Change, *Rev. Geophys.*, **26**, 63-80.

Rodhe, H., 1976: An atmospheric sulphur budget for NW Europe, in *Nitrogen, Phosphorus and Sulphur-Global Cycles*, B.H. Svensson and R. Soderlund (eds.) SCOPE Report 7, Ecol. Bull. (Stockholm) **22**: 123-134.

Ronen, D., M. Mordeckai, and E. Almon, 1988: Contaminated aquifers are a forgotten component of the global N_2O budget, *Nature*, **335**, 57-59.

Rotty, R. M., and G. Marland, 1986: Production of CO_2 from fossil fuel burning by fuel type, 1860-1982, *Report NDP-006*, Carbon Dioxide Information Center, Oak Ridge National Laboratory,USA.

Ryaboshapko, A.G., 1983: The atmospheric sulphur cycle, In *The Global Biogeochemical Sulphur Cycle. M.V. Ivanov and G.R. Freney (eds.), SCOPE 19*, 203-296, John Wiley & Sons, Chichester, New York.

Ryden, J. C., 1981: N_2O exchange between a grassland soil and the atmosphere, *Nature*, **292**, 235-237.

Sarmiento, J.L., J.C. Orr, and U. Siegenthaler,1990: A perturbation simulation of CO_2 uptake in an ocean general circulation model. Submitted to *J. Geophys. Res.*

Sebacher, D. I., R. C. Harriss, K. B. Bartlett, S. M. Sebacher, and S. S. Grice, 1986: Atmospheric methane sources: Alaskan tundra bogs, an alpine fen, and a subarctic boreal marsh, *Tellus*, **38B**, 1-10.

Schmidt, J., W. Seiler, and R. Conrad, 1988: Emission of nitrous oxide from temperate forest soils into the atmosphere, *J. Atm. Chem.*, **6**, 95-115.

Seiler, W. and J. Fishman, 1981: The distribution of carbon monoxide and ozone in the free troposphere. *J. Geophys. Res.*, **86**, 7255-7265.

Seiler, W., R. Conrad, and D. Scharffe, 1984: Field studies of methane emission from termite nests into the atmosphere and measurements of methane uptake by tropical soils, *J. Atmos. Chem.*, **1**, 171-186.

Seiler, W., and R. Conrad, 1987: Contribution of tropical ecosystems to the global budget of trace gases, especially CH_4, H_2, CO, and N_2O, *The Geophysiology of Amazonia, Ed. R. Dickinson, 133-160*, John Wiley, New York.

Siegenthaler, U., 1983: Uptake of excess CO_2 by an outcrop-diffusion model of the ocean, *J. Geophys. Res.*, **88**, 3599-3608.

Siegenthaler, U., 1986: Carbon dioxide: it's natural cycle and anthropogenic perturbation. In: *The Role of Air-Sea Exchange in Geochemical Cycling* (P.Buat-Menard, ed.), 209-247, Reidel.

Siegenthaler, U., and H. Oeschger, 1987: Biospheric CO_2 emissions during the past 200 years reconstructed by deconvolution of ice core data, *Tellus*, **39B**, 140-154.

Siegenthaler, U., H. Friedli, H. Loetscher, E. Moor, A. Neftel, H. Oeschger, and B. Stauffer, 1988: Stable-isotope ratios and concentrations of CO_2 in air from polar ice cores, *Annals of Glaciology*, **10**.

Slemr, F., R. Conrad, and W. Seiler, 1984: Nitrous oxide emissions from fertilized and unfertilized soils in a subtropical region (Andalusia, Spain), *J. Atmos. Chem.*, **1**, 159-169.

Stauffer, B., H. Hofer, H. Oeschger, J. Schwander, and U. Siegenthaler, 1984: Atmospheric CO_2 concentrations during the last glaciation, *Annals of Glaciology*, **5**, 760-764.

Stauffer, B., G. Fischer, A. Neftel, and H. Oeschger, 1985: Increases in atmospheric methane recorded in Antarctic ice core, *Science*, **229**, 1386-1388.

Stauffer, B., E. Lochbronner, H. Oeschger, and J. Schwander, 1988: Methane concentration in the glacial atmosphere was only half that of the preindustrial Holocene, *Nature*, **332**, 812-814.

Steele, L. P., P. J. Fraser, R. A. Rasmussen, M. A. K. Khalil, T. J. Conway, A. J. Crawford, R. H. Gammon, K. A. Masarie, and K. W. Thoning, 1987: The global distribution of methane in the troposphere, *J. Atmos. Chem.*, **5**, 125-171.

Stevens, C. M., A. E. Engelkemeir, R. A. Rasmussen, 1990: The contribution of increasing fluxes of CH_4 from biomass burning in the southern hemisphere based on measurements of the temporal trends of the isotopic composition of atmospheric CH_4, *Chapman Conference on Global Biomass Burning: Atmospheric, Climatic, and Biospheric Implications*, Williamsburgh, Virginia, USA, March 19-23, J. S. Levine, Editor.

Strain, B. R., and J. D. Cure, eds., 1985: *Direct Effect on Increasing Carbon Dioxide on Vegetation*, DOE/ER-0238, U.S. Department of Energy, Washington, DC.

Stuiver, M. and P. D. Quay, 1981: Atmospheric ^{14}C changes resulting from fossil fuel CO_2 release and cosmic ray flux variability, *Earth Planet. Sci. Lett.*, **53**, 349-362.

Sundquist, E.T., 1985: Geological perspectives on carbon dioxide and the carbon cycle, in: The Carbon Cycle and Atmospheric CO_2: Natural Variations Archean to Present, E.T. Sundquist and W.S. Broecker (eds.), *Geophysical Monograph*, **32**, 5-60, AGU.

Svensson, B. H., and T. Rosswall, 1984: In situ methane production from acid peat in plant communities with different moisture regimes in a subarctic mire, *Oikos*, **43**, 341-350.

Sze, N. D., 1977: Anthropogenic CO emissions: Implications for the atmospheric CO-OH-CH_4 cycle, *Science*, **195**, 673-675.

Tans, P. P., T.J.Conway and T. Nakasawa, 1989: Latitudinal distribution of sources and sinks of atmospheric carbon dioxide, *J. Geophys Res*, **94**, 5151-5172.

Tans, P.P., I. Y. Fung and T. Takahashi, 1990: Observational constraints on the global atmospheric carbon dioxide budget, *Science*, **247**, 1431-1438.

Teramura, A.H., 1983: Effects of ultraviolet-B radiation on the growth and yield of crop plants. *Physiol. Plant.*, **58**, 415-427.

Thompson, A. M., and R. J. Cicerone, 1986: Possible perturbations to atmospheric CO, CH4, and OH, *J. Geophys. Res.*, **91**, 10,853-10, 864.

Thompson, A.M., R.W. Stewart, M. A. Owens and J.A. Herwehe, 1989: Sensitivity of tropospheric oxidants to global chemical and climate change, *Atmos. Environ.*, **23**, 519-532.

Thompson, M. L., I. G. Enting, G. I. Pearman, and P. Hyson, 1986: Interannual variation of atmospheric CO_2 concentration, *J. Atmos. Chem.*, **4**, 125-155.

Tiao, G.C., G.C. Reinsel, J.H. Pedrick, G.M. Allenby, C.L. Mateer, A.J. Miller, and J.J. DeLuisi, 1986: A statistical trend analysis of ozonesonde data. *J. Geophys. Res.*, **91**, 133121-13136.

Trabalka, J.R., (ed.), 1985: Atmospheric Carbon Dioxide and the Global Carbon Cycle, *U.S. Department of Energy, DOE/ER-0239*, Washington D.C..

UNEP, 1987: Montreal Protocol on Substances that Deplete the Ozone Layer", UNEP conference services number 87-6106.

UNEP, 1989: Report of the Technology Review Panel, ICN number 92-807-1247-0.

Volz, A., and D. Kley, 1988: Ozone measurements made in the 19th century: An evaluation of the Montsouris series, *Nature*, **332**, 240-242.

Wagenbach, D., K. O. Munnich, U. Schotterer, and H. Oeschger, 1988: The anthropogenic impact on snow chemistry at Colle Gnifetti, Swiss Alps, *Ann. Glaciol.*, **10**, 183-187.

Wahlen, M., N. Takata, R. Henry, B. Deck, J. Zeglen, J. S. Vogel, J. Southon, A. Shemesh, R. Fairbanks, and W. Broecker, 1989: Carbon-14 in methane sources and in atmospheric methane: The contribution from fossil carbon, *Science*, **245**, 286-290.

Weiss, R. F., 1981: The temporal and spatial distribution of tropospheric nitrous oxide, *J. Geophys. Res.*, **86(C8)**, 7185-7195.

Weiss, R. F. and H. Craig, 1976: Production of atmospheric nitrous oxide by combustion, *Geophys. Res. Lett.*, **3(12)**, 751-753.

Wenk, T., 1985: Einflüsse der Ozeanzirkulation und der marinen biologie auf die atmosphärische CO_2-konzentration, Ph.D. thesis, University of Bern, Switzerland.

Whalen, S. C., and W. S. Reeburgh, 1988: A methane flux time series for tundra environments, *Global Biogeochem. Cycles*, **2**, 399-410.

Winkler, P.and U. Kaminski, 1988: Increasing submicron particle mass concentration at Hamburg, 1. Observations, *Atmos. Environ.*, **22**, 2871-2878.

Winstead, E. L., K. G. Hoffman, W. R. Coffer III, and J. S. Levine, 1990: Nitrous oxide emissions from biomass burning, *Chapman Conference on Global Biomass Burning: Atmospheric, Climatic, and Biospheric Implications*, Williamsburgh, Virginia, USA, March 19-23, J. S. Levine, Editor.

WMO, 1985: Atmospheric ozone 1985: Assessment of our understanding of the processes controlling its present distribution and change, Global Ozone Research and Monitoring Project, report *16*, Geneva.

WMO, 1989a: Report of the NASA/WMO Ozone Trends Panel, 1989, Global Ozone Research and Monitoring Project, Report *18*, Geneva.

WMO, 1989b: Scientific assessment of stratospheric ozone: 1989, Global Ozone Research and Monitoring Project, Report *20*, Geneva.

Wofsy, S. C., M. B. McElroy, and Y. L. Yung, 1975: The chemistry of atmospheric ozone, *Geophys. Res. Lett.*, **2**, 215-218.

Woodwell, G. M., 1983: Biotic effects on the concentration of atmospheric carbon dioxide: A review and projection, *Changing Climate*, National Academy Press, 216-241.

Yagi, K., and K. Minami, 1990: Effects of mineral fertilizer and organic matter application on the emission of methane from some Japanese paddy fields, *submitted to Soil Sci. Plant Nutr.*.

Zander, R., Ph. Demoulin, D. H. Ehhalt, and U. Schmidt, 1990: Secular increases of the vertical abundance of methane derived from IR solar spectra recorded at the Jungfraujoch Station, *J. Geophys. Res.*, in press.

Zander, R., Ph. Demoulin, D. H. Ehhalt, U. Schmidt, and C.P. Rinsland, 1989: Secular increases in the total vertical abundance of carbon monoxide above central Europe since 1950, *J. Geophys. Res.*, **94**, 11,021-11,028.

Zardini, D., D. Raynaud, D. Scharffe, and W. Seiler, 1989: N_2O measurements of air extracted from Antarctic ice cores: Implications on atmospheric N_2O back to the last glacial-interglacial transition, *J. Atmos. Chem.*, **8**, 189-201.

Zimmerman, P. R., J. P. Greenberg, S. O. Wandiga, and P. J. Crutzen, Termites: 1982: A potentially large source of atmospheric methane, carbon dioxide and molecular hydrogen, *Science*, **218**, 563-565.

2

Radiative Forcing of Climate

K.P. SHINE, R.G. DERWENT, D.J. WUEBBLES, J-J. MORCRETTE

Contributors:
A.J. Apling; J.P. Blanchet; R.J. Charlson; D. Crommelynck; H. Grassl; N. Husson; G.J. Jenkins; I. Karol; M.D. King; V. Ramanathan; H. Rodhe; G-Y. Shi; G. Thomas; W-C. Wang; T.M.L. Wigley; T. Yamanouchi

CONTENTS

EXECUTIVE SUMMARY

1. The climate of the Earth is affected by changes in radiative forcing due to several sources (known as radiative forcing agents): these include the concentrations of radiatively active (greenhouse) gases, solar radiation, aerosols and albedo. In addition to their direct radiative effect on climate, many gases produce indirect effects on global radiative forcing.

2. The major contributor to increases in radiative forcing due to increased concentrations of greenhouse gases since pre-industrial times is carbon dioxide (CO_2) (61%), with substantial contributions from methane (CH_4) (17%), nitrous oxide (N_2O) (4%) and chlorofluorocarbons (CFCs) (12%). Stratospheric water vapour increases, which are expected to result from methane emissions, contribute 6%, although evidence for changes in concentration is based entirely on model calculations.
The contribution from changes in tropospheric and stratospheric ozone is difficult to estimate; increased levels of tropospheric ozone may have caused 10% of the total forcing since pre-industrial times. Decreases in lower stratospheric ozone may have decreased radiative forcing in recent decades.

3. The most recent decadal increase in radiative forcing is attributable to CO_2 (56%), CH_4 (11%), N_2O (6%) and CFCs(24%); stratospheric H_2O is estimated to have contributed 4%.

4. Using the scenario A ("business-as-usual" case) of future emissions derived by IPCC WG3, calculations show the following forcing from pre-industrial values (and percentage contribution to total) by the year 2025:

CO_2: 2.9 Wm^{-2} (63%); CH_4: 0.7 Wm^{-2} (15%); N_2O: 0.2 Wm^{-2} (4%); CFCs and HCFCs: 0.5 Wm^{-2} (11%); stratospheric H_2O: 0.2Wm^{-2} (5%)

The total, 4.6 Wm^{-2}, corresponds to an effective CO_2 amount of more than double the pre-industrial value.

5. An index is developed which allows the climate effects of the *emissions* of greenhouse gases to be compared. This is termed the Global Warming Potential (GWP). The GWP depends on the position and strength of the absorption bands of the gas, its lifetime in the atmosphere, its molecular weight and the time period over which the climate effects are of concern. A number of simplifications are used to derive values for GWPs and the values presented here should be considered as preliminary. It is quoted here as relative to CO_2.

Over a 500 year time period, the GWP of equal mass emissions of the gases is as follows:

CO_2: 1; CH_4: 9; $_2O$: 190; CFC-11: 1500; CFC-12: 4500; HCFC-22: 510.

Over a 20 year time period, the corresponding figures are:

CO_2: 1; CH_4: 63; N_2O: 270; CFC-11: 4500; CFC-12: 7100; HCFC-22: 4100 .

Values for other gases are given in the text. There are many uncertainties associated with this analysis; for example the atmospheric lifetime of CO_2 is not well characterized. The GWPs can be applied by considering actual emissions of the greenhouse gases. For example, considering anthropogenic emissions of all gases in 1990, and integrating their effect over 100 years, shows that 60% of the greenhouse forcing from these emissions comes from CO_2.

6. Although potential CFC replacements are less (or, in some cases, not at all) damaging to the ozone layer, the GWPs of several of them are still substantial; however, over periods greater than about 20 years most of the substitutes should have a markedly smaller impact on global warming than the CFCs they replace, assuming the same emissions.

7. Changes in climate forcing over the last century due to greenhouse gas increases are likely to have been much greater than that due to solar radiation. Although decadal variations of solar radiation can be comparable with greenhouse forcing, the solar forcing is not sustained and oscillates in sign. This limits the ability of the climate system to respond to the forcing. In contrast, the enhanced greenhouse effect causes a sustained forcing.

8. Stratospheric aerosols resulting from volcanic eruptions can cause a significant radiative forcing. A large eruption, such as El Chichón, can cause a radiative forcing, averaged over a decade, about one-third of (but the opposite sign to) the greenhouse gas forcing between 1980 and 1990. Regional and short-term effects of volcanic eruptions can be even larger.

9. Man-made sulphur emissions, which have increased in the Northern Hemisphere over the last century, affect radiative forcing by forming aerosols and influencing the radiative properties of clouds so as to cool the Earth. It is very difficult to

estimate the size of this effect, but it is conceivable that this radiative forcing has been of a comparable magnitude, but of opposite sign, to the greenhouse forcing earlier in this century; regional effects could even have been larger. The change in forcing due to sulphur emissions in the future could be of either sign, as it is not known whether the emissions will increase or decrease.

2.1 Introduction

The climate of the Earth has the potential to be changed on all timescales by the way in which shortwave radiation from the Sun is scattered and absorbed, and thermal infrared radiation is absorbed and emitted by the Earth-atmosphere system. If the climate system is in equilibrium, then the absorbed solar energy is exactly balanced by radiation emitted to space by the Earth and atmosphere. Any factor that is able to perturb this balance, and thus potentially alter the climate, is called a radiative forcing agent.

Of particular relevance to concerns about climate change are the changes in radiative forcing which arise from the increases in the concentration of radiatively active trace gases ("greenhouse gases") in the troposphere and stratosphere described in Section 1. These changes in concentration will come about when their emissions or removal mechanisms are changed, so that the atmospheric concentrations are no longer in equilibrium with the sources and sinks of the gas. The growing concentrations of greenhouse gases such as carbon dioxide, methane, chlorofluorocarbons and nitrous oxide are of particular concern. In addition, indirect effects on radiative forcing can result from molecules that may not themselves be greenhouse gases but which lead to chemical reactions which create greenhouse gases. For example, indirect effects are believed to be altering the distribution of stratospheric and tropospheric ozone.

Although water vapour is the single most important greenhouse gas, the effect of changes in its tropospheric concentration (which may arise as a natural consequence of the warming) is considered as a feedback to be treated in climate models; similarly changes in cloud amount or properties which result from climate changes will be considered as feedbacks. Both these factors are discussed in Section 3. Possible feedbacks between ocean temperature and dimethyl sulphide emissions, which may alter sulphate aerosol amounts, are also considered to be a feedback and will be considered in Section 3.

Other factors can alter the radiative balance of the planet. The most obvious of these is the amount of solar radiation reaching the Earth and this is known to vary on a wide range of time-scales. The amount of solar radiation absorbed by the Earth-atmosphere system is determined by the extent to which the atmosphere and Earth's surface reflect the radiation (their albedo) and by the quantities of gases such as ozone and water vapour in the atmosphere. The albedo of the Earth's surface can be affected by changes in the land surface, e.g., desertification. The planetary albedo can be altered by changes in the amount of aerosol particles in the atmosphere; in the stratosphere, the dominant source is from volcanic eruptions, while in the troposphere the source can be either natural or man-made.

The planetary albedo will also change if the properties of clouds are changed, for instance, if additional cloud condensation nuclei are provided by natural or man-made changes in aerosol concentrations. Changes in aerosol concentrations can also affect radiative forcing by their ability to absorb thermal infrared radiation.

Although all of the above factors will be considered in this section, the emphasis will be very strongly on the greenhouse gases, as they are likely to change radiative forcing over the next few decades by more than any other factor, natural or anthropogenic. They are also candidates for any policy action which may be required to limit global climate change. Obviously factors such as those related to emissions from volcanoes and the effects of solar variability are completely outside our control.

The purpose of this section is to use the information described in Section 1, on how the forcing agents themselves have changed in the past and how, based on a number of emission scenarios, they may change in the future. This information will then be used in climate models, later in the report, to show the climate and sea level consequences of the emission scenarios.

However, we can also use the estimates of radiative forcing from this section in their own right, by looking at the relative contribution from each of the agents - and in particular the greenhouse gases. The advantage of dealing with radiative forcing, rather than climate change itself, is that we can estimate the former with a great deal more certainty than we can estimate the latter. In the context of policy formulation, the relative importance of these agents is of major significance in assessing the effectiveness of response strategies. The radiative forcing is expressed as a change in flux of energy, in Wm^{-2}.

In order to formulate policy on the possible limitations of greenhouse gas emissions (undertaken within IPCC by Working Group III), it is essential to know how abatement of the emissions of each of the trace gases will affect global climate forcing in the future. This information can then be used for calculations of the cost-effectiveness of reductions, e.g., CO_2 emissions compared to CH_4 emissions. There is no ideal index that can be used for each gas, but values of one index, the Global Warming Potential, are derived in this section. Research now under way will enable such indices to be refined.

2.2 Greenhouse Gases

2.2.1 Introduction

A typical global-average energy budget for the climate system shows that about half of the incident solar radiation (at wavelengths between 0.2 and 4.0 μm) is absorbed at the Earth's surface. This radiation warms the Earth's surface which then emits energy in the thermal infrared region (4-

100μm); constituents in the Earth's atmosphere are able to absorb this radiation and subsequently emit it both upwards to space and downwards to the surface. This downward emission of radiation serves to further warm the surface; this warming is known as the greenhouse effect.

The strength of the greenhouse effect can be gauged by the difference between the effective emitting temperature of the Earth as seen from space (about 255K) and the globally-averaged surface temperature (about 285K). The principal components of the greenhouse effect are the atmospheric gases (Section 2.2.2); clouds and aerosols also absorb and emit thermal infrared radiation, but they also increase the planetary albedo, and it is believed that their net effect is to cool the surface (see Sections 3.3.4 and 2.3.2). Of the atmospheric gases, the dominant greenhouse gas is water vapour. If H_2O was the only greenhouse gas present, then the greenhouse effect of a clear-sky mid-latitude atmosphere, as measured by the difference between the emitted thermal infrared flux at the surface and the top of the atmosphere, would be about 60-70% of the value with all gases included; by contrast, if CO_2 alone was present, the corresponding value would be about 25% (but note that because of overlap between the absorption bands of different gases, such percentages are not strictly additive).

Here we are primarily concerned with the impacts of changing concentrations of greenhouse gases. A number of basic factors affect the ability of different greenhouse gases to force the climate system.

The absorption strength and the wavelength of this absorption in the thermal infrared are of fundamental importance in dictating whether a molecule can be an important greenhouse forcing agent; this effect is modified by both the existing quantities of that gas in the atmosphere and the overlap between the absorption bands and those of other gases present in the atmosphere.

The ability to build up significant quantities of the gas in the atmosphere is of obvious importance and this is dictated not only by the emissions of the gas, but also by its lifetime in the atmosphere. Further, these gases, as well as those that are not significant greenhouse gases can, via chemical reactions, result in products that are greenhouse gases.

In addition, the relative strength of greenhouse gases will depend on the period over which the effects of the gases are to be considered. For example, a short-lived gas which has a strong (on a kg-per-kg basis) greenhouse effect may, in the short term, be more effective at changing the radiative forcing than a weaker but longer-lived gas; over longer periods, however, the integrated effect of the weaker gas may be greater as a result of its persistence in the atmosphere.

From this introduction, it is clear that an assessment of the strength of greenhouse gases in influencing radiative forcing depends on how that strength is measured. There are many possible approaches and it is important to distinguish between them.

Some of the more important indices that have been used as measures of the strength of the radiative forcing by greenhouse gases include:

i) **Relative molecular forcing.** This gives the relative forcing on a molecule-per-molecule basis of the different species. It is normally quoted relative to CO_2. Since the forcing of some atmospheric species (most notably CO_2, methane and nitrous oxide) is markedly non-linear in absorber amount, this relative forcing will be dependent on the concentration changes for which the calculations are performed. A small change in current atmospheric concentrations is generally used. This measure emphasises that the contributions of individual gases *must not* be judged on the basis of concentration alone. The relative molecular forcing will be considered in Section 2.2.4.

ii) **Relative mass forcing.** This is similar to the relative molecular forcing but is relative on a kilogram-per-kilogram basis. It is related to the relative molecular forcing by the molecular weights of the gases concerned. It will also be considered in Section 2.2.4.

iii) **Contribution of past, present and future changes in trace gas concentration.** This measure, which can either be relative or absolute, calculates the contribution to radiative forcing over some given period due to observed past or present changes, or scenarios of future changes in trace gas concentration. This is an important baseline. The relative measures (i) and (ii) above, can belittle the influence of carbon dioxide since it is relatively weak on a molecule-per-molecule basis, or a kg-per-kg basis. This measure accounts for the fact that the concentration changes for CO_2 are between two and four orders of magnitude greater than the changes of other important greenhouse gases. This measure will be considered in sections 2.2.5 and 2.2.6. Care must be taken in interpreting this measure as it is sometimes presented as the total change in forcing since pre-industrial times and sometimes as the change in forcing over a shorter period such as a decade or 50 years.

iv) **Global Warming Potential (GWP).** All the above measures are based on *concentration* changes in the atmosphere, as opposed to *emissions*. Assessing the potential impact of future emissions may be far more important from a policy point of view. Such measures combine calculations of the absorption strength of a molecule with assessments of its atmospheric lifetime; it can also include the indirect greenhouse effects due to chemical changes in the atmosphere caused by the gas. The development of an index is still at an early stage, but progress has been made and preliminary values are given in Section 2.2.7.

A detailed assessment of the climatic effects of trace gases was made by WMO (1985) (see also Ramanathan et al., 1987). The effect of halocarbons has been considered in detail in the recent Scientific Assessment of Stratospheric Ozone (UNEP, 1989) (see also Fisher et al., 1990). This section should be considered as building on these assessments and bringing them up to date.

2.2.2 Direct Effects

Many molecules in the atmosphere possess pure-rotation or vibration-rotation spectra that allow them to emit and absorb thermal infrared radiation (4-100 μm); such gases include water vapour, carbon dioxide and ozone (but not the main constituents of the atmosphere, oxygen or nitrogen). These absorption properties are directly responsible for the greenhouse effect.

It is not the change in thermal infrared flux at the surface that determines the strength of the greenhouse warming. The surface, planetary boundary layer and the free troposphere are tightly coupled via air motions on a wide range of scales, so that in a global-mean sense they must be considered as a single thermodynamic system. As a result it is the change in the radiative flux at the *tropopause*, and not the surface, that expresses the radiative forcing of climate system (see e.g., Ramanathan et al., 1987).

A number of factors determine the ability of an added molecule to affect radiative forcing and in particular the spectral absorption of the molecule in relation to the spectral distribution of radiation emitted by a black-body. The distribution of emitted radiation with wavelength is shown by the dashed curves for a range of atmospheric temperatures in Figure 2.1. Unless a molecule possesses strong absorption bands in the wavelength region of significant emission, it can have little effect on the net radiation.

These considerations are complicated by the effect of naturally occurring gases on the spectrum of net radiation at the tropopause. Figure 2.1 shows the spectral variation of the net flux at the tropopause for a clear-sky mid-latitude profile. For example, the natural quantities of carbon dioxide are so large that the atmosphere is very opaque over short distances at the centre of its 15 μm band. At this wavelength, the radiation reaching the tropopause, from both above and below, comes from regions at temperatures little different to the tropopause itself. The net flux is thus close to zero. The addition of a small amount of gas capable of absorbing at this wavelength has negligible effect on the net flux at the tropopause. The effect of added carbon dioxide molecules is, however, significant at the

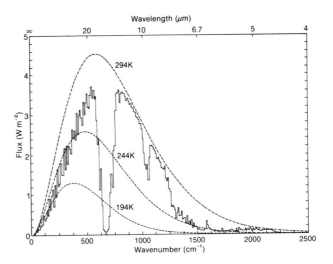

Figure 2.1: The dashed lines show the emission from a black body (Wm^{-2} per 10 cm^{-1} spectral interval) across the thermal infrared for temperatures of 294K, 244K and 194K. The solid line shows the net flux at the tropopause (Wm^{-2}) in each 10 cm^{-1} interval, using a standard narrow-band radiation scheme and a clear-sky mid-latitude summer atmosphere with a surface temperature of 294K (Shine, pers. comm.). In general, the closer this line is to the dashed line for 294K, the more transparent the atmosphere.

edges of the 15 μm band, and in particular around 13.7 and 16 μm. At the other extreme, in more transparent regions of the atmosphere (for example between 10 and 12 μm), much of the radiation reaching the tropopause from beneath is, for clear skies, from the warm surface and the lower troposphere; this emission is not balanced by downward emission of radiation from the overlying stratosphere. A molecule able to absorb in such a transparent spectral region is able to have a far larger effect.

The existing concentrations of a particular gas dictate the effect that additional molecules of that gas can have. For gases such as the halocarbons, where the naturally occurring concentrations are zero or very small, their forcing is close to linear in concentration for present-day concentrations. Gases such as methane and nitrous oxide are present in such quantities that significant absorption is already occurring and it is found that their forcing is approximately proportional to the square root of their concentration. Furthermore, there is significant overlap between some of the infrared absorption bands of methane and nitrous oxide which must be carefully considered in calculations of forcing. For carbon dioxide, as has already been mentioned, parts of the spectrum are already so opaque that additional molecules of carbon dioxide are even less effective; the forcing is found to be logarithmic in concentration. These effects are reflected in the empirical expressions used to calculate the radiative forcing that are discussed in Section 2.2.4.

A further consideration is the spectroscopic strength of the bands of molecules which dictates the strength of the infra-red absorption. Molecules such as the halocarbons have bands with intensities about an order of magnitude or more greater, on a molecule-per-molecule basis, than the 15 μm band of carbon dioxide. The actual absorptance by a band is, however, a complicated function of both absorber amount and spectroscopic strength so that these factors cannot be considered entirely in isolation.

2.2.3 *Indirect Effects*

In addition to their direct radiative effects, many of the greenhouse gases also have indirect radiative effects on climate through their interactions with atmospheric chemical processes. Several of these interactions are shown in Table 2.1.

For example, both atmospheric measurements and theoretical models indicate that the global distribution of ozone in the troposphere and stratosphere is changing as a result of such interactions (UNEP, 1989; also see Section 1).

Ozone plays an important dual role in affecting climate. While CO_2 and other greenhouse gases are relatively well-mixed in the atmosphere, the climatic effect of ozone depends strongly on its vertical distribution throughout the troposphere and stratosphere, as well as on its total amount in the atmosphere. Ozone is a primary absorber of solar radiation in the stratosphere where it is directly responsible for the increase in temperature with altitude. Ozone is also an important absorber of infrared radiation. It is the balance between these radiative processes that determines the net effect of ozone on climate. Changes in ozone in the upper troposphere and lower stratosphere (below 25 km) are most effective in determining the change in radiative forcing, with increased ozone leading to an increased radiative forcing which would be expected to warm the surface (e.g., Wang and Sze, 1980; Lacis et al., 1990). This is because the greenhouse effect is directly proportional to the temperature contrast between the level of emission and the levels at which radiation is absorbed. This contrast is greatest near the tropopause where temperatures are at a minimum compared to the surface. Above about 30 km, added ozone causes a decrease in surface temperature because it absorbs extra solar radiation, effectively robbing the troposphere of direct solar energy that would otherwise warm the surface (Lacis et al, 1990).

Table 2.1: *Direct radiative effects and indirect trace gas chemical-climate interactions (based on Wuebbles et al., 1989)*

Gas	Greenhouse Gas	Is its tropospheric concentration affected by chemistry?	Effects on tropospheric chemistry? *	Effects on * stratospheric chemistry?
CO_2	Yes	No	No	Yes, affects O_3 (see text)
CH_4	Yes	Yes, reacts with OH	Yes, affects OH, O_3 and CO_2	Yes, affects O_3 and H_2O
CO	Yes, but weak	Yes, reacts with OH	Yes, affects OH, O_3 and CO_2	Not significantly
N_2O	Yes	No	No	Yes, affects O_3
NOx	Yes	Yes, reacts with OH	Yes, affects OH and O_3	Yes, affects O_3
CFC-11	Yes	No	No	Yes, affects O_3
CFC-12	Yes	No	No	Yes, affects O_3
CFC-113	Yes	No	No	Yes, affects O_3
HCFC-22	Yes	Yes, reacts with OH	No	Yes, affects O_3
CH_3CCl_3	Yes	Yes, reacts with OH	No	Yes, affects O_3
CF_2ClBr	Yes	Yes, photolysis	No	Yes, affects O_3
CF_3Br	Yes	No	No	Yes, affects O_3
SO_2	Yes, but weak	Yes, reacts with OH	Yes, increases aerosols	Yes, increases aerosols
CH_3SCH_3	Yes, but weak	Yes, reacts with OH	Source of SO_2	Not significantly
CS_2	Yes, but weak	Yes, reacts with OH	Source of COS	Yes, increases aerosols
COS	Yes, but weak	Yes, reacts with OH	Not significant	Yes, increases aerosols
O_3	Yes	Yes	Yes	Yes

* - Effects on atmospheric chemistry are limited to effects on constituents having a significant influence on climate.

Stratospheric water vapour is an important greenhouse gas. A major source of stratospheric water vapour is the oxidation of methane (e.g., WMO 1985); it is anticipated that increased atmospheric concentrations of methane will lead to increases in stratospheric water vapour. It is also possible that changes in climate will affect the transfer of water vapour from the troposphere to the stratosphere, although the sign of the net effect on stratospheric water vapour is unclear. Unfortunately, observations of stratospheric water vapour are inadequate for trend detection. In this section the impact of increased emissions of methane on stratospheric water vapour will be included as an indirect radiative forcing due to methane.

The oxidation of fossil-based methane and carbon monoxide in the atmosphere lead to the production of additional carbon dioxide. Although CO_2 has no known chemical interactions of consequence within the troposphere or stratosphere, its increasing concentrations can affect the concentrations of stratospheric ozone through its radiative cooling of the stratosphere. In the upper stratosphere the cooling slows down catalytic ozone destruction and results in a net increase in ozone; where heterogeneous ozone destruction is important, as in the Antarctic lower stratosphere, ozone destruction may be accelerated by this cooling (UNEP, 1989). The combination of these indirect effects, along with their direct radiative effects, determines the actual changes in radiative forcing resulting from these greenhouse gases.

The hydroxyl radical, OH, is not itself a greenhouse gas, but it is extremely important in the troposphere as a chemical scavenger. Reactions with OH largely control the atmospheric lifetime, and, therefore, the concentrations of many gases important in determining climate change. These gases include CH_4, CO, the non-methane hydrocarbons (NMHCs), the hydrochlorofluorocarbons (HCFCs), the hydrofluorocarbons (HFCs), CH_3CCl_3, H_2S, SO_2 and dimethyl sulphide (DMS). Their reaction with OH also affects the production of tropospheric ozone, as well as determining the amounts of these compounds reaching the stratosphere, where these species can cause changes in the ozone distribution. In turn, the reactions of these gases with OH also affects its atmospheric concentration. The increase in tropospheric water vapour concentration expected as a result of global warming would also increase photochemical production of OH. It is important that effects of interaction between OH and the greenhouse gases, along with the resulting impact on atmospheric lifetimes of these gases, be accounted for in analysing the possible state of future climate.

The indirect effects can have a significant effect on the total forcing; these effects will be detailed later in the section.

2.2.4 Relationship Between Radiative Forcing and Concentration

To estimate climate change using simple energy balance climate models (see Section 6) and in order to estimate the relative importance of different greenhouse gases in past, present and future atmospheres (e.g., using Global Warming Potentials, see Section 2.2.7), it is necessary to express the radiative forcing for each particular gas in terms of its concentration change. This can be done in terms of the changes in net radiative flux at the tropopause

$$\Delta F = f(C_0, C)$$

where ΔF is the change in net flux (in Wm^{-2}) corresponding to a volumetric concentration change from C_0 to C.

Direct-effect ΔF-ΔC relationships are calculated using detailed radiative transfer models. Such calculations simulate the complex variations of absorption and emission with wavelength for the gases included, and account for the overlap between absorption bands of the gases; the effects of clouds on the transfer of radiation are also accounted for.

As was discussed in Section 2.2.2, the forcing is given by the change in net flux at the tropopause. However, as is explained by Ramanathan et al. (1987) and Hansen et al. (1981), great care must be taken in the evaluation of this change. When absorber amount varies, not only does the flux at the tropopause respond, but also the overlying stratosphere is no longer in radiative equilibrium. For some gases, and in particular CO_2, the concentration change acts to cool the stratosphere; for others, and in particular the CFCs, the stratosphere warms (see e.g., Table 5 of Wang et al. (1990)). Calculations of the change in forcing at the tropopause should allow the stratosphere to come into a new equilibrium with this altered flux divergence, while tropospheric temperatures are held constant. The consequent change in stratospheric temperature alters the downward emission at the tropopause and, hence, the forcing. The ΔF-ΔC relationships used here implicitly account for the stratospheric response. If this point is ignored, then the same change in flux at the tropopause from different forcing agents can lead to a different tropospheric temperature response. Allowing for the stratospheric adjustment means that the temperature response for the same flux change from different causes are in far closer agreement (Lacis, personal communication).

The form of the ΔF-ΔC relationship depends primarily on the gas concentration. For low/moderate/high concentrations, the form is well approximated by a linear/square-root/logarithmic dependence of ΔF on concentration. For ozone, the form follows none of these because of marked vertical variations in absorption and concentration. Vertical variations in concentration change

Table 2.2: *Expressions used to derive radiative forcing for past trends and future scenarios of greenhouse gas concentrations*

TRACE GAS	RADIATIVE FORCING APPROXIMATION GIVING ΔF IN Wm^{-2}	COMMENTS
Carbon dioxide	$\Delta F = 6.3 \ln (C/C_O)$ where C is CO_2 in ppmv for C < 1000 ppmv	Functional form from Wigley (1987); coefficient derived from Hansen et al. (1988).
Methane	$\Delta F = 0.036 (\sqrt{M} - \sqrt{M_O}) - (f(M, N_O)-f(M_O, N_O))$ where M is CH_4 in ppbv and N is N_2O in ppbv. Valid for M <5ppmv.	Functional form from Wigley (1987); coefficient derived from Hansen et al. (1988). Overlap term, f(M, N) from Hansen et al (1988)[*]
Nitrous Oxide	$\Delta F = 0.14 (\sqrt{N} - \sqrt{N_O}) - (f(M_O, N) - f(M_O, N_O))$ with M and N as above. Valid for N <5ppmv.	Functional form from Wigley (1987); coefficient derived from Hansen et al.(1988). Overlap term from Hansen et al. (1988)[*]
CFC-11	$\Delta F = 0.22 (X - X_O)$ where X is CFC-11 in ppbv Valid for X <2ppbv.	Based on Hansen et al. (1988)
CFC-12	$\Delta F = 0.28 (Y - Y_O)$ where Y is CFC-12 in ppbv Valid for Y <2ppbv.	Based on Hansen et al. (1988)
Stratospheric water vapour	$\Delta F = 0.011 (\sqrt{M} - \sqrt{M_O})$ where M is CH_4 in ppbv	Stratospheric water vapour forcing taken to be 0.3 of methane forcing without overlap based on Wuebbles et al. (1989)
Tropospheric ozone	$\Delta F = 0.02 (O - O_O)$ where O is ozone in ppbv	Very tentative illustrative parameterization based on value from Hansen et al. (1988).
Other CFCs, HCFCs and HFCs	$\Delta F = A (Z - Z_O)$ where A based on forcing relative to CFC-11 in Table 2.4 and Z is constituent in ppbv	Coefficients A derived from Fisher et al. (1990)

[*] Methane-Nitrous Oxide overlap term

f (M, N) = $0.47 \ln [1 + 2.01 \times 10^{-5} (MN)^{0.75} + 5.31 \times 10^{-15} M (MN)^{1.52}]$; M and N are in ppbv.

Note typographical error on page 9360 of Hansen et al. (1988): 0.014 should be 0.14.

for ozone make it even more difficult to relate ΔF to concentration in a simple way.

The actual relationships between forcing and concentration derived from detailed models can be used to develop simple expressions (e.g., Wigley, 1987; Hansen et al 1988) which are then more easily used for a large number of calculations. Such simple expressions are used in this Section. The values adopted and their sources are given in Table 2.2. Values derived from Hansen et al. have

been multiplied by 3.35 (Lacis, personal communication) to convert forcing as a temperature change to forcing as a change in net flux at the tropopause after allowing for stratospheric temperature change. These expressions should be considered as global mean forcings; they implicitly include the radiative effects of global mean cloud cover.

Significant spatial variations in ΔF will exist because its value for any given ΔC depends on the assumed temperature and water vapour profiles. Variations will also occur due to spatial variations in mean cloudiness. These factors can produce marked differences in the relative contributions of different greenhouse gases to total radiative forcing in different regions but these are not accounted for here.

Uncertainties in ΔF-ΔC relationships arise in three ways. First, there are still uncertainties in the basic spectroscopic data for many gases. In particular, data for CFCs, HFCs and HCFCs are probably only accurate to within ± 10-20%. Part of this uncertainty is related to the temperature dependence of the intensities, which is generally not known. For some of these gases, only cross-section data are available. For the line intensity data that do exist, there have been no detailed intercomparisons of results from

different laboratories. Further information on the available spectroscopic data is given by Husson (1990).

Second, uncertainties arise through details in the radiative transfer modelling. Intercomparisons made under the auspices of WCRP (Luther and Fouquart, 1984) suggest that these uncertainties are around $\pm 10\%$ (although schemes used in climate models disagreed with detailed calculations by up to 25% for the flux change at the tropopause on doubling CO_2).

Third, uncertainties arise through assumptions made in the radiative model with regard to the following:

(i) the assumed or computed vertical profile of the concentration change. For example, for CFCs and HCFCs, results can depend noticeably on the assumed change in stratospheric concentration (see e.g., Ramanathan et al., 1985).

(ii) the assumed or computed vertical profiles of temperature and moisture.

Table 2.3: *Radiative forcing relative to CO_2 per unit molecule change, and per unit mass change in the atmosphere for present day concentrations. CO_2, CH_4 and N_2O forcings from 1990 concentrations in Table 2.5.*

TRACE GAS	ΔF for ΔC per molecule relative to CO_2	ΔF for ΔC per unit mass relative to CO_2
CO_2	1	1
CH_4	21	58
N_2O	206	206
CFC-11	12400	3970
CFC-12	15800	5750
CFC-113	15800	3710
CFC-114	18300	4710
CFC-115	14500	4130
HCFC-22	10700	5440
CCl_4	5720	1640
CH_3CCl_3	2730	900
CF_3Br	16000	4730
Possible CFC substitutes:		
HCFC-123	9940	2860
HCFC-124	10800	3480
HFC-125	13400	4920
HFC-134a	9570	4130
HCFC-141b	7710	2900
HCFC-142b	10200	4470
HFC-143a	7830	4100
HFC-152a	6590	4390

Table 2.4: *Radiative forcing of a number of CFCs, possible CFC substitutes and other halocarbons relative to CFC-11 per unit molecule and per unit mass change. All values, except CF_3Br, from Fisher et al., 1990: CF_3Br from Ramanathan et al., 1985.*

TRACE GAS	$\Delta F/\Delta C$ per molecule relative to CFC11	$\Delta F/\Delta C$ per unit mass relative to CFC11
CFC-11	1.00	1.00
CFC-12	1.27	1.45
CFC-113	1.27	0.93
CFC-114	1.47	1.18
CFC-115	1.17	1.04
HCFC-22	0.86	1.36
HCFC-123	0.80	0.72
HCFC-124	0.87	0.88
HFC-125	1.08	1.24
HFC-134a	0.77	1.04
HCFC-141b	0.62	0.73
HCFC-142b	0.82	1.12
HFC-143a	0.63	1.03
HFC-152a	0.53	1.10
CCl_4	0.46	0.45
CH_3CCl_3	0.22	0.23
CF_3Br	1.29	1.19

(iii)   assumptions made with regard to cloudiness. Clear sky ΔF values are in general 20% greater than those using realistic cloudiness.

(iv)   the assumed concentrations of other gases (usually, present-day values are used). These are important because they determine the overall IR flux and because of overlap between the absorption lines of different gases.

(v)   the indirect effects on the radiative forcing due to chemical interactions as discussed in Section 2.2.3.

The overall effect of this third group of uncertainties on ΔF is probably at least ±10%.

Direct radiative forcing changes for the different greenhouse gases can be easily compared using the above ΔF-ΔC relationships. There are two ways in which these comparisons may be made, per unit volumetric concentration change (equivalent to per molecule) or per unit mass change. Comparison for the major greenhouse gases are given in Table 2.3. The relative strength of the CFCs, HFCs and HCFCs, relative to CFC-11, are shown in Table 2.4 (from Fisher et al., 1990). It can be seen that, by these measures, many of the potential CFC substitutes are strong infrared absorbers.

2.2.5 Past and Present Changes in Radiative Forcing

Based on the expressions given in Table 2.2 the radiative forcing between 1765 and 1990 was calculated using observed variations of the greenhouse gases. The concentrations are given in Table 2.5; they are updated values from Wigley (1987) and Section 1. Values for 1990 have been extrapolated from recent values. In addition to the well-observed variations in the gases given in Table 2.5, it is assumed that increased concentrations of methane have led to increases in stratospheric water vapour, although such changes are based entirely on model estimates (see Section 2.2.3).

Table 2.6 gives the contributions to the forcing for a number of periods. This is shown diagrammatically in Figure 2.2 as the change in total forcing from 1765 concentrations; it is shown as a change in forcing per decade in Figure 2.3

Table 2.5: *Trace gas concentrations from 1765 to 1990, used to construct Figure 2.2*

YEAR	CO_2 (ppmv)	CH_4 (ppbv)	N_2O (ppbv)	CFC-11 (ppbv)	CFC-12 (ppbv)
1765	279.00	790.0	285.00	0	0
1900	295.72	974.1	292.02	0	0
1960	316.24	1272.0	296.62	0.0175	0.0303
1970	324.76	1420.9	298.82	0.0700	0.1211
1980	337.32	1569.0	302.62	0.1575	0.2725
1990	353.93	1717.0	309.68	0.2800	0.4844

Table 2.6: *Forcing in Wm^{-2} due to changes in trace gas concentrations in Table 2.5. All values are for changes in forcing from 1765 concentrations. The change due to stratospheric water vapour is an indirect effect of changes in methane concentration (see text).*

YEAR	SUM	CO_2	CH_4 direct	Strat H_2O	N_2O	CFC-11	CFC-12	Other CFCs
1765 - 1900	0.53	0.37	0.10	0.034	0.027	0.0	0.0	0.0
1765 - 1960	1.17	0.79	0.24	0.082	0.045	0.004	0.008	0.005
1765 - 1970	1.48	0.96	0.30	0.10	0.054	0.014	0.034	0.021
1765 - 1980	1.91	1.20	0.36	0.12	0.068	0.035	0.076	0.048
1765 - 1990	2.45	1.50	0.42	0.14	0.10	0.062	0.14	0.085

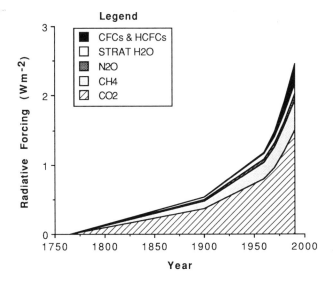

Figure 2.2: Changes in radiative forcing (Wm^{-2}) due to increases in greenhouse gas concentrations between 1765 and 1990. Values are changes in forcing from 1765 concentrations.

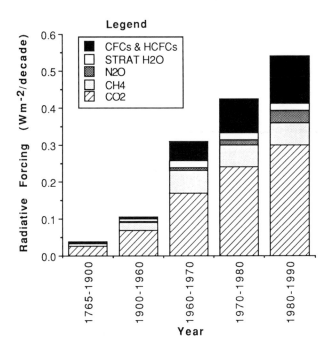

Figure 2.3: Decadal contributions to radiative forcing (Wm^{-2}) due to increases in greenhouse gas concentrations for periods between 1765 and 1990. The changes for the periods 1765-1900 and 1900-1960 are the total changes during these periods divided by the number of decades.

Changes in halocarbons other than CFC-11 and CFC-12 have been accounted for by using concentration changes from Section 1, and the forcing-versus-concentration changes given in Tables 2.2 and 2.4. It is found that they contribute an extra 43% of the sum of the forcing from CFC-11 and CFC-12; most of this contribution results from

changes in HCFC-22, CFC-113, carbon tetrachloride and methyl chloroform. This is in reasonable agreement with Hansen et al. (1989), who, using less recent spectroscopic data, find these halocarbons contribute an extra 60% of the combined CFC-11 and CFC-12 forcing.

For the period 1765 to 1990, CO_2 has contributed about 61% of the forcing, methane 17% plus 6% from stratospheric water vapour, N_2O 4% and the CFCs 12%. For the decade 1980-1990, about 56% of the forcing has been due to changes in CO_2, 11% due to the direct effects of CH_4 and 4% via stratospheric water vapour, 6% from N_2O and 24% from the CFCs.

As discussed in Section 1, the distribution of tropospheric ozone has almost certainly changed over this period, with a possible impact on radiative forcing. Difficulties in assessing the global changes in ozone, and in calculating the resultant radiative forcing, prevent a detailed assessment of the effect. Estimates of tropospheric ozone change driven by changing methane and NO_X emissions are highly model-dependent partly because of the inherent spatial averages used in current two-dimensional models. Estimates of changes in tropospheric ozone from pre-industrial values (e.g., Hough and Derwent 1990) and simplified estimates of the radiative forcing (Table 2.2) suggest that tropospheric ozone may have contributed about 10% of the total forcing due to greenhouse gases since pre-industrial times.

Decreases in lower stratospheric ozone, particularly since the mid-70s, may have led to a decreased radiative forcing; this may have compensated for the effects of tropospheric ozone (Hansen et al.,1989; Lacis et al. 1990). This compensation should be considered as largely fortuitous, as the mechanisms influencing ozone concentrations in the troposphere and stratosphere are somewhat different.

2.2.6 Calculations of Future Forcing

Using the radiative forcing expressions described in Section 2.2.4, and the four scenarios developed by Working Group III, possible changes in radiative forcing over the next century can be calculated. The four scenarios are intended to provide insight into policy analysis for a range of potential changes in concentrations; Scenario A is a "Business-as-Usual" case, whilst Scenarios B,C and D represent cases of reduced emissions. These four scenarios are considered in more detail in the Appendix 1. As in the previous section, the indirect effect of methane on forcing via stratospheric water vapour changes is included, whilst the effects of possible changes in ozone are neglected.

It must be stressed here that the gas referred to as HCFC-22 as given in the scenarios is used as a surrogate for all the CFC substitutes. Since all HCFCs and HFCs are of similar radiative strength, on a molecule-per-molecule basis (see Table 2.4), the error from this source in using HCFC-22 as

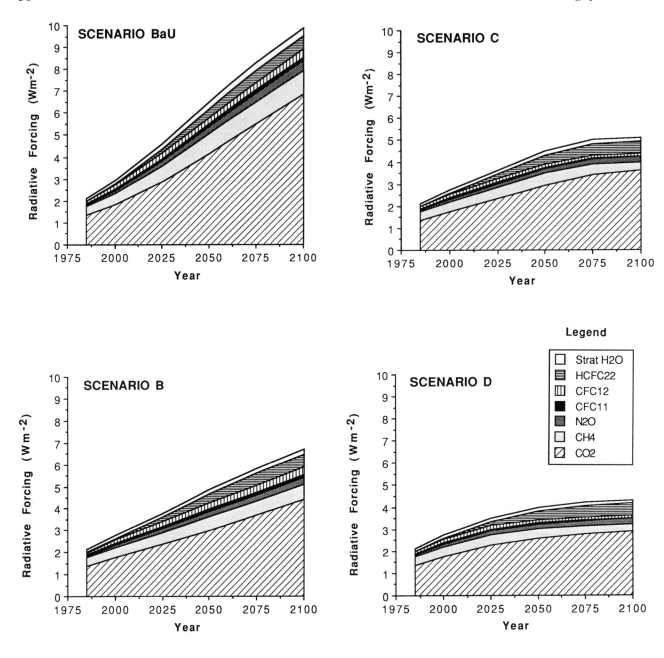

Figure 2.4: Possible future changes in radiative forcing (Wm^{-2}) due to increases in greenhouse gas concentrations between 1985 and 2100 using the four policy scenarios given in the Appendix 1. Values are changes in forcing from 1765 concentrations.

a proxy for the other gases will be small. However, since the concentrations, as specified in the scenarios, were calculated assuming the HCFC-22 lifetime and molecular weight, considerable errors in the forcing may result from errors in the concentrations. Since some of the CFC substitutes have a longer lifetime than HCFC-22, and some shorter, it is not possible to calculate the sign of the error without knowing the precise mix of substitutes used.

Figure 2.4 shows the radiative forcing change (from pre-industrial) for each gas from the four scenarios; the results are tabulated in Table 2.7.

For these scenarios, CO$_2$ remains the dominant contributor to change throughout the period. In the

Business-as-Usual Scenario, for example, its contribution to the change always exceeds 60%. For the scenarios chosen for this analysis, the contribution of HCFC-22 becomes significant in the next century. It is contributing 11% of the 25 year forcing change between 2025-2050 in the Business-as-Usual Scenario and 18% in Scenario B.

Since the concentration of chlorine can be anticipated to increase in the stratosphere for at least the next decade (Section 1.6.2; see also Prather and Watson, 1990), further decreases in stratospheric ozone can be anticipated. Decreases in upper stratospheric ozone will lead to a small warming effect; decreases in the lower stratosphere would cause a cooling effect. A 1% loss in ozone in the lower

Table 2.7: *Changes in radiative forcing in Wm^{-2} for the 4 policy scenarios. The change due to stratospheric water vapour is an indirect effect of changes in methane concentration (see text). All values are changes in forcing from 1765 concentrations.*

SCENARIO A (Business-as-Usual)

YEAR	SUM	CO$_2$	CH$_4$ direct	Strat H$_2$O	N$_2$O	CFC-11	CFC-12	HCFC- 22
1765-2000	2.95	1.85	0.51	0.18	0.12	0.08	0.17	0.04
1765-2025	4.59	2.88	0.72	0.25	0.21	0.11	0.25	0.17
1765-2050	6.49	4.15	0.90	0.31	0.31	0.12	0.30	0.39
1765-2075	8.28	5.49	1.02	0.35	0.40	0.13	0.35	0.55
1765-2100	9.90	6.84	1.09	0.38	0.47	0.14	0.39	0.59

SCENARIO B (Low Emissions)

YEAR	SUM	CO$_2$	CH$_4$ direct	Strat H$_2$O	N$_2$O	CFC-11	CFC-12	HCFC- 22
1765-2000	2.77	1.75	0.45	0.16	0.11	0.08	0.17	0.04
1765-2025	3.80	2.35	0.56	0.19	0.18	0.10	0.24	0.17
1765-2050	4.87	2.97	0.65	0.22	0.23	0.11	0.29	0.39
1765-2075	5.84	3.69	0.66	0.23	0.28	0.12	0.33	0.53
1765-2100	6.68	4.43	0.66	0.23	0.33	0.12	0.36	0.56

SCENARIO C (Control Policies)

YEAR	SUM	CO$_2$	CH$_4$ direct	Strat H$_2$O	N$_2$O	CFC-11	CFC-12	HCFC- 22
1765-2000	2.74	1.75	0.44	0.15	0.11	0.08	0.17	0.05
1765-2025	3.63	2.34	0.51	0.17	0.17	0.07	0.17	0.20
1765-2050	4.49	2.96	0.53	0.18	0.22	0.05	0.14	0.41
1765-2075	5.00	3.42	0.47	0.16	0.25	0.03	0.12	0.55
1765-2100	5.07	3.62	0.37	0.13	0.27	0.02	0.10	0.57

SCENARIO D (Accelerated Policies)

YEAR	SUM	CO$_2$	CH$_4$ direct	Strat H$_2$O	N$_2$O	CFC-11	CFC-12	HCFC- 22
1765-2000	2.74	1.75	0.44	0.15	0.11	0.08	0.17	0.04
1765-2025	3.52	2.29	0.47	0.16	0.17	0.07	0.17	0.20
1765-2050	3.99	2.60	0.43	0.15	0.21	0.05	0.14	0.40
1765-2075	4.22	2.77	0.39	0.13	0.24	0.03	0.12	0.53
1765-2100	4.30	2.90	0.34	0.12	0.26	0.02	0.10	0.56

stratosphere would cause a change of about 0.05 Wm^{-2}, so that changes could be significant on a decadal time-scale. Possible decreases in chlorine content as a result of international agreements (Prather and Watson, 1990) would be expected to lead to a slow recovery of stratospheric ozone over many decades, which would then result in a small positive forcing over the period of that recovery.

2.2.7 A Global Warming Potential Concept for Trace Gases

In considering the policy options for dealing with greenhouse gases, it is necessary to have a simple means of describing the relative abilities of emissions of each greenhouse gas to affect radiative forcing and hence climate. A useful approach could be to express any estimates relative to the trace gas of primary concern, namely carbon dioxide. It would follow on from the concept of relative Ozone Depletion Potential (ODP) which has become an integral part of the Montreal Protocol and other national and international agreements for controlling emissions of halocarbons (e.g., UNEP, 1989). The long lifetime of some greenhouse gases implies some commitment to possible climate impacts for decades or centuries to come, and hence the inclusion of 'potential' in the formulation of the concept.

Estimates of the relative greenhouse forcing based on atmospheric concentrations have been detailed in Section 2.2.3; these are relatively straightforward to evaluate. Relative forcings based on emissions are of much greater intrinsic interest to policy makers, but require a careful consideration of the radiative properties of the gases, their lifetimes and their indirect effects on greenhouse gases. Wuebbles (1989) has reviewed various approaches to the design of relative forcings based on emissions using past and current trends in global emissions and concentrations.

It must be stressed that there is no universally accepted methodology for combining all the relevant factors into a single global warming potential for greenhouse gas emissions. In fact there may be no single approach which will represent all the needs of policy makers. A simple approach has been adopted here to illustrate the difficulties inherent in the concept, to illustrate the importance of some of the current gaps in understanding and to demonstrate the current range of uncertainties. However, because of the importance of greenhouse warming potentials, a preliminary evaluation is made.

The Global Warming Potential (GWP) of the emissions of a greenhouse gas, as employed in this report, is the time integrated commitment to climate forcing from the instantaneous release of 1 kg of a trace gas expressed relative to that from 1 kg of carbon dioxide:

$$GWP = \frac{\int_0^n a_i \, c_i \, dt}{\int_0^n a_{CO_2} \, c_{CO_2} \, dt}$$

where a_i is the instantaneous radiative forcing due to a unit increase in the concentration of trace gas, i, c_i is concentration of the trace gas, i, remaining at time, t, after its release and n is the number of years over which the calculation is performed. The corresponding values for carbon dioxide are in the denominator.

Fisher et al. (1990) have used a similar analysis to derive a global warming potential for halocarbons taken relative to CFC-11. In their work it is implicitly assumed that the integration time is out to infinity.

Early attempts at defining a concept of global warming potentials (Lashof and Ahuja, 1990; Rodhe, 1990; Derwent, 1990) are based on the instantaneous emissions into the atmosphere of a quantity of a particular trace gas. The trace gas concentration then declines with time and whilst it is present in the atmosphere, it generates a greenhouse warming. If its decline is due to atmospheric chemistry processes, then the products of these reactions may generate an additional greenhouse warming. A realistic emissions scenario can be thought of as due to a large number of instantaneous releases of different magnitudes over an extended time period and some emission abatement scenarios can be evaluated using this concept.

Particular problems associated with evaluating the GWP are:

   the estimation of atmospheric lifetimes of gases (and in particular CO_2), and the variation of that lifetime in the future;

   the dependence of the radiative forcing of a gas on its concentration and the concentration of other gases with spectrally overlapping absorption bands;

   the calculation of the indirect effects of the emitted gases and the subsequent radiative effects of these indirect greenhouse gases (ozone poses a particular problem);

   the specification of the most appropriate time period over which to perform the integration.

The full resolution of the above problems must await further research. The assumptions made in the present assessment are described below.

For some environmental impacts, it is important to evaluate the cumulative greenhouse warming over an extended period after the instantaneous release of the trace gas. For the evaluation of sea-level rise, the commitment to greenhouse warming over a 100 year or longer time horizon may be appropriate. For the evaluation of short-term effects, a time horizon of a few decades could be taken; for example, model studies show that continental areas are able to respond rapidly to radiative forcing (see e.g., Section 6) so that the relative effects of emissions on such timescales are relevant to predictions of near-term climate change. This consideration alone dramatically changes the emphasis between the different greenhouse gases, depending on their persistence in the atmosphere. For this reason, global warming potentials in Table 2.8 have been evaluated over 20, 100 and 500 years. These three different time horizons are presented as candidates for discussion and should not be considered as having any special significance.

The figures presented in Table 2.8 should be considered preliminary only. Considerable uncertainty exists as to the lifetimes of methane and many of the halocarbons, due to difficulties in modelling the chemistry of the troposphere. The specification of a single lifetime for carbon dioxide also presents difficulties; this is an approximation of the actual lifetime due to the transfer of CO_2 amongst the different reservoirs. The detailed time behaviour of a pulse of carbon dioxide added to the atmosphere has been described using an ocean-atmosphere-biosphere carbon dioxide model (Siegenthaler, 1983). The added carbon dioxide declines in a markedly non-exponential manner; there is an initial fast decline over the first 10 year period, followed by a more gradual decline over the next 100 years and a rather slow decline over the thousand year time-scale. The time period for the first half-life is typically around 50 years, for the second, about 250 years (see Section 1.2.1 for details). A single lifetime figure defined by the decline to $1/e$ is about 120 years. Indeed, the uncertainties associated with specifying the lifetime of CO_2 means that presentation of the GWP relative to CO_2 may not be the ideal choice; relative GWPs of gases other than CO_2 to each other are not affected by this uncertainty.

In performing the integration of greenhouse impacts into the future, a number of simplifications have been made. The neglect of the dependence of the radiative term on the trace gas concentration implies small trace gas concentration changes. Further, the overlap of the infrared absorption bands of methane and nitrous oxide may be significant and this restricts the application of the GWP to small perturbations around present day concentrations.

An assumption implicit in this simple approach is that the atmospheric lifetimes of the trace gases remain constant over the integration time horizon. This is likely to be a poor assumption for many trace gases for a variety of different reasons. For those trace gases which are removed by tropospheric OH radicals, a significant change in lifetime could be anticipated in the future, depending on the impact of human activities on methane, carbon monoxide and oxides of nitrogen emissions. For some scenarios, as much as a 50% increase in methane and HCFC-22 lifetimes has been estimated. Such increases in lifetime have a dramatic influence on the global warming potentials in Table 2.8, integrated over the longer time horizons. Much more work needs to be done to determine global warming potentials which will properly account for the processes affecting atmospheric composition and for the possible non-linear

feedbacks influencing the impacts of trace gases on climate.

It is recognised that the emissions of a number of trace gases, including NO_X, carbon monoxide, methane and other hydrocarbons, have the potential to influence the distribution of tropospheric ozone. It is not straightforward to estimate the greenhouse warming potential of these indirect effects because changes in tropospheric ozone depend, in a complex and non-linear manner on the concentrations of a range of species. The limited spatial resolution in current tropospheric chemistry models means that estimates of increased tropospheric ozone production are highly model-dependent. Furthermore, the radiative impacts of tropospheric ozone changes depend markedly on their spatial distribution. As a result, the GWP values for the secondary greenhouse gases have been provided as first order estimates only, using results from a tropospheric two-dimensional model of global atmospheric chemistry (Hough and Derwent, 1990) and the radiative forcing given in Table 2.2 (see Derwent (1990) for further details). Evaluation of the radiative forcing resulting from changes in concentrations of stratospheric ozone (as a result of CFC, N_2O, and CH_4 emissions) have not been included due to insufficient time to undertake the analysis this requires.

Bearing in mind the uncertainties inherent in Table 2.8, a number of important points are raised by the results. Firstly, over a twenty-year period a kilogram of all the proposed CFC-substitutes, with the exception of the relatively short-lived HCFC-123 and HFC-152a, cause more than a three-order of magnitude greater warming than 1 kg of CO_2. However, for a number of these gases (but *not* the five CFCs themselves) the global warming potential reduces markedly as the integration time is increased; this implies that over the long term, the replacement compounds should have a much lower global warming effect than the CFCs they replace, for the same levels of emissions. In addition, the shorter lifetimes imply that abrupt changes in total emissions would impact on the actual global warming relatively quickly. A further important point is that in terms of radiative forcing over the short-term, the effect of the CFC substitutes is considerably greater than indicated by the halocarbon global warming potential (GWP) of Fisher et al. (1990). For example, over a 20 year period, the effect of 1 kg emission of HCFC-22 contributes only slightly less to the radiative forcing than the same amount of CFC-11, even though its 'infinite' GWP is about 0.35. This is because, on a kg-per-kg basis, HCFC-22 is a stronger greenhouse gas than CFC-11 (Table 2.4).

The indirect greenhouse warmings listed in Table 2.8 are potentially very significant. The production of CO_2, stratospheric water vapour and tropospheric ozone as a result of emissions of methane leads to an indirect effect

Table 2.8: *Global warming potentials following the instantaneous injection of 1 kg of each trace gas, relative to carbon dioxide. A specific example of an application of these potentials is given in Table 2.9.*

Trace Gas	Estimated Lifetime, years	Global Warming Potential Integration Time Horizon, Years		
		20	100	500
Carbon Dioxide	*	1	1	1
Methane - inc indirect	10	63	21	9
Nitrous Oxide	150	270	290	190
CFC-11	60	4500	3500	1500
CFC-12	130	7100	7300	4500
HCFC-22	15	4100	1500	510
CFC-113	90	4500	4200	2100
CFC-114	200	6000	6900	5500
CFC-115	400	5500	6900	7400
HCFC-123	1.6	310	85	29
HCFC-124	6.6	1500	430	150
HFC-125	28	4700	2500	860
HFC-134a	16	3200	1200	420
HCFC-141b	8	1500	440	150
HCFC-142b	19	3700	1600	540
HFC-143a	41	4500	2900	1000
HFC-152a	1.7	510	140	47
CCl_4	50	1900	1300	460
$CH_3 CCl_3$	6	350	100	34
CF_3Br	110	5800	5800	3200

INDIRECT EFFECTS Source Gas	Greenhouse Gas Affected			
CH_4	Tropospheric O_3	24	8	3
CH_4	CO_2	3	3	3
CH_4	Stratospheric H_2O	10	4	1
CO	Tropospheric O_3	5	1	0
CO	CO_2	2	2	2
NO_x	Tropospheric O_3	150	40	14
NMHC	Tropospheric O_3	28	8	3
NMHC	CO_2	3	3	3

CFCs and other gases do not include effect through depletion of stratospheric ozone.

Changes in lifetime and variations of radiative forcing with concentration are neglected. The effects of N_2O forcing due to changes in CH_4 (because of overlapping absorption), and vice versa, are neglected.

* The persistence of carbon dioxide has been estimated by explicitly integrating the box-diffusion model of Siegenthaler (1983); an approximate lifetime is 120 years.

almost as large as the direct effect for integration times of a century or longer. The potential for emissions of gases, such as CO, NO_x and the non-methane hydrocarbons, to contribute indirectly to global warming is also significant. It must be stressed that these indirect effects are highly model dependent and they will need further revision and evaluation. An example of uncertainty concerns the impact of NO_x emissions; these emissions generate OH which leads to increased destruction of gases such as methane (e.g., Thompson et al., 1989). This would constitute a

Table 2.9: Example of use of Global Warming Potentials. The table shows the integrated effects over a 100-year time horizon of total emissions in 1990, given as a fraction of the total effect.

Trace Gas	Current Man Made Emissions Tg yr^{-1}	Proportion of total effects %
CO_2	26000	61
CH_4	300	15
N_2O	6	4
CFC-11	0.3	2
CFC-12	0.4	7
HCFC-22	0.1	0.4
CFC-113	0.15	1.5
CFC-114	0.015	0.2
CFC-115	0.005	0.1
CCl_4	0.09	0.3
CH_3CCl_3	0.81	0.2
CO	200	1
NO_x	66	6
NMHCs	20	0.5

Carbon dioxide emissions given on CO_2 basis; equivalent to 7 GtC yr^{-1}. Nitrous oxide emissions given on N_2O basis; equivalent to 4 MtN yr^{-1}. NO_x emissions given on NO_2 basis; equivalent to 20 MtN yr^{-1}

negative indirect effect of NO_x emissions which would oppose the forcing due to increased tropospheric ozone formation.

As an example of the use of the Global Warming Potentials, Table 2.9 shows the integrated effects over a 100 year time horizon for the estimated human-related greenhouse gas emissions in 1990. The derived cumulative effects, derived by multiplying the appropriate GWP by the 1990 emissions rate, indicates that CO_2 will account for 61% of the radiative forcing over this time period. Emissions of NO_x, whose effect is entirely indirect, is calculated to contribute 6% to the total forcing.

2.3 Other Radiative Forcing Agents

2.3.1 Solar Radiation
The Sun is the primary source of energy for the Earth's climate system. Variations in the amount of solar radiation received by the Earth can affect our climate. There are two distinct sources of this variability. The first, which acts

with greatest impact on time-scales of 10,000 to 100,000 years is caused by changes in the Sun-Earth orbital parameters. The second comes from physical changes on the Sun itself; such changes occur on almost all time-scales.

2.3.1.1 Variability due to orbital changes
Variations in climate on time-scales ranging from 10,000 to 100,000 years, including the major glacial/interglacial cycles during the Quaternary period, are believed to be initiated by variations in the Earth's orbital parameters which in turn influence the latitudinal and seasonal variation of solar energy received by the Earth (the Milankovitch Effect). Although the covariation of these orbital parameters and the Earth's climate provides a compelling argument in favour of this theory, internal feedback processes have to be invoked to explain the observed climatic variations, in particular the amplitude of the dominating 100,000 year period; one such feedback could be the changes to the carbon cycle and the greenhouse effect of atmospheric CO_2 (see Section 1).

The radiative forcing associated with the Milankovitch Effect can be given for particular latitudes and months to illustrate that the rate of change of forcing is small compared to radiative forcing due to the enhanced greenhouse effect; of course, the climatic impact of the Milankovitch Effect results from the redistribution of solar energy, latitudinally and seasonally, so that a comparison is necessarily rather rough. As an example, in the past 10,000 years, the incident solar radiation at 60°N in July has decreased by about 35 Wm^{-2} (e.g., Rind et al., 1989); the average change in one decade is -0.035 Wm^{-2}, compared with the estimate, in Section 2.2.5, that the greenhouse forcing over the most recent decade increased by 0.6 Wm^{-2}, more than 15 times higher than the Milankovitch forcing.

2.3.1.2 Variability due to changes in total solar irradiance
Variations in the short-wave and radio-frequency outputs of the Sun respond to changes in the surface activity of the star and follow in phase with the 11-year sunspot cycle. The greatest changes, in terms of total energy, occur in the short-wave region, and particularly the near ultraviolet. At 0.3 µm, the solar cycle variation is less than 1%; since only about 1% of the Sun's radiation lies at this or shorter wavelengths, solar-cycle variations in the ultraviolet will by themselves induce variations of no more than 0.01% in total irradiance, although these may be important for atmospheric chemistry in the middle atmosphere.

Of greater potential importance, in terms of direct affects on climate, are changes integrated over all wavelengths, the total solar irradiance or the so-called "solar constant". Continuous, spaceborne measurements of total irradiance have been made since 1978. These have shown that, on time-scales of days to a decade, there are irradiance

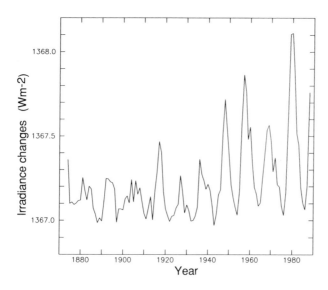

Figure 2.5: Reconstructed solar irradiance (Wm^{-2}) from 1874 to 1988 using the model of Foukal and Lean (1990); The model was calibrated using direct observations of solar irradiance from satellites between 1980 and 1988. Data from J.Lean (pers. comm.). Note that the solar forcing is only 0.175 times the irradiance, due to area and albedo effects.

variations that are associated with activity in the Sun's outer layer, the photosphere - specifically, sunspots and bright areas known as faculae. The very high frequency changes are too rapid to affect the climate noticeably. However, there is a lower frequency component that follows the 11-year sunspot cycle which may have a climatic effect. It has been found that the increased irradiance due to faculae more than offsets the decreases due to the cooler sunspots - consequently, high sunspot numbers are associated with high solar output (Foukal and Lean, 1990). Over the period 1980-86, there was a decline in irradiance of about 1 Wm^{-2} corresponding to a globally-averaged forcing change at the top of the atmosphere of a little less than 0.2 Wm^{-2}. Since then irradiance has increased, following the sunspot cycle (e.g., Willson and Hudson, 1988).

This is comparable with the greenhouse forcing which, over the period 1980-86, increased by about 0.3 Wm^{-2}. However, over longer periods these solar changes would have contributed only minimally towards offsetting the greenhouse effect on global-mean temperature because of the different time-scales on which the two mechanisms operate. Because of oceanic thermal inertia (see Section 6), and because of the relatively short time scale of the forcing changes associated with the solar cycle, only a small fraction of possible temperature changes due to this source can be realised (Wigley and Raper, 1990). In contrast, the *sustained* nature of the greenhouse forcing allows a much greater fraction of the possible temperature change to be realised, so that the greenhouse forcing dominates.

Because the satellite record of solar irradiance began so recently, we cannot say with absolute certainty what past variations may have been. However, a physically-based statistical model has been developed by Foukal and Lean (1990), which attempts to reconstruct the solar-cycle-related changes back more than 100 years (see Figure 2.5). This figure illustrates that the changes from 1980 to 1986 were probably the largest in the past century.

While the model of Foukal and Lean (1990) indicates that the direct effects of solar-cycle-related irradiance changes may have been very small this does not rule out the possibility of larger, lower-frequency effects. Three possibilities have been hypothesized; they are not supported by direct observational evidence of solar irradiance variations, and their magnitudes are derived by assuming that observed or inferred temperature variations are responses to solar forcing. The first idea is that on the time scale of about a century, some underlying variation exists that parallels the envelope of sunspot activity, i.e., the smooth curve joining the peaks of successive sunspot maxima (Eddy 1977; Reid,1987). The envelope curve shows a quasi-cyclic behaviour with period about 80-90 years referred to as the Gleissberg cycle (e.g., Gilliland 1982; Gilliland and Schneider, 1984).

There is no reason why one should expect the envelope curve to be related to solar irradiance variations beyond those associated with the Foukal-Lean mechanism. Reid's study appears to have been spurred by the visual similarity between the Folland et al. (1984) global marine temperature curve and the envelope curve. This similarity is less apparent when more recently compiled temperatures are considered (see e.g., Section 7) and is much less apparent in the Southern Hemisphere than in the Northern. With no way to estimate the range of irradiance variation *a priori,* Reid tuned this to obtain a best match between modelled and observed temperatures. Assuming solar change as the sole forcing mechanism, the implied decadal time-scale irradiance range is about 0.6%, or 1.5 Wm^{-2} at the top of the atmosphere, for an assumed climate sensitivity of 2.5°C for a CO_2 doubling. This value is about thirty times that inferred by direct satellite data.

Reid emphasizes that his work is mainly an exercise in curve-fitting, so that the results should be used with extreme caution. Nevertheless, it has been taken seriously by the Marshall Institute (1989) so a brief analysis is in order. Kelly and Wigley (1990) have performed a similar analysis to Reid's incorporating a greenhouse forcing history (Section 2.2.5) and using more recent temperature compilations (Section 7). The amplitude A of the radiative forcing due to solar variability (which is tied to sunspot number) is evaluated so as to give the best agreement between observed and modelled temperatures between 1861 and 1989. The value of A which gives the best fit is found to depend critically on the assumed climate sensitivity. For values of equilibrium change due to doubled CO_2 (see Section 5.2.1) of greater than 2°C, it is found that the best fit is obtained if there is a negative correlation between solar output and sunspot number (which contradicts recent observations). At the lower end of the range of climate sensitivity suggested in Section 5.2.1 (1.5°C), the best fit is obtained for a value of A about one-fifth that derived by Reid. However, even for this value, the percentage variance explained is only marginally better when the solar and greenhouse effects are considered together, than when greenhouse forcing is considered alone. This analysis provides no evidence for low-frequency irradiance variations larger than the small changes that have been directly inferred from satellite-based irradiance observations.

The second suggested solar effect makes use of the relationship between solar radius variations and irradiance changes. Radius variations have been observed over the past few centuries, but whether these could have significant irradiance changes associated with them is unknown. The proportionality constant relating radius and irradiance changes is so uncertain that it could imply an entirely negligible or a quite noticeable irradiance variation (Gilliland, 1982; Wigley, 1988). Gilliland (1982) therefore attempted to estimate the solar effect empirically by comparing modelled and observed data. Gilliland concluded that solar-induced, quasi-cyclic temperature changes (~80 year cycle) with range about 0.2°C might exist - but to obtain a reasonable fit he had to invoke a phase lag between radius and irradiance changes. Most theories relating radius and irradiance changes do not allow such a phase lag, although an exception has been noted by Wigley (1988). While the physical basis for the radius effect is at least reasonable, these results are far from being convincing in a statistical sense, as Gilliland himself noted. Nevertheless, we cannot completely rule out the possibility of solar forcing changes related to radius variations on an 80-year time-scale causing global-mean temperature fluctuations with a range of up to 0.2°C. Hansen and Lacis (1990) regard about 0.8 Wm^{-2} as a probable upper limit for the change in forcing due to variations in solar output over such periods.

The third suggested solar effect is that related to the minima in sunspot activity such as the Maunder Minimum, for which the associated changes in atmospheric radiocarbon content are used as a proxy. These ideas were revived by Eddy (1977). The hypothesis has some credence in that the sunspot minima are manifestations of solar change (although irradiance changes associated with them would be only a few tenths Wm^{-2} based on the Foukal-Lean model) as are radiocarbon fluctuations. But neither is direct evidence of solar *irradiance* changes. Indirect evidence of irradiance changes comes from the climate record; specifically the observation that during the Holocene, the timing of the neoglacial (i.e., "Little Ice Age" type) events show some correspondence with times of anomalous atmospheric radiocarbon content. Wigley (1988) and Wigley and Kelly (1990) found the correlation over a 10,000 year period to be statistically significant but far from convincing. Nevertheless, if one accepts its reality, the magnitude of the solar forcing changes required to cause the observed neoglacial events can be shown to have been up to 1.3 Wm^{-2} at the top of the atmosphere, averaged over 100-200 years. These results have also been used by the Marshall Institute (1989) who suggest that another Little Ice Age is imminent and that this may substantially offset any future greenhouse-gas-induced warming. While one might expect such an event to occur some time in the future, the timing cannot be predicted. Further, the 1.3 Wm^{-2} solar change (which is an upper limit) is small compared with greenhouse forcing and even if such a change occurred over the next few decades, it would be swamped by the enhanced greenhouse effect.

2.3.2 Direct Aerosol Effects

The impact of aerosol particles, i.e., solid or liquid particles in the size range 0.001-10 µm radius, on the radiation budget of the Earth-atmosphere system is manifold, either *directly* through scattering and absorption in the solar and thermal infrared spectral ranges or *indirectly* by the modification of the microphysical properties of clouds which affects their radiative properties. There is no doubt that aerosol particles influence the Earth's climate. However, their influence is far more difficult to assess than that of the trace gases, because they constitute their own class of substances with different size distributions, shape, chemical compositions and optical properties, and because their concentrations vary by orders of magnitude in space and time and because observations of their temporal and spatial variation are poor (Section 1).

It is not easy to determine the sign of changes in the planetary radiation budget due to aerosols. Depending on absorption-to-backscattering ratio, surface albedo, total aerosol optical depth and solar elevation - if ordered approximately according to importance - additional aerosol particles may either increase or decrease local planetary albedo (e.g., Coakley and Chylek, 1975; Grassl and Newiger, 1982). A given aerosol load may increase the planetary albedo above an ocean surface and decrease it above a sand desert. The effect of aerosol particles on terrestrial radiation cannot be neglected; in conditions where the albedo change is small, the added greenhouse effect can dominate (Grassl 1988).

While it is easy to demonstrate that aerosol particles measurably reduce solar irradiance in industrial regions, the lack of data and inadequate spatial coverage preclude extending this demonstration to larger spatial scales. For example, Ball and Robinson (1982) have shown for the eastern U.S. an average annual depletion of solar irradiance of 7.5% at the surface. Some of this depleted radiation will, however, have been absorbed within the troposphere, so that the perturbation to the net flux at the tropopause will be somewhat less and the impact on the thermal infrared is not quantified. Most of this perturbation is anthropogenic. The depletion is regionally very significant; for example, for a daily mean surface irradiance of 200 Wm^{-2}, if about half of the depleted irradiance is lost to space, the change in forcing would be 7.5 Wm^{-2}.

Carbon black (soot) plays an especially important role for the local heating rate in the air as it is the only strong absorber in the visible and near infrared spectrum present in aerosol particles. Soot incorporated into cloud particles can also directly affect the radiative properties of clouds by decreasing cloud albedo and hence lead to a positive forcing (e.g., Grassl 1988).

In view of the above uncertainties on the sign, the affected area and the temporal trend of the direct impact of aerosols, we are unable to estimate the change in forcing due to tropospheric aerosols.

Concentrations of stratospheric aerosols may be greatly enhanced over large areas for a few years following large explosive volcanic eruptions although there is no evidence for any secular increase in background aerosol (Section 1).

Major volcanic eruptions can inject gaseous sulphur dioxide and dust, among other chemicals, into the stratosphere. The sulphur dioxide is quickly converted into sulphuric acid aerosols. If present in sufficient quantities in the stratosphere, where the half-life is about 1 year, these aerosols can significantly affect the net radiation balance of the Earth.

These aerosols can drastically reduce (by up to tens of percent) the direct solar beam, although this is, to some extent, compensated by an increase in diffuse radiation, so that decreases in total radiation are smaller (typically 5-10%) (e.g., Spaenkuch, 1978; Coulson, 1988). This decrease in insolation, coupled with the warming due to the thermal infrared effects of the aerosols, leaves only a small deficit in the radiative heating at the surface, for even a major volcanic eruption. Furthermore, volcanic aerosol clouds usually cover only a limited portion of the globe and they exist for a time (1-3 years) that is short compared to the response time of the ocean-atmosphere system (which is of order decades). Thus their climatic effects should be relatively short-lived. Because the size distribution and the optical properties of the particles are very important in determining whether the Earth's surface warms or cools, theoretical estimates of their effect on the surface climate are strongly dependent on the assumptions made about the aerosols (e.g., Mass and Portman (1989) and references therein).

A number of empirical studies have been carried out to detect the impact of volcanic eruptions on surface temperatures over the last 100 years or more (e.g., Bradley, 1988; Mass and Portman, 1989). Generally these studies have concluded that major volcanic events, of which there were only about 5 during the past century, may cause a global-mean cooling of 0.1 to 0.2°C for a one to two year period after the event. A direct calculation of the radiative impact of a major volcanic eruption (Ramanathan, 1988) shows that the decadal radiative forcing may be 0.2 - 0.4 Wm^{-2}, indicating that they can have a significant climatic impact on decadal time-scales.

There have also been claims of longer time-scale effects. For example, Hammer et al. (1980) and Porter (1987) have claimed that the climate fluctuations of the last millenium, including events like the Little Ice Age, were due largely to variations in explosive volcanic activity, and various authors have suggested that decadal time-scale trends in the twentieth century were strongly influenced by the changing frequencies of large eruptions (SCOPE, 1986). These claims are highly contentious and generally based on debatable evidence. For instance, a major problem in such studies is that there is no agreed record of past volcanic forcing - alternative records published in the literature correlate poorly. In consequence, the statistical evidence for a low frequency volcanic effect is poor (Wigley et al 1986) but not negligible (Schönwiese, 1988); since the lifetime of the aerosols in the stratosphere is only a few years, such an effect would require frequent explosive eruptions to cause long time-scale fluctuations in aerosol loading.

In summary, there is little doubt that major volcanic eruptions contribute to the interannual variability of the global temperature record. There is no convincing evidence, however, of longer time-scale effects. In the future, the effects of volcanic eruptions will continue to impose small year-to-year fluctuations on the global mean temperature. Furthermore, a period of sustained intense volcanic activity could partially offset or delay the effects of warming due to increased concentrations of greenhouse gases. However, such a period would be plainly evident and readily allowed for in any contemporary assessment of the progress of the greenhouse warming.

2.3.3 *Indirect Aerosol Effects*

Cloud droplets form exclusively through condensation of water vapour on cloud condensation nuclei (CCN): i.e., aerosol particles. Therefore, the size, number and the chemical composition of aerosol particles, as well as updraughts, determine the number of cloud droplets. As a consequence, continental clouds, especially over populated regions, have a higher droplet concentration (by a factor of order 10) than those in remote marine areas. Clouds with the same vertical extent and liquid water content are calculated to have a higher short-wave albedo over continents than over the oceans (e.g., Twomey, 1977). In other words, the more polluted an area by aerosol particles the more reflective the clouds. This effect is most pronounced for moderately thick clouds such as marine stratocumulus and stratus clouds which cover about 25% of the Earth's surface. Hence, an increased load of aerosol particles has the potential to increase the albedo of the planet and thus to some extent counteract the enhanced greenhouse effect.

The strongest confirmation of this aerosol/cloud albedo connection stems from observations of clouds in the wake of ship-stack effluents. Ships enhance existing cloud cover (Twomey et al., 1984), and measurably increase the reflectivities (albedo) of clouds in overcast conditions (Coakley et al., 1987). While the *in-situ* observations (Radke et al., 1990) have shown the expected increase in droplet numbers and decrease in droplet sizes for the contaminated clouds, they have also shown an increase in cloud liquid water content (LWC) in contradiction to the suggestion by Twomey et al., (1984) that the changes in the droplet size distribution will leave the LWC nearly unchanged. Albrecht (1989) has suggested that the LWC-increase could be due to the suppression of drizzle in the contaminated clouds. An increase of the number of CCN therefore may have an even more complicated influence than has been analysed.

The increase in aerosol sulphate caused by anthropogenic SO_2 emissions (Section 1, Figure 1.16) may have caused an increase in the number of CCN with possible subsequent influence on cloud albedo and climate.

Cess (personal communication) has reported changes in planetary albedo over cloudy skies that are consistent with a larger-scale effect of sulphate emissions. Measurements from the Earth Radiation Budget Experiment satellite instruments indicate, after other factors have been taken into account, that the planetary albedo over low clouds decreases by a few per cent between the western and eastern North Atlantic. The implication is that sulphate emissions from the east coast of North America are affecting cloud albedos downwind. A similar effect can be seen in the North Pacific, off the coast of Asia.

There are important gaps in our understanding, and too little data, so that a confident assessment of the influence of sulphur emissions on radiative forcing cannot be made. Wigley (1989) has estimated a global-mean forcing change of between -0.25 and -1.25 Wm^{-2} from 1900 to 1985 (with all of it actually occurring in the Northern Hemisphere). Deriving a forcing history during this period presents even further difficulties, so that we use, for a typical decadal forcing, the average change of -0.03 to -0.15 Wm^{-2} per decade.

Reference to Figure 2.3 shows that this forcing may have contributed significantly to the total forcing, particularly earlier in the century; at these times it may have been of a similar size, but of opposite sign, to the forcing caused by

the enhanced greenhouse effect. Indeed, it has been suggested that the increase in CCN of industrial origin (see Section 1.7.1) might explain why the Northern Hemisphere has not been warming as rapidly as the Southern Hemisphere over the last 50 years. Wigley (1989) estimates that each 0.1°C increase in the twentieth century warming of the Southern Hemisphere relative to the Northern Hemisphere corresponds to a mean forcing differential of around -0.5 Wm^{-2}, or a CCN increase of about 10%.

Sulphur emissions are actively being reduced in many countries. Hence, even if some compensation in the total forcing is occurring because of changes in sulphate and greenhouse gases, it is not clear whether that compensation will continue in the future. Because of the limited atmospheric residence time of the sulphur compounds, their possible effects on climate will be reduced as soon as their emissions are decreased. A decrease in sulphur emissions would, via this theory, cause a decrease in cloud albedo. The change in forcing over a decade could then be positive (although the total change from pre-industrial times would remain less than or equal to zero). Hence we are unable to estimate even the sign of future changes in forcing due to this sulphate effect.

A further important point is that even if the cloud albedo increases exactly offset the forcing due to increased concentrations of greenhouse gases, this would not necessarily imply zero climate change. The sulphate effect would tend to act only regionally, whilst the greenhouse forcing is global. Hence regional climate change would still be possible even if the global mean perturbation to the radiation balance were to be zero.

2.3.4 Surface Characteristics

The effects of desertification, salinization, temperate and tropical deforestation and urbanization on the surface albedo have been calculated by Sagan et al., (1979). They calculated an absolute change in surface albedo of 6 x 10^{-3} over the last 1,000 years, and 1 x 10^{-3} over as short a time as the last 25 years. Henderson-Sellers and Gornitz (1984) updated these latter calculations to a maximum albedo change over the last 25-30 years of between 3.3 and 6.4 x 10^{-4}. From Hansen et al., (1988) the radiative forcing (in Wm^{-2}) for a change in a land surface albedo is about:

$$\Delta F = -43 \, \Delta x \qquad (\Delta x \leq 0.1)$$

where Δx is the change (as a decimal fraction) in the land albedo. (The expression implicitly accounts for the fact that the land surface occupies only 30% of the total surface area of the globe).

Thus the albedo change over the last few decades will have produced a radiative forcing of 0.03 Wm^{-2} at most; i.e., the effects of surface albedo changes on the planetary radiation budget are very small. The effects of changes in surface characteristics on water balance and surface roughness are likely to be far more important for the regional climate; the changes are discussed in Section 5.6.

2.4 The Relative Importance of Radiative Forcing Agents in the Future

The analyses of past trends and future projections of the changes in concentrations of greenhouse gases indicate that the radiative forcing from these gases may increase by as much as 0.4-0.6 Wm^{-2} per decade over the next several decades. As discussed in Section 2.3, decadal-scale changes in the radiative forcing can also result from other causes. Natural effects on the forcing as a result of solar variability and volcanic eruptions are particularly relevant on decadal timescales. Other potentially important anthropogenic effects may result from increases in the aerosol content of the lower atmosphere, particularly as a result of sulphur emissions. It is important to consider how these additional forcings may modify the atmospheric radiative forcing from that expected from greenhouse gases on both decadal and longer timescales.

Over the period of a decade, the other radiative forcings could extensively modify the expected radiative forcing from greenhouse gases. The additional forcing could either add to, subtract from or even largely negate the radiative forcing from greenhouse gases, with the effect over any given decade possibly being quite different from that over other decades. Figure 2.6(a) estimates the range of possible effects from solar variability, volcanic eruptions, and man-made sulphur emissions over a decade as compared with the results using the four policy scenarios which give, over the next decade, changes ranging from 0.41 to 0.56 Wm^{-2}. For solar flux variations, it is assumed that the variability over a decade, when averaged over the eleven year solar cycle, should be less than the longer-term change. The earlier discussion indicates that over a decade the solar flux variability could modify the radiative forcing by ± 0.1 Wm^{-2} and one large volcanic eruption in a decade could cause a decrease of 0.2 Wm^{-2}. The global-mean effect of sulphur emissions on cloud albedos was estimated to be up to 0.15 Wm^{-2} per decade, but, on a decadal scale, not even the sign of the effect is certain. Since both the volcanic and sulphate effects do not act globally, the possible compensations between increased greenhouse forcing and possible decreases from the other effects may be even greater regionally, whilst in other regions, such as in the southern hemisphere, the impact of sulphur emissions may be very small.

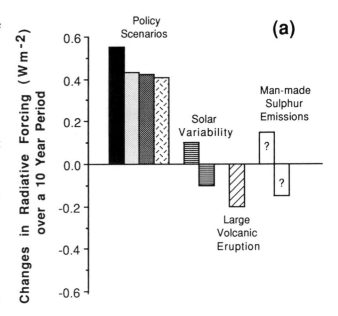

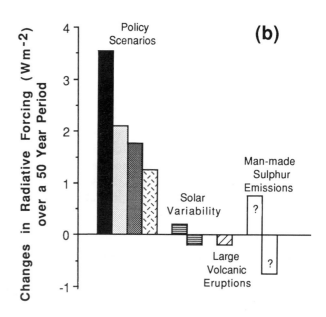

Figure 2.6: Comparison of different radiative forcing mechanisms for: (a) a 10-year period, and (b) a 50-year period in the future. The greenhouse gas forcings are for the periods 1990-2000 and 2000-2050 respectively, using the four policy scenarios. Forcings due to changes in solar radiation and sulphur emissions could be either positive or negative over the two periods.

While other effects could greatly amplify or negate the greenhouse-gas-induced radiative forcing over any given decade, the effects of such forcings over a longer time period should generally be much smaller than the forcing expected from the greenhouse gases. This is shown in Figure 2.6(b) for the changes in radiative forcing over a 50-year period. The four policy scenarios lead to changes in forcing of between 1.3 and 3.5 Wm^{-2} for the period 2000-2050. The effects from solar variability, volcanic eruptions, and man-made sulphur emissions are likely to be much smaller. The prior discussion suggests a change in radiative forcing of 0.2 Wm^{-2} from solar variability could occur over several decades. In the unlikely case of one major volcanic eruption per decade, a resulting net decrease in radiative forcing of 0.2 Wm^{-2} could be sustained over a 50 year period. The effect of man-made sulphur emissions is again highly uncertain, but using the earlier estimates, it could be up to 0.75 Wm^{-2}, of either sign. Effects on radiative forcing from changes in surface characteristics should be less than 0.1 Wm^{-2} over this time period.

In addition to the effects from other forcings that oppose or reinforce the greenhouse gas forcing, there are also decadal-scale climate changes that can occur without any changes in the radiative forcing. Non-linear interactions in the Earth-ocean-atmosphere system can result in "unforced" internal climatic variability (see e.g., Section 6.5.2). As a result of the combined effects of forced and unforced effects on climate, a range of unpredictable variations of either sign will be superimposed on a trend of rising temperature.

References

Albrecht, B.A., 1989: Aerosols, cloud microphysics and fractional cloudiness. *Science*, **242**, 1227-1330.

Ball, R.J. and G.D. Robinson, 1982: The origin of haze in the Central United States and its effect on solar radiation. *J. Appl. Meteorol*, **21**, 171-188.

Bradley, R.S.,1988: The explosive volcanic eruption signal in northern hemisphere continental temperature records. *Clim.Change*, **12**, 221-243.

Coakley, J.A.Jr., and P. Chylek, 1975: The two stream approximation in radiative transfer: including the angle of the incident radiation. *J.Atmos. Sci..*, **46**, 249-261.

Coakley, J.A.Jr., R.L. Bernstein and P.A. Durkee, 1987: Effect of ship-stack effluents on cloud reflectivity . *Science*, **237**, 1020-1022.

Coulson, K.L., 1988: *Polarization and intensity of light in the atmosphere*. A.Deepak Publishing, Hampton, VA, USA.

Derwent, R.G., 1990: Trace gases and their relative contribution to the greenhouse effect. Atomic Energy Research Establishment, Harwell, Oxon, Report *AERE* - **R13716**.

Eddy, J.A., 1977: Climate and the changing sun. *Clim. Change*, **1**, 173-190

Fisher, D.A., C.H. Hales, W-C. Wang, M.K.W. Ko and N.D. Sze, 1990: Model calculations of the relative effects of CFCs and their replacements on global warming. *Nature*, **344**, 513-516.

Folland, C.K., D.E. Parker and F.E. Kates, 1984: Worldwide marine temperature fluctuations 1856-1981. *Nature*, **310**, 670-673.

Foukal, P. and J. Lean, 1990: An empirical model of total solar irradiance variations between 1874 and 1988. *Science*, **247**, 556-558.

Gilliland, R.L., 1982: Solar,volcanic and CO$_2$ forcing of recent climatic changes. *Clim.Change*, **4**, 111-131.

Gilliland, R.L. and S.H. Schneider, 1984: Volcanic, CO$_2$ and solar forcing of northern and southern hemisphere surface temperatures. *Nature*, **310**, 38-41.

Grassl, H., 1988: What are the radiative and climatic consequences of the changing concentration of atmospheric aerosol particles. In *The Changing Atmosphere* , eds F.S.Rowland and I.S.A.Isaksen, pp 187-199, John Wiley and Sons Ltd.

Grassl, H. and M. Newiger ,1982: Changes of local planetary albedo by aerosol particles. In: *Atmospheric Pollution , Studies of Environmental Science,*, **20**, 313-320.

Hansen, J.E. and A.A. Lacis, 1990: Sun and dust versus the greenhouse. *Clim. Change* (submitted).

Hansen, J., A. Lacis and M. Prather, 1989: Greenhouse effect of chlorofluorocarbons and other trace gases. *J.Geophys.Res.*, **94**, 16417-16421.

Hansen,J., D. Johnson, A. Lacis, S. Lebedeff, P. Lee, D. Rind and G. Russell, 1981: Climate impacts of increasing carbon dioxide. *Science*, **213**, 957-966.

Hansen, J., I. Fung, A. Lacis, D. Rind, S. Lebedeff, R. Ruedy and G. Russell, 1988: Global climate changes as forecast by Goddard Institute for Space Studies Three-Dimensional Model. *J. Geophys.Res.* **93**, 9341-9364.

Hammer, C.U., H.B. Clausen and W. Dansgaard, 1980: Greenland ice sheet evidence of postglacial volcanism and its climatic impact. *Nature*, **288**, 230-235.

Hough, A.M. and R.G. Derwent ,1990: Changes in the global concentration of tropospheric ozone due to human activities. *Nature* , **344**, 645-648.

Henderson-Sellers, A. and V. Gornitz, 1984: Possible climatic impacts of land cover transformations with particular emphasis on tropical deforestation. *Clim.Change*, **6**, 231-257.

Husson, N., 1990: Compilation of references to the spectroscopic data base for greenhouse gases. WMO (To be published).

Kelly, P.M. and T.M.L. Wigley, 1990: The relative contribution of greenhouse and solar forcing to observed trends in global-mean temperature. Submitted to *Nature*.

Lacis, A.A., D.J. Wuebbles and J.A. Logan, 1990: Radiative forcing of global climate changes in the vertical distribution of ozone. *J. Geophys. Res.* (to appear).

Lashof, D.A. and D.R. Ahuja, 1990: Relative contributions of greenhouse gas emissions to global warming. *Nature* , **344**, 529-531.

Luther, F.M. and Y. Fouquart, 1984: The Intercomparison of Radiation Codes in Climate Models (ICRCCM). *World Climate Programme Report* **WCP-93**, WMO, Geneva.

Marshall Institute, 1989: Scientific Perspectives on the Greenhouse Problem. Ed. F.Seitz. Marshall Institute, Washington D.C.

Mass, C.F. and D.A. Portman, 1989: Major volcanic eruptions and climate: A critical evaluation. *J.Climate*, **2**, 566-593.

Porter, S.C., 1987: Pattern and forcing of the northern hemisphere glacier variations during the last millenium. *Quart. Res.*, **26**, 27-48.

Prather, M.J. and R.T. Watson, 1990: Stratospheric ozone depletion and future levels of atmospheric chlorine and bromine. *Nature*, **344**, 729-734.

Radke, L.F., J.A. Coakley, Jr., and M.D. King, 1990: Direct and remote sensing observations of the effects of ships on clouds. *Science* (submitted for publication).

Ramanathan, V., 1988: The greenhouse theory of climate change: A test by inadvertent global experiment. *Science, 240,* 293-299.

Ramanathan, V., R.J. Cicerone, H.B. Singh and J.T. Kiehl, 1985: Trace gas trends and their potential role in climate change. *J.Geophys. Res., 90,* 5547-5566.

Ramanathan, V., L. Callis, R. Cess, J. Hansen, I. Isaksen, W. Kuhn, A. Lacis, F. Luther, J. Mahlman, R. Reck and M. Schlesinger, 1987: Climate-chemical interactions and effects of changing atmospheric trace gases. *Rev. Geophys., 25,* 1441-1482.

Reid, G.C., 1987: Influence of solar variability on global sea surface temperature. *Nature, 329,* 142-143.

Rind, D., D. Peteet and G. Kukla, 1989: Can Milankovitch orbital variations initiate the growth of ice sheets in a general circulation model? *J.Geophys.Res., 94,* 12851-12871.

Rodhe, H. , 1990: A comparison of the contribution of various gases to the greenhouse effect, *Science* (to appear).

Sagan, C., O.B. Toon and J.B. Pollack, 1979: Anthropogenic albedo changes and the Earth's climate. *Science, 206,* 1363-1368.

Schneider, S.H. and C. Mass, 1975: Volcanic dust, sunspots and temperature trends. *Science, 190,* 741-746.

Schönwiese, C.D., 1988: Volcanic activity parameters and volcanism-climate relationships within the recent centuries. *Atmosfera, 1,* 141-156.

SCOPE, 1986: *The Greenhouse Effect, Climatic Change and Exosystems.* Edited by B.Bolin, B.R.Döös, J.Jäger and R.A.Warrick, SCOPE 29, John Wiley and Sons, Chichester.

Seigenthaler, U., 1983: Uptake of excess CO_2 by an outcrop-diffusion model of the ocean. *J.Geophys.Res., 88,* 3599-3608.

Spaenkuch, D., 1978: The variation of the shortwave radiation balance of the Earth-atmosphere system with increasing turbidity. *Z.Meteorol., 28,* 199-207.

Thompson, A.M., R.W. Stewart, M.A. Owens and J.A.Herwehe, 1989: Sensitivity of tropospheric oxidants to global chemical and climate change. *Atmos. Environ., 23,* 519-532.

Twomey, S.A., 1977: *Atmospheric Aerosols.* Elsevier, Amsterdam.

Twomey, S.A., M. Piepgrass and T.L. Wolfe , 1984: An assessment of the impact of pollution on global cloud albedo. *Tellus* **36B,** 356-366.

UNEP, 1989: *Scientific Assessment of Stratospheric Ozone,* Nairobi.

Wang, W-C. and N.D. Sze, 1980: Coupled effects of atmospheric N_2O and O_3 Earth's climate. *Nature, 286,* 589-590

Wang, W-C., G-Y. Shi and J.T. Kiehl, 1990: Incorporation of the thermal radiative effect of CH_4, N_2O, CF_2Cl_2 and $CFCl_3$ into the NCAR Community Climate Model. *J.Geophys.Res* (to appear).

Wigley, T.M.L., 1987: Relative contributions of different trace gases to the greenhouse effect. *Climate Monitor,* **16,** 14-29.

Wigley, T.M.L., 1988: The climate of the past 10000 years and the role of the Sun. In *Secular solar and geomagnetic variations in the last 10000 years* edited by F.R.Stephenson and A.W.Wolfendale, 209-224, Kluwer.

Wigley, T.M.L., 1989: Possible climate change due to SO_2-derived cloud condensation nuclei. *Nature ,* **339,** 365-367.

Wigley, T.M.L. and P.M. Kelly, 1990: Holocene climatic change, ^{14}C wiggles and variations in solar irradiance. *Phil.Trans.R.Soc.London.* **A330,** 547-560

Wigley, T.M.L. and S.C.B. Raper, 1990: Climatic change due to solar irradiance changes. *Geophys.Res.Lett.* (to appear).

Wigley, T.M.L., P.D. Jones and P.M. Kelly, 1986: Warm World Scenarios and the Detection of Climatic Changed Induced by Radiatively Active Gases. In *The Greenhouse Effect, Climatic Change and Ecosystems,* Edited by B.Bolin, B.R.Döös, J.Jäger and R.A.Warrick, SCOPE 29, John Wiley and Sons, Chichester.

Willson, R.C. and H.S. Hudson, 1988: Solar luminosity variations in Solar Cycle 21. *Nature,* **332,** 810-812.

WMO 1985: Atmospheric ozone 1985: World Meteorological Organisation Global Ozone Research and Monitoring Project - *Report* **No 16.** WMO Geneva.

Wuebbles, D.J, 1989: Beyond CO_2 - the other greenhouse gases. Lawrence Livermore National Laboratory , California, *Report UCRL* **99883;** also Air and Waste Management Assoc. paper 89-119.4.

Wuebbles, D.J., K.E. Grant, P.S. Connell, and J. E. Penner, 1989: The role of atmospheric chemistry in climate change. *J. Air Poll. Control Assoc.,* **39,** 22-28.

3

Processes and Modelling

U. CUBASCH, R.D. CESS

Contributors:
F. Bretherton; H. Cattle; J.T. Houghton; J.F.B. Mitchell; D. Randall; E. Roeckner; J. D. Woods; T. Yamanouchi.

CONTENTS

EXECUTIVE SUMMARY

The climate system consists of the five components:

    atmosphere
    ocean
    cryosphere (ice)
    biosphere
    geosphere

The fundamental process driving the global climate system is heating by incoming short-wave solar radiation and cooling by long-wave infrared radiation into space. The heating is strongest at tropical latitudes, while cooling predominates at the polar latitudes of each winter hemisphere. The latitudinal gradient of heating drives the large-scale circulations in the atmosphere and in the ocean, thus providing the heat transfer necessary to balance the system.

Many facets of the climate system are not well understood, and a significant number of the uncertainties in modelling atmospheric, cryospheric and oceanic interactions are directly due to the representation or knowledge of interactive climate feedback mechanisms. Such feedback mechanisms can either amplify or reduce the climate response resulting from a given change of climate forcing.

In order to predict changes in the climate system, numerical models have been developed which try to simulate the different feedback mechanisms and the interaction between the different components of the climate system.

So far most climate simulations have been carried out with numerical Atmospheric General Circulation Models (AGCMs) which have been developed or derived from weather forecast models. For investigations of climate change due to increased greenhouse gas concentrations, they have generally been run coupled with simple representations of the upper ocean and, in some cases, with more detailed, but low resolution, dynamical models of the ocean to its full depth. Relatively simple schemes for interactive land surface temperature and soil moisture are also usually included. Representations of the other elements of the climate system (land-ice, biosphere) are usually included as non-interactive components. The resolution of these models is as yet too coarse to allow more than a limited regional interpretation of the results.

Unfortunately, even though this is crucial for climate change prediction, only a few models linking all the main components of the climate system in a comprehensive way have been developed. This is mainly due to a lack of computer resources, since a coupled system has to take the different timescales of the sub-systems into account. An atmospheric general circulation model on its own can be integrated on currently available computers for several model decades to give estimates of the variability about its equilibrium response; when coupled to a global ocean model (which needs millennia to reach an equilibrium) the demands on computer time are increased by several orders of magnitude. The inclusion of additional sub-systems and the refinement of resolution needed to make regional predictions demands computer speeds several orders of magnitude faster than is available on current machines.

It should be noted that current simulations of climate change obtained by incomplete models may be expected to be superseded as soon as more complete models of the climate system become available.

An alternative to numerical model simulations is the palaeo-analogue method (the reconstruction of past climates). Although its usefulness for climate prediction is questioned because of problems involving data coverage and the validity of past climate forcing compared with future scenarios, the method gives valuable information about the possible spectrum of climate change and it provides information for the broader calibration of atmospheric circulation models in different climate regimes.

3.1 Introduction

The aim of this section is to provide background understanding of the climate system, to explain some of the technical terms used in climate research (i.e., what is a transient and what is an equilibrium response), and to describe how climate change can be predicted. In the limited space available to this Section it is impossible to give more than a brief description of the climate system and its prediction. The discussion will therefore be limited to the most relevant aspects. More detailed description are found in the references and in, for example, the books of Gates (1975) and Houghton (ed) (1984).

A section has been devoted to feedback processes which introduce the non-linearities into the climate system, and which account for many of the difficulties in predicting climate change. Climate models and their technical details are discussed where relevant to subsequent Sections of the Report. For more detailed information the reader is referred to the book by Washington and Parkinson (1986).

To illustrate some of the difficulties and uncertainties which arise in climate change predictions from numerical models we compare results from two independent numerical simulations at the end of the Section.

3.2 Climate System

The climate system (see Figure 3.1) consists of the five components:

    atmosphere
    ocean
    cryosphere
    biosphere
    geosphere

The fundamental processes driving the global climate system are heating by incoming short wave solar radiation and the cooling by long-wave radiation into space. The heating is strongest at tropical latitudes, while cooling predominates in the polar regions during the winter of each hemisphere. The latitudinal gradient of heating drives the atmosphere and ocean circulations; these provide the heat transfer necessary to balance the system (see Simmons and Bengtsson, 1984).

3.2.1 The Atmosphere

The bulk of the incoming solar radiation is absorbed not by the atmosphere but by the Earth's surface (soil, ocean, ice). Evaporation of moisture and direct heating of the surface generate a heat transfer between the surface and the atmosphere in the form of latent and sensible heat. The atmosphere transports this heat meridionally, mainly via transient weather systems with a timescale of the order of days.

The following processes are important in determining the behaviour of the atmospheric component of the climate system:

Turbulent transfer of heat , momentum and moisture at the surface of the Earth;

The surface type (i.e., its albedo), which determines the proportion of incoming to reflected solar radiation.

Latent heat release when water vapour condenses; clouds, which play an important role in reflecting

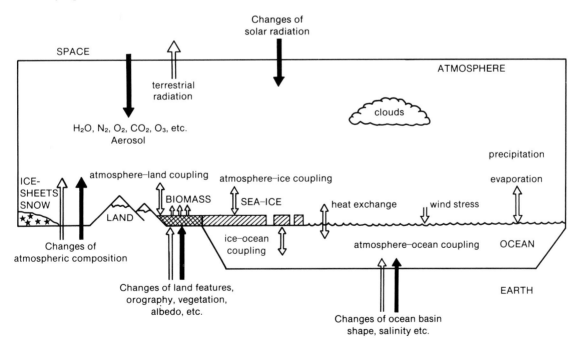

Figure 3.1: Schematic illustration of the components of the coupled atmosphere-ocean-ice-land climatic system. The full arrows are examples of external processes, and the open arrows are examples of internal processes in climatic change (from Houghton, 1984).

incoming solar short-wave radiation and in absorbing and emitting long-wave radiation;

The radiative cooling and heating of the atmosphere by CO_2, water vapour, ozone and other trace gases;

Aerosols (such as volcanic dust), the orbital parameters, mountain ranges and the land-sea distribution.

Atmospheric processes are also influenced by a number of feedback mechanisms which involve interactions between the atmospheric processes themselves (radiation and clouds, for example) and between these processes and the underlying surface. Such feedback mechanisms are discussed in more detail in 3.3.1.

The problems concerning the impact of human activities on the greenhouse effect has broadened in scope from a CO_2 climate problem to a trace gas climate problem (Ramanathan et al., 1987). The climatic effects of trace gases are strongly governed by interactions between chemistry, radiation and dynamics. The nature of the trace gas radiative heating and the importance of chemical-radiative interactions has been already discussed in Section 2.

3.2.2 The Ocean

The ocean also plays an essential role in the global climate system. Over half of the solar radiation reaching the Earth's surface is first absorbed by the ocean, where it is stored and redistributed by ocean currents before escaping to the atmosphere, largely as latent heat of evaporation, but also as long-wave radiation. The currents are driven by the exchange of momentum, heat and water between the ocean and atmosphere. They have a complicated horizontal and vertical structure determined by the pattern of winds blowing over the sea and the distribution of continents and submerged mountain ranges. The vertical structure of the ocean comprises three layers:

The Seasonal Boundary Layer, mixed annually from the surface, is less than 100 metres deep in the tropics and reaches hundreds of metres in the sub-polar seas (other than the North Pacific) and several kilometres in very small regions of the polar seas in most years;

The Warm Water Sphere (permanent thermocline), ventilated (i. e., exchanging heat and gases) from the seasonal boundary layer, is pushed down to depths of many hundreds of metres in gyres by the convergence of surface (Ekman) currents driven directly by the wind; and

The Cold Water Sphere (deep ocean), which fills the bottom 80% of the ocean's volume, ventilated from the seasonal boundary layer in polar seas.

The ocean contains chemical and the biological mechanisms which are important in controlling carbon dioxide in the climate system. Carbon dioxide is transferred from the atmosphere into the interior of the ocean by the physical pump mechanism (described in the previous Section) caused by differences in the partial pressure of carbon dioxide in the ocean and the lowest layers of the atmosphere. Furthermore the annual ventilation of the seasonal boundary layer from the surface mixed-layer controls the efficiency of the biological pump by which ocean plankton convert dissolved carbon dioxide into particulate carbon, which sinks into deep water. These two pumps are responsible for extracting carbon dioxide from the global carbon cycle for periods in excess of a hundred years. The ocean branch of the carbon cycle involves a flux of carbon dioxide from the air into the sea at locations where the surface mixed layer has a partial pressure of CO_2 lower than the atmosphere and vice versa. Mixed-layer partial pressure of CO_2 is depressed by enhanced solubility in cold water and enhanced plankton production during the spring bloom. The rate of gas exchange depends on the air-sea difference in partial pressure of CO_2 and a coefficient which increases with wind speed.

The following processes control the the climate response of the ocean:

The small-scale (of order 50 km) transient eddies inside the ocean influence the structure of permanent gyres and streams and their interaction with submerged mountain ranges. The eddies also control the horizontal dispersion of chemicals (such as CO_2) dissolved in seawater.

The small-scale (tens of kilometres) patches of deep winter convection in the polar seas and the northernmost part of the North Atlantic, which transport heat and dissolved carbon dioxide below one kilometre into the deep reservoir of the cold water sphere, and the slow currents which circulate the newly implanted water around the world ocean.

The more extensive mechanism of thermocline ventilation by which some of the water in the surface mixed-layer flows from the seasonal boundary layer into the warm water sphere reservoir of the ocean, which extends for several hundreds of metres below most of the ocean's surface area.

The global transport of heat, freshwater and dissolved chemicals carried by ocean currents which dictate the global distributions of temperature, salinity, sea-ice and chemicals at the sea surface. Fluctuations in the large-scale circulation have modulated these patterns over years and decades. They also control the regional variations in sea surface properties which affect climate at this scale.

The biological pump in the seasonal boundary layer by which microscopic plants and animals (the plankton) consume some of the carbon dioxide dissolved in the seawater and sequester the carbon in the deep ocean away from the short term (up to a hundred years) interactions between ocean and atmosphere.

3.2.3 *The Cryosphere*
The terrestrial cryosphere can be classified as follows (Untersteiner, 1984):

Seasonal snow cover, which responds rapidly to atmospheric dynamics on timescales of days and longer. In a global context the seasonal heat storage in snow is small. The primary influence of the cryosphere comes from the high albedo of snow covered surfaces.

Sea ice, which affects climate on time scales of seasons and longer. This has a similar effect on the surface heat balance as snow on land. It also tends to decouple the ocean and atmosphere, since it inhibits the exchange of moisture and momentum. In some regions it influences the formation of deep water masses by salt extrusion during the freezing period and by the generation of fresh water layers in the melting period.

Ice sheets of Greenland and the Antarctic, which can be considered as quasi-permanent topographic features. They contain 80% of the existing fresh water on the globe, thereby acting as a long-term reservoir in the hydrological cycle. Any change in size will therefore influence the global sea level.

Mountain glaciers are a small part of the cryosphere. They also represent a freshwater reservoir and can therefore influence the sea level. They are used as an important diagnostic tool for climate change since they respond rapidly to changing environmental conditions.

Permafrost affects surface ecosystems and river discharges. It influences the thermohaline circulation of the ocean.

3.2.4 *The Biosphere*
The biosphere on land and in the oceans (discussed above) controls the magnitude of the fluxes of several greenhouse gases, including CO_2 and methane, between the atmosphere, the oceans and the land. The processes involved are sensitive to climatic and environmental conditions, so any change in the climate or the environment (e.g., increases in the atmospheric abundance of CO_2) will influence the atmospheric abundance of these gases. A detailed description of the feedbacks and their respective magnitudes can be found in Section 10.

3.2.5 *The Geosphere*
The land processes play an important part in the hydrological cycle. These concern the amount of fresh water stored in the ground as soil moisture (thereby

interacting with the biosphere) and in underground reservoirs, or transported as run-off to different locations where it might influence the ocean circulation, particularly in high latitudes. The soil interacts with the atmosphere by exchanges of gases, aerosols and moisture, and these are influenced by the soil type and the vegetation, which again are strongly dependent on the soil wetness. Our present knowledge about these strongly interactive processes is limited and will be the target of future research (see Section 11).

3.2.6 *Timescales*
While the atmosphere reacts very rapidly to changes in its forcing (on a timescale of hours or days), the ocean reacts more slowly on timescales ranging from days (at the surface layer) to millennia in the greatest depths. The ice cover reacts on timescales of days for sea ice regions to millennia for ice sheets. The land processes react on timescales of days up to months, while the biosphere reacts on time scales from hours (plankton growth) to centuries (tree-growth).

3.3 Radiative Feedback Mechanisms

3.3.1 *Discussion of Radiative Feedback Mechanisms*
Many facets of the climate system are not well understood, and a significant number of the uncertainties in modelling atmospheric, cryospheric and oceanic interactions are directly due to interactive climate feedback mechanisms. They can either amplify or damp the climate response resulting from a given climate forcing (Cess and Potter, 1988). For simplicity, emphasis will here be directed towards global-mean quantities, and the interpretation of climate change as a two-stage process: forcing and response. This has proved useful in interpreting climate feedback mechanisms in general circulation models. It should, in fact, be emphasized that the conventional concept of climate feedback applies only to global mean quantities and to changes from one equilibrium climate to another.

As discussed in Section 2, the radiative forcing of the surface-atmosphere system, ΔQ, is evaluated by holding all other climate parameters fixed, with $G = 4$ Wm^{-2} for an instantaneous doubling of atmospheric CO_2. It readily follows (Cess et al., 1989) that the change in surface climate, expressed as the change in global-mean surface temperature ΔTs, is related to the radiative forcing by $\Delta Ts = \lambda \times \Delta Q$, where λ is the climate sensitivity parameter

$$\lambda = \frac{1}{\Delta F/\Delta Ts - \Delta S/\Delta Ts}$$

where F and S denote respectively the global-mean emitted infrared and net downward solar fluxes at the Top Of the

Atmosphere (TOA). Thus ΔF and ΔS are the climate-change TOA responses to the radiative forcing ΔQ. An increase in λ thus represents an increased climate change due to a given radiative forcing ΔQ (= ΔF - ΔQ).

The definition of radiative forcing requires some clarification. Strictly speaking, it is defined as the change in net downward radiative flux at the tropopause, so that for an instantaneous doubling of CO_2 this is approximately 4 Wm^{-2} and constitutes the radiative heating of the surface-troposphere system. If the stratosphere is allowed to respond to this forcing, while the climate parameters of the surface-troposphere system are held fixed, then this 4 Wm^{-2} flux change also applies at the top of the atmosphere. It is in this context that radiative forcing is used in this section.

A doubling of atmospheric CO_2 serves to illustrate the use of λ for evaluating feedback mechanisms. Figure 3.2 schematically depicts the global radiation balance. Averaged over the year and over the globe, there is 340 Wm^{-2} of incident solar radiation at the TOA. Of this, roughly 30%, or 100 Wm^{-2}, is reflected by the surface - atmosphere system. Thus the climate system absorbs 240 Wm^{-2} of solar radiation, so that under equilibrium conditions it must emit 240 Wm^{-2} of infrared radiation. The CO_2 radiative forcing constitutes a reduction in the emitted infrared radiation, since this 4 Wm^{-2} forcing represents a heating of the climate system. Thus the CO_2

doubling results in the climate system absorbing 4 Wm^{-2} more energy than it emits, and global warming then occurs so as to increase the emitted radiation in order to re-establish the Earth's radiation balance. If this warming produced no change in the climate system other than temperature, then the system would return to its original radiation balance, with 240 Wm^{-2} both absorbed and emitted. In this absence of climate feedback mechanisms, $\Delta F/\Delta Ts$ = 3.3 Wm^{-2} K^{-1} (Cess et al., 1989) while $\Delta S/\Delta Ts$ = 0, so that λ = 0.3 Km^2 W^{-1}. It in turn follows that ΔTs = λ x ΔQ = 1.2°C. If it were not for the fact that this warming introduces numerous interactive feedback mechanisms, then ΔTs = 1.2°C would be quite a robust global-mean quantity. Unfortunately, such feedbacks introduce considerable uncertainties into ΔTs estimates. Three of the commonly discussed feedback mechanisms are described in the following sub-sections.

3.3.2 *Water Vapour Feedback*

The best understood feedback mechanism is water vapour feedback, and this is intuitively easy to comprehend. For illustrative purposes a doubling of atmospheric CO_2 will again be considered. The ensuing global warming is, of course, the result of CO_2 being a greenhouse gas. This warming, however, produces an interactive effect; the warmer atmosphere contains more water vapour, itself a greenhouse gas. Thus an increase in one greenhouse gas (CO_2) induces an increase in yet another greenhouse gas (water vapour), resulting in a positive (amplifying) feedback mechanism.

To be more specific on this point, Raval and Ramanathan (1989) have recently employed satellite data to quantify the temperature dependence of the water vapour greenhouse effect. From their results it readily follows (Cess, 1989) that water vapour feedback reduces $\Delta F/\Delta Ts$ from the prior value of 3.3 Wm^{-2} K^{-1} to 2.3 Wm^{-2} K^{-1}. This in turn increases λ from 0.3 Km^2 W^{-1} to 0.43 Km^2 W^{-1} and thus increases the global warming from ΔTs = 1.2°C to ΔTs = 1.7°C. There is yet a further amplification caused by the increased water vapour. Since water vapour also absorbs solar radiation, water vapour feedback leads to an additional heating of the climate system through enhanced absorption of solar radiation. In terms of $\Delta S/\Delta Ts$ as appears within the expression for λ, this results in $\Delta S/\Delta Ts$ = 0.2Wm^{-2} K^{-1} (Cess et al., 1989), so that λ is now 0.48 Km^2 W^{-1} while ΔTs = 1.9°C. The point is that water vapour feedback has amplified the initial global warming of 1.2°C to 1.9°C, i.e., an amplification factor of 1.6.

3.3.3 *Snow-Ice Albedo Feedback*

An additional well-known positive feedback mechanism is snow-ice albedo feedback, by which a warmer Earth has less snow and ice cover, resulting in a less reflective planet

Global Radiation Budget

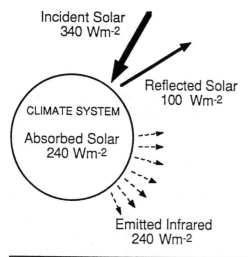

	Absorbed	Emitted
Instantaneous CO2 doubling	240 Wm^{-2}	236 Wm^{-2}
New Equilibrium with no other change	240 Wm^{-2}	240 Wm^{-2}

Figure 3.2: Schematic illustration of the global radiation budget at the top of the atmosphere.

which in turn absorbs more solar radiation. For simulations in which the carbon dioxide concentration of the atmosphere is increased, general circulation models produce polar amplification of the warming in winter, and this is at least partially ascribed to snow-ice albedo feedback. The real situation, however, is probably more complex as, for example, the stability of the polar atmosphere in winter also plays a part. Illustrations of snow-ice albedo feedback, as produced by general circulation models, will be given in Section 3.5. It should be borne in mind, however, that there is a need to diagnose the interactive nature of this feedback mechanism more fully.

3.3.4 Cloud Feedback

Feedback mechanisms related to clouds are extremely complex. To demonstrate this, it will be useful to first consider the impact of clouds upon the present climate. Summarized in Table 3.1 are the radiative impacts of clouds upon the global climate system for annual mean conditions. These radiative impacts refer to the effect of clouds relative to a "clear-sky" Earth; as will shortly be described, this is termed cloud-radiative forcing.

The presence of clouds heats the climate system by 31 Wm^{-2} through reducing the TOA infrared emission. Note the similarity to trace-gas radiative forcing, which is why this impact is referred to as cloud- radiative forcing. Although clouds contribute to the greenhouse warming of the climate system, they also produce a cooling through the reflection and reduction in absorption of solar radiation. As demonstrated in Table 3.1, the latter process dominates over the former, so that the net effect of clouds on the annual global climate system is a 13 Wm^{-2} radiative cooling. As discussed below with respect to cloud feedback components, cloud-radiative forcing is an integrated effect governed by cloud amount, cloud vertical distribution, cloud optical depth and possibly the cloud droplet distribution (Wigley, 1989; Charlson et al, 1987).

Although clouds produce net cooling of the climate system, this must not be construed as a possible means of offsetting global warming due to increasing greenhouse gases. As discussed in detail by Cess et al. (1989), cloud feedback constitutes the *change* in net CRF associated with a change in climate. Choosing a hypothetical example, if climate warming caused by a doubling of CO_2 were to result in a change in net CRF from -13 Wm^{-2} to -11 Wm^{-2}, then this increase in net CRF of 2 Wm^{-2} would amplify the 4 Wm^{-2} initial CO_2 radiative forcing and would so act as a positive feedback mechanism. It is emphasized that this is a hypothetical example, and there is no *a priori* means of determining the sign of cloud feedback. To emphasize the complexity of this feedback mechanism, three contributory processes are summarized as follows:

Cloud Amount: If cloud amount decreases because of global warming, as occurs in typical general circulation model simulations, then this decrease reduces the infrared greenhouse effect attributed to clouds. Thus as the Earth warms it is able to emit infrared radiation more efficiently, moderating the warming and so acting as a negative climate feedback mechanism. But there is a related positive feedback; the solar radiation absorbed by the climate system increases because the diminished cloud amount causes a reduction of reflected solar radiation by the atmosphere. There is no simple way of appraising the sign of this feedback component.

Cloud Altitude: A vertical redistribution of clouds will also induce feedbacks. For example, if global warming displaces a given cloud layer to a higher and colder region of the atmosphere, this will produce a positive feedback because the colder cloud will emit less radiation and thus have an enhanced greenhouse effect.

Cloud Water Content: There has been considerable recent speculation that global warming could increase cloud water content, thereby resulting in brighter clouds and hence a negative component of cloud feedback. Cess et al. (1989) have recently suggested that this explanation is probably an oversimplification. In one case, they demonstrated that this negative solar feedback induces a compensating positive infrared feedback. In a more recent study they further indicate that in some models the net effect might thereby be that of positive feedback (see also Schlesinger and Roeckner, 1988, Roeckner et al., 1987).

The above discussion clearly illustrates the multitude of complexities associated with cloud feedback, and the uncertainties due to this feedback will further be emphasized in Section 3.5. In that both cloud and snow-ice albedo feedbacks are geographical in nature, then these feedback mechanisms can only be addressed through the use of three-dimensional numerical circulation models.

Table 3.1: Infrared, solar and net cloud-radiative forcing (CRF). These are annual-mean values.

Infrared CRF	31 Wm^{-2}
Solar CRF	- 44 Wm^{-2}
Net CRF	- 13 Wm^{-2}

3.4 Predictability Of The Climate System

The prediction of change in the climate system due to changes in the forcing is called "climate forecasting". In the climate system the slow components (for example the oceanic circulation) are altered by the fast components (for example the atmosphere) (Hasselmann, 1976; Mikolajewicz and Maier Reimer, 1990), which again are influenced by the slow components, so that the complete system shows a considerable variance just by an interaction of all components involved. This effect is an illustration of "natural variability".

Taking the climate system as a whole, we note that some elements of the system are chaotic, viewed on a century to millennium time scale, while other parts are remarkably stable on those time scales. The existence of these (in the time frame considered) stable components allows prediction of global change despite the existence of the chaotic elements. The chaotic elements of the climate system are the weather systems in the atmosphere and in the ocean.

The weather systems in the atmosphere have such a large horizontal scale that it is necessary to treat the whole of the atmospheric circulation as chaotic; nevertheless there are stable elements in the atmosphere as witnessed by the smooth seasonal cycle in such phenomena as the temperature distributions over the continents, the monsoon, storm tracks, inter-tropical convergence zone, etc. That stability gives us hope that the response of the atmospheric climate (including the statistics of the chaotic weather systems) to greenhouse forcing will itself be stable and that the interactions between the atmosphere and the other elements of the climate system will also be stable, even though the mechanisms of interaction depend on the weather systems.

This leads to the common assumption used in climate prediction, that the climate system is in equilibrium with its forcing. That means, as long as its forcing is constant and the slowly varying components alter only slightly in the time scale considered, the mean state of the climate system will be stable and that if there is a change in the forcing, the mean state will change until it is again in balance with the forcing. This state is described as an equilibrium state; the transition between one mean and another mean state is called a transient state.

The time-scale of the transition period is determined by the adjustment time of the slowest climate system component, i. e., the ocean. The stable ("quasi stationary") behaviour of the climate system gives us the opportunity to detect changes by taking time averages. Because the internal variability of the system is so high, the averaging interval has to be long compared to the chaotic fluctuations to detect a statistically significant signal which can be attributed to the external forcing.

A number of statistical test have been devised to optimize the detection of climate change signals (v. Storch & Zwiers, 1988; Zwiers, 1988; Hasselmann, 1988; Santer and Wigley, 1990) (see Section 8).

Studies of the completed change from one mean state to another are called "equilibrium response" studies. Studies of the time evolution of the climate change due to an altered forcing, which might also be time dependent, are called "transient response" experiments.

The "weather systems" in the ocean have much smaller horizontal scales (less than one hundred kilometres) than in the atmosphere leaving the large-scale features of the world ocean circulation to be non-chaotic. The success of classical dynamical oceanography depends on that fact. Observations of the penetration of transient tracers into the ocean show that the large-scale ocean currents are stable over periods of several decades. Palaeo-oceanographic evidence shows that the currents and gyres adjusted smoothly to the ice age cycle. That evidence and theoretical understanding of the large-scale ocean circulation suggests that we are indeed dealing, in the ocean, with a predictable system, at least on timescales of decades. The question is whether the existence of predictability in the ocean component of the Earth's climate system makes the system predictable as a whole. However, this seems to be a reasonable working hypothesis, which receives some support from the smooth transient response simulated by coupled ocean-atmosphere models (see Section 6).

3.5 Methods Of Predicting Future Climate

Two approaches have been taken to predict the future climate:

a)   the "analogue method", which tries to estimate future climate change from reconstructions of past climates using palaeo-climatic data,

b)   climate simulations with numerical models (GCM's) of the atmospheric general circulation, which have been derived from weather forecast models. They include representations of the other elements of the climate system (using ocean models, land surface models, etc).which have varying degrees of sophistication. A comprehensive list of the models employed and the research groups involved can be found in Table 3.2(a) and (b).

Table 3.2(a): Summary of results from global mixed-layer ocean-atmosphere models used in equilibrium 2 x CO_2 experiments

ENTRY	Group	Investigators	Year	RESOLUTION No. of waves, or °lat. x °long.	No. of Vertical Layers	Diurnal Cycle	Convection	Ocean Heat Transport	Cloud	Cloud Properties	ΔT (°C)	ΔP (%)	COMMENTS
A.	**Fixed, zonally averaged cloud; no ocean heat transport**												
1.	GFDL	Manabe & Stouffer	1980	R15	9	N	MCA	N	FC	F	2.0	3.5	Based on 4 x CO_2 simulation
2.	GFDL	Wetherald & Manabe	1986, 8	R15	9	N	MCA	N	FC	F	3.2	n/a	
B.	**Variable cloud; no ocean heat transport**												
3.	OSU	Schlesinger & Zhao	1989	4° x 5°	2	N	PC	N	RH	F	2.8	8	
4.			1989	4° x 5°	2	N	PC	N	RH	F	4.4	11	As (3), but with revised clouds.
5.	MRI	Noda & Tokioka	1989	4° x 5°	5	Y	PC	N	RH	F	4.3 *	7 *	* Equilibrium not reached .
6.	NCAR	Washington & Meehl	1984	R15	9	N	MCA	N	RH	F	3.5 *	7 *	* Excessive ice. Estimate ΔT = 4°C at equilibrium.
7.			1989	R15	9	N	MCA	N	RH	F	4.0	8	As (6), but with revised albedos for sea-ice, snow.
8.	GFDL	Wetherald & Manabe	1986, 8	R15	9	N	MCA	N	RH	F	4.0	9	As (2), but with variable cloud.
C.	**Variable cloud; prescribed oceanic heat transport**												
9.	AUS	Gordon & Hunt	1989	R21	4	Y	MCA	Y	RH	F	4.0	7	
10.	GISS	Hansen et al.	1981	8° x 10°	7	Y	PC	Y	RH	F	3.9	n/a	
11.		Hansen et al.	1984	8° x 10°	9	Y	PC	Y	RH	F	4.2	11	
12.		Hansen et al.	1984	8° x 10°	9	Y	PC	Y	RH	F	4.8	13	As (11), but with more sea-ice control.
13.	GFDL	Wetherald & Manabe	1989 †	R15	9	N	MCA	Y	RH	F	4.0	8	
14.	MGO	Meleshko et al.	1990	T21	9	N	PC	Y	RH	F	n/a	n/a	Simulation in progress.
15.	UKMO	Wilson & Mitchell	1987	5° x 7.5°	11	Y	PC	Y	RH	F	5.2	15	
16.		Mitchell & Warrilow	1987	5° x 7.5°	11	Y	PC	Y	RH	F	5.2	15	As (15), but with four revised surface schemes.
17.		Mitchell et al.	1989	5° x 7.5°	11	Y	PC	Y	CW	F	2.7	6	As (16), but with cloud water scheme.
18.			1989	5° x 7.5°	11	Y	PC	Y	CW	F	3.2	8	As (17), but with alternative ice formulation.
19.			1989	5° x 7.5°	11	Y	PC	Y	CW	V	1.9	3	As (17), but with variable cloud radiative properties.
D.	**High Resolution**												
20.	CCC	Boer et al.	1989	T32	10	Y	MCA	Y	RH	V	3.5	4	* "Soft" convective adjustment.
21.	GFDL	Wetherald & Manabe	1989 †	R30	9	N	MCA	*	RH	F	4.0	8	* SSTs prescribed, changes prescribed from (13).
22.	UKMO	Mitchell et al.	1989	2.5°x3.75°	11	Y	PC	Y	CW	F	3.5	9	As (18), but with gravity wave drag.

All models are global, with realistic geography, a mixed-layer ocean, and a seasonal cycle of insolation. Except where stated, results are the equilibrium response to doubling CO_2.

R, T	= Rhomboidal/Triangular truncation in spectral space;	ΔT	= Equilibrium surface temperature change on doubling CO_2;	ΔP	= Percentage change in precipitation;

R, T = Rhomboidal/Triangular truncation in spectral space;
N = Not included;
PC = Penetrative convection;
FC = Fixed cloud;
F = Fixed cloud radiative properties;
GFDL = Geophysical Fluid Dynamics Laboratory, Princeton, USA;
MGO = Main Geophysical Observatory, Leningrad, USSR;
AUS = CSIRO, Australia;

ΔT = Equilibrium surface temperature change on doubling CO_2;
Y = Included;
CA = Convective adjustment;
RH = Condensation or relative humidity based cloud;
† = Personal communication.
NCAR = National Center for Atmospheric Research, Boulder, CO, USA;
CCC = Canadian Climate Center;

ΔP = Percentage change in precipitation;
MCA = Moist convective adjustment;
CW = Cloud water;
V = Variable cloud radiative properties;
n/a = Not available
MRI = Meteorological Research Institute, Japan;
UKMO = Meteorological Office, United Kingdom;

Table 3.2(b): *Summary of experiments carried out with global coupled ocean-atmosphere models*

ENTRY	Group	Investigators	Year	RESOLUTION No. of Spectral Waves	Atmos. Levels	Diurnal Cycle	Conv- ection	Ocean Levels	Cloud	COMMENTS
1.	GFDL	Stouffer et al.	1989	R15	9	N	RH	12	MCA	100 Years, 1% CO_2 increase compounded.
2.	NCAR	Washington & Meehl	1989	R15	9	N	RH	4	FC	30 Years, 1% CO_2 increased linear.
3.	MPI	Cubasch et al.	1990	T21	19	Y	PC	11	CW	25 Years, instantaneous CO_2 doubling.
4.	UHH	Oberhuber et al.	1990	T21	19	Y	PC	9	CW	25 Years, instantaneous CO_2 doubling.

All models are global, with realistic geography and a seasonal cycle of insolation.

R, T	= Number of waves in spectral space;	Y	= Included;
N	= Not included;	PC	= Penetrative convection;
MCA	= Moist convective adjustment;	CW	= Cloud water;
CA	= Convective adjustment ;	V	= Variable cloud radiative properties.
FC	= Fixed cloud;		

GFDL	= Geophysical Fluid Dynamics Laboratory,Princeton, USA;	UHH	= Met Institute, University of Hamburg, FRG;
MPI	= Max Planck Institut für Meteorologie, Hamburg, FRG;	NCAR	= National Center for Atmospheric Research, Boulder, Co, USA.

3.5.1 The Palaeo-Analogue Method

This method has two distinct and rather independent parts. The first derives an estimate of global temperature sensitivity to atmospheric CO_2 concentrations based on estimates of CO_2 concentrations at various times in the past and the corresponding global average temperatures, adjusted to allow for past changes in albedo and solar constant. In the second part regional patterns of climate are reconstructed for selected past epochs, and they are regarded as analogues of future climates under enhanced greenhouse conditions. For a further discussion of the method, see for example Budyko and Izrael (1987).

3.5.1.1 Estimate of temperature sensitivity to CO_2 changes

There are three stages (Budyko et al., 1987)

i)   determining the global mean changes for past palaeo-climates. This is done for four periods (Early Pliocene, early and middle Miocene, Palaeocene-Eocene and the Cretaceous). The temperature changes are based on isotopic temperatures obtained by Emiliani (1966) and maps derived by Sinitsyn (1965,1967) (see Budyko, 1982).

ii) subtracting the temperature change attributed to changes in the solar constant which is assumed to have increased by 5% every billion years, and to changes in surface albedo. A 1% increase in solar constant is assumed to raise the global mean surface temperature by $1.4^\circ C$. The changes in albedo are derived from the ratio of land to ocean, and each 0.01 reduction in albedo is assumed to have raised global mean temperature by $2^\circ C$. These corrections contribute to between 25% and 50% of the total change.

iii) relating the residual warming to the estimated change in atmospheric CO_2 concentrations. The CO_2 concentrations are derived from a carbon cycle model. The concentrations during the Eocene are estimated to be more than five times greater than present, and for the Cretaceous nine times greater (Budyko et al., 1987). On the other hand Shackleton (1985) argued that it is possible to constrain the total CO_2 in the ocean, and suggests that atmospheric CO_2 concentrations were unlikely to have been more than double today's value.

The result is a sensitivity of $3.0^\circ C$ for a doubling of CO_2, with a possible range of $\pm 1^\circ C$, which is very similar to that obtained on the basis of numerical simulations (Section 5).

3.5.1.2 Construction of the analogue patterns

In their study, Budyko et al. (1987) used the mid-Holocene (5-6 kbp), the Last Interglacial (Eemian or Mikulino,125 kbp) and the Pliocene (3-4 mbp) as analogues for future climates. January, July and mean annual temperatures, and mean annual precipitation were reconstructed for each of the above three epochs (see Figures 7.3, 7.4 and 7.5). Estimates of the mean temperatures over the Northern Hemisphere exceeded the temperature at the end of the pre-industrial period (the 19th century) by approximately 1°, 2° and 3-4°C during the mid-Holocene, Eemian and Pliocene respectively. These periods were chosen as analogues of future climate for 2000, 2025 and 2050 respectively.

Although the nature of the forcing during these periods was probably different, the relative values of the mean latitudinal temperature change in the Northern Hemisphere for each epoch were similar in each case (Figure 3.3). Note however that the observational coverage was rather limited, especially for the Eemian when the land-based data came essentially from the Eastern Hemisphere (see Section 7.2.2). Correlations were also calculated between estimated temperature anomalies for 12 regions of the Northern Hemisphere in each of the three epochs. These were found to be statistically significant in most cases, despite the limited quality and quantity of data in the earlier epochs.

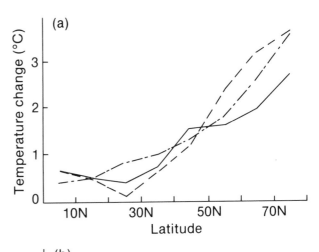

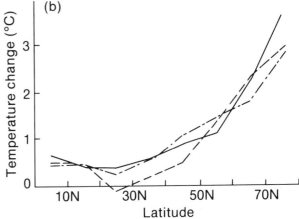

Figure 3.3: Relative surface air temperature changes in different latitudes of the Northern Hemisphere during the palaeo-climatic warm epochs: **(a)** winter, **(b)** summer. Full line = Holocene; Dashed line = last interglacial; Dash-dotted line = Pliocene.

The considerable similarity between the temperature anomaly maps for the three different epochs suggests that the regional temperature anomaly changes are, to a first approximation, directly proportional to increasing mean global temperature. If this is true, then the regional distributions of surface air temperature anomalies are analogous to each other and the similarities between these maps also suggest that the empirical methods for estimating the spatial temperature distribution with global warming may be relatively robust.

Similarly, annual mean precipitation changes have been reconstructed, though the patterns in the mid-Holocene differ from those found for the other two periods (see Section 5.4, Section 7.2.2).

When reconstructions of past climate conditions are accurate and thorough, they can provide relatively reliable estimates of self-consistent spatial patterns of climatic changes. Weaknesses in developing these relationships can arise because of uncertainties

i)   in reconstructing past climates,
ii)   in extending limited areal coverage to global scales,
iii) in interpreting the effects of changing orography and equilibrium versus non-equilibrium conditions,
iv)   in determining the relative influences of the various factors that have caused the past climatic changes.

3.5.2 *Atmospheric General Circulation Models*

General circulation models are based on the physical conservation laws which describe the redistribution of momentum, heat and water vapour by atmospheric motions. All of these processes are formulated in the "primitive" equations, which describe the behaviour of a fluid (air or water) on a rotating body (the Earth) under the influence of a differential heating (the temperature contrast between equator and pole) caused by an external heat source (the Sun). These governing equations are non-linear partial differential equations, whose solution cannot be obtained except by numerical methods. These numerical methods subdivide the atmosphere vertically into discrete layers, wherein the variables are "carried" and computed. For each layer the horizontal variations of the predicted quantities are determined either at discrete grid points over the Earth, as in grid point (finite difference) models, or by a finite number of prescribed mathematical functions as in spectral models. The horizontal resolution of a typical atmospheric model used for climate studies is illustrated by its representation of land and sea shown in Figure 3.4.

The values of the predicted variables (wind, temperature, humidity, surface pressure, rainfall etc.) for each layer (including the surface) and grid point (or mathematical function) are determined from the governing equation by "marching" (integrating) forward in time in discrete time steps starting from some given initial conditions. To

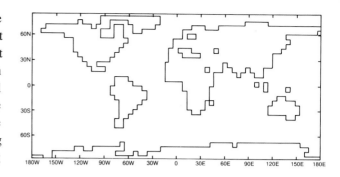

Figure 3.4: The model land-sea mask for a typical climate model (T21, ECHAM, after Cubasch et al, 1989)

prevent the solution from becoming numerically unstable, the time step must be made smaller than a value that depends on the speed of the fastest moving disturbance (wave), the the grid size (or smallest resolved wavelength), and the integration method.

The spatial resolution of GCM's is constrained for practical reasons by the speed and memory capacity of the computer used to perform the numerical integrations. Increasing the resolution not only increases the memory required (linearly for vertical resolution, quadratically for horizontal resolution), but also generally requires a reduction in the integration time step. Consequentially, the computer time required increases rapidly with increasing resolution. Typical models have a horizontal resolution of 300 to 1000 km and between 2 and 19 vertical levels. These resolutions are sufficient to represent large-scale features of the climate, but allow only a limited interpretation of results on the regional scale.

3.5.2.1 *Physical parameterizations*

Due to their limited spatial resolution, GCM's do not (and will not with any foreseeable increase of resolution) resolve several physical processes of importance to climate. However, the statistical effects of these sub grid-scale processes on the scales resolved by the GCM have to be incorporated into the model by relating them to the resolved scale variables (wind, temperature, humidity and surface pressure) themselves. Such a process is called parametrization, and is based on both observational and theoretical studies. Figure 3.5 shows the physical processes parameterized in a typical GCM, and their interactions.

3.5.2.2 *Radiation and the effect of clouds*

The parametrization of radiation is possibly the most important issue for climate change experiments, since it is through radiation that the effects of the greenhouse gases are transferred into the general circulation. A radiation parametrization scheme calculates the radiative balance of

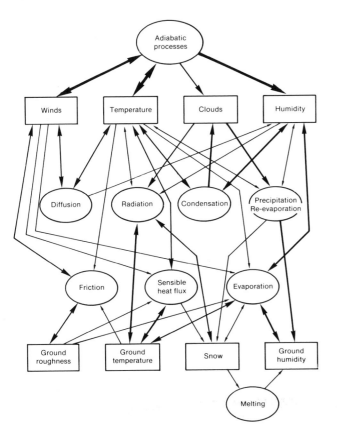

Figure 3.5: The processes parametrized in a numerical atmosphere model (ECMWF) and their interaction. The thickness of the arrows indicates the strength of the interaction (from Houghton, 1984)

many simulations. Climate experiments run without a seasonal cycle are limited in scope and their reliability for climate change experiments is therefore doubtful. The inclusion of the diurnal cycle improves the realism of some feedback mechanisms and therefore the quality of the climate simulations.

3.5.2.3 Sub grid-scale transports

Most of the solar radiation absorbed by the climate system is absorbed at the surface. This energy becomes available for driving the atmospheric general circulation only after it has been transferred to the atmosphere through the planetary boundary layer (PBL), primarily by small-scale turbulent and convective fluxes of sensible and latent heat, but also by net long-wave radiative exchange. On the other hand, the general circulation is slowed down by frictional dissipation which basically takes place in the PBL through vertical transport of momentum by turbulent eddies.

In most GCMs the turbulent fluxes of heat, water vapour and momentum at the surface are calculated from empirical "bulk formulae" with stability dependent transfer coefficients. The fluxes at the PBL top (at a fixed height generally) are either neglected or parametrized from simple mixed-layer theory. In GCMs that resolve the PBL, the eddy diffusion approach is generally employed. Considerable efforts are made to incorporate into the PBL parametrizations the effects of cloud, vegetation and sub grid-scale terrain height.

Cumulus convection in a vertically unstable atmosphere is one of the main heat-producing mechanisms at scales which are unresolvable in GCMs. A common procedure is to adjust the temperature and water vapour profile to a conditionally stable state (Moist Convective Adjustment: MCA). The second class of cumulus parameterizations often employed in GCMs is based on a "moisture convergence" closure (KUO). Other GCMs use Penetrative Convection (PC) schemes to mix moist conditionally unstable air from lower model layers with dry air aloft. The question of how sophisticated convective para- meterizations in GCMs need be, and how much the sensitivity of climate change experiments depends on their formulation, is still open.

3.5.2.4 Land surface processes

Another important parametrization is the transfer of heat and water within the soil, for instance the balance between evaporation and precipitation, snow melt, storage of water in the ground and river runoff. This parametrization is of extreme relevance for climate change predictions, since it shows how local climates may change from humid to arid and vice versa depending on global circulation changes. It furthermore reflects, in some of the more sophisticated schemes, the changes that could occur through alterations in surface vegetation and land-use.

the incoming solar radiation and the outgoing terrestrial long-wave radiation and, as appropriate, the reflection, emission and absorption of these fluxes in the atmosphere. Absorption and emission are calculated in several broad spectral bands (for reasons of economy) taking into account the concentration of different absorbers and emitters like CO_2, water vapour, ozone and aerosols.

One sensitive part in any radiation scheme is the calculation of the radiative effect of clouds. In early GCM experiments clouds were prescribed using observed cloud climatologies (fixed cloud (FC) experiments), and were not allowed to alter during the experiments with (for example) changed CO_2 concentration. Later schemes contained interactive cloud parametrizations of various soph- istication, but mostly based on an estimate of the cloud amount from the relative humidity (RH experiments). Only the most advanced schemes calculate the variation of cloud optical properties by the cloud water content (CW experiments). Capital letters in brackets indicate abbreviations used in Table 3.2.(a) and 3.2.(b).

The seasonal variation of the solar insolation is included in almost all experiments, but a diurnal cycle is omitted in

Most soil moisture schemes used to date are based either on the so-called "bucket" method or the force-restore method. In the former case, soil moisture is available from a single reservoir, or thick soil layer. When all the moisture is used up, evaporation ceases. In the latter method, two layers of soil provide moisture for evaporation, a thin, near-surface layer which responds rapidly to precipitation and evaporation, and a thick, deep soil layer acting as a reservoir. If the surface layer dries out, deep soil moisture is mostly unavailable for evaporation and evaporation rates fall to small values. However, in the presence of vegetation, realistic models use the deep soil layer as a source of moisture for evapotranspiration.

At any given grid-point over land, a balance between precipitation, evaporation, runoff and local accumulation of soil moisture is evaluated. If precipitation exceeds evaporation, then local accumulation will occur until saturation is achieved. After this, runoff is assumed and the excess water is removed. The availability of this runoff as fresh-water input to the ocean has been allowed for in ocean models only recently (Cubasch et al., 1990). Most models differ in the amount of freshwater required for saturation, and few treat more than one soil type. The force-restore method has recently been extended to include a range of soil types by Noilham and Planton (1989).

3.5.2.5 *Boundary conditions*
To determine a unique solution of the model equations, it is necessary to specify a set of upper and lower boundary conditions. These are:

  input of solar radiation (including temporal variation) at
     the top of the atmosphere;
  orography and land sea distribution;
  albedo of bare land;
  surface roughness;
  vegetation characteristics.

The lower boundary over the sea is either prescribed from climatological data or, as this is not very appropriate for climate change experiments, it has to be calculated by an ocean model. As comprehensive ocean models are expensive to run (see Section 3.5.3), the most commonly used ocean model coupled to atmosphere models is the "mixed-layer" model. This model describes the uppermost layer of the ocean where the oceanic temperature is relatively uniform with depth. It is frequently modelled as a simple slab for which a fixed depth of the mixed layer is prescribed and the oceanic heat storage is calculated; the oceanic heat transport is either neglected, or is carried only within the mixed layer, or is prescribed from climatology. Sea ice extents are determined interactively, usually with a variant of the thermodynamic sea ice model due to Semtner (1976). Such an ocean model evidently has strong limitations for studies of climate change, particularly as it does not allow for the observed lags in heat storage of the upper ocean to be represented. Variations of mixed-layer depth, oceanic heat flux convergence, and exchanges with the deep ocean, which would entail an additional storage and redistribution of heat, are all neglected as well. Attempts have been made to couple atmospheric models to ocean models of intermediate complexity. Thus, for example, Hansen et al. (1988) have used a low resolution atmospheric model run with a mixed layer model coupled diffusively to a deep ocean to simulate the time dependent response to a gradual increase in trace gases.

3.5.3 *Ocean Models*
To simulate the role of the ocean more adequately, a number of dynamical ocean models have been developed (Bryan, 1969; Semtner, 1974; Hasselmann, 1982; Cox, 1984; Oberhuber, 1989). The typical ocean model used for climate simulations follows basically the same set of equations as the atmosphere if the equation defining the water vapour balance is replaced by one describing salinity. As with atmospheric GCMs, numerical solutions can be obtained by applying finite difference techniques and specifying appropriate surface boundary conditions (i.e., fluxes of heat, momentum and fresh-water) either from observations (uncoupled mode) or from an atmospheric GCM (coupled mode, see Section 3.5.6). The vertical and horizontal exchange of temperature, momentum and salinity by diffusion or turbulent transfers is parametrized.

The formation of sea-ice is generally treated as a purely thermodynamic process. However, some models already include dynamical effects such as sea ice drift and deformation caused by winds and ocean currents (Oberhuber et al., 1989).

One of the problems of simulating the ocean is the wide range of time and length scales involved. The models for climate sensitivity studies resolve only the largest time and length scales (horizontal resolution: 200 to 1000 km; time scale: hours to 10,000 years; vertical resolution: 2 to 20 levels). High resolution models, which can resolve eddies, are now being tested (Semtner and Chervin, 1988), but with the currently available computer power cannot be run sufficiently long enough to simulate climate changes.

3.5.4 *Carbon Cycle Models*
The exchange of carbon dioxide between the ocean and atmosphere can be simulated by adding equations to the ocean component for the air-sea gas flux, the physics of gas solubility, the chemistry of carbon dioxide buffering in sea water, and the biological pump (Maier Reimer and Hasselmann, 1989). This extension of the coupled ocean-atmosphere model will permit diagnosis of the fractionation of carbon dioxide between the atmosphere and ocean in the last hundred years, and changes to that fractionation in the future as the ocean begins to respond to

global warming, in particular through changes in the ocean mixed layer depth, which affects both the physical uptake of carbon dioxide and the efficiency of the biological pump. The physical and chemical equations are well established, but more work is needed to establish equations for the biological pump. The latter must parametrize the biological diversity, which varies regionally and seasonally and is likely to vary as the climate changes; ideally the equations themselves must cope with such changes without introducing too many variables. Candidate sets of such "robust biological equations" have been tested in one-dimensional models and are now being used in ocean circulation models with encouraging results. It seems likely that they will have to be incorporated into eddy-resolving ocean circulation models in order to avoid biases due to the patchy growth of plankton. They will also have to pay special attention to the seasonal boundary layer (the biologist's euphotic zone) and its interaction with the permanent thermocline in order to deal with nutrient and carbon dioxide recirculation. Such models are computationally expensive and complete global models based on these equations will have to await the arrival of more powerful supercomputers later in the 1990s. Besides the biological organic carbon pump the biological calcium carbonate counter pump and interactions between the seawater and carbon sediment pools must be considered. First results with models which include the organic carbon pump with a sediment reservoir indicate the importance of these processes (Heinze and Maier Reimer, 1989).

3.5.5 Chemical Models
Due to the increasing awareness of the importance of trace gases other than CO_2 a number of research groups have now started to develop models considering the chemical interactions between a variety of trace gases and the general circulation (Prather et al, 1987). At the time of writing, these models have not yet been used in the models discussed so far to estimate the global climate change. It will be interesting to see their impact on future climate change modelling.

3.5.6 Coupled Models of the Atmosphere and the Ocean
Due to the dominating influence of the ocean - atmosphere link in the climate system, realistic climate change experiments require OGCM's and AGCM's to be coupled together by exchanging information about the sea surface temperature, the ice cover, the total (latent ,sensible and net longwave radiative) heat flux, the solar radiation and the wind stress.

One basic problem in the construction of coupled models arises from the wide range of time scales from about one day for the atmosphere to 1000 years for the deep ocean. Synchronously coupled atmosphere-ocean models are extremely time consuming, and limited computer resources

prohibit equilibrium being reached except with mixed-layer models. Various asynchronous coupling techniques have been suggested to accelerate the convergence of a coupled model. However, the problem is far from being solved and can only really be tackled by using faster computers.

A second basic problem that arises through such coupling is "model drift". The coupled model normally drifts to a state that reflects the systematic errors of each respective model component because each sub-model is no longer constrained by prescribed observed fluxes at the ocean-atmosphere interface. Therefore flux correction terms are sometimes introduced to neutralize the climate drift and to obtain a realistic reference climate for climate change experiments (Sausen et al, 1988, Cubasch, 1989) (c.f. Section 4.9). However, these terms are additive and do not guarantee stability from further drift, They are also prescribed from present-day conditions and are not allowed to change with altered forcing from increased CO_2.

Carbon cycle models have already been coupled to ocean models, but coupling to an AGCM-OGCM has not yet been carried out.

3.5.7 Use of Models
Despite their shortcomings, models provide a powerful facility for studies of climate and climate change. A review of such studies is contained in Schlesinger (1983). They are normally used for investigations of the sensitivity of climate to internal and external factors and for prediction of climate change by firstly carrying out a "control" integration with parameters set for present day climate in order to establish a reference mean model climatology and the necessary statistics on modelled climatic variability. These can both be verified against the observed climate and used for examination and assessment of the subsequently modelled climate change. The climate change (perturbation) run is then carried out by repeating the model run with appropriately changed parameters (a doubling of CO_2 for example) and the differences between this and the parallel control run examined. The difference between the control and the perturbed experiments is called the "response". The significance of the response must be assessed against the model's natural variability (determined in the control run) using appropriate statistical tests. These enable an assessment to be made (usually expressed in terms of probabilities) of the confidence that the changes obtained represent an implied climatic change, rather than simply a result of the natural variability of the model.

Typical integration times range from 5 to 100 years, depending on the nature of the investigation. Until now, most effort to study the response to increased levels of greenhouse gas concentrations has gone into determining the equilibrium response of climate to a doubling of CO_2, using atmospheric models coupled to slab ocean models. A comparatively small number of attempts have been made to determine the "transient" (i.e. time-dependent) climate response to anthropogenic forcing using coupled atmosphere and ocean circulation models.

3.5.7.1 *Equilibrium response experiments*

In an equilibrium response experiment both simulations, i.e., the control experiment with the present amount of atmospheric CO_2 and the perturbation experiment with doubled CO_2, are run sufficiently long to achieve the respective equilibrium climates. A review of such experiments is given in Schlesinger and Mitchell (1987). For a mixed-layer ocean the response time to reach equilibrium amounts to several decades, which is feasible with present day computers. For a fully coupled GCM the equilibrium response time would be several thousand years and cannot be achieved with present day computers. A comprehensive list of equilibrium response experiments can be found in Table 3.2(a).

3.5.7.2 *Time-dependent response experiments*

Equilibrium response studies for given CO_2 increases are required as standard benchmark calculations for model intercomparison. The results may be misleading, however, if applied to actual climate change caused by man's activities, because the atmospheric CO_2 concentrations do not change abruptly but have been growing by about 0.4% per year. Moreover, the timing of the atmospheric response depends crucially on the ocean heat uptake which might delay the CO_2 induced warming by several decades. Thus, for realistic climate scenario computations, not only have the atmospheric processes to be simulated with some fidelity but also the oceanic heat transport which is largely governed by ocean dynamics. First experiments with coupled dynamical atmosphere-ocean models have been performed (Table 3.2(b)) and will be discussed later in Section 6.

3.6 Illustrative Equilibrium Experiments

In this section climate sensitivity results, as produced by a large number of general circulation models, are summarized for two quite different climate change simulations. The first refers to a simulation that was designed to suppress snow-ice albedo feedback so as to concentrate on the water vapour and cloud feedbacks. The second consists of a summary of global warming due to a CO_2 doubling.

The first case, addressing only water vapour and cloud feedbacks, consists of a perpetual July simulation in which the climate was changed by imposing a 4°C perturbation on the global sea surface temperature while holding sea ice fixed. Since a perpetual July simulation with a general circulation model results in very little snow cover in the Northern Hemisphere, this effectively eliminates snow-ice albedo feedback. The details of this simulation are given by

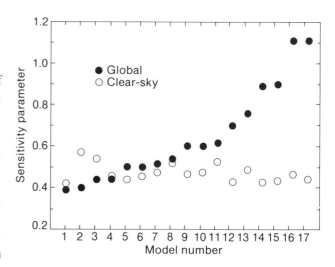

Figure 3.6: Summary of the clear-sky and global sensitivity parameters for 17 general circulation models.

Cess et al. (1989); the main point is that it was chosen so as to minimize computer time and thus allow a large number of modelling groups to participate in the intercomparison. This procedure is in essence an inverse climate change simulation. Rather than introducing a radiative forcing into the models and then letting the model climates respond to this forcing, the climate change was instead prescribed and the models in turn produced their respective radiative forcings.

Cess et al. (1989) have summarized climate sensitivity parameters as produced by 14 atmospheric general circulation models (most of them are referenced in Table 3.2(a) and 3.2(b). This number has since risen to 17 models, and their sensitivity parameters (λ as defined in Section 3.3.1) are summarized in Figure 3.6. The important point here is that cloud effects were isolated by separately averaging the models' clear-sky TOA fluxes, so that in addition to evaluating the climate sensitivity parameter for the globe as a whole (filled circles in Figure 3.6), it was also possible to evaluate the sensitivity parameter for an equivalent "clear-sky" Earth (open circles).

Note that the models are in remarkable agreement with respect to the clear-sky sensitivity parameter, and the model-average $\lambda = 0.47$ Km^2W^{-1} is consistent with the discussion of water vapour feedback (Section 3.3.2), for which it was suggested that $\lambda = 0.48$ Km^2W^{-1}. There is, however, a nearly threefold variation in the global sensitivity parameter, and since the clear-sky sensitivity parameters are in good agreement, then this implies that most of the disagreements can be attributed to differences in cloud feedback. A more detailed demonstration of this is given by Cess et al. (1989). The important conclusions from this intercomparison are that the 17 models agree well with an observational determination of water vapour feedback, whereas improvements in the treatment of cloud feedback are needed if general circulation models are ultimately to be used as reliable climate predictions.

The second type of simulation refers to a doubling of atmospheric CO_2, so that in proceeding from one equilibrium climate to another, snow-ice albedo feedback is additionally activated in the general circulation models. It must be cautioned however, that cloud feedback in this type of simulation should not be expected to be similar to that for the perpetual July simulation. Furthermore, one should anticipate interactive effects between cloud feedback and snow-ice albedo feedback.

Summarized in Table 3.2(a) are ΔTs results, as well as the related changes in global precipitation, for CO_2 doubling simulations using a number of general circulation models. All models show a significant increase in global-mean temperature which ranges from 1.9°C to 5.2°C. As in the perpetual July simulations, cloud feedback probably introduces a large uncertainty, although here it is difficult to quantify this point.

Most results lie between 3.5°C and 4.5°C, although this does not necessarily imply that the correct value lies in this range. Nor does it mean that two models with comparable ΔTs values likewise produce comparable individual feedback mechanisms. For example, consider the Wetherald and Manabe (1988) and Hansen et al. (1984) simulations for which the respective ΔTs values are 4.0°C and 4.2°C. Summarized in Table 3.3 are their diagnoses of individual feedback mechanisms. These two models (labelled GFDL and GISS respectively) produce rather similar warmings in the absence of both cloud feedback and snow-ice albedo feedback. The incorporation of cloud feedback, however, demonstrates that this is a stronger feedback in the GISS model, as is consistent with the perpetual July simulations. But curiously the additional incorporation of snow-ice albedo feedback compensates for

their differences in cloud feedback. Thus, while the two models produce comparable global warming, they do so for quite different reasons.

It should be emphasized that Table 3.3 should not be used to appraise the amplification factor due to cloud feedback since feedback mechanisms are interactive. For example, from Table 3.3 the cloud feedback amplifications for the GFDL and GISS models might be inferred to be 1.2 and 1.6 respectively. But, these are in the ***absence*** of snow-ice albedo feedback. Conversely, if snow-ice albedo feedback is incorporated before cloud feedback, then the respective amplification factors are 1.3 and 1.8. These larger values are due to an amplification of cloud feedback by snow-ice albedo feedback.

3.7 Summary

Many aspects of the global climate system can now be simulated by numerical models. The feedback processes associated with these aspects are usually well represented, but there appear to be considerable differences in the strength of the interaction of these processes in simulations using different models. Section 4 examines results from various models in more detail.

Unfortunately, even though this is crucial for climate change prediction, only a few models linking all the main components of the climate system in a comprehensive way have been developed. This is mainly due to a lack of computer resources, since a coupled system has to take the different timescales of the sub-systems into account, but also the task requires interdisciplinary cooperation.

An atmospheric general circulation model on its own can be integrated on currently available computers for several model decades to give estimates of the variability about its equilibrium response; when coupled to a global ocean model (which needs millennia to reach an equilibrium) the demands on computer time are increased by several orders of magnitude. The inclusion of additional sub-systems and a refinement of resolution needed to make regional predictions demands computer speeds several orders of magnitude faster than is available on current machines.

We can only expect current simulations of climate change to be broadly accurate and the results obtained by existing models may become obsolete as more complete models of the climate system are used. Results from fully coupled atmosphere ocean models are now beginning to emerge; these are given in Section 6.

The palaeo-analogue method, although of limited use for detailed climate prediction (see Section 5), nevertheless gives valuable information about the spectrum of past and future climate changes and provides data for the calibration of circulation models in climate regimes differing from the present. Results from these calibrations are shown in Section 4.

Table 3.3: *Comparison of ΔTs (°C) for the GFDL and GISS models with the progressive addition of cloud and snow-ice feedbacks.*

FEEDBACKS	GFDL	GISS
No cloud or snow-ice	1.7	2.0
Plus cloud	2.0	3.2
Plus snow-ice	4.0	4.2

References

Boer, G. et al., 1989: Equilibrium response studies with the CCC climate model. Pers. Com.

Bryan, K., 1969: A numerical method for the study of the circulation of the world ocean. *J. Comp. Phys.*, **4**, 347-376.

Budyko, M.I., 1982: The Earth's climate ,past and future.(English translation) Academic Press.

Budyko, M.I., A.B. Ronov and A.L.Yanshin, 1987: History of the Earth's atmosphere (English translation) Springer-Verlag.

Budyko, M.I. and Y. Izrael, 1987: "Anthropogenic climate change", Gidrometeorizdat, 406pp.

Cess, R.D. and G.L. Potter, 1988: A methodology for understanding and intercomparing atmospheric climate feedback processes in general circulation models. *J. Geophys. Res.*, **93**, 8305-8314.

Cess, R.D., G L. Potter, J.P. Blanchet, G.J. Boer, S.J. Ghan, J.T. Kiehl, H. Le Treut, Z.X. Li, X.Z. Liang, J.F.B. Mitchell, J.-J. Morcrette, D.A. Randall, M.R. Riches, E. Roeckner, U. Schlese, A. Slingo, K.E. Taylor, W.M. Washington, R.T. Wetherald and I. Yagai, 1989: Interpretation of cloud climate feedback as produced by 14 atmospheric general circulation models. *Science*, **245**, 513-516.

Cess, R.D., 1989: Gauging water vapour feedback. *Nature*, **342**, 736-737.

Cox, M.D., 1984: A primitive equation, three dimensional model of the ocean. GFDL ocean group, *Techn. Rep.* No. **1**. 250pp.

Charlson, R.J., J.E. Lovelock, M.O. Andrae and S.G. Warren, 1987: Oceanic plankton, atmospheric sulphur, cloud albedo and climate. *Nature*, **326**, 655-661.

Cubasch, U., 1989: Coupling a global atmosphere model with a global ocean model using the flux correction method, In *"Aspects of coupling atmosphere and ocean models"*, Rep. No. 6. , Met. Inst. Univ. Hamburg, FRG.

Cubasch, U., E. Maier Reimer, B. Santer, U. Mikolajewicz, E. Roeckner and M. Boettinger, 1990: The response of a global coupled O-AGCM to CO_2 doubling. MPI report, MPI fuer Meteorologie, Hamburg, FRG.

Emiliani C, 1966: Isotopic palaeotemperatures, *Science*, **154**, 851-857.

Gates, L, 1975: The physical basis of climate and climate modelling; GARP Publication Series No. 16, WMO, Geneva.

Gordon and Hunt, 1989: Australian equilibrium response experiments. Pers. Com.

Hansen, J., D. Johnston, A. Lacis, S. Lebedeff, P. Lee, D. Rind, and G. Russel, 1981: Climate impact of increasing atmospheric carbon dioxide. *Science*, **213**, 957-966.

Hansen, J., A. Lacis, D. Rind, L. Russel, P.Stone, I. Fung, R. Ruedy, J. Lerner, 1984: Climate sensitivity analysis of feedback mechanisms. In: *Climate processes and climate sensitivity*, (ed. J. Hansen and T. Takahashi), Geophys. Monogr. Ser. **29**, 130-163. AGU, Washington

Hansen, J., I. Fung, A. Lacis, D. Rind, S. Lebedeff, R. Ruedy and G. Russel, 1988: Global climate changes as forecast by GISS's three-dimensional model. *J. Geophys. Res.*, **93**, 9341-9364.

Hasselmann, K., 1976: Stochastic climate models, Part I: Theory. *Tellus*, **28**, 6, 473-485.

Hasselmann, K., 1982: An ocean model for climate variability studies. *Prog. Oceanogr.*, **11**, 69-92.

Hasselmann, K., 1988: PIPs and POPs - A general formalism for the reduction of dynamical systems in terms of Principal Interaction and Patterns and Principal Oscillation Patterns. *J. Geophys. Res.*, **93**, 11,015 - 11,022 .

Heinze, C. and E. Maier Reimer, 1989: Glacial pCO_2 reduction by the world ocean: Experiments with the Hamburg Carbon Cycle model; In: *3. World conference on Analysis and Evaluation of atmospheric CO_2 data, present and past*. WMO, Environmental Pollution Monitoring and Research Program, Rep. No. **59**, 9-14.

Houghton, J. (ed), 1984: The Global Climate. Cambridge University Press, Cambridge, UK, 233pp.

Manabe, S. and R.J. Stouffer, 1980: Sensitivity of a global climate model to an increase in the CO_2 concentration in the atmosphere. *J. Geophys. Res.*, **85**, 5529-5554.

Manabe, S., K. Bryan and M. Spelman, 1989: Transient response of a global ocean-atmosphere model to a doubling of atmospheric carbon dioxide. submitted to *J. Geophys. Res.*

Maier Reimer, E. and K. Hasselmann, 1987: Transport and storage of CO_2 in the ocean - an inorganic ocean circulation carbon cycle model; Climate Dynamics, **2**, 63-90.

Maier Reimer, E. and U. Mikolajewicz, 1989: Experiments with an OGCM on the cause of the "Younger Dryas"; MPI Report No. 39, MPI für Meteorologie, Hamburg.

Meehl, G.A. and W.M. Washington, 1988: A comparison of soil moisture sensitivity in two global climate models. *JAS*, **45**, 1476-1492.

Meleshko, 1990: CO_2 equilibrium response studies with the CCCP climate model. Pers. Com.

Mikolajewicz, U. and E. Maier Reimer, 1990: Internal secular variability in an ocean general circulation model; MPI Report No. 50, MPI für Meteorologie, Hamburg.

Mitchell, J.F.B and D.A. Warrilow, 1987: Summer dryness in northern mid-latitude due to increased CO_2. *Nature*, **330**, 238-240.

Mitchell, J.F.B., C.A. Senior and W.J. Ingram, 1989: CO_2 and climate: A missing feedback?. *Nature*, **341**, 132-134.

Noda, A. and T. Tokioka, 1989: The effect of doubling CO_2 concentration on convective and non-convective precipitation in a general circulation model coupled with a simple mixed layer ocean. *J. Met. Soc. Japan*, **67**, 95-110.

Noilham, J. and S. Planton, 1989: A simple parametrization of land surface processes for meteorological models. *MWR*, **117**, 536-549

Oberhuber, J., 1989: An isopycnic global ocean model. MPI report, MPI fuer Meteorologie, Hamburg

Oberhuber, J., F. Lunkeit and R. Sausen, 1990: CO_2 doubling experiments with a coupled global isopycnic ocean-atmosphere circulation model. Rep. d. Met. Inst. Univ. Hamburg, Hamburg, FRG.

Prather, M. , M. McElroy, S. Wolfsy, G. Russel and D. Rind, 1987: Chemistry of the global troposphere: Fluorocarbons as tracers of air motion. *JGR*, **92**, 6579-6611.

Ramanathan, V., L. Callis, R. Cess, J. Hansen, F. Isaksen, W. Kuhn, A. Lacis, F. Luther, J. Mahlman, R. Reck and M. Schlesinger, 1987: Climate-chemical Interactions and effects of changing atmospheric trace gases. *Rev. of Geophysics*, **25**, 1441-1482.

Raval, A. and V. Ramanathan, 1989: Observational determination of the greenhouse effect. *Nature*, **342**, 758-761.

Roeckner E., U. Schlese, J. Biercamp and P. Loewe, 1987: Cloud optical depth feedbacks and climate modelling. *Nature*, **329**, 138-140.

Santer, B. and T.M.L. Wigley, 1990: Regional validation of means, variances and spatial patterns in general circulation model control runs. *JGR*, **95**, 829-850.

Sausen, R., K.Barthel and K. Hasselmann, 1988: Coupled ocean-atmosphere models with flux corrections. *Climate Dynamics*, **2**, 154-163.

Schlesinger, M.E., 1983: A review of climate model simulations of CO_2 induced climatic change. OSU Report 41, Climatic Research Institute and Department of Atmospheric Sciences.

Schlesinger, M.E. and J.F.B. Mitchell, 1987: Climate Model Simulations of the Equilibrium Climatic Response to Increased Carbon Dioxide. *Reviews of Geophysics*, **25**, 760-798.

Schlesinger, M.E. and E. Roeckner, 1988: Negative or positive cloud optical depth feedback. *Nature*, **335**, 303-304.

Schlesinger, M.E. and Z.C. Zhao, 1989: Seasonal climatic change introduced by doubled CO_2 as simulated by the OSU atmospheric GCM/mixed-layer ocean model. *J. Climate*, **2**, 429-495.

Semtner, A.J., Jr., 1974: An oceanic general circulation model with bottom topography. Numerical Simulation of Weather and Climate, Technical Report No. 9, Department of Meteorology, University of California, Los Angeles.

Semtner, A.J., Jr., 1976: A model for the thermodynamic growth of sea ice in numerical investigations of climate. *J. Phys. Oceanogr.*, **6**, 379-389.

Semtner, A.J., Jr. and R.M. Chervin, 1988: A simulation of the global ocean circulation with resolved eddies. *J. Geophys. Res.*, **93**, 15,502 - 15,522.

Shackleton, N.J., 1985: Oceanic carbon isotope constraints on oxygen and carbon dioxide in the Cenozoic atmosphere. "The carbon cycle and atmospheric CO_2; natural variations Archean to present. *Geophysical monograph*, **32**, 412-417

Simmons, A.J. and L. Bengtsson, 1984: Atmospheric general circulation models, their design and use for climate studies. In: *The Global Climate* (ed. J. Houghton), Cambridge University Press, 233 pp.

Sinitsyn, V.M., 1965: "Ancient climates in Eurasia, Part 1:Palaeogene and Neogene" Izd. LGU, Leningrad(R)

Sinitsyn, V.M. 1967: "Introduction to Palaeo-climatology".Nedra,Leningrad. (R)

Stouffer, R.J., S. Manabe and K. Bryan, 1989: On the climate change induced by a gradual increase of atmospheric carbon dioxide. *Nature*, **342**, 660-662.

Untersteiner, N., 1984: The Cryosphere. In: *The Global Climate* (ed. J. Houghton), Cambridge University Press, 233 pp.

v. Storch, H. and F. Zwiers, 1988: Recurrence analysis of climate sensitivity experiments. *Journal of Climate*, **1**, 157-171.

Washington, W.M. and G.A. Meehl, 1984: Seasonal Cycle experiments on the climate sensitivity due to a doubling of CO_2 with an atmospheric GCM coupled to a simple mixed layer ocean model. *J. Geophys. Res.*, **89**, 9475-9503.

Washington, W.M. and G.A. Meehl, 1989: Climate sensitivity due to increased CO_2: experiments with a coupled atmosphere and ocean general circulation model. *Climate Dynamics*, **4**, 1-38.

Washington, W.M. and C.L. Parkinson, 1986: An introduction to three dimensional climate modeling. University Science Books, 20 Edgehill Road, Mill Valley, CA. 94941 , 422pp.

Wetherald, R.T. and S. Manabe, 1986: An investigation of cloud cover change in response to thermal forcing. *Climatic Change*, **10**, 11-42.

Wetherald, R.T. and S. Manabe, 1988: Cloud feedback processes in a general circulation model. *J. Atmos. Sci.*, **45**, 1397-1415.

Wigley, T.M.L., 1989: Possible climatic change due to SO_2-derived cloud condensation nuclei. *Nature*, **339**, 365-367.

Wilson, C.A. and J.F.B. Mitchell, 1987: Simulated climate and CO_2 induced climate change over western Europe. *Climatic Change*, **10**, 11-42.

Zwiers, F.W., 1988: Aspects of the statistical analysis of climate experiments with multiple integrations. MPI Report No. 18, MPI für Meteorologie, Hamburg.

4

Validation of Climate Models

W.L. GATES, P.R. ROWNTREE, Q.-C. ZENG

Contributors:
P. Arkin; A. Baede; L. Bengtsson; A. Berger; C. Blondin; G. Boer; K. Bryan;
R. Dickinson; S. Grotch; D. Harvey; E. Holopainen; R. Jenne; J. Kutzbach;
H. Le Treut; P. Lemke; B. McAvaney; G. Meehl; P Morel; T. Palmer; L. Prahm;
S. Schneider; K. Shine; I. Simmonds; J. Walsh; R. Wetherald, J. Willebrand.

CONTENTS

Executive Summary

1. The validation of the present day climate simulated by atmospheric general circulation models shows that there is considerable skill in the portrayal of the large-scale distribution of the pressure, temperature, wind and precipitation in both summer and winter, although this success is in part due to the constraints placed on the sea surface temperature and sea-ice.

2. On regional scales there are significant errors in these variables in all models. Validation for five selected regions shows mean surface air temperature errors of 2 to 3°C, compared with an average seasonal variation of 15°C. Similarly, the simulation of precipitation exhibits errors on sub-continental scales (1000-2000 km) which differ in location between models. Validation on these scales for the five selected regions shows mean errors of from 20% to 50% of the average rainfall depending on the model.

3. The limited soil moisture data available show that the simulated middle latitude summer and winter distributions qualitatively reflect most of the observed large-scale characteristics.

4. Snow cover can be well simulated in winter apart from errors in regions where the temperature is poorly simulated. Though comparison is difficult in other seasons because of the different forms of model and observed data, it is evident that the broad seasonal variation can be simulated, although there are significant errors on regional scales.

5. The radiative fluxes at the top of the atmosphere, important for the response of climate to radiative perturbations, are simulated well in some models, indicating some skill in cloud parameterization. Errors averaged around latitude circles are mostly less than 20 Wm^{-2} with average error magnitudes as low as 5 Wm^{-2} or about 2% of the unperturbed values; however, there are substantial discrepancies in albedo, particularly in middle and high latitudes due to the sensitivity of the parameterization schemes.

6. There has been a general reduction in the errors in more recent models as a result of increased resolution, changes in the parameterization of convection, cloudiness and surface processes, and the introduction of parameterizations of gravity wave drag.

7. Although the daily and interannual variability of temperature and precipitation have been examined only to a limited extent, there is evidence that they are overestimated in some models, especially during summer. The daily variability of sea-level pressure can be well simulated, but the eddy kinetic energy in the upper troposphere tends to be underestimated.

8. Our confidence that changes simulated by the latest atmospheric models used in climate change studies can be given credence is increased by their generally satisfactory portrayal of aspects of low-frequency variability, such as the atmospheric response to sea surface temperature anomalies associated with the El Niño and with wet and dry periods in the Sahel, and by their ability to simulate aspects of the climate at selected times during the last 18,000 years.

9. Models of the oceanic general circulation simulate many of the observed large-scale features of ocean climate, especially in lower latitudes, although their solutions are sensitive to resolution and to the parameterization of sub-gridscale processes such as mixing and convective overturning.

10. Atmospheric models have been coupled with simple mixed-layer ocean models in which a flux adjustment is often made to compensate for the omission of heat advection by ocean currents and for other deficiencies. Confidence in these models is enhanced by their ability to simulate aspects of the climate of the last ice age.

11. Atmospheric models have been coupled with multi-layer oceanic general circulation models, in which an adjustment is sometimes made to the surface heat and salinity fluxes. Although so far such models are of relatively coarse resolution, the large-scale structure of the ocean and atmosphere can be simulated with some skill.

12. There is an urgent need to acquire further data for climate model validation on both global and regional scales, and to perform validation against data sets produced in the course of operational weather forecasting.

4.1 INTRODUCTION

Climate models, and those based on general circulation models (GCMs) in particular, are mathematical formulations of atmosphere, ocean and land surface processes that are based on classical physical principles. They represent a unique and potentially powerful tool for the study of the climatic changes that may result from increased concentrations of CO_2 and other greenhouse gases in the atmosphere. Such models are the only available means to consider simultaneously the wide range of interacting physical processes that characterize the climate system, and their objective numerical solution provides an opportunity to examine the nature of both past and possible future climates under a variety of conditions. In order to evaluate such model estimates properly, however, it is necessary to validate the simulations against the observed climate, and thereby to identify their systematic errors, particularly errors common to several models. These errors or model biases must be taken into account in evaluating the estimates of future climate changes. Additional caution arises from the GCMs' relatively crude treatment of the ocean and their neglect of other potentially-important elements of the climate system, such as the upper atmosphere and atmospheric chemical and surface biological processes. While it is to be expected that GCMs will gradually improve, there will always be a range of uncertainty associated with their results; the scientific challenge to climate modelling is to make these uncertainties as small as possible.

The purpose of this section is to present an authoritative overview of the accuracy of current GCM-based climate models, although space limitations have not allowed consideration of all climate variables. We have also not considered the simpler climate models since they do not allow assessment of regional climate changes and have to be calibrated using the more complex models. We begin this task by evaluating the models' ability to reproduce selected features of the observed mean climate and the average seasonal climate variations, after which we consider their ability to simulate climate anomalies and extreme events. We also consider other aspects which increase our overall confidence in models, such as the performance of atmospheric models in operational weather prediction, and of atmospheric and coupled atmosphere-ocean models in the simulation of low-frequency variability and palaeo-climates.

4.1.1 Model Overview

The models that have been used for climate change experiments have been described in Section 3.5 and are discussed further below. Because of limitations in computing power, the higher resolution atmospheric models have so far been used only in conjunction with the simple mixed-layer ocean models, as in the "equilibrium" experiments in Section 5. Many of these models give results similar to those from experiments with prescribed sea surface temperature (SST) and sea-ice, because these variables are constrained to be near the observed values by use of prescribed advective heat fluxes. This assessment places an upper bound on the expected performance of models with more complete representations of the ocean, whose results are discussed in Section 6.

Although the atmospheric models that have been developed over the past several decades have many differences in their formulations, and especially in their physical parameterizations, they necessarily have a strong family resemblance. It can therefore be understood that, though they all generate simulations which are to a substantial degree realistic, at the same time they display a number of systematic errors in common, such as excessively low temperatures in the polar lower stratosphere and excessively low levels of eddy kinetic energy in the upper troposphere. On a regional basis, atmospheric GCMs display a wide variety of errors, some of which are related to the parameterizations of sub-grid-scale processes, and some to the models' limited resolution. Recent numerical experimentation with several models has revealed a marked sensitivity of simulated climate (and climate change) to the treatment of clouds, while significant sensitivity to the parameterization of convection, soil moisture and frictional dissipation has also been demonstrated. These model errors and sensitivities, and our current uncertainty over how best to represent the processes involved, require a serious consideration of the extent to which we can have confidence in the performance of models on different scales.

As anticipated above, the first part of this validation of climate models focuses on those models that have been used for equilibrium experiments with increased CO_2, as discussed in Section 5. Many of the models considered (Table 3.2(a) - see caption to Figure 4.1(a) for models' reference numbers) are of relatively low resolution, since until recently it is only such models that could be integrated for the long periods required to obtain a clear signal. To represent the seasonal cycle realistically, and to estimate equilibrium climate change, the ocean must be represented in such a way that it can respond to seasonal forcing with an appropriate amplitude. All the models in Table 3.2(a) have been run with a coupled mixed-layer or slab ocean. The period used for validation of these models is typically about ten years; this is believed to be sufficient to define the mean and standard deviation of atmospheric variables for validation purposes. Additionally, some of these models (versions listed in Table 3.2(b)) have been coupled to a dynamic model of the deeper ocean (see Section 6).

4.1.2 Methods and Problems of Model Validation

The questions we need to answer in this assessment concern the suitability of individual models for estimating climate change. The response of modelled climate to a perturbation of the radiative or other forcing has been shown to depend on the control climate. How serious can a model's errors be for its response to a perturbation still to be credible? Mitchell et al. (1987) pointed out that it may be possible to allow for some discrepancies between simulated and observed climates, provided the patterns are sufficiently alike that relevant physical mechanisms can be identified. For example, they found that with increased CO_2 and increased surface temperatures, precipitation tended to increase where it was already heavy, so that if a rainbelt was differently located in two models, the response patterns could differ but still have the same implications for the real climate change. However, although a perfect simulation may not be required, it is clear that the better the simulation the more reliable the conclusions concerning climate change that may be made. Also, since, as discussed in Section 3, the magnitude of the response depends on the feedbacks, these feedbacks should be realistically represented in the models. Thus, in selecting model variables with which to validate atmospheric components of the climate models, we considered the following:

a) Variables that are important for the description of the atmospheric circulation and which therefore ought to be realistically portrayed in the control simulations if the modelled changes are to be given credence. Examples include sea-level pressure and atmospheric wind and temperature, and their variability as portrayed by the kinetic energy of eddies.

b) Variables that are critical in defining climate changes generated by greenhouse gases. These data also need to be realistic in control simulations for the present climate if the model predictions are to be credible. Examples include surface air temperature, precipitation, and soil moisture, along with their day-to-day and year-to-year variability.

c) Variables that are important for climate feedbacks. If they are poorly simulated, we cannot expect changes in global and regional climate to be accurately estimated. Examples are snow cover, sea-ice, and clouds and their radiative effects.

In general, the assessments made can only be relative in character. It is not usually possible to specify a critical value that errors must not exceed. Thus, temperature changes may be realistically simulated even when the modelled temperatures are in error by several degrees; for example, the error may be due to excessive night-time cooling of air near the surface, which may have little effect on other aspects of the simulation. On the other hand, it may be obvious that an error is too serious for much

credence to be given to changes in a particular region - for example, a prediction of changes in temperature in a coastal region with observed winds off the ocean, if the simulated winds blow off the land. To allow the reader to make such assessments, maps are shown for a number of key variables and models for which detailed changes are depicted in Section 5.

The validation of climate models requires, of course, the availability of appropriate observed data. For some variables of interest, observed data are unavailable, or are available for only certain regions of the world. In addition to traditional climatological data, useful compilations of a number of variables simulated by climate models have been provided by Schutz and Gates (1971, 1972), Oort (1983) and Levitus (1982), and more recent compilations of atmospheric statistics have been made using analyses from operational weather prediction (see, for example, Trenberth and Olson, 1988). Rather than attempting to provide a comprehensive summary of observed climatic data, we have used what appear to be the best available data in each case, even though the length and quality of the data are uneven. Satellite observing systems also provide important data sets with which to validate some aspects of climate models, and when fully incorporated in the data assimilation routines of operational models such data are expected to become an important new source of global data for model validation.

A wide range of statistical methods has been used to compare model simulations with observations (Livezey, 1985; Katz, 1988; Wigley and Santer, 1990; Santer and Wigley, 1990). No one method, however, is "ideal" in view of the generally small samples and high noise levels involved and the specific purposes of each validation. Other factors that can complicate the validation process include variations in the form in which variables are represented in different models; for example, soil moisture may be expressed as a fraction of soil capacity or as a depth, and snowcover may be portrayed by the fractional cover or by the equivalent mass of liquid water. Another problem is the inadequate representation of the distribution of some climatic variables obtainable with available observations, such as precipitation over the oceans and soil moisture.

Another method of validation which should be considered is internal validation where the accuracy of a particular model process or parameterization is tested by comparison with observations or with results of more detailed models of the process. This approach has only been applied to a very limited extent. The best example is the validation of radiative transfer calculations conducted under the auspices of the WCRP programme for Intercomparison of Radiation Codes in Climate Models (ICRCCM). This intercomparison established the relative accuracy of radiation codes for clear sky conditions against line-by-line calculations, and has led to improvements in several climate models. Similarly, the intercomparison of simulated cloud-radiative forcing with satellite observations from the Earth Radiation Budget Experiment (ERBE) should result in improvements in the representation of clouds in climate models.

In addition to validation of the present climate, it is instructive to consider the evidence that climate models are capable of simulating climate changes. Important evidence comes from atmospheric models when used in other than climate simulations, since the ocean surface temperature is often constrained in similar ways. Relevant experiments in this regard for atmospheric models are those with variations of tropical sea-surface temperature (SST) in the El Niño context. Numerical weather prediction, which uses atmospheric models that are similar in many respects to those used in climate simulation, provides an additional source of validation. The simulation of climate since the last glacial maximum, for which we have some knowledge of the land ice, trace gas concentrations and ocean surface

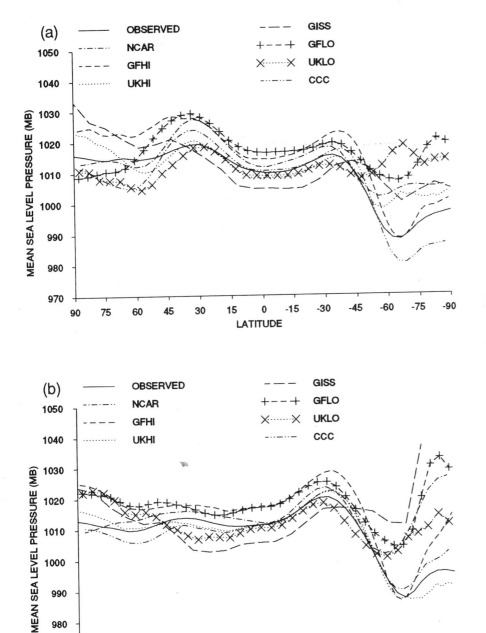

Figure 4.1: *Zonally averaged sea-level pressure (hPa) for observed (Schutz and Gates, 1971, 1972) and models: (a) December-January-February, (b) June-July-August. Model reference numbers (Table 3.2 (a)) are CCC (20), GFHI (21), GFLO (13), GISS (11), NCAR (7), UKHI (22), UKLO (15).*

temperatures, also provides a useful test of climate models. Finally, for validation of the experiments on transient climate change discussed in Section 6, it is important to consider the validation of ocean and coupled ocean-atmosphere models.

4.2 Simulation of the Atmospheric Circulation

In this section we consider a number of basic atmospheric variables for which validation data are readily available, and whose satisfactory simulation is a prerequisite for confidence in the models' ability to portray climate change.

4.2.1 Sea-Level Pressure

The sea-level pressure pattern provides a useful characterization of the atmospheric circulation near the surface and is closely related to many aspects of climate. A simple but revealing measure of the pressure pattern is the north-south profile of the zonal average (average around a latitude circle) (Figure 4.1). In both solstitial seasons the structure is rather similar, with a deep Antarctic trough, subtropical ridges with a near-equatorial trough between, and a rather weak and asymmetric pattern in northern middle and high latitudes. The models approximate the observed pattern with varying degrees of success; all simulate to some extent the subtropical ridges and Antarctic trough. The ridges are in some cases displaced poleward and there is a considerable range in their strength, particularly in the NH (Northern Hemisphere). In the lower resolution models, the Antarctic trough is generally too weak and sometimes poorly located.

The dependence of the simulation of the Antarctic trough on resolution, evident here for the GFDL and UKMO models, has been found in several previous studies (Manabe et al., 1978; Hansen et al., 1983; Dyson, 1985). While Manabe et al. found it to be marked in the GFDL spectral models only in July, the similar GFDL model considered here shows it clearly in January also. The earlier GFDL result is consistent with experiments with the CCC model by Boer and Lazare (1988), showing only a slight deepening of an already deep Antarctic trough as resolution was increased. The important result in the present context is that the more realistic models used in CO_2 experiments are those with higher resolution (CCC, GFHI and UKHI).

The NH winter subtropical ridge is too strong in most models, while the decrease in pressure from this ridge to the mid-latitude trough is generally excessive; this is associated with excessively deep oceanic lows in some models and with spurious westerlies over the Rockies in others. The westerlies in high resolution models are very sensitive to the representation of the drag associated with gravity waves induced by mountains; without it, a ring of strong westerlies extends around middle latitudes during northern winter (Slingo and Pearson, 1987).

The models simulate the seasonal reversal from northern summer to winter (for example Figure 4.2). To some degree this reflects the dominance of thermal forcing of the pressure pattern, and the fact that most of the models have ocean temperatures which are kept close to climatology. Summer temperatures over land are strongly affected by the availability of soil moisture (see Section 4.3.3); the absence of evaporative cooling generates higher surface temperature and lower pressure as in the "dry land" experiment of Shukla and Mintz (1982). This effect is evident in some of the models' simulations of pressure over land in the summer. A serious shortcoming of the lower resolution models in the northern summer is the tendency to develop too strong a ridge between the Azores high and the Arctic; this error shows up as the absence of a trough near 60°N in Figure 4.1(b), and was also found in previous assessments (e.g., Manabe et al., 1978; Dyson, 1985).

The variability of the pressure pattern can usefully be separated into the daily variance within a month and the interannual variability of monthly means. Both are simulated with some skill by models, especially the daily variance (e.g., Figure 4.3); in particular, the variability maxima over the eastern Atlantic and northeast Pacific are well simulated, and in the Southern Hemisphere high values are simulated near 60°S as observed. These results indicate that the models can successfully simulate the major storm tracks in middle latitudes. On smaller scales, however, there are regionally important errors, associated for example with the displacements of the variability maxima in the Northern Hemisphere.

In summary, the recent higher resolution models are capable of generally realistic simulations of the time averaged sea-level pressure and of the temporal pressure variability.

4.2.2 Temperature

While models successfully simulate the major features of the observed temperature structure of the atmosphere, all models contain systematic errors such as those shown in the simulated zonally averaged temperatures in Figure 4.4. Of errors common to many models, the most notable is the general coldness of the simulated atmosphere; simulated temperatures in the polar upper troposphere and lower stratosphere are too low in summer by more than 10°C while the lower troposphere in tropical and middle latitudes is too cold in both summer and winter. The latter error may in some cases be alleviated by increasing the horizontal resolution (Boer and Lazare, 1988; Ingram, personal communication). The existence of such common deficiencies, despite the considerable differences in the models' resolution, numerical treatments and physical parameterizations, implies that all models may be misrepresenting (or indeed omitting) some physical mechanisms. In contrast, in some regions of the

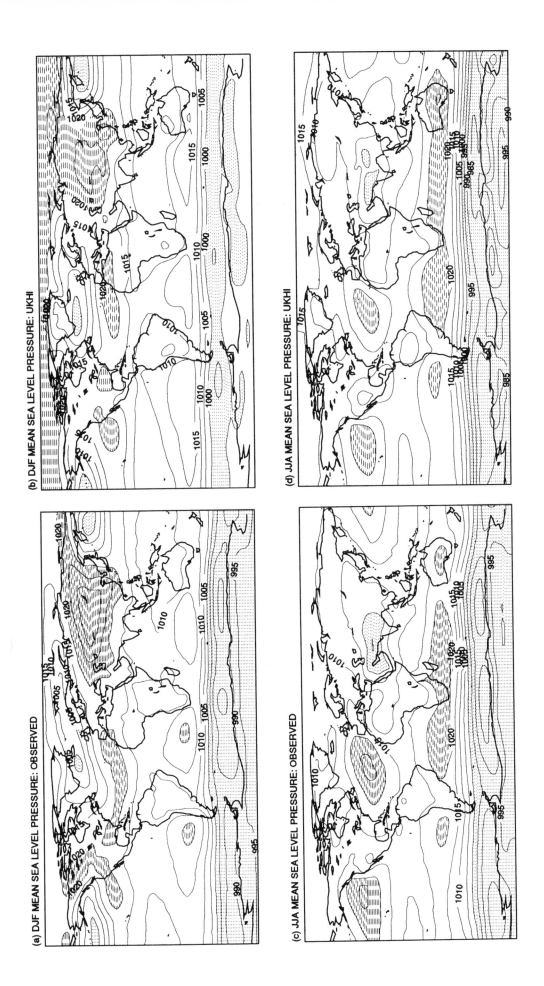

Figure 4.2: Sea-level pressure (hPa) for December-January-February (a,b) and June-July-August (c,d) for: (a,c) Observed (Schutz and Gates 1971, 1972) and (b,d) the UKHI model (No. 22, Table 3.2(a))

(a)

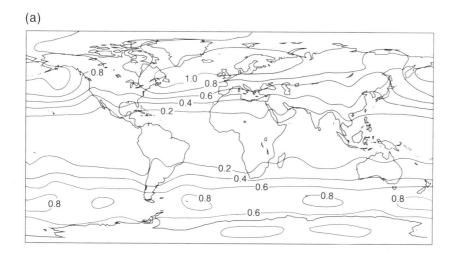

(b)

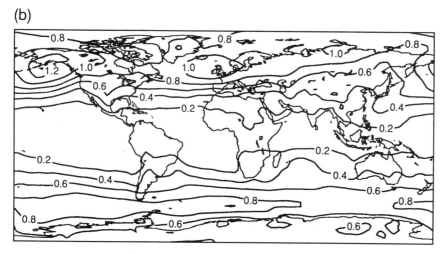

Figure 4.3: Daily standard deviation of 1000 hPa height (hm) for December-January-February for: (a) Observed (Trenberth and Olson, 1988), (b) For CCC model (No. 20 Table 3.2(a)).

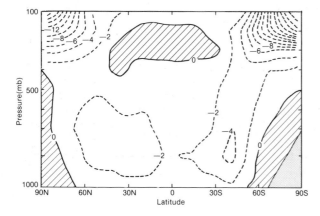

Figure 4.4: Zonally averaged temperature errors (°C) for an early version of the CCC model (not shown in Table 3.2(a)) for December-January-February. Contours every 2°C with negative errors dashed.

atmosphere, the simulated temperatures do not have a consistent bias and are warmer than observed in some models and cooler than observed in others.

In summary, the major features of the observed zonally averaged temperature structure are successfully simulated by modern general circulation models. There are, however, characteristic errors in specific regions, notably the polar upper troposphere and lower stratosphere in summer and much of the lower troposphere where temperatures are colder than observed in all models.

4.2.3 Zonal Wind

The winds are closely linked to the pressure and temperature distributions. The general poleward decrease of temperature in most of the troposphere leads to westerly flow, at least in the zonal mean (Figure 4.5). The major features in both solstitial seasons are the closed-off jets in each hemisphere and the easterlies in the tropics, especially near the surface and in the lower stratosphere. While most of the models represent these features to some extent, there

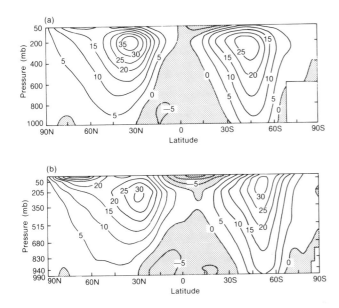

Figure 4.5: Zonally averaged zonal wind (ms⁻¹) for December-January-February for: (a) Observed (Arpe, personal communication), based on ECMWF analyses; (b) GFHI model (No. 21 Table 3.2(a)).

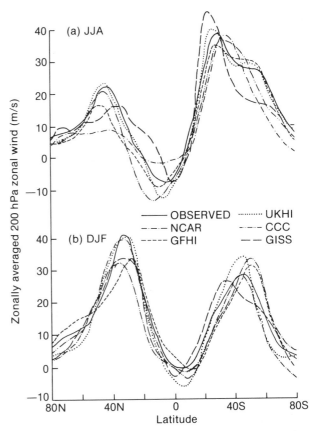

Figure 4.6: Zonally averaged 200 hPa zonal wind (ms⁻¹) for various models and as observed (Trenberth, personal communication) for: (a) June-July-August, (b) December-January-February.

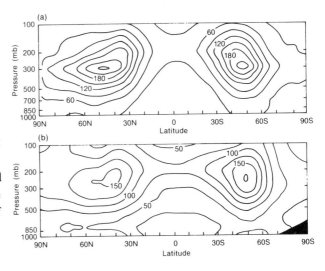

Figure 4.7: Zonally averaged transient eddy kinetic energy (m²s⁻²) for December-January-February for (a) Observed (Trenberth and Olson, 1988) based on ECMWF analyses; (b) T30 version of the CCC model (not listed in Table 3.2(a)) (from Boer and Lazare, 1988).

are a number of errors in common. The most prominent of these are excessive westerlies above the summer jet, especially in the SH, and above and poleward of the winter jet. In consequence, some models fail to close the winter subtropical jet (i.e., separate it from the stratospheric polar night jet). This problem can also be alleviated by improving the vertical resolution (Cariolle et al., (1990)). In the NH the closure error is smallest in the more recent simulations using higher resolution models (e.g., Figure 4.5(b)), although this improvement may owe more to the inclusion of gravity wave drag than to the improvement in resolution. Dyson (1985) found that improved resolution helped to intensify the NH summer jet; this result is also evident here, as shown by Figure 4.6, which allows comparison of the observed and modelled meridional profiles of wind at 200 hPa. At this level the June-July-August jet in the SH is realistically simulated in most models, but again most fail to close it.

In summary, although most models represent the broad features of the observed zonal wind structure, only the more recent models with the more realistic simulations of sea-level pressure succeed in closing the winter jets and in providing a sufficiently strong NH summer jet.

4.2.4 Eddy Kinetic Energy

A realistic climate model should simulate correctly not only the zonally-averaged atmospheric variables but also their space-time variability. Nearly half the atmospheric kinetic energy resides in "eddies", by which meteorologists understand deviations of the flow from zonally-averaged

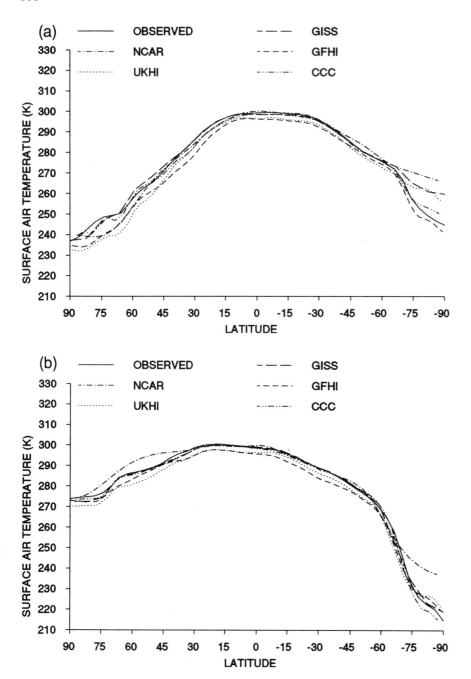

Figure 4.8: Zonally averaged surface air temperatures (K) for various models and as observed (Schutz and Gates, 1971, 1972) for: (a) December-January-February, (b) June-July-August

conditions, as caused, for example, by cyclones and meanders of the jet stream. A measure of the intensity of the eddies is the "eddy kinetic energy" (EKE), which is observed to be largest in the extratropical latitudes in the upper troposphere.

A persistent error of atmospheric general circulation models is their tendency to underestimate the EKE, particularly the "transient" part representing variations about the time-averaged flow. When integrated from real initial conditions, the models tend to lose EKE in the course of the integration until they reach their own

characteristic climate. For example, Figure 4.7 compares the transient EKE for December-January-February in a version of the CCC model (Boer and Lazare, 1988) with the corresponding observed EKE (Trenberth, personal communication), and shows a significant underestimate of the EKE maxima in middle latitudes. There is a suggestion of overestimation in the tropics. Similar results have been obtained for forecast models (WGNE, 1988; see also Section 4.7). Since recent experiments have indicated that resolution is not the basic reason for this systematic EKE error (Boer and Lazare, 1988; Tibaldi et al., 1989), it is

(a) DJF GISS - OBSERVED SURFACE AIR TEMPERATURE

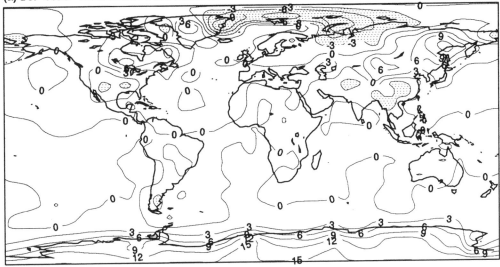

(b) JJA GISS - OBSERVED SURFACE AIR TEMPERATURE

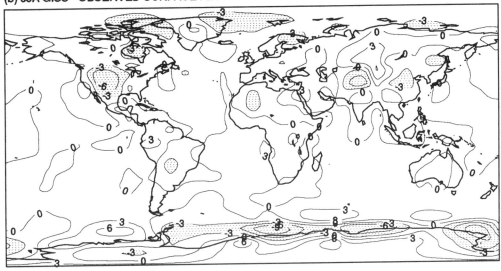

Figure 4.9: Surface air temperature errors for December-January-February (a) and June-July-August (b) for GISS model (No. 11, Table 3.2(a)). Errors calculated relative to Schutz and Gates (1971, 1972). Areas less than -3°C shaded.

probably caused by the models' treatment of physical processes.

In summary, models suffer from a deficiency of transient eddy kinetic energy, an error which appears most marked in the upper extratropical troposphere and which may be reversed in the tropics.

4.3 Simulation of Other Key Climate Variables

In this section we assess the global distribution of variables involved in the energy and hydrological balances whose satisfactory simulation is important for determining the climate's response to increased greenhouse gases.

4.3.1 Surface Air Temperature

The temperature of the air near the surface is an important climatic parameter. The global pattern is dominated by large pole to equator gradients which models simulate well (Figure 4.8) though it should be recalled that in most of the models shown, the ocean surface temperatures are maintained near the correct level by 'flux adjustment' techniques. Because of the dominance of the pole-to-equator gradient, the maps shown (Figure 4.9) are of departures of temperature from the observed. Validation of this quantity from atmospheric GCMs is complicated by the models' low vertical resolution and by the relatively large diurnal variation of temperature near the surface.

A principal conclusion from the comparison of simulated with observed near-surface air temperature is that while each model displays systematic errors, there are few errors common to all models. One characteristic error is that temperatures over eastern Asia are too cold in winter (e.g., Figure 4.9(a) over southeast Asia). Another, common to most models, is that temperatures are too high over the Antarctic ice sheet (Figure 4.8); in winter at least, this can be attributed to the models' difficulty in resolving the shallow cold surface layer. In summer, errors in the simulated ground wetness appear to be responsible for many of the temperature errors over the continents (compare Figures 4.8 and 4.12), this error being less marked in models with more complete representations of the land surface (e.g., Figure 4.9). (See Section 4.4 for a detailed assessment for five selected areas).

The variability of surface air temperature can, like sea-level pressure, be considered in terms of the day-to-day variations within a month or season, and the interannual variations of monthly or seasonal means. Detailed validations of these quantities for selected regions in North America in relatively low resolution models have been made by Rind et al. (1989) and Mearns et al., (199). Both found variances to be too high on a daily time-scale, while for interannual timescales, the results differed between the models. An earlier study by Reed (1986) with a version of the UKMO model (not in Table 3.2(a)) also revealed too high variability on a daily time-scale in eastern England. On the other hand, for the models reviewed in this assessment for which data were available, the daily variance appeared to be capable of realistic simulation though it tended to be deficient over northern middle latitudes, especially in summer.

In summary, the patterns of simulated surface air temperature are generally similar to the observed. Errors common to most models include excessively cold air over eastern Asia in winter and too warm conditions over Antarctica. Errors over the continents in summer are often associated with errors in ground wetness.

4.3.2 Precipitation

A realistic simulation of precipitation is essential for many if not all studies of the impact of climate change. A number of estimates of the distributions of precipitation from observations are available; some of these are derived from station observations, which are generally considered adequate over land; one is derived from satellite measurements of outgoing longwave radiation (Arkin and Meisner, 1987; Arkin and Ardanuy, 1989), while another is from ship observations of current weather coupled with estimated equivalent rainfall rates (e.g., Dorman and Bourke, 1979). The differences among these analyses are not insignificant, but are mostly smaller than the differences between the analyses and model simulations.

While all models simulate the broad features of the observed precipitation pattern, with useful regional detail in some regions, (see, for example the zonally averaged patterns in Figure 4.10 and the patterns for the higher resolution models in Figure 4.11), significant errors are present, such as the generally inadequate simulation of the southeast Asia summer monsoon rainfall (see also Section 4.5.3), the zonal rainfall gradient across the tropical Pacific, and the southern summer rains in the Zaire basin. These errors reduce the correlation between observed and modelled patterns over land to about 0.75 (model 22, Table 3.2(a)). A similar level of skill is evident in the assessment in Section 4.4.2. Some of the earlier models underestimate the dryness of the subtropics (Figure 4.10), while several models are much too wet in high latitudes in winter. Models that do not use a flux correction to ensure an approximately correct SST fail to simulate the eastern and central equatorial Pacific dry zone (not shown). There are also large differences between the recent model simulations of the intensity of the tropical oceanic rainbelts though the (inevitable) uncertainties in the observed ocean precipitation can make it unclear which models are nearer reality.

Rind et al. (1989) and Mearns et al. (1990) have found that over the USA interannual variability of simulated precipitation in the GISS and NCAR models is generally excessive in both summer and winter, while daily variability is not seriously biased in either season, at least relative to the mean precipitation. The NCAR study also revealed considerable sensitivity of the daily precipitation to the model formulation, particularly to aspects of the parameterization of evaporation over land. Analyses of the UKMO model (Reed, 1986; Wilson and Mitchell, 1987) over western Europe showed that there were too many rain days, although occurrences of heavy rain were underpredicted.

In summary, current atmospheric models are capable of realistic simulations of the broadscale precipitation pattern provided the ocean surface temperatures are accurately represented. All the models assessed, however, have some important regional precipitation errors.

4.3.3 Soil Moisture

Soil moisture is a climatic variable that has a significant impact on ecosystems and agriculture. Some model experiments on the impact of increasing greenhouse gases on climate have shown large decreases in soil moisture over land in summer; this can provide a positive feedback with higher surface temperatures and decreased cloud cover. Since there is no global coverage of observations of soil moisture its validation is difficult. Estimates of soil moisture have been made by Mintz and Serafini (1989) from precipitation data and estimates of evaporation but comparison with a recent analysis of observations over the

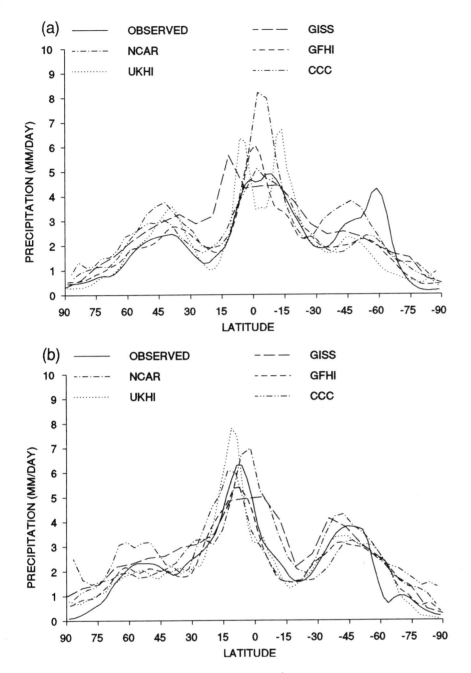

Figure 4.10: Zonally averaged precipitation (mm day^{-1}) for various models and as observed (Jaeger, 1976) for: (a) December-January-February, (b) June-July-August.

western USSR (Vinnikov and Yeserkepova, 1990) suggests the estimates are too low in high latitudes in summer (Table 4.1).

Compared with the Mintz-Serafini observational estimates, model simulations generally show a greater seasonal variation, especially in the tropics. In general the simulations' errors in soil moisture resemble those in precipitation, and vary considerably among models, as is evident from the zonally averaged June-August data (Figure 4.12). The simulations over Eurasia and northern Africa are quite close to the "observed" in most models. However, the zonally averaged data and also comparison with the USSR data (e.g., Table 4.1), suggests that some

models become too dry in summer in middle latitudes. All models have difficulty with the extent of the aridity over Australia, especially in the (southern) summer.

In summary, the limited soil moisture data available show that the simulated middle latitude summer and winter distributions qualitatively reflect most of the observed large-scale characteristics. However, there are large differences in the models' simulations of soil moisture, as expected from the precipitation simulations, and it should be emphasized that the representation (and validation) of soil moisture in current climate models is still relatively crude.

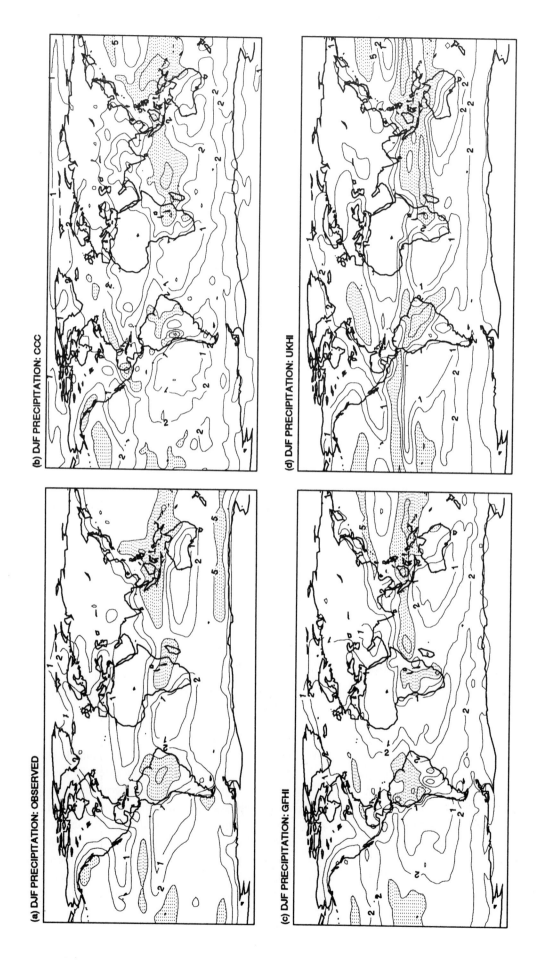

Figure 4.11: Precipitation (mm day^{-1}) for December-January-February (a, b, c, d) and June-July-August (e, f, g, h); observed (Jaeger, 1976) (a, e) and CCC model (No. 20) (b, f, GFHI model (No. 21) (c, g) and UKHI model (No. 22) (d, h) (see Table 3.2(a) for model reference numbers).

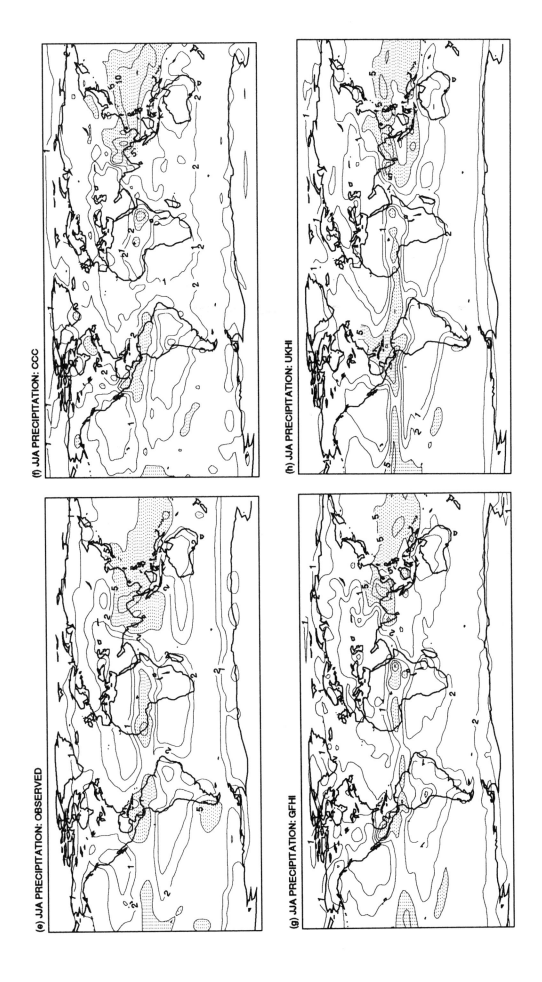

Figure 4.11   (continued)

Table 4.1 *Averages of available soil moisture as percentage of capacity for 30-60°E as observed (Vinnikov and Yeserkepova, 1990), estimated from observations (Mintz and Serafini, 1989), and for three models (CCC, GFHI and UKHI) for December-February (DJF) and June-August (JJA).*

Model or Data	Vinnikov & Yeserkepova		Mintz & Serafini		CCC		GFHI		UKHI	
Seasons	DJF	JJA	DJF	JJA	DJF	JJA	DJF	JJA	DJF	JJA
62°N	100	100	100	55	>90	88	56	54	93	97
58°N	100	98	97	41	>90	85	32	22	90	93
54°N	85	57	83	27	78	77	25	13	81	67
50°N	63	24	57	12	35	22	28	10	70	30

NOTE: The capacities vary between the data sets. For Mintz and Serafini and GFHI, they are 15cm. For UKHI, there is no fixed capacity, but the runoff parameterization leads to an effective capacity close to 15cm, as used for the above results. CCC and Vinnikov and Yeserkepova have larger capacities; their data were provided as actual fractions of capacity.

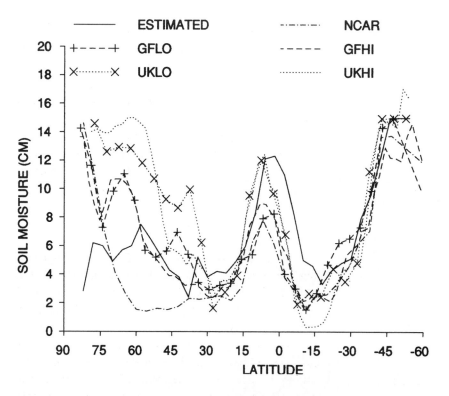

Figure 4.12: Zonally averaged soil moisture (cm) for land points as estimated by Mintz and Serafini (1989) for July and as modelled for June-July-August.

4.3.4 Snow Cover

Snow is an important climate element because of its high reflectivity for solar radiation and because of its possible involvement in a feedback with temperature. The correct simulation of snow extent is thus critical for accurate prediction of the response to increasing greenhouse gases.

Snow cover observations used for model validation were a 15 year satellite-derived data set of the frequency of cover (Matson et al., 1986) and an earlier snow depth data set (Arctic Construction and Frost Effects Laboratory, 1954). These observations document the expected maximum snow cover in the Northern Hemisphere winter, with Southern Hemisphere snow confined mainly to Antarctica.

Detailed assessments of the simulations, especially for seasons other than winter, are hindered by the different forms of the model data (mostly seasonal mean liquid water content) and the observed data (either frequency of cover or maps of depth at ends of months). While all models capture the gross features of the seasonal cycle of snow cover, some models exhibit large errors. Otherwise, except over eastern Asia where snow extents are mostly excessive (consistent with the low simulated temperatures (e.g., Figure 4.9)), the models' average winter snow depths can be near those observed; this is illustrated for North America in Figure 4.13, which compares the observed 5cm snowdepth contour at the end of January with the modelled 1cm liquid water equivalent contour averaged for December to February. Comparable results are obtained over Europe and western Asia.

In summary, several models achieve a broadly realistic simulation of snow cover. Provided snow albedos are realistic, the simulated snow extent should thus not distort simulated global radiative feedbacks. However, there are significant errors in the snow cover on regional scales in all models.

4.3.5 Sea-Ice
An accurate simulation of sea-ice is important for a model's ability to simulate climate change by virtue of its profound effect on the surface heat flux and radiative feedbacks in high latitudes. An attempt is made to ensure a good simulation of sea-ice extent by including a prescribed ocean heat flux in many current models; this flux is assumed to be unchanged when the climate is perturbed. Without it, models tend to simulate excessive temperature gradients between pole and equator, particularly in the Northern Hemisphere winter, with a consequent excess of sea-ice. Sea-ice in the Arctic Ocean is constrained to follow the coast in winter, but in summer and autumn the ice separates from the coast in many places; this behaviour is simulated by some models. Experiments with relatively sophisticated dynamic-thermodynamic sea-ice models (Hibler, 1979; Hibler and Ackley, 1983; Owens and Lemke, 1989) indicate that a realistic simulation of sea-ice variations may require the inclusion of dynamic effects, although the optimal representation for climate applications has not yet been determined. Although the thickness of sea-ice is not readily validated due to the inadequacy of observational data, models display substantial differences in simulated sea-ice thickness.

In summary, considerable improvement in the representation of sea-ice is necessary before models can be expected to simulate satisfactorily high-latitude climate changes.

4.3.6 Clouds and Radiation
The global distribution of clouds has been analysed from satellite data over recent years in ISCCP, and the diagnosed cloud cover can be compared with modelled cloud. For example, Li and Letreut (1989) showed that the patterns of cloud amounts in a 10-day forecast were similar to those diagnosed in ISCCP over Africa in July, but were deficient over the southeast Atlantic. However, the definition of cloud may differ from model to model and between model and observations so, as discussed by Li and Letreut, it is often easier and more satisfactory to compare measurable radiative quantities. A useful indication of cloud cover can be obtained from top-of-the-atmosphere satellite measurements of the outgoing longwave radiation (OLR) and planetary albedo, as provided by Nimbus 7 Earth Radiation budget measurements (Hartmann et al., 1986; Ardanuy et al., 1989). These quantities describe the exchange of energy between the whole climate system (ocean, ground, ice, and atmosphere) and outer space, and in that sense constitute the net forcing of the climate. At the same time

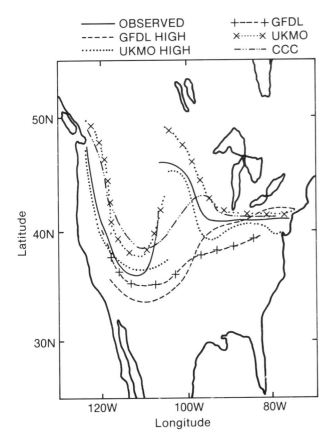

Figure 4.13: Winter snow cover over North America as defined for various models by the minimum latitude at which the December-January-February simulated snow-cover had a 1 cm liquid water equivalent contour, and, as observed, by the minimum latitude of the end-of-January average 2 inch (5 cm) depth contour. (GFDL HIGH = GFHI; UKMO HIGH = UKHI; GFDL = GFLO; UKMO = UKLO).

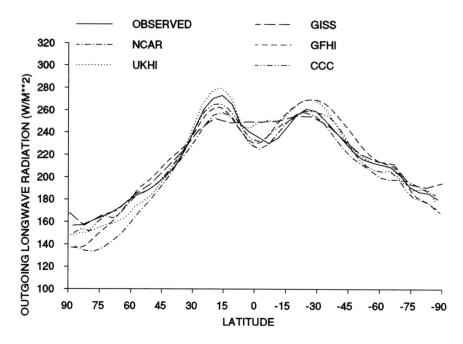

Figure 4.14: Zonally averaged OLR (Wm^{-2}) for December-January-February (models) and for January (observed, Nimbus-7 NFOV).

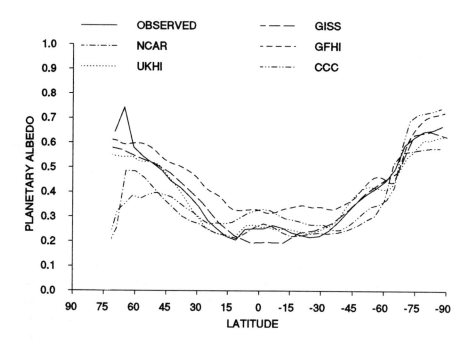

Figure 4.15: Zonally averaged planetary albedo for December-January-February (models) and for January (observed, Nimbus-7 NFOV).

they depend critically on many mechanisms that are internal to the climate system, and in particular the hydrological cycle. As a result, the ability of models to simulate the OLR and planetary albedo properly depends not only on the algorithms used to compute the radiative transfers within the atmosphere, but also on the simulated snow cover, surface temperature and clouds.

The zonally averaged OLR (Figure 4.14) is dominated by maxima in low latitudes and minima in high latitudes.

Clouds generate minimum OLR near the tropical convergence zones, which are also evident as maxima in the albedo (Figure 4.15). The increase in albedo toward high latitudes, on the other hand, can be associated with clouds, snow and ice, or changes in land surface type and solar zenith angle.

Because of the important radiative effects of clouds and their association with precipitation, observed tropical and subtropical OLR extremes are highly correlated with those of precipitation. In general, in tropical regions the models' OLR values are realistic, and models successfully simulate the correlation of precipitation with planetary albedo. At higher latitudes in winter, there is considerable disparity among models in the simulated values of planetary albedo, evidently due to the differing simulations of snow cover and/or clouds and the different specifications of albedo for particular surfaces or cloud types. In general, the models simulate polar OLR minima which are below the observed values, probably because of the temperature errors there (Figure 4.4, 4.8). Apart from these high latitude regions, the zonally averaged OLR is generally within 20 Wm^{-2} of the observed. The mean error magnitudes for individual models are as low as 5 Wm^{-2} (Model 20, Table 3.2(a)) or 2% of the climatological values. For absorbed solar radiation, errors are mostly below 20 Wm^{-2} with albedo errors less than 0.1 except in northern middle and high latitudes.

In summary, this assessment has shown that although the latitudinal variation of top of the atmosphere radiative parameters can be well simulated, there are some discrepancies, particularly in the albedo in middle and high latitudes due to the sensitivity of the parameterization schemes. Most models underestimate the OLR in high latitudes.

4.4 Simulation of the Regional Seasonal Cycle

The seasonal cycle constitutes the largest regularly observed change of the atmosphere-ocean system, and provides an important opportunity for model validation. In general, all GCMs simulate a recognizable average seasonal variation of the principal climate variables, as measured by the phase and amplitude of the annual harmonic. The seasonal variation of the amplitudes of the transient and stationary waves can also be simulated with reasonable fidelity (e.g., the GLA GCM; see Straus and Shukla, 1988 and Section 4.2.1).

A more detailed summary of GCMs' simulations of the seasonal cycle and a comparison with observational estimates for five selected regions is given in Table 4.2 in terms of the surface air temperature and precipitation. In this statistical summary, each model's grid-point data over land areas within the selected region have been averaged without interpolation or area-weighting; the areas are bounded as follows: Region 1 (35-50N, 85-105W), Region 2 (5-30N, 70-105E), Region 3 (10-20N, 20W-40E), Region 4 (35-50N, 10W-45E), Region 5 (10-45S, 110-155E). Similar areas are used in analysing regional changes in Section 5.

4.4.1 Surface Air Temperature
The surface temperature data in Table 4.2 are for the bottom model layer except for the CCC model, for which

an estimate of 'screen' temperature at 2m was supplied. For the UKMO models, 00GMT data were adjusted to daily means using detailed data for selected points. The differences between the model simulations are generally much larger than those between the observed data sets, with the best agreement among the models' surface air temperature occurring over southeast Asia in summer, and the poorest agreement over the Sahel in winter. In general, the seasonal differences for each of the regions show that the models are, on average, capable of a good representation of the seasonal variation of surface air temperature.

The magnitudes of the average errors of the individual models lie in the range 2.6±0.8°C, with larger values in winter (3.1°C) than in summer (2.1°C). For the high resolution models, the average is 2.3°C. These figures may be compared with the mean seasonal variation of 15.5°C. There appears to be no surface air temperature bias common to all the models, although the models with higher resolution (of the eight assessed) show an average temperature below that observed. Average regional errors are generally small, with only southeast Asia in winter having a mean error (-2.6°C) of more than 1.5°C; the models' average estimates of the seasonal range are within 1°C of that observed for each region except southeast Asia.

In summary, climate models simulate the regional seasonal cycle of surface air temperature with an error of 2 to 3°C, though this error is in all cases a relatively small fraction of the seasonal temperature range itself.

4.4.2 Precipitation
Average values of the simulated and observed precipitation over the five regions are presented in Table 4.2. Most models succeed in identifying southeast Asia in summer as the wettest and the Sahel in winter as the driest seasonal precipitation regimes of those assessed; the region and season which gives the most difficulty appears to be southeast Asia in winter, where several of the models are much too wet. Indeed, all four northern winter validations reveal a preponderance of positive errors, and Australia also tends to be too wet. The mean magnitude of model error varies quite widely between models, from 0.5 to 1.2 mm day^{-1}, or from 20 to 50% of the observed; the three higher resolution models have the smallest mean errors. The relatively large differences among the models indicate the difficulty of accurately simulating precipitation in a specific region (even on a seasonal basis), and underscores the need for improved parameterization of precipitation mechanisms.

In summary, as for temperature, the range of model skill in simulating the seasonal precipitation is substantial, the mean errors being from 20% to 50% of the average precipitation. The models tend to overestimate precipitation in winter.

Table 4.2 *Regional unweighted averages of seasonal surface air temperature ($°C$, upper portion) and precipitation (mm day^{-1}, lower portion) as simulated in model control runs and as observed over five selected regions (see text). Here DJF is December-January-February and JJA is June-July-August (see Table 3.2(a) for model identification, where different from Figure 4.1).*

Model or Data	Region 1 Gt. Plains		Region 2 S E Asia		Region 3 Sahel		Region 4 S Europe		Region 5 Australia	
	DJF	JJA	DJF	JJA	DJF	JJA	DJF	JJA	DJF	JJA
CCC	-8.4	21.2	10.9	25.3	13.5	27.5	2.3	20.7	26.9	11.3
NCAR (#6)	-3.5	29.9	10.5	27.4	25.2	31.8	2.3	26.4	29.3	17.0
GFDL R15 (#8)	-5.7	25.9	14.1	27.3	25.9	31.7	2.0	26.7	31.9	16.1
GFHI	-7.3	23.7	9.0	25.5	18.3	26.0	-3.8	20.9	24.9	11.8
GISS	-1.2	19.5	14.7	25.4	21.3	28.6	7.5	22.9	26.5	14.3
OSU (#3)	-4.8	20.4	13.9	28.2	30.5	32.8	-1.0	20.2	30.7	22.4
UKLO	-1.7	19.5	17.7	25.8	26.0	26.9	3.7	20.1	25.2	16.3
UKHI	-11.4	20.2	13.1	25.2	21.1	28.5	-2.0	18.5	25.5	15.3
Oort	-6.3	20.8	16.2	25.9	22.7	28.8	1.5	20.8	27.3	15.3
Schutz	-7.7	22.1	15.0	25.6	22.1	28.2	0.3	21.9	27.6	14.4
CCC	1.4	3.8	2.0	8.6	0.1	2.9	2.4	1.7	2.0	0.8
NCAR (#6)	1.6	1.0	3.1	9.3	0.5	4.3	2.9	0.8	3.4	2.7
GFDL R15 (#8)	1.9	3.3	3.3	9.5	1.0	3.9	2.8	1.1	2.3	2.5
GFHI	1.3	2.1	1.6	8.6	0.5	4.5	1.6	1.4	2.9	1.0
GISS	2.0	3.1	5.9	6.0	0.9	3.2	3.0	2.0	2.6	1.6
OSU (#3)	1.4	1.7	0.8	1.4	0.2	1.5	2.2	1.1	1.3	0.9
UKLO	1.2	4.0	1.5	4.1	0.3	3.8	2.2	3.5	3.0	0.8
UKHI	1.0	2.7	0.5	4.3	0.0	2.8	2.8	1.5	4.1	1.0
Jaeger	1.1	2.5	0.6	9.0	0.1	4.4	2.1	2.0	2.4	0.8
Schutz	1.1	2.4	0.6	6.3	0.2	3.4	1.7	1.8	2.1	1.1

4.5 Simulation of Regional Climate Anomalies

4.5.1 *Response to El Niño SST Anomalies*

The El Niño Southern Oscillation (ENSO) phenomenon is now recognized to be an irregular oscillation of the coupled ocean/atmosphere system in the tropical Pacific, occurring approximately every three to five years. During the peak of an El Niño, sea surface temperatures (SSTs) in the eastern tropical Pacific can be several degrees warmer than the climatological mean. The convective rainfall maximum is shifted towards the warm SST anomalies, and the associated anomalous latent heat release forces changes in the large scale atmospheric circulation over the Pacific basin (and this in turn helps to maintain the anomalous SST). In addition, there is evidence that the extratropical

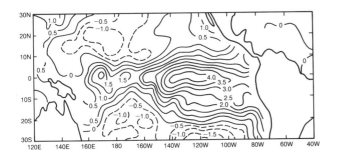

Figure 4.16: Sea-surface temperature anomaly (°C) for January 1983. Dashed contours are negative. (Fennessy and Shukla, 1988b)

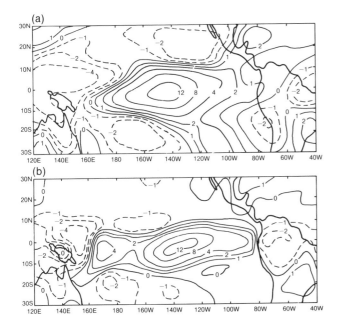

Figure 4.17: Anomalies of mean convective precipitation (mm day^{-1}) from mid-December 1982 to mid-February 1983: (a) Observed, (b) Simulated. Observed precipitation anomalies are calculated from OLR data; simulated anomalies are the average of three 60-day integrations with the Goddard Laboratory for Atmospheres GCM starting from 15, 16 and 17 December. (Fennessy and Shukla, 1988b)

jet streams are significantly displaced from their climatological positions during strong El Niño events, particularly over the North Pacific and North America (Fennessy and Shukla, 1988a).

Before the capability of coupled atmosphere/ocean GCMs to simulate El Niño and its teleconnections can be assessed, it is first necessary to assess whether the atmospheric component of these models can respond realistically to observed SST anomalies. There has been

considerable research on this problem in recent years as part of the WCRP TOGA programme (see for example, Nihoul, 1985; WMO, 1986, 1988) following Rowntree's (1972) initial studies. Figure 4.16 shows the SST anomaly in the Pacific for January 1983, with a maximum of 4°C in the eastern Pacific. An example of the observed and simulated precipitation anomalies from mid-December 1982 to mid-February 1983 is shown in Figure 4.17 (Fennessy and Shukla, 1988b), where a close correspondence across the central and eastern Pacific as well as over Indonesia and northern Australia can be seen. The observed and simulated anomalies in the zonal departure of the 200mb stream function for this same period are shown in Figure 4.18. The strong anticyclonic couplet straddling the equator in the central and eastern Pacific and the weaker couplet to the east in the tropics are well simulated in the model though their magnitudes are too weak. In the extratropics over the Pacific and North America, an eastward-shifted PNA-like pattern (Wallace and Gutzler, 1981), with cyclonic anomaly over the North Pacific and anticyclonic anomaly over Canada, is present in both the observed and simulated anomaly fields.

While the results above indicate that atmospheric GCMs can respond realistically to El Niño SST patterns, comparison of different GCMs' responses to identical SST anomalies underscores the importance of a realistic model control climate (Palmer and Mansfield, 1986). For example, the responses of models to an identical El Niño SST anomaly are significantly different in those regions where the models' control climates differ markedly. In such experiments there are also errors in the simulated anomalous surface heat flux and wind stress, which would give rise to quite different SST if used to force an ocean model.

Comparative extended-range forecast experiments using initial data from the El Niño winter of 1982/3 with both observed and climatological sea surface temperatures, showed that in the tropics the use of the observed SST led to consistent improvements in forecast skill compared with runs with climatological SST, while in the extratropics the improvements were more variable (WMO, 1986). These results suggest that, in the winter-time extratropics, the internal low-frequency variability of the atmosphere is as large as the signal from tropical forcing by El Niño, while in the tropics the influence of the El Niño forcing is dominant.

In summary, we may conclude that, given a satisfactory estimate of anomalous SST in the tropical Pacific, atmospheric GCMs can provide a realistic simulation of seasonal tropical atmospheric anomalies at least for intense El Niño episodes. This success serves to increase our confidence in these models and in their response to surface forcing. Problems associated with climate drift, particularly in relation to fluxes at the ocean-atmosphere interface,

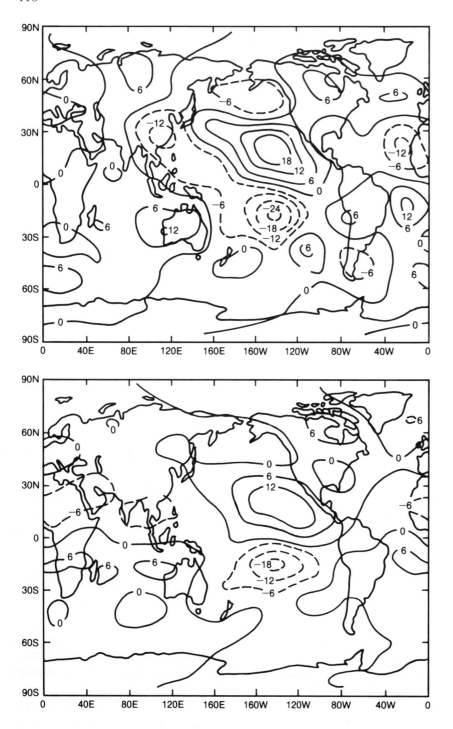

Figure 4.18: Mean departure from the zonal mean of anomalies of the 200 hPa stream-function ($10^6 m^2 s^{-1}$). Dashed contours are negative. Other details as in Figure 4.17. (Upper panel shows observed, lower panel shows simulated).

however, have so far inhibited consistently successful El Niño prediction with coupled ocean-atmosphere GCMs, although recent simulations have reproduced some aspects of observed El Niño phenomena (Sperber et al., 1987; Meehl, 1990).

4.5.2 Sahelian Drought

Over much of the 1970s and 1980s, sub-Saharan Africa experienced persistent drought, while in the 1950s rainfall was relatively plentiful. Climate models have been useful in determining the mechanisms responsible for the drought, although the successful prediction of seasonal rainfall anomalies requires further model development.

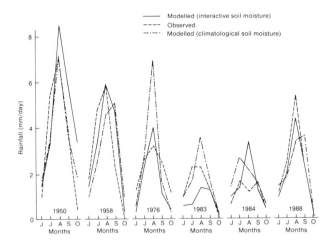

Figure 4.19: Simulated and observed Sahel rainfall in the six years for which simulations have been made. Simulations were made both with climatological and interactive soil moisture in the UKMO GCM (version not shown in Table 3.2(a)) (Folland, 1990 - personal communication).

A number of GCM studies have indicated that local changes in land surface conditions have an important influence on rainfall. For example, Charney et al. (1977), Sud and Fennessy (1982) and Laval and Picon (1986) have shown that an increase in land albedo over the Sahel can inhibit rainfall, while Rowntree and Sangster (1986) have shown that restriction of soil moisture storage (as well as albedo increases) can also have a substantial impact on rainfall in the Sahel. Other experiments indicate that the climate of the Sahel is sensitive to changes in local vegetation cover.

Further GCM experimentation has been described by Folland et al. (1989); the UKMO GCM has now been run from observed initial conditions in March and forced with the observed SST for seven months of each of 1950, 1958, 1976, 1983, 1984 and 1988 (Folland, personal communication). For each of these years, two experiments were performed: one with an interactive soil moisture parameterization, and one with fixed climatological soil moisture. Figure 4.19 shows a comparison of the observed and simulated rainfall over the Sahel. It is clear that the decadal time-scale trend in Sahel rainfall has been well captured. Moreover, the results suggest that soil moisture feedback is not the main cause of the large modelled differences in rain between the wet and dry years, though it does contribute to the skill of the simulations. Insofar as these decadal timescale fluctuations in large-scale SST are associated with internal variability of the ocean-atmosphere system, it would appear that Sahel drought is part of the natural variability of the climate, although the physical mechanisms whereby SST influences Sahel rain clearly involve remote dynamical processes (Palmer, 1986).

In summary, atmospheric model experiments exhibit an ability to simulate some of the observed interannual variations in Sahel rainfall, given the correct SST patterns.

4.5.3 Summer Monsoon

The monsoon, especially the Asian monsoon, displays significant seasonal variation and interannual variability, and the onset and retreat of the summer monsoons in Asia and Australia are associated with abrupt changes in the atmospheric general circulation (Yeh et al., 1959; McBride, 1987). An earlier or later monsoon onset, or a longer or shorter duration, usually causes flood or drought. Therefore, not only the accuracy of the seasonal monsoon precipitation but also the accuracy of the monsoon timing are important aspects of a model's ability to simulate regional climate anomalies.

In general, most atmospheric GCMs simulate the gross features of summer monsoon precipitation patterns though there are significant deficiencies (see Section 4.3.2); although this aspect of model performance has not been extensively examined, some models have been shown to simulate the monsoon onset and associated abrupt changes (Kitoh and Tokioka, 1987; Zeng et al., 1988). Part of the interannual variability of the summer monsoon has been found to be associated with anomalies in SST, both local (Kershaw, 1988) and remote. For example, there is an apparent correlation between the strength of the Indian monsoon and SST in the eastern tropical Pacific, in the sense that a poor monsoon is generally associated with a warm east Pacific (Gregory, 1989).

4.6 Simulation of Extreme Events

The occurrence of extreme events is an important aspect of climate, and is in some respects more important than the mean climate. Many relatively large-scale extreme events such as intense heat and cold, and prolonged wet and dry spells, can be diagnosed from climate model experiments (e.g., Mearns et al., 1984). The ability of climate models to simulate smaller-scale extreme events is not well established, and is examined here only in terms of tropical storm winds and small-scale severe storms.

Krishnamurti and Oosterhof (1989) made a five-day forecast of the Pacific typhoon Hope (July 1979) using a 12-layer model, with different horizontal resolutions. At a resolution of 75 km the model's forecast of strong winds was close to the observed maximum, while with a resolution of 400 km the maximum wind was less than half of the correct value and was located much too far from the centre of the storm. Since most climate models have been run with a resolution of 300 km or more, they do not adequately resolve major tropical storms and their associated severe winds. It may be noted, however, that by using appropriate criteria, Manabe et al. (1970),

Haarsma et al. (1990), and Broccoli and Manabe (1990), have reported that climate models can simulate some of the geographical structure that is characteristic of tropical cyclones.

Neither models with resolutions of 300 - 1000 km nor current numerical weather prediction models simulate individual thunderstorms, which are controlled by mesoscale dynamical processes. However, they do simulate variables that are related to the probability and intensity of severe weather such as thunderstorms, hail, wind gusts and tornadoes. If the appropriate variables are saved from a climate model, it should therefore be possible to determine whether the frequency and intensity of severe convective storms will change in an altered climate.

In summary, while changes in the occurrence of some types of extreme events, such as the frequency of high temperatures, can be diagnosed directly from climate model data, special techniques are needed for inferring changes in the occurrence of extreme events such as intense rainfall or severe local windstorms.

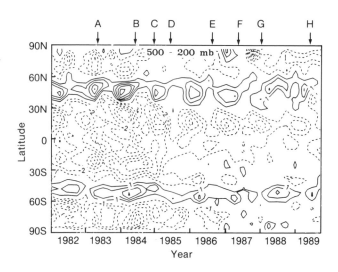

Figure 4.20: Zonal and vertical means (500-200 hPa) of zonal wind errors (forecast-analysis) of ECMWF day-10 forecasts. Contours drawn from seasonal mean values at intervals of 1 ms^{-1} (zero line suppressed). The lettering at the top of the diagram indicates times of major model changes.

4.7 Validation from Operational Weather Forecasting

As was recognized at the outset of numerical modelling, the climatological balance of a weather forecast model becomes of importance after a few days of prediction, while an extension of the integration domain to the whole globe becomes necessary. This means that modelling problems in numerical weather prediction (NWP), at least in the medium and extended range, have become similar to those in modelling the climate on timescales of a few months (Bengtsson, 1985).

The development of numerical models over the past several decades has led to a considerable improvement in forecast skill. This advance can be seen in the increased accuracy of short-range predictions, in the extension of the time range of useful predictive skill, and in the increase in the number of useful forecast products. A systematic evaluation of the quality of short-range forecasts in the Northern Hemisphere has been carried out by the WMO/CAS Working Group on Weather Prediction Research covering the 10-year period 1979 - 1988 (Lange 1989). Under this intercomparison project operational forecasts from several centres have been verified on a daily basis; considerable improvement has taken place, especially in the tropics and in the Southern Hemisphere (Bengtsson, 1989).

Of particular importance for climate modelling are model errors of a systematic (or case-independent) nature. Such model deficiencies give rise to a "climate drift", in which the model simulations generally develop significant differences from the real climate. Although there has been a progressive reduction of systematic errors in NWP models as noted above, a tendency to zonalization of the

flow is still present. Figure 4.20 shows the evolution of 10-day errors of the zonal wind component in the ECMWF model, and illustrates the global character of the model's errors (Arpe, 1989). The systematic errors typical of forecast models include the tendency for reduced variability in large-scale eddy activity, which shows up synoptically as a reduction in the frequency of blocking and quasi-stationary cut-off lows (see also Section 4.2.4).

In summary, the development and sustained improvement of atmospheric models requires long periods of validation using a large ensemble of different weather situations. Confidence in a model used for climate simulation will therefore be increased if the same model is successful when used in a forecasting mode.

4.8 Simulation of Ocean Climate

The ocean influences climate change on seasonal, decadal and longer timescales in several important ways. The large-scale transports of heat and fresh water by ocean currents are important climate parameters, and affect both the overall magnitude as well as the regional distribution of the response of the atmosphere-ocean system to greenhouse warming (Spelman and Manabe, 1985). The circulation and thermal structure of the upper ocean control the penetration of heat into the deeper ocean and hence also the timescale by which the ocean can delay the atmospheric response to CO_2 increases (Schlesinger et al., 1985). Vertical motions and water mass formation processes in high latitudes are controlling factors (besides chemical and biological interactions) for the oceanic uptake of carbon dioxide through the sea surface, and thus influence the radiative forcing in the atmosphere. To be a credible tool for the prediction of climate change, ocean models must therefore be capable of simulating the present circulation and water mass distribution, including their seasonal variability.

4.8.1 Status of Ocean Modelling

The main problems in ocean modelling arise from uncertainties in the parameterization of unresolved motions, from insufficient spatial resolution, and from poor estimates of air-sea fluxes. In general, ocean modelling is less advanced than atmospheric modelling, reflecting the greater difficulty of observing the ocean, the much smaller number of scientists/institutions working in this area, and the absence until recently of adequate computing resources and of an operational demand equivalent to numerical weather prediction. Global ocean models have generally followed the work by Bryan and Lewis (1979); they mostly have horizontal resolutions of several hundred kilometres and about a dozen levels in the vertical. Coarse grid models of this type have also been used in conjunction with atmospheric GCMs for studies of the coupled ocean-atmosphere system (see Section 4.9).

The performance of ocean models on decadal and longer timescales is critically dependent upon an accurate parameterization of sub-gridscale mixing. The main contribution to poleward heat transport in ocean models arises from vertical overturning, whereas the contribution associated with the horizontal circulation is somewhat smaller. Most models underestimate the heat transport, and simulate western boundary currents which are less intense and broader than those observed. The need for eddy-resolving models (e.g., Semtner and Chervin, 1988) in climate simulations is not yet established. Coarse vertical resolution, on the other hand, can significantly alter the effective mixing and thus influence the overturning and heat transport in a model. The main thermocline in most coarse-resolution simulations is considerably warmer and more diffuse than observed, a result probably due to a deficient representation of lateral and vertical mixing.

An important component of the deeper ocean circulation is driven by fluxes of heat and fresh water at the sea surface. In the absence of reliable data on the surface fluxes of heat and fresh water, many ocean modellers have parameterized these in terms of observed sea-surface temperature and salinity. The fluxes diagnosed in this way vary considerably among models. While surface heat flux and surface temperature are strongly related, there is no correspondingly strong connection between surface fresh water flux and surface salinity; a consequence is the possible existence of multiple equilibrium states with significant differences in oceanic heat transport (Manabe and Stouffer, 1988).

4.8.2 Validation of Ocean Models

The distribution of temperature, salinity and other water mass properties is the primary information for the validation of ocean models. Analysed data sets (e.g., Levitus, 1982) have been very useful although the use of original hydrographic data is sometimes preferred. The distributions of transient tracers, in particular tritium/helium-3 and C-14 produced by nuclear bomb tests, and CFCs, place certain constraints on the circulation and are also useful diagnostics for model evaluation (Sarmiento 1983, Toggweiler et al., 1989). The poleward transport of heat in the ocean zonal hydrographic sections (e.g., the annual mean of 1.0-1.2 PW at 25N in the North Atlantic found by Bryden and Hall, 1980) appears to have high reliability, and zonal sections planned in the World Ocean Circulation Experiment could significantly improve ocean heat transport estimates.

Direct observations of the fluxes of heat, fresh water and momentum at the sea surface are not very accurate, and would not appear to be viable in the near future. However, monitoring upper ocean parameters, in particular the heat and fresh water content in connection with ocean circulation models, can contribute to an indirect determination of the surface fluxes. The validation of ocean models using large data sets can in general be made more efficient if appropriate inverse modelling and/or data assimilation procedures are employed.

The validation of ocean models may conveniently be considered separately on the time-scales of a season, a decade and a century. The goal is a model which will correctly sequester excess heat produced by greenhouse warming, and produce the right prediction of changing sea surface temperatures when it is run in coupled mode with an atmospheric model.

Seasonal timescales in the upper ocean are important for a simulation of greenhouse warming both because seasonal variations are a fundamental component of climate, and because of the seasonal variation of vertical mixing in the ocean. Sarmiento (1986) has demonstrated that the seasonal variation of mixed layer depths can be simulated with a sufficiently detailed representation of the upper ocean.

Perhaps the best way to check ocean models on decadal time-scales is to simulate the spread of transient tracers. Data sets based on these tracers provide a unique picture of the downward paths from the surface into the ocean thermocline and deep water. The invasion of tritium into the western Atlantic about a decade after the peak of the bomb tests is compared with calculations by Sarmiento (1983) in Figure 4.21. The data show that the tritium very rapidly invades the main thermocline but only a small fraction gets into deep water. The model succeeds in reproducing many of the important features of the data such as the shallow equatorial penetration and the deep penetration in high latitudes. The main failure is the lack of

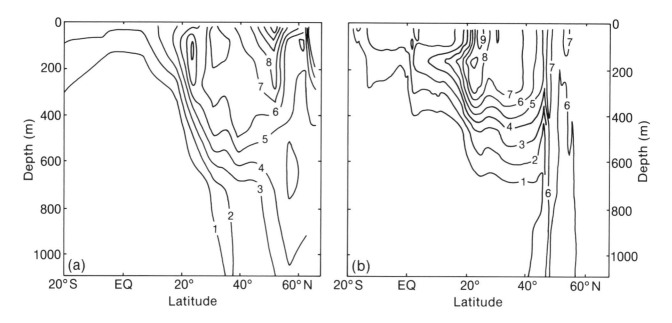

Figure 4.21: Tritium in the GEOSECS section in the western North Atlantic approximately one decade after the major bomb tests. (a) GEOSECS observations; (b) as predicted by a 12-level model (Sarmiento, 1983). In Tritium units.

penetration at 30-50°N, which may be related to some inadequacy in simulating cross-Gulf Stream/North Atlantic Current exchange (Bryan and Sarmiento, 1985). A notable result is the importance of seasonal convection for vertical mixing. An obvious difficulty in using transient tracer data to estimate the penetration of excess heat from greenhouse warming is the feedback caused by changes in the density field. A very small temperature perturbation should behave like a tracer, but as the amplitude increases, the perturbation will affect the circulation (Bryan and Spelman 1985).

In many parts of the ocean, salinity is an excellent tracer of ocean circulation. The salinity field of the ocean is extremely difficult to simulate. The reason for this is that sources and sinks of fresh water at the ocean surface have a rather complex distribution, much more so than the transient tracers. Water masses with distinctive salinity signatures lie at the base of the thermocline in all the major oceans. In the Southern Ocean and the Pacific, these water masses are characterized by salinity minima and relatively weak stability. The characteristic renewal timescale of these water masses is greater than a decade but less than a century, a range very important for greenhouse warming. At present, these water masses have not been simulated in a satisfactory way in an ocean circulation model.

In summary, oceanic processes are expected to play a major role in climate change. The satisfactory representation of vertical and horizontal transport processes (and of sea-ice) are thus of particular importance. There is encouraging evidence from tracer studies that at least some aspects of these mixing processes are captured by ocean

models. However, at present, ocean models tend to underestimate heat transport.

4.9 Validation of Coupled Models

While much has been learned from models of the atmosphere and ocean formulated as separate systems, a more fundamental approach is to treat the ocean and atmosphere together as a coupled system. This is unlikely to improve on the simulation of the time-meaned atmosphere and ocean when treated as separate entities (with realistic surface fluxes), since the average SST can only become less realistic; however, it is the only way in which some of the climate system's long-term interactions, including the transient response to progressively increasing CO_2, can be realistically studied (see Section 6).

Typical of the current generation of coarse-grid coupled GCMs is the simulation shown in Figure 4.22 from Washington and Meehl (1989). The general pattern of zonal mean temperature is reproduced in both atmosphere and ocean in this freely interacting coupled model, although during the time period simulated the temperature in the deeper ocean is still strongly related to the initial conditions. On closer inspection, comparison of the simulated near-surface temperatures in the ocean with observed values from Levitus (1982) shows warmer-than-observed temperatures in the high-latitude southern oceans, colder-than-observed temperatures in the tropics, and colder-than-observed temperatures at high northern latitudes. The latter can be traced to the North Atlantic where the lack of a well-defined Gulf Stream and associated thermohaline circulation inhibits the transport of

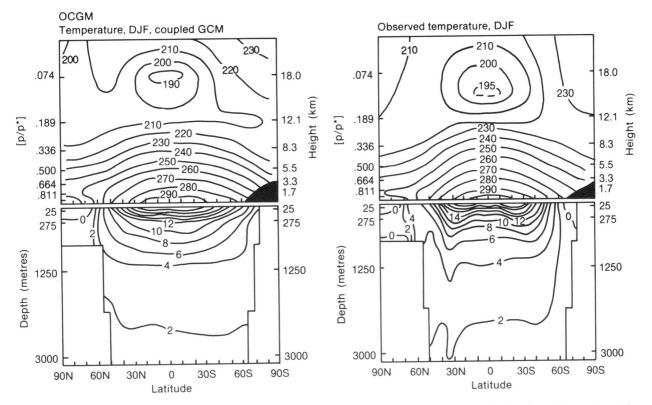

Figure 4.22: Zonal mean temperatures for December-January-February for atmosphere and ocean as simulated in a 30 year integration with (left) the NCAR coupled model (Washington and Meehl, 1989) and (right) from observations. Observed atmospheric temperatures are from Newell et al. (1972); observed ocean temperatures are from Levitus (1982). The unlabelled contour in the observed near the tropical surface is 295K. The maximum ocean temperatures in this same region are 27°C (observed) and 25°C (computed).

heat to those latitudes. Similar patterns of systematic sea-surface temperature errors have been found in other coupled models (Gates et al., 1985; Manabe and Stouffer 1988), and their effects on the simulated surface heat flux have been examined by Meehl (1989).

In view of such errors, a practical decision faces those designing coupled models. On the one hand, they can decide that the systematic errors, while serious in terms of the control integration, do not prevent the useful interpretation of results from sensitivity experiments. On the other hand, they may consider that the systematic errors represent a significant bias in the control run and would affect the results of sensitivity experiments to an unacceptable degree. The alternative then is to somehow adjust for the errors in the control run to provide a more realistic basic state for sensitivity experiments. Such techniques have been devised and are variously called "flux correction" (Sausen et al., 1988) or "flux adjustment" (Manabe and Stouffer, 1988). These methods effectively remove a large part of the systematic errors and such coupled simulations are closer to observed conditions. However, since the correction terms are additive, the coupled model can still exhibit drift, and the flux correction

terms cannot change during the course of a climate change experiment, (i.e., it is effectively assumed that the model errors are the same for both the control and perturbed climates).

One way to validate coupled models is to analyse the simulated interannual variability, a fundamental source of which is associated with the El Niño - Southern Oscillation (ENSO) (see also Section 4.5.1). The current generation of coarse-grid coupled models has been shown to be capable of simulating some aspects of the ENSO phenomenon (Sperber et al., 1987; Meehl, 1990; Philander et al., 1989), although the simulated intensity is in general too weak. Ultimately a coupled climate model should be verified by its simulation of the observed evolution of the atmosphere and ocean over historical times. For hypothetical future rates of CO_2 increase, current coupled GCMs at least indicate that the patterns of the climate's transient response are likely to be substantially different in at least some ocean regions from those given in equilibrium simulations without a fully interacting ocean (Washington and Meehl 1989; Stouffer et al., 1989; see also Section 6).

In summary, coupled models of the ocean-atmosphere system are still in an early stage of development and have so far used relatively coarse resolution. Nevertheless, the large-scale structures of the ocean and atmosphere can be simulated with some skill using such models and current simulations give results that are generally similar to those of equilibrium models (see Sections 5 and 6).

4.10 Validation from Palaeo-Climate

Studies of palaeo-climatic changes are an important element in climate model validation for two reasons:

1) they improve our physical understanding of the causes and mechanisms of large climatic changes so that we can improve the representation of the appropriate processes in models; and

2) they provide unique data sets for model validation.

4.10.1 Observational Studies of the Holocene

The changes of the Earth's climate during the Holocene and since the last glacial maximum (the last 18,000 years) are the largest and best-documented in the palaeo-climatic record, and are therefore well-suited for model validation; the data sets are near-global in distribution, the time control (based upon radiocarbon dating) is good, and estimates of palaeo-climatic conditions can be obtained from a variety of palaeo-environmental records, such as lake sediment cores, ocean sediment cores, ice cores, and soil cores. At the last glacial maximum there were large ice sheets in North America and northern Eurasia, sea-level was about 100m below present, the atmospheric CO_2 concentration was around 200ppm, sea-ice was more extensive than at present, and the patterns of vegetation and lake distribution were different from now. During the last 18,000 years, we

therefore have the opportunity to observe how the climate system evolved during the major change from glacial to present (interglacial) conditions. CLIMAP Project Members (1976, 1981) and COHMAP Members (1988) have assembled a comprehensive data set for the climate of the last 18,000 years as summarized in Figure 4.23.

The period from 5.5-6kbp (thousand years before present) is probably the earliest date in the Holocene when the boundary conditions of ice-sheet extent and sea level were analogous to the present. There is also general agreement that vegetation was close to equilibrium with the climate at this time. Radiocarbon dating of most of the sources of stratigraphic data allows an accuracy of better than plus-or-minus 1000 years in the selection of data for the purposes of making reconstructions. The earlier period around 9kbp is of particular interest because the differences of the radiative forcing from the present were particularly large (Berger, 1979), although there was still a substantial North American ice sheet.

4.10.2 Model Studies of Holocene Climate

Several atmospheric GCMs have been used to simulate the climate of the 18kbp glacial maximum, and have helped to clarify the relative roles of continental ice sheets, sea-ice, ocean temperature and land albedo in producing major shifts in circulation, temperature and precipitation patterns (Gates, 1976a, b; Manabe and Hahn, 1977; Kutzbach and Guetter, 1986; Rind 1987). In addition to specifying land-based ice sheets and changed land albedos, these models also prescribed SSTs and sea-ice extents. Manabe and Broccoli (1985) successfully simulated the SSTs and sea-ice during the Last Glacial Maximum using an atmospheric GCM coupled to a mixed-layer ocean model; this

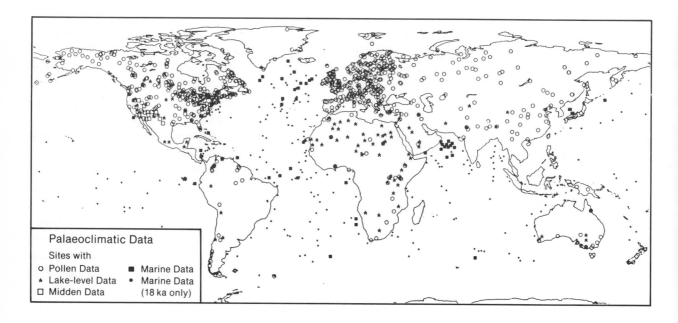

Figure 4.23: Data sites in the CLIMAP/COHMAP global palaeo-climatic data base (COHMAP Members, 1988).

SPRUCE POLLEN (as observed)

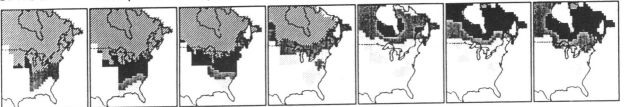

SPRUCE POLLEN (as simulated by GCM output)

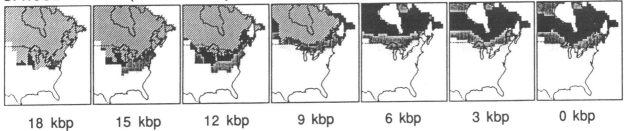

| 18 kbp | 15 kbp | 12 kbp | 9 kbp | 6 kbp | 3 kbp | 0 kbp |

Figure 4.24: Maps showing observed (upper row) and simulated (lower row) percentages of spruce pollen for each 3000-year interval from 18,000 YBP to the present. The region with diagonal lines in the north is a digital representation of the location of the Laurentide ice sheet. Area of spruce abundance as determined by spruce pollen is shown by dark stippling for >20%, intermediate stippling for 5 to 20%, and light stippling for 1 to 5%. Simulation values were produced by applying the observed (present) association between spruce pollen and climatic variables to the temperature and precipitation values simulated by the climate model. (From COHMAP Members, 1988).

experiment also demonstrated the sensitivity of glacial-age climate simulations to the lowered level of glacial-age atmospheric CO_2 (Broccoli and Manabe, 1987). The climate's sensitivity to orbital parameter changes has been confirmed through comparisons of model simulations with palaeo-climatic data both using atmospheric GCMs (Kutzbach and Guetter, 1986; Royer et al., 1984) and using atmospheric GCMs coupled to mixed-layer ocean models (Kutzbach and Gallimore, 1988; Mitchell et al., 1988). Manabe and Stouffer (1988), moreover, report evidence of two stable equilibria of a coupled atmosphere-ocean GCM that may be of importance for explaining abrupt short-term climate changes such as the cooling which occurred about 10,500 years ago. Rind et al. (1986), Overpeck et al. (1989) and Oglesby et al. (1989) have sought understanding of this cooling from model experiments in which cooling of the North Atlantic or the Gulf of Mexico was specified.

In general, palaeo-climate modelling studies have found encouraging agreement between simulations and observations on continental scales. For example, the COHMAP comparisons using the NCAR model show temperature and precipitation changes, 18kbp to present, that are generally consistent with observations in North America, at least as interpreted by the movement of spruce populations (Figure 4.24), while the simulated enhancement of northern tropical monsoons around 9kbp is also supported by palaeo-climatic data (COHMAP Members, 1988). On the other hand, palaeo-climate

modelling studies, like their modern counterparts, also reveal regions and times where model and data disagree. For example, Figure 4.24 shows errors over southeastern North America from 18 to 12kbp associated with simulated summer temperatures that are too high, while Manabe and Broccoli (1985) obtain larger cooling of the tropical oceans at 18kbp than palaeo-climatic data suggest.

Because the mid-Holocene may have been warmer than now, at least during northern summer, the question arises whether or not this period might be in some sense analogous to the climate with doubled CO_2. Gallimore and Kutzbach (1989) and Mitchell (1990) have discussed the differences in forcing (orbital parameters versus CO_2) and differences in the climatic response as simulated by GCMs. Even though the two types of forcing are very different, we can learn a great deal about our models by determining how well model experiments with orbitally-caused changes in solar radiation simulate the observed extent of increase in northern mid-continent summer warmth and dryness, the decrease in Arctic sea-ice, and the increase in northern tropical precipitation.

4.10.3 Other Validation Opportunities

Studies of the previous interglacial around 125kbp by CLIMAP Project Members (1984) and Prell and Kutzbach (1987), among others, show evidence for warmer conditions, especially in high latitudes, reduced sea-ice extent, and enhanced northern tropical monsoons. At this time CO_2 levels were above pre-industrial levels, sea-level was somewhat higher than now, the Greenland ice sheet was perhaps smaller, and orbital parameters favoured greatly enhanced Northern Hemisphere seasonality. Because of the indications of warmth and relatively high CO_2 levels (relative to before and after 125 kbp) this period is also of interest for modelling and model validation studies. Modelling experiments by Royer et al. (1984) have emphasised the strong cooling from the equilibrium climate with orbital parameters for 125kbp to that for 115kbp. However, data sets are not nearly as extensive or as well-dated as for the mid-Holocene.

There is strong evidence that the first growth of ice sheets and the development of glacial/interglacial cycles began in North America and northern Eurasia around 2.4 million years ago; prior to this time the climate was presumably significantly warmer than at present. This period may well be our only geologically-recent example of a climate that was significantly warmer than now over large areas. However, the period poses many problems, including the marked differences from the present day in major global topographic features and the uncertainties in forcing conditions; these factors make it unsuitable for detailed model validation at the present time, although such simulations would be of considerable scientific interest.

In summary, palaeo-climatic data have provided encouraging evidence of the ability of climate models to simulate climates different from the present, especially during the Holocene. This indicates that further such data would be useful for climate model validation.

4.11 Conclusions and Recommendations

This somewhat selective review of the performance of current global climate models has shown that there is considerable skill in the simulation of the present day climate by atmospheric general circulation models in the portrayal of the large-scale distribution of the pressure, temperature, wind and precipitation in both summer and winter. As discussed in Section 4.1.2, the responses to perturbations can be given credence, provided simulated and observed patterns are sufficiently similar for corresponding features and mechanisms to be identified. Recent models appear to satisfy this condition over most of the globe. Although quantification of this conclusion is difficult, it is supported by the skill demonstrated by atmospheric models in simulating, firstly, the circulation and rainfall changes associated with the El Niño ocean temperature anomalies and, secondly, the rainfall anomalies characteristic of wet and dry periods in the Sahel region of Africa when the observed sea surface temperature anomalies are used.

On regional scales there are significant errors in these variables in all models. Validation for selected regions

shows mean surface air temperature errors of 2 to 3°C, compared with an average seasonal variation of 15°C. The large-scale distribution of precipitation can be realistically simulated apart from some errors on sub-continental scales (1000-2000 km) whose locations differ between models. Validation on these scales for selected regions shows mean errors of from 20% to 50% of the average rainfall depending on the model.

The limited data available show that the simulated summer and winter soil moisture distributions in middle latitudes qualitatively reflect most of the large-scale characteristics of observed soil wetness. Snow cover can be well simulated in winter except in regions where the temperature is poorly simulated. The radiative fluxes at the top of the atmosphere, important for the response of climate to radiative perturbations, are simulated with average errors in the zonal mean as small as 5 Wm^{-2}. There are, however, substantial discrepancies in albedo, particularly in middle and high latitudes.

There has been a general reduction in the errors in more recent models as a result of increased resolution, changes in the parameterization of convection, cloudiness and surface processes, and the introduction of parameterizations of gravity wave drag. In addition to the conclusions drawn from the validation of atmospheric model control simulations, our overall confidence in the models is increased by their relatively high level of accuracy when used for short and medium-range weather prediction, by their portrayal of low-frequency atmospheric variability such as the atmospheric response to realistic sea surface temperature anomalies (also referred to above), and by their ability to simulate aspects of the climate at selected times during the last 18,000 years. Further confidence in atmospheric models would be obtained by their successful simulation of the climate changes shown by the observed instrumental record.

Other opportunities for validation not considered in detail here include the simulation of variations in stratospheric temperature and circulation. Models have been successful in simulating the impact on temperatures of the Antarctic ozone hole (Kiehl et al., 1988; Cariolle et al., 1990), although they have not successfully simulated the quasi-biennial oscillation in stratospheric wind and temperature.

The latest atmospheric models, while by no means perfect, are thus sufficiently close to reality to inspire some confidence in their ability to predict the broad features of a doubled CO_2 climate at equilibrium, provided the changes in sea-surface temperature and sea-ice are correct. The models used in simulating the equilibrium responses to increased greenhouse gases employ simple mixed-layer ocean models, in which adjustments to the surface fluxes have usually been made to maintain realistic present day sea-surface temperatures and sea-ice in the control experiments. Our confidence in the ability of these models to simulate changes in the climate, including ocean temperatures and sea-ice, is enhanced by their successful simulation of aspects of the climates during and since the most recent ice age.

Despite the present computational constraints on resolution, the performance of ocean models lends credence to our ability to simulate many of the observed large-scale features of ocean climate, especially in lower latitudes. However, coupled ocean-atmosphere models exhibit characteristic errors which as yet can only be removed by empirical adjustments to the ocean surface fluxes. This is due in part to the use of atmospheric and oceanic models of relatively low resolution, and in part to inadequate parameterizations of fluxes at the air-sea interface. Nevertheless, the latest long runs with such models, discussed in Section 6, exhibit variability on decadal timescales which is similar to that observed (compare Figure 6.2 and Figure 7.10).

There is a clear need for further improvement of the accuracy of climate models through both increased resolution and improved parameterization of small-scale processes, especially the treatment of convection, clouds and surface effects in atmospheric GCMs, and mixing and sea-ice behaviour in oceanic GCMs. Much further experience needs to be gained in the design of coupled models in order to avoid the equally unsatisfactory choices of accepting a progressive climatic drift or of empirically correcting the behaviour of the upper ocean. These improvements and the associated extended simulations will require substantial amounts of computer time, along with increased coordination and cooperation among the world's climate modelling community. Data from satellite programmes, such as EOS, and from field experiments are needed to provide more complete data sets for specifying land surface characteristics, for initialisation and validation of ocean simulations and to improve parameterizations. Of particular value should be ERBE, ISCCP and FIRE data for radiation and cloud, the ISLSCP and the HAPEX for land surface processes, the GEWEX for energy and water balances, and WOCE and TOGA data for the oceans.

The validation of a number of atmospheric model variables has been handicapped by limitations in the available observed and model data. In particular, future model assessments would benefit from improved estimates of precipitation and evaporation over the oceans, and of evaporation, soil moisture and snow depth over land, and by more uniform practices in the retention of model data such as snow-cover frequency and depth, and daily near-surface temperature extremes or means. The generation of data suitable for validating cloud simulations deserves continuing attention, as does the assembly of palaeo-climatic data sets appropriate for climate model validation over the Earth's recent geological history. The lack of

appropriate data has also severely hindered the validation of ocean models. Adequate data on the seasonal distribution of ocean currents and their variability and on salinity and sea-ice thickness are especially needed.

Although the ten-year atmospheric data set for 1963-73 compiled by Oort (1983) and the oceanographic set assembled by Levitus (1982) have been of great value in model assessment, the subsequent availability of 4-dimensional assimilation techniques and the expansion of observing platforms provide the opportunity for considerably improved data sets. Indeed, many data sets now used by modellers for validation have been produced by global forecasting centres as a by-product of their operational data assimilation, although changes in forecast and assimilation techniques have led to temporal discontinuities in the data. The proposal by Bengtsson and Shukla (1988) for a re-analysis of observations over a recent decade (e.g., 1979-1988) with a frozen up-to-date assimilation system is therefore of great potential value for climate model validation. If carried out over additional decades, such a data set could also contribute to our understanding of how to distinguish between natural climate fluctuations and changes caused by increased greenhouse gases.

References

Arctic Construction and Frost Effects Laboratory, 1954: Depth of snow cover in the Northern Hemisphere. New England Division, Corps of Engineers, Boston, Mass.

Ardanuy, P. E., L. L. Stowe, A. Gruber, M. Weiss and C. S. Long, 1989: Longwave cloud radiative forcing as determined from Nimbus-7 observations. *J. Clim.*, **2**, 766-799.

Arkin, P. A., and P. E. Ardanuy, 1989: Estimating climatic-scale precipitation from space: a review. *J. Clim.*, **2**, 1229-1238.

Arkin, P. A., and B. N. Meisner, 1987: The relationship between large-scale convective rainfall and cold cloud over the western hemisphere during 1982-1984, *Mon. Weath. Rev.*, **115**, 51-74.

Arpe, K., 1989: Systematic errors of the ECMWF model. ECMWF Internal Report, SAC (89).

Bengtsson, L., 1985: Medium-range forecasting at the ECMWF. *Adv. Geoph*, **28**, 3-54.

Bengtsson, L., 1989: Numerical weather prediction at the Southern Hemisphere. Third International Conference on Southern Hemisphere Meteorology and Oceanography, Buenos Aires, Argentina, November 13-17, 1989.

Bengtsson, L., and J. Shukla, 1988: Integration of space and *in situ* observations to study global climate change. *Bull. Amer. Meteor. Soc.*, **69**, 1130-1143.

Berger, A., 1979: Insolation signatures of Quaternary climatic changes. *Il Nuovo Cimento*, **20(1)**, 63-87.

Boer, G. J., and M. Lazare, 1988: Some results concerning the effect of horizontal resolution and gravity-wave drag on simulated climate. *J. Clim.*, **1**, 789-806.

Broccoli, A.J. and S. Manabe, 1987: The influence of continental ice, atmospheric CO_2 and land albedo on the climate of the last glacial maximum. *Clim. Dyn.*, **1**, 87-99.

Broccoli, A.J. and S. Manabe, 1990: Will global warming increase the frequency and intensity of tropical cyclones? (submitted for publication).

Bryan, K., and L. J. Lewis, 1979: A water mass model of the world ocean. *J. Geophys. Res.*, **84**, 2503-2517.

Bryan, K. and J. L. Sarmiento, 1985: Modelling ocean circulation. In Issues in atmospheric and oceanic modelling (Part A) (Ed. S. Manabe), *Adv. Geophys,.* **28**, 433-459.

Bryan, K. and M. J. Spelman, 1985: The ocean's response to a CO_2-induced warming. *J. Geophys. Res.*, **90**, 11679-11688.

Bryden, H. L., and M. Hall, 1980: Heat transport by currents across 25N latitude in the Atlantic Ocean. *Science,* **207**, 884-885.

Cariolle, D., A. Lasserre-Bigorry, J.-F. Royer and J.-F. Geleyn, 1990: A general circulation model simulation of the springtime Antarctic ozone decrease and its impact on mid-latitudes. *J. Geophys. Res.*, **95**, 1883-1898.

Charney, J. G., W. J. Quirk, S. H. Chow, and J. Kornfeld, 1977: A comparative study of the effects of albedo change on drought in semi-arid regions. *J. Atmos. Sci.*, **34**, 1366-1385.

CLIMAP Project Members, 1976: The surface of the ice-age Earth. *Science*, **191**, 1131-1136.

CLIMAP Project Members, 1981: Seasonal reconstructions of the Earth's surface at the last glacial maximum. *Geological Society of America Map and Chart Series*, MC-36.

CLIMAP Project Members, 1984: The last interglacial ocean. *Quat. Res.*, **21**, 123-224.

COHMAP Members, 1988: Climatic changes of the last 18,000 years: Observations and model simulations. *Science* , **241**, 1043-1052.

Dorman, C. E., and R. H. Bourke, 1979: Precipitation over the Pacific Ocean, 30°S to 60°N. *Mon. Wea. Rev.*, **107**, 896-910.

Dyson, J. F., 1985: The effect of resolution and diffusion on the simulated climate. Research Activities in Atmospheric and Oceanic Modelling, Report No. **8**, WGNE, WMO, Geneva, pp. 4.4-4.6.

Fennessy, M.J., and J. Shukla, 1988a: Numerical simulation of the atmospheric response to the time-varying El Niño SST anomalies during May 1982 through October 1983. *J. Clim.*, **1**, 195-211.

Fennessy, M. J., and J. Shukla, 1988b: Impact of the 1982/3 and 1986/7 Pacific SST anomalies on time-mean prediction with the GLAS GCM. WCRP-**15**, WMO, pp. 26-44.

Folland, C. K., J. A. Owen and K. Maskell, 1989: Physical causes and predictability of variations in seasonal rainfall over sub-Saharan Africa. IAHS Publ. No. **186**, 87-95.

Gallimore, R.G., and J.E. Kutzbach, 1989: Effects of soil moisture on the sensitivity of a climate model to Earth orbital forcing at 9000 yr BP. *Climatic Change,* **14**, 175-205.

Gates, W.L., 1976a: Modeling the ice-age climate. *Science*, **191**, 1138-1144.

Gates, W.L., 1976b: The numerical simulation of ice-age climate with a global general circulation model. *J. Atmos. Sci.*, **33**, 1844-1873.

Gates, W. L., Y. J. Han, and M. E. Schlesinger, 1985: The global climate simulated by a coupled atmosphere-ocean general circulation model: Preliminary results. *In*: Nihoul, J. C. J. (ed.), *Coupled Ocean-Atmosphere Models*, Elsevier Oceanogr. Ser. 40, 131-151.

Gregory, S., 1989: Macro-regional definition and characteristics of Indian summer monsoon rainfall, 1871-1985. *Int. J. Clim.*, **9**, 465-483.

Haarsma, R., J.F.B. Mitchell and C.A. Senior, 1990: Tropical cyclones in a warmer climate. (submitted for publication).

Hansen, J., et al., 1983: Efficient three-dimensional global models for climate studies: Models I and II. *Mon. Wea. Rev.*, **111**, 609-662.

Hartmann, D. L., V. Ramanathan, A. Berroir and G. E. Hunt, 1986: Earth radiation budget data and climate research. *Rev. Geophys.*, **24**, 439-468.

Hibler, W. D. III, 1979: A dynamic thermodynamic sea ice model. *J. Phys. Oceanogr.*, **9**, 815-846.

Hibler, W. D. III, and S. F. Ackley, 1983: Numerical simulations of the Weddell Sea pack ice. *J. Geophys. Res.*, **88**, 2873-2887.

Jaeger, L., 1976: Monatskarten des Niederschlags fur die ganze Erde. *Bericht Deutscher Wetterdienst,* **18**, Nr. 139, 38pp.

Katz, R. W., 1988: Statistical procedures for making inferences about changes in climate variability. *J. Clim.*, **1**, 1057-1068.

Kershaw, R., 1988: The effect of a sea surface temperature anomaly on a prediction of the onset of the south-west monsoon over India. *Quart. J. R. Meteor. Soc.*, **114**, 325-345.

Kiehl, J.T., B.A. Boville and B.P. Briegleb, 1988: Response of a general circulation model to a prescribed Antarctic ozone hole. *Nature*, **332**, 501-504.

Kitoh, A. and T. Tokioka, 1987: A simulation of the tropospheric general circulation with the MRI atmospheric general circulation model, Part III: The Asian summer monsoon. *J. Meteor. Soc. Japan,* **65**, 167-187.

Krishnamurti, T.N., and D. Oosterhof, 1989: Prediction of the life cycle of a supertyphoon with a high-resolution global model. *Bull. Amer. Meteor. Soc.*, **70**, 1218 - 1230.

Kutzbach, J.E., and P.J. Guetter, 1986: The influence of changing orbital parameters and surface boundary conditions on climate simulations for the past 18,000 years. *J. Atmos. Sci.*, **43**, 16, 1726-1759.

Kutzbach, J.E., and R.G. Gallimore, 1988: Sensitivity of a coupled atmosphere/mixed-layer ocean model to changes in orbital forcing at 9000 yr BP. *J. Geophys. Res.*, **93**, 803-821.

Lange, A., 1989: Results of the WMO/CAS NWP data study and intercomparison project for forecasts for the Northern Hemisphere in 1988. World Meteorological Organization, World Weather Watch, Technical Report No. **7**.

Laval, K. and L. Picon, 1986: Effect of a change of the surface albedo of the Sahel on climate. *J. Atmos. Sci.*, **43**, 2418-2429.

Levitus, S., 1982: Climatological Atlas of the World Ocean. NOAA Professional Paper 13, National Oceanic and Atmospheric Administration, Washington, DC, 173 pp, 17 microfiche.

Li, Z.-X. and Letreut, H., 1989: Comparison of GCM results with data from operational meteorological satellites, *Ocean-Air Interactions*, **1**, 221-237.

Livezey, R. E., 1985: Statistical analysis of general circulation model climate simulations: Sensitivity and prediction experiments. *J. Atmos. Sci.*, **42**, 1139-1149.

Manabe, S. and A. J. Broccoli, 1985: A comparison of climate model sensitivity with data from the Last Glacial Maximum. *J. Atmos. Sci.*, **42**, 2643-2651.

Manabe, S., and D.G. Hahn, 1977: Simulation of the tropical climate of an ice age. *J. Geophys. Res.*, **82**, 3889-3911.

Manabe, S., and R.J. Stouffer, 1988: Two stable equilibria of a coupled ocean-atmosphere model. *J. Clim.*, **1**, 841-866.

Manabe, S., D. G. Hahn and J. L. Holloway, 1978: Climate simulations with GFDL spectral models of the atmosphere. Report of the JOC Study Conference on Climate Models: Performance, Intercomparison and Sensitivity Studies, GARP Publications Series No. 22, Vol. 1, 41-94, WMO, Geneva.

Manabe, S., J. L. Holloway and H. M. Stone, 1970: Tropical circulation in a time integration of a global model of the atmosphere. *J. Atmos. Sci.*, **27**, 580-613.

Matson, M., C. F. Ropelewski and M. S. Varnadore, 1986: An Atlas of Satellite-Derived Northern Hemisphere Snow Cover Frequency. U.S. Govt. Printing Office Publ. 1986-151-384, NOAA.

McBride, J. L., 1987: The Australian summer monsoon. In: *Monsoon Meteorology*, Oxford University Press, 203-232.

Mearns, L.O., R.W. Katz and S.H. Schneider, 1984: Extreme high-temperature events: changes in their probabilities with changes in mean temperature. *J. Clim. Appl. Met.*, **23**, 1601-1613.

Mearns, L.O., S.H. Schneider, S.L. Thompson and L.R. McDaniel, 1990: Analysis of climate variability in general circulation models: comparison with observations and changes in variability in 2 x CO_2 experiments. (submitted for publication)

Meehl, G. A., 1989: The coupled ocean-atmosphere modeling problem in the tropical Pacific and Asian monsoon regions. *J. Clim.*, **2**, 1122-1139.

Meehl, G. A., 1990: Seasonal cycle forcing of El Nino-Southern Oscillation in a global coupled ocean-atmosphere GCM. *J. Clim.*, **3**, 72-98.

Mintz, Y., and Y. V. Serafini, 1989: Global monthly climatology of soil moisture and water balance. LMD Internal Report No. 148, LMD, Paris.

Mitchell, J. F. B., 1990: Greenhouse warming: Is the Holocene a good analogue? *J Clim.* (in press)

Mitchell, J.F.B., N.S. Grahame, and K.H. Needham, 1988: Climate simulations for 9000 years before present: Seasonal variations and the effect of the Laurentide Ice Sheet, *J. Geophys. Res.*, **93**, 8283-8303.

Mitchell, J. F. B., C.A. Wilson and W. M. Cunnington, 1987: On CO_2 climate sensitivity and model dependence of results. *Q. J. R. Meteor. Soc.*, **113**, 293-322.

Newell, R. E., J. W. Kidson, D. G. Vincent, and G. J. Boer, 1972: The General Circulation of the Tropical Atmosphere and Interactions with Extra-Tropical Latitudes, Vol. 1., M.I.T. Press, Cambridge, MA, 258 pp.

Nihoul, J.C.J. (Ed.), 1985: Coupled ocean-atmosphere models. Elsevier Oceanography Series, **40**, Elsevier, Amsterdam, 767pp.

Oglesby, R.J., K.A. Maasch and B. Saltzman, 1989: Glacial meltwater cooling of the Gulf of Mexico: GCM implications for Holocene and present-day climates. *Clim. Dyn.*, **3**, 115-133.

Oort, A. H., 1983: Global atmospheric circulation statistics, 1958-1973. NOAA Professional Paper 14, U.S. Department of Commerce.

Overpeck, J.T., L.C. Peterson, N. Kipp, J. Imbrie, and D. Rind, 1989: Climate change in the circum-North Atlantic region during the last deglaciation. *Nature*, **338**, 553-557.

Owens, W. B. and P. Lemke, 1989: Sensitivity studies with a sea ice - mixed layer - pycnocline model in the Weddell Sea. *J. Geophys. Res.* (in press).

Palmer, T. N., 1986: Influence of the Atlantic, Pacific and Indian Oceans on Sahel rainfall. *Nature*, **322**, 236-238.

Palmer, T.N., and D.A. Mansfield, 1986: A study of wintertime circulation anomalies during past El Niño events using a high resolution general circulation model. I: Influence of model climatology. *Quart. J. R. Meteor. Soc.*, **112**, 613-638.

Philander, S. G. H., N. C. Lau, R. C. Pacanowski, M. J. Nath, 1989: Two different simulations of Southern Oscillation and El Nino with coupled ocean-atmosphere general circulation models. *Phil. Trans. Roy. Soc.*, in press.

Prell, W.L., and J.E. Kutzbach, 1987: Monsoon variability over the past 150,000 years. *J. Geophys. Res.*, **92**, 8411-8425.

Reed, D.N., 1986: Simulation of time series of temperature and precipitation over eastern England by an atmospheric general circulation model. *J. Climatology*, **6**, 233-253.

Rind, D., 1987: Components of the Ice Age circulation. *J. Geophys. Res*, **92**, 4241-4281.

Rind, D., D. Peteet,W. Broecker, A. McIntyre, and W. Ruddiman , 1986: The impact of cold North Atlantic sea surface temperatures on climate; Implications for the Younger Dryas cooling (ll-l0K): *Clim. Dyn.*, **1**, 3-34.

Rind, D., R. Goldberg and R. Ruedy, 1989: Change in climate variability in the 21st century. *Climatic Change*, **14**, 5-37.

Rowntree, P. R., and A. B. Sangster, 1986: Remote sensing needs identified in climate model experiments with hydrological and albedo changed in the Sahel. Proc. ISLSCP Conference, Rome, Italy, ESA SP-248, 175-183.

Rowntree, P.R., 1972: The influence of tropical east pacific ocean temperatures on the atmosphere. *Quart. J. R. Meteor. Soc.*, **98**, 290--312.

Royer, J.F., M. Deque and P. Pestiaux, 1984: A sensitivity experiment to astronomical forcing with a spectral GCM: Simulation of the annual cycle at 125,000 BP and 115,000 BP. *In -Milankovitch and Climate*, (A. Berger, J. Imbrie, J. Hays, G. Kukla, and B. Saltzman, eds.), Part 2, Reidel Publ., Dordrecht, Netherlands, pp. 801-820.

Santer, B. D. and T. M. L. Wigley, 1990: Regional validation of means, variances and spatial patterns in general circulation model control runs. *J. Geophys. Res.* (in press).

Sarmiento, J. L., 1983: A simulation of bomb-tritium entry into the Atlantic Ocean. *J. Phys. Oceanogr.*, **13**, 1924-1939.

Sausen, R., R. K. Barthels, and K. Hasselmann, 1988: Coupled ocean-atmosphere models with flux correction. *Clim. Dyn.*, **2**: 154-163.

Schlesinger, M. E., W. L. Gates, and Y. -J. Han, 1985: The role of the ocean in carbon-dioxide-induced climate warming. Preliminary results from the OSU coupled atmosphere-ocean model. In: *Coupled Ocean-Atmosphere Models*, J. C. J. Nihoul (ed.), Elsevier, 447-478.

Schutz, C. and W. L. Gates, 1971: Global climatic data for surface, 800 mb, 400 mb: January. Rand, Santa Monica, R-915-ARPA, 173 pp.

Schutz, C. and W. L. Gates, 1972: Global climatic data for surface, 800 mb, 400 mb: July. Rand, Santa Monica, R-1029-ARPA, 180 pp.

Semtner, A. J., and R. Chervin, 1988: A simulation of the global ocean circulation with resolved eddies. *J. Geophys., Res.*, **93**, 15, 502-15, 522.

Shukla, J., and Y. Mintz, 1982: Influence of land surface evapotranspiration on the Earth's climate. *Science*, **215**, 1498-1501.

Slingo, A. and D. W. Pearson, 1987: A comparison of the impact of an envelope orography and of a parameterization of orographic gravity wave drag on model simulations. *Quart. J. R Meteor. Soc.*, **113**, 847-870.

Spelman, M. J., and S. Manabe, 1985: Influence of oceanic heat transport upon the sensitivity of a model climate. *J. Geophys. Res.*, **89**, 571-586.

Sperber, K. R., S. Hameed, W. L. Gates, and G. L. Potter, 1987: Southern Oscillation simulated in a global climate model. *Nature* , **329**: 140-142.

Stouffer, R. J., S. Manabe, and K. Bryan, 1989: Interhemispheric asymmetry in climate response to a gradual increase of atmospneric CO_2. *Nature*, **342**, 660-662.

Straus, D. M. and J. Shukla, 1988: A comparison of a GCM simulation of the seasonal cycle of the atmosphere with observations. *Atmos-Ocean*, **26**, 541-604.

Sud, Y. C., and M. Fennessy, 1982: A study of the influence of surface albedo on July circulation in semi-arid regions using the GLAS GCM. *J. Clim.*, **2**, 105-125.

Tibaldi, S., T. Palmer, C. Brankovic, and U. Cubasch, 1989: Extended-range predictions with ECMWF models. II: Influence of horizontal resolution on systematic error and forecast skill. *ECMWF Technical Memorandum* , **No.152**, 39 pp.

Toggweiler, J. R., K. Dixon and K. Bryan, 1989: Simulations of radiocarbon in a coarse-resolution world ocean model, 2. Distributions of bomb-produced carbon-14. *J. Geophys. Res.*, **94**, 8243-8264.

Trenberth, K.E. and J.G. Olson, 1988: ECMWF global analyses 1979-86: Circulation statistics and data evalation. NCAR Technical Note NCAR/TN-300+STR, 94pp. plus 12 fiche

Vinnikov, K. Ya. and I. B. Yeserkepova, 1990: Soil moisture: empirical data and model results. (submitted for publication)

Wallace, J.M., and Gutzler, D. S. 1981: Teleconnections in the geopotential height field during the northern hemisphere winter. *Mon. Wea. Rev.*, **109**, 784-812.

Washington, W. M., and G. A. Meehl, 1989: Climate sensitivity due to increased CO_2: Experiments with a coupled atmosphere and ocean general circulation model. *Clim. Dyn.*, **4** 1-38.

WGNE, 1988: Proceedings of the Workshop on Systematic Errors in Models of the Atmosphere, Toronto, 19-23 September 1988. CAS/JSC Working Group on Numerical Experimentation, WNN/TD - No.2 & 3, 382 pp.

Wigley, T. M. L., and B. D. Santer, 1990: Statistical comparison of spatial fields in model validation, perturbation and predictability experiments. *J. Geophys. Res.* (in press).

Wilson, C. A., and J. F. B. Mitchell, 1987: Simulated climate and CO_2- induced climate change over western Europe. *Climatic Change*, **10**, 11-42.

WMO, 1986: Workshop on Comparison of Simulations by Numerical Models of the Sensitivity of the Atmospheric Circulation to Sea Surface Temperature Anomalies. WMO/TD-No. 138, WCP-121, World Meteorological Organization, Geneva, 188 pp.

WMO, 1988: Modelling the Sensitivity and Variations of the Ocean-Atmosphere System. WMO/TD-No. 254, WCRP-15, World Meteorological Organization, Geneva, 289 pp.

Yeh, T. C., S. Y. Tao and M. C. Li, 1959: The abrupt change of circulation over the Northern Hemisphere during June and October. In: *The Atmosphere and the Sea in Motion*, 249-267.

Zeng, Qing-cun et al., 1988: Numerical simulation of monsoon and the abrupt change in atmospheric general circulation. *Chinese Journal of the Atmospheric Sciences*, Special Issue.

5

Equilibrium Climate Change - and its Implications for the Future

J. F. B. MITCHELL, S. MANABE, V. MELESHKO, T. TOKIOKA

Contributors:
A. Baede; A. Berger; G. Boer; M. Budyko; V. Canuto; H. Cao; R. Dickinson;
H. Ellsaesser; S. Grotch; R. Haarsma; A. Hecht; B. Hunt; B. Huntley;
R. Keshavamurty; R. Koerner; C. Lorius; M. MacCracken; G. Meehl; E. Oladipo;
B. Pittock; L. Prahm; D. Randall; P. Rowntree; D. Rind; M. Schlesinger; S. Schneider;
C.Senior; N. Shackleton; W. Shuttleworth; R. Stouffer; F. Street-Perrott; A. Velichko;
K. Vinnikov, R. Wetherald.

CONTENTS

EXECUTIVE SUMMARY

1. All models show substantial changes in climate when CO_2 concentrations are doubled, even though the changes vary from model to model on a sub-continental scale.

2. The main equilibrium changes in climate due to doubling CO_2 deduced from models are given below. The number of *'s indicates the degree of confidence determined subjectively from the amount of agreement between models, our understanding of the model results and our confidence in the representation of the relevant process in the model. Five *'s indicate virtual certainties, one * indicates low confidence.

Temperature:

*****	the lower atmosphere and Earth's surface warm;
*****	the stratosphere cools;
***	near the Earth's surface, the global average warming lies between +1.5°C and +4.5°C, with a "best guess" of 2.5°C;
***	the surface warming at high latitudes is greater than the global average in winter but smaller than in summer. (In time dependent simulations with a deep ocean, there is little warming over the high latitude southern ocean);
***	the surface warming and its seasonal variation are least in the tropics.

Precipitation:

****	the global average increases (as does that of evaporation), the larger the warming, the larger the increase;
***	increases at high latitudes throughout the year;
***	increases globally by 3 to 15% (as does evaporation);
**	increases at mid-latitudes in winter;
**	the zonal mean value increases in the tropics although there are areas of decrease. Shifts in the main tropical rain bands differ from model to model, so there is little consistency between models in simulated regional changes;
**	changes little in subtropical arid areas.

Soil moisture:

***	increases in high latitudes in winter;
**	decreases over northern mid-latitude continents in summer.

Snow and sea-ice:

****	the area of sea-ice and seasonal snow-cover diminish.

The results from models become less reliable at smaller scales, so predictions for smaller than continental regions should be treated with great caution. The continents warm more than the ocean. Temperature increases in southern Europe and central North America are greater than the global mean and are accompanied by reduced precipitation and soil moisture in summer. The Asian summer monsoon intensifies.

3. Changes in the day-to-day variability of weather are uncertain. However, episodes of high temperature will become more frequent in the future simply due to an increase in the mean temperature. There is some evidence of a general increase in convective precipitation.

4. The direct effect of deforestation on global mean climate is small. The indirect effects (through changes in the CO_2 sink) may be more important. However, tropical deforestation may lead to substantial local effects, including a reduction of about 20% in precipitation.

5. Improved predictions of global climate change require better treatment of processes affecting the distribution and properties of cloud, ocean-atmosphere interaction, convection, sea-ice and transfer of heat and moisture from the land surface. Increased model resolution will allow more realistic predictions of global-scale changes, and some improvement in the prediction of regional climate change.

5.1 Introduction

5.1.1 Why Carry Out Equilibrium Studies ?

Climate is in equilibrium when it is in balance with the radiative forcing (Section 3.2). Thus, as long as greenhouse gas concentrations continue to increase, climate will not reach equilibrium. Even if concentrations are eventually stabilised at constant levels and maintained there, it would be many decades before full equilibrium is reached. Thus equilibrium simulations cannot be used directly as forecasts. Why carry out equilibrium studies?

First, approximate equilibrium simulations using atmosphere-oceanic mixed layer models which ignore both the deep ocean and changes in ocean circulation (Section 3, Section 6.4.4.1) require less computer time than time-dependent simulations which must include the influence of the deep ocean to be credible.

Second, equilibrium experiments are easier to compare than time-dependent experiments. This, combined with the fact that they are relatively inexpensive to carry out, makes equilibrium simulations ideal for sensitivity studies in which the effect of using alternative parameterizations (for example, of cloud) can be assessed.

Third, it appears that apart from areas where the oceanic thermal inertia is large, as in the North Atlantic and in high southern latitudes, equilibrium solutions can be scaled and used as approximations to the time-dependent response (see Section 6).

Most equilibrium experiments consider the effect of doubling the concentration of atmospheric carbon dioxide, since the effect of increases in other trace gases can be calculated in terms of an effective CO_2 increase (see Section 2). Note that only the radiative effects of increases in gases are taken into account, and not the effects of related factors such as deforestation and possible changes in cloud albedo due to sulphur emissions.

Simulated changes in climate are known to be dependent on the simulation of the undisturbed climate [1] (see, for example, Mitchell et al., 1987). The simulation of present day climate is discussed in more detail in Section 4.

5.1.2 What Are The Limitations Of Equilibrium Climate Studies ?

Firstly, most equilibrium studies use models which exclude possible changes in ocean circulation. Nearly all the equilibrium studies which do allow changes in ocean circulation have been simplified in other ways such as ignoring the seasonal cycle of insolation (Manabe et al., 1990), or using idealised geography (Manabe and Bryan, 1985). The effects of the ocean and the differences between

non-equilibrium and equilibrium climate simulations are discussed further in Section 6.

Secondly, different areas of the world respond at different rates to the gradual increase in greenhouse gases. Over most of the ocean, the response to the increase in radiative heating will be relatively rapid, as little of the extra heat will penetrate below the thermocline at about 500m (see Section 6). On the other hand, in parts of the northern North Atlantic and the high latitude southern ocean, particularly in winter, the extra heat will be mixed down to several kilometres, significantly reducing the rate of warming and consequently the warming reached at any given time. In other words, the geographical patterns of the equilibrium warming may differ from patterns of the time-dependent warming as it evolves in time. This applies both to model simulations and palaeo-climatic analogues.

5.1.3 How Have The Equilibrium Experiments Been Assessed ?

Over 20 simulations of the equilibrium response to doubling CO_2 using general circulation models (GCMs) coupled to mixed-layer oceans have been carried out by 9 modelling groups (Table 3.2a). All those cited involved global models with realistic geography, a mixed layer ocean and a seasonal cycle of insolation. The more recent studies include a prescribed seasonally-varying oceanic heating (Section 3). Models 13, 20 and 21 in Table 3.2a also prescribe a heat convergence under sea ice. Clearly, it is not possible here to show results from all 20 or so experiments, so some way of condensing the available data must be chosen. We have chosen not to average the results, as there are aspects of each model which are misleading. Nor is it reasonable to choose a "best" model, as a particular model may be more reliable than another for one climatic parameter, but not for another. Moreover, a result which is common to most models is not necessarily the most reliable - it may merely reflect the fact that many models use similar (possibly erroneous) representations of complex atmospheric processes.

In this section, the climate changes which are common to all models, **and which are physically plausible** are emphasised, and illustrated by typical results. Where there is disagreement among model results, those which are probably unreliable (for example, because of large errors in the simulation of present climate) have been eliminated, and examples illustrating the range of uncertainty are included. The reasons for the discrepancies (if known) are stated, and an assessment of what seems most likely to be the correct result **in the light of current knowledge**, including evidence from time-dependent simulations (Section 6), is given.

The contents of the remainder of this section are as follows. First we consider the large-scale changes in

[1] *For example, if the snowline is misplaced in the simulation of present climate, then the large warming associated with the retreat of the snowline will be misplaced in the simulated climate change.*

temperature, precipitation and other climatic elements in equilibrium simulations of the effect of doubling CO_2, with the emphasis on new results. Several comprehensive reviews have been published recently (for example Dickinson, 1986; Schlesinger and Mitchell, 1985,1987; Mitchell, 1989) to which the reader is referred for further discussion of earlier studies. The purpose here is to describe the changes and to assess the realism of the mechanisms producing them. The possible changes in climatic variability are then discussed. Next, we consider simulated seasonal mean changes from three different models in five selected regions. These results have been scaled to give a "best estimate" of the warming which would occur at 2030 (at about the time of effective doubling of CO_2 in the IPCC Business-as-Usual Scenario. This is followed by an assessment of forecasts using the palaeo-analogue method and a review of attempts to model the direct climatic effects of deforestation. Finally, the main uncertainties are discussed.

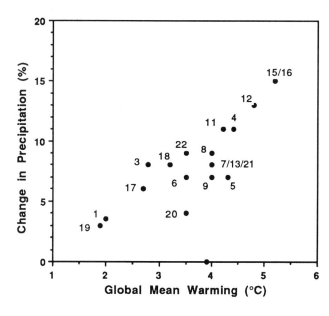

Figure 5.1: Percentage change in globally and annually averaged precipitation as a function of global mean warming from 17 models. The numbers refer to the entries describing the models in Table 3.2a.

5.2 Equilibrium Changes in Climatic Means Due to Doubling CO_2

5.2.1 *The Global Mean Equilibrium Response*

All models show a significant equilibrium increase in global average surface temperature due to a doubling of CO_2 which ranges from 1.9 to 5.2°C (Table 3.2a). Most results lie between 3.5 and 4.5°C, although this does not necessarily imply that the correct value lies in this range. The main uncertainty arises from the problems of simulating cloud. With no changes in cloud, a warming of 2 to 3°C is obtained (Table 3.2a, entries 1 and 2 - see also Hansen et al., 1984; Schlesinger, 1985), whereas models in which cloud amount is calculated interactively from relative humidity, but in which radiative properties are fixed, give a warming of 4 to 5°C (Table 3.2a, entries 3 to 16, 21).

Even amongst those models which calculate cloud from relative humidity, there is a wide variation in sensitivity (Table 3.2a, entries 3 to 16, 20, 22) and cloud feedback (Cess et al., 1989 - Table 3.2a, entries 3 and 4, see also Section 3.4.4). Sensitivity also varies for many other reasons, including the extent of sea ice in the control climate (Table 3.2a, entries 11 and 12; Manabe and Stouffer, 1980; Ingram et al., 1989).

Specifying cloud from relative humidity with fixed radiative properties ignores the possible effects of changes in cloud microphysics, such as changes in total cloud water content, the partition between cloud ice and cloud water, and changes in the effective radius of cloud particles. Because of this, attempts have been made to model cloud water content explicitly (Roeckner et al., 1987; Smith, 1990). Climate warming may produce an increase in cloud water content and hence in the reflectivity of cloud (a

negative feedback), but also an increase in the long-wave emissivity of cloud (increasing the greenhouse effect, a positive feedback especially for thin high cloud). Models disagree about the net effect which depends crucially on the radiative properties at solar and infrared wavelengths. One general circulation experiment (Mitchell et al., 1989 - Table 3.2a, entries 18 and 19) and experiments with simple one-dimensional radiative convective models (Somerville and Remer, 1984; Somerville and Iacobellis, 1988), suggest a negative feedback. A further possible negative feedback due to increases in the proportion of water cloud at the expense of ice cloud has been identified (Table 3.2a, entries 16, 17 - Mitchell et al., 1989).

On the basis of evidence from the more recent modelling studies (Table 3.2a, entries 3, 4, 7-9, 17-22) it appears that the equilibrium change in globally averaged surface temperature due to doubling CO_2 is between 1.9 and 4.4°C. The model results to not provide any compelling reason to alter the previously accepted range 1.5 to 4.5°C (U.S. National Academy of Sciences, 1979; Bolin et al., 1986). The clustering of estimates around 4°C (Table 3.2a, Figure 5.1) may be largely due to the neglect of changes in cloud microphysics in the models concerned. In the 2xCO_2 simulations in which some aspects of cloud microphysics are parameterized (Table 3.2a, entries 17-20, 22), the warming is less than 4°C. However, in idealised simulations (Cess et al., 1990), changes in cloud microphysics produced positive feedbacks in some models and negative feedbacks in others. Thus we cannot reduce the upper limit of the range. The modelling studies do not on their own provide a basis for choosing a most likely value.

Dickinson (in Bolin et al., 1986) attempted to quantify the uncertainty in the sensitivity of global mean temperature ΔTs to doubling CO_2 by considering the uncertainties in individual feedback processes as determined from climate model experiments. The climate sensitivity parameter Λ (the reciprocal of that defined in Section 3.3.1) is the sum of the individual feedback strengths, and the range of ΔTs is deduced from the range of Λ through

$$\Delta Ts = \Delta Q / \Lambda$$

where ΔQ is the change in radiative heating due to doubling CO_2 (See Section 3.3.1). Repeating this analysis with revised estimates [2] gives a range of 1.7 to 4.1°C. The mid-range value of Λ gives a sensitivity of 2.4°C which is less than the mid-range value of ΔTs. This is because for high sensitivity (small Λ), a given increment in Λ gives a bigger change in ΔTs than for low sensitivity. Similarly, taking the value of Λ corresponding to the middle of the range of Λ implied by 1.5 to 4.5°C gives a value of 2.3°C for ΔTs.

One can attempt to constrain the range of model sensitivities by comparing predictions of the expected warming to date with observations. This approach is fraught with uncertainty. Global mean temperatures are subject to considerable natural fluctuations (Section 7) and may have been influenced by external factors other than the greenhouse effect. In particular, the effect of aerosols on cloud (Section 2.3.3) may have suppressed the expected warming. There is also some uncertainty concerning the extent to which the thermal inertia of the oceans slows the rate of warming (Section 6). Hence, observations alone cannot be used to reduce the range of uncertainty, though assuming that factors other than the greenhouse effect remain unchanged, they are more consistent with a value in the lower end of the range 1.5 to 4.5°C (Section 8.1.3).

The evidence from the modelling studies, from observations and the sensitivity analyses indicate that the sensitivity of global mean surface temperature to doubling CO_2 is unlikely to lie outside the range 1.5 to 4.5°C. There is no compelling evidence to suggest in what part of this range the correct value is most likely to lie. There is no particular virtue in choosing the middle of the range, and both the sensitivity analysis and the observational evidence

neglecting factors other than the greenhouse effect, indicate that a value in the lower part of the range may be more likely. Most scientists declined to give a single number, but for the purpose of illustrating the IPCC Scenarios, a value of 2.5°C is considered to be the "best guess" **in the light of current knowledge.**

The simulated global warming due to a doubling of CO_2 is accompanied by increases in global mean evaporation and precipitation, ranging from 3 to 15%, (Table 3.2a, Figure 5.1). In general, the greater the warming, the greater the enhancement of the hydrological cycle. Since evaporation increases as well as precipitation, increased precipitation does not necessarily imply a wetter land surface.

5.2.2 What Are The Large-Scale Changes On Which The Models Agree ?

Although globally averaged changes give an indication of the likely magnitude of changes in climate due to increases in greenhouse gases, the geographical and seasonal distribution of the changes are needed to estimate the economic and social impacts of climate change. Despite the large range of estimates for the global annual average warming, there are several large-scale features of the simulated changes which are common to all models. These are outlined below. Where appropriate, results from the high resolution models (Table 3.2a, entries 20-22) are quoted to give the reader a rough indication of the size of the changes.

5.2.2.1 Temperature changes
The results from equilibrium simulations shown here are averaged over periods of 5 to 15 years. Because there is considerable interannual variability in simulated surface temperatures particularly in high latitudes in winter, some of the smaller-scale features may be random fluctuations due to the short sampling period rather than persistent changes due to doubling atmospheric CO_2.

1. All models produce a warming of the Earth's surface and troposphere (lower atmosphere) and a cooling of the stratosphere (Figure 5.2):
The warming of the surface and troposphere are due to an enhancement of the natural greenhouse effect. The stratospheric cooling is due to enhanced radiative cooling to space and increases with height, reaching 3 to 6°C at about 25 mb. Note that the models considered have at most two levels in the tropical stratosphere and so cannot resolve the details of the stratospheric cooling. Models with high resolution produce a cooling of up to 11°C in the stratosphere on doubling CO_2 (Fels et al., 1980). The indirect effects of stratospheric cooling are discussed in Section 2.2.3. Note that other greenhouse gases (for

[2] *water vapour and lapse rate feedback, $2.4 \pm 0.1 Wm^{-2}K^{-1}$ (Raval and Ramanathan, 1989); surface albedo feedback, $0.3 \pm 0.2 Wm^{-2} K^{-1}$ (Ingram et al., 1989); cloud feedback, $-0.3 \pm 0.7 Wm^{-2} K^{-1}$ (as originally estimated by Dickinson, consistent with range of ΔTs in Mitchell et al., 1989); the sensitivity $\Lambda = 1.8 \pm 0.7 Wm^{-2} K^{-1}$, assuming the errors are independent of one another.*

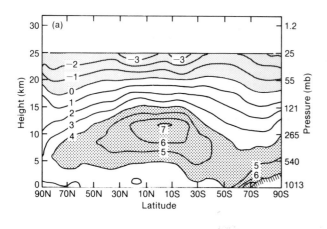

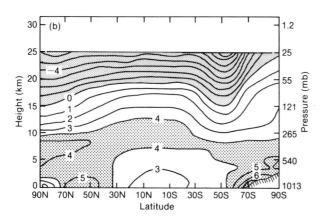

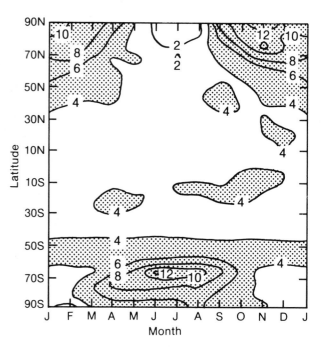

Figure 5.3: Time-latitude diagram of the zonally averaged increase in surface temperature due to doubling CO_2 in the GISS model (Hansen et al, 1984; from Schlesinger and Mitchell, 1987). Warming > 4°C stippled.

Figure 5.2: Height latitude diagram of the zonally averaged change in air temperature due to doubling CO_2 for the months of June, July and August in two models giving a global mean warming of 4°C. Cooling, and warming by > 4°C, stippled. (from Schlesinger and Mitchell, 1987): **(a)** with penetrative convection (Hansen et al, 1984), **(b)** with moist convection adjustment (Manabe and Wetherald, 1988).

example, methane, chlorofluorocarbons) produce a weak radiative warming of the stratosphere (Wang et al., 1990).

2. All models produce an enhanced warming in higher latitudes in late autumn and winter (Figures 5.2, 5.3):
This enhancement of the warming in higher latitudes is the result of a variety of processes.

First, in the warmer 2 x CO_2 climate, sea-ice forms later in autumn giving a pronounced warming in the Arctic (Figures 5.3, 5.4 a, b, c - see over page) and around Antarctica in the corresponding Southern Hemisphere season (Figures 5.3, 5.4 d, e, f - over page). (In high latitudes over the southern ocean, there is little or no warming at any time of year in time-dependent simulations (Section 6)). The reduction in the extent of sea-ice, which is highly reflective, leads to greater solar heating of the surface, mainly in summer, and further warming (a positive "temperature albedo" feedback, see Section 3.3.3). Second,

the warming leads to thinner sea ice allowing a greater flux of heat through the ice from the ocean in winter, enhancing the warming of the surface (Manabe and Stouffer, 1980). Third, there is a further temperature albedo feedback over the northern extra-tropical continents in spring due to the reduced extent and earlier melting of highly reflective snow-cover. Fourth, the warming is confined to near the surface. Thus the increase in outgoing long-wave radiation (at the top of the atmosphere) for each degree of warming is small relative to lower latitudes, where the warming is mixed throughout the troposphere. As a result, a larger warming is required in high latitudes to counterbalance the increase in downward radiation due to the increase in greenhouse gases. Finally, there is increased latent-heat release in high latitudes because of the stronger flux of moisture from the tropics (see Section 3).

In the more recent high resolution simulations (Table 3.2a, entries 20 - 22), the warming over North America in winter is about 4°C, rising to 8°C in the northeast of the continent (for example, Figure 5.4 a, b, c - over page). Similarly, over Europe and northern Asia, the warming is of order 4°C, with some areas of much larger warming as for example in eastern Siberia.

The comparatively large warming over sea-ice in autumn and early winter, and over the northern continents in spring, is physically plausible. Much of the variation in these features from model to model can be attributed to

(a) DJF 2 X CO2 - 1 X CO2 SURFACE AIR TEMPERATURE: CCC

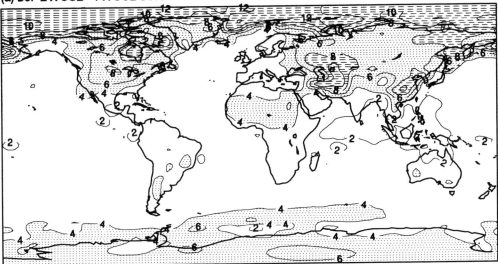

(b) DJF 2 X CO2 - 1 X CO2 SURFACE AIR TEMPERATURE: GFHI

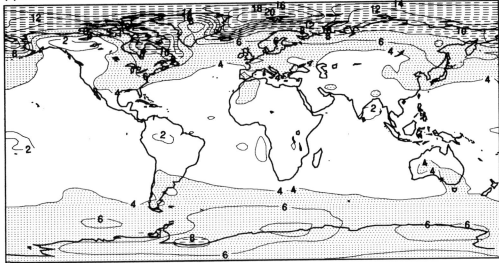

(c) DJF 2 X CO2 - 1 X CO2 SURFACE AIR TEMPERATURE: UKHI

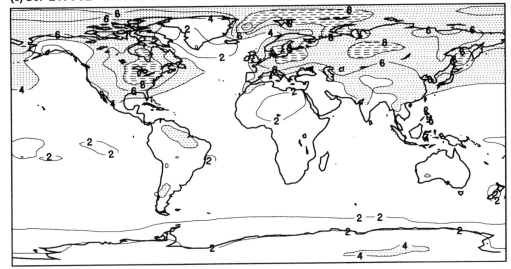

Figure 5.4: Change in surface air temperature (ten year means) due to doubling CO_2, for months December-January-February, as simulated by three high resolution models: (a) CCC: Canadian Climate Centre (Boer, pers. comm., 1989), (b) GFHI: Geophysical Fluids Dynamics Laboratory (Manabe and Wetherald, pers. comm., 1990), and (c) UKHI: United Kingdom Meteorological Office (Mitchell and Senior, pers. comm., 1990). Contours every 2°C, light stippling where the warming exceeds 4°C, dashed shading where the warming exceeds 8°C. Also shown in the colour section.

(d) JJA 2 X CO2 - 1 X CO2 SURFACE AIR TEMPERATURE: CCC

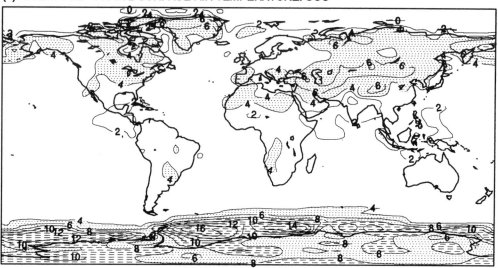

(e) JJA 2 X CO2 - 1 X CO2 SURFACE AIR TEMPERATURE: GFHI

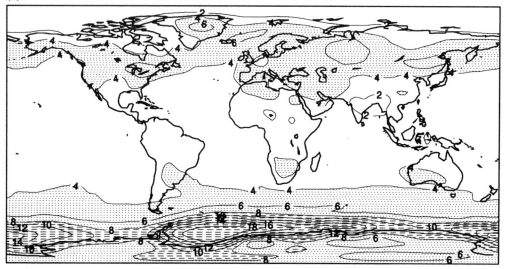

(f) JJA 2 X CO2 - 1 X CO2 SURFACE AIR TEMPERATURE: UKHI

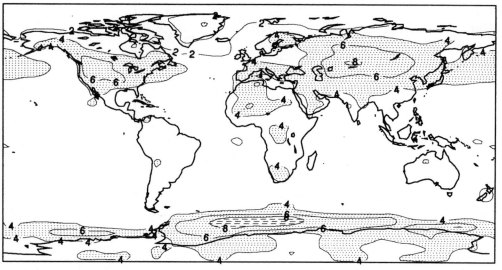

Figure 5.4 continued: Change in surface air temperature (ten year means) due to doubling CO_2, for months June-July-August, as simulated by three high resolution models: (d) CCC, (e) GFHI, and (f) UKHI. Other details as Figure 5.4. Also shown in the colour section.

differences in the sea-ice extents and snow cover in the simulation of present climate.

3. The warming is smaller than the global mean over sea-ice in the Arctic in summer (Figure 5.4 d, e, f) and around Antarctica in the corresponding season (Figure 5.4 a, c):
In summer the temperature of the surface of permanent sea-ice reaches melting point in both $1 \times CO_2$ (present day) and $2 \times CO_2$ simulations (Figure 5.3). Even in models where sea-ice disappears in summer in the $2 \times CO_2$ simulation, the large heat capacity of the oceanic mixed layer inhibits further warming above 0°C during the few months when it is ice free. Thus, the winter and annual average warmings are largest in high latitudes, but the summer warming is smaller than the annual average warming.

4. In all models, the tropical warming is both smaller than the global mean and varies little with season being typically 2 to 3°C (for example, Figures 5.3, 5.4):
The saturation vapour pressure of water increases non-linearly with temperature, so that at higher temperature, proportionally more of the increase in radiative heating of the surface is used to increase evaporation rather than to raise surface temperature. As a result, the surface warming is reduced relative to the global mean because of enhanced evaporative cooling. The enhanced evaporation is associated with increased tropical precipitation (see Section 5.2.2.2). Thus the warming of the upper troposphere in the tropics is greater than the global mean due to increased latent-heat release (Figure 5.2). Note that the magnitude of the warming in the tropics in those models with a similar global mean warming varies by a factor of almost 2 (Figure 5.2). The reasons for this are probably differences in the treatment of convection (Schlesinger and Mitchell, 1987; Cunnington and Mitchell, 1990) (the vertical transfer of heat and moisture on scales smaller than the model grid), in the choice of cloud radiative parameters (Cess and Potter, 1988) and the distribution of model layers in the vertical (Wetherald and Manabe, 1988). Some of these factors have been discussed further in Section 3.

5. In most models, the warming over northern mid-latitude continents in summer is greater than the global mean (for example, Figure 5.4 d, e, f):
Where the land surface becomes sufficiently dry to restrict evaporation, further drying reduces evaporation and hence evaporative cooling, leading to further warming of the surface (Figure 5.5). The reduction in evaporation may also produce a reduction in low cloud (for example, Manabe and Wetherald, 1987), further enhancing the surface warming (Figure 5.5). In one model, (Washington and Meehl, 1984; Meehl and Washington, 1989) the land surface becomes generally wetter in these latitudes in summer, reducing the warming. This is probably due to the

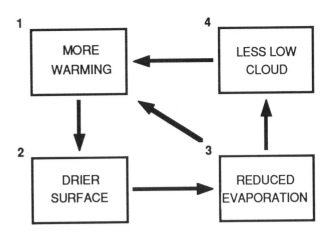

Figure 5.5: Schematic representation of soil moisture temperature feedback through changes in evaporation and low cloud.

land surface being excessively dry in the control simulation (see Section 5.2.2.3).

The summer warming in the more recent simulations (Table 3.2a, entries 20-22) is typically 4 to 5°C over the St. Lawrence-Great Lakes region and 5 to 6°C over central Asia (for example, Figure 5.4 d, e, f). Many of the inter-model differences in the simulated warming over the summer continents can be attributed to differences in the simulated changes in soil moisture and cloud.

5.2.2.2 Precipitation changes
1. All models produce enhanced precipitation in high latitudes and the tropics throughout the year, and in mid-latitudes in winter (see for example, Figure 5.6):
All models simulate a substantially moister atmosphere (increased specific humidity). Precipitation occurs in regions of lower level convergence, including the mid-latitude storm tracks and the inter-tropical convergence zone (ITCZ), where moist inflowing air is forced to ascend, cool and precipitate to remove the resulting super-saturation. The increases in atmospheric moisture will lead to a greater flux of moisture into these regions and hence increased precipitation provided there are no large compensational changes in circulation. The high resolution models (Table 3.2a, entries 20-22) give an increase of 10 to 20% in precipitation averaged over land between 35 and 55°N.

(a) DJF 2 X CO2 - 1 X CO2 PRECIPITATION: CCC

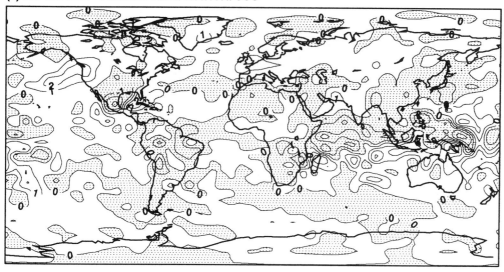

(b) DJF 2 X CO2 - 1 X CO2 PRECIPITATION: GFHI

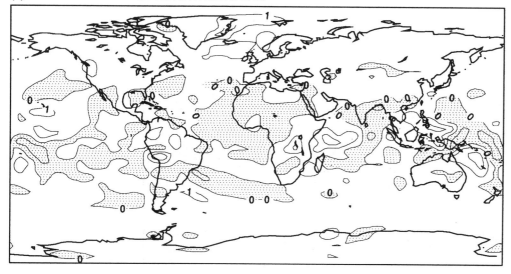

(c) DJF 2 X CO2 - 1 X CO2 PRECIPITATION: UKHI

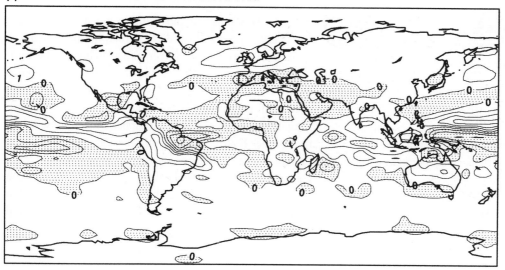

Figure 5.6: Change in precipitation (smoothed 10-year means) due to doubling CO_2, for months December-January-February, as simulated by three high resolution models: (a) CCC, (b) GFHI, and (c) UKHI. Contours at $\pm$ 0, 1, 2, 5 mm day^{-1}, areas of decrease stippled. Also shown in the colour section.

(d) JJA 2 X CO2 - 1 X CO2 PRECIPITATION: CCC

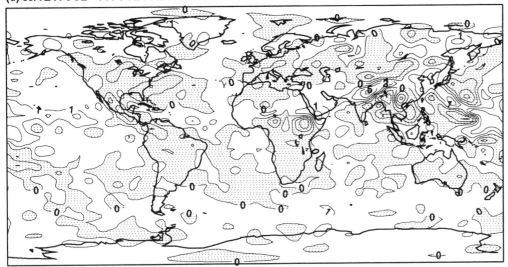

(e) JJA 2 X CO2 - 1 X CO2 PRECIPITATION: GFHI

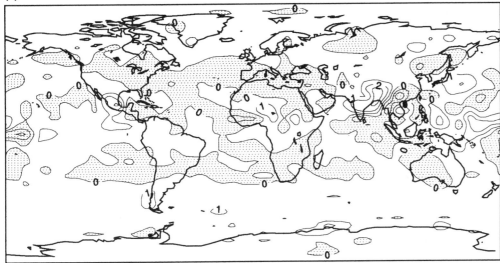

(f) JJA 2 X CO2 - 1 X CO2 PRECIPITATION: UKHI

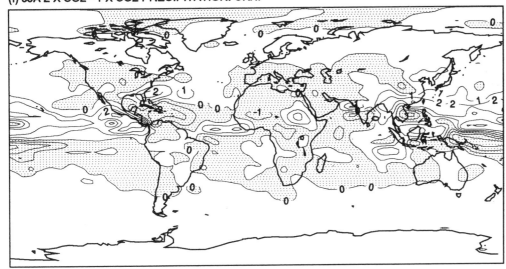

Figure 5.6 continued: Change in precipitation (smoothed 10-year means) due to doubling CO_2, for months June-July-August, as simulated by three high resolution models: (d) CCC, (e) GFHI, and (f) UKHI. Contours at $\pm 0, 1, 2, 5$ mm day^{-1}, areas of decrease stippled. Also shown in the colour section.

2. Changes in the dry subtropics are generally small and with both increases and decreases:

The interannual variability of precipitation is large relative to its mean value in these regions, so many of the changes indicated by the models cannot be demonstrated to be statistically significant. Note that even small changes of precipitation in arid regions can have substantial impacts.

3. There are considerable discrepancies regarding changes in precipitation on sub-continental scales, especially in the tropics although most models simulate an enhancement of the precipitation associated with a strengthening of the Southwest Asian monsoon (Table 3.2a, entries 3, 7, 8, 13, 15-22, Figure 5.6):

The inter-model agreement concerning changes in precipitation is less than in the case of temperature changes for two reasons. First, precipitation is changed ndirectly by a wider variety of different processes, many of them not resolved on the model's grid, and so is inherently more difficult to model, whereas the warming is primarily a direct response to increased radiative heating. Second, the changes in precipitation are relatively small compared with the natural variations and so are more difficult to detect in the short sampling period available.

In many of the models, summer rainfall decreases slightly over much of the northern mid-latitude continents and there is a tendency for the tropical maximum of the precipitation to shift further into the summer hemisphere. In other models, the tropical rain belt tends to shift into the winter hemisphere (Table 3.2a, entries 11, 12) or southwards throughout the year (Table 3.2a, entry 6). In some models, enhancement of the Asian monsoon appears to be associated with strong positive cloud feedback whereby decreases in cloud cover over Eurasia in summer enhance the solar heating of the surface (Wilson and Mitchell 1987a). This increases the land-sea temperature contrast which drives the summer monsoon.

Some of the precipitation over mid-latitude continents in summer originates from local evaporation (Mintz, 1984) so the simulated changes in precipitation are likely to be sensitive to the wetness of the surface and the formulation of evapotranspiration. Changes in soil moisture are considered in more detail in the following sub-section.

5.2.2.3 Soil moisture changes.

In all simulations with enhanced CO_2, both evaporation and precipitation increase. In regions where precipitation increases, increases in evaporation may be even greater. Current models represent the availability of water in the upper soil layers by a soil moisture variable, which is augmented by precipitation and snow melt, and depleted by evaporation and runoff. The representations of surface hydrology in models used so far in $2 \times CO_2$ experiments are highly simplified, though some models make allowance for

the type of soil and/or vegetation in a simple manner (Table 3.2a, entries 11, 12, 20). Note that none of the models considered allow for the direct effect of CO_2 on vegetation. Of particular importance is the expected increase in water efficiency (Section 10) which, in the absence of other changes, would reduce evapotranspiration from the surface.

The main findings from equilibrium simulations are:

1. All models simulate a general increase in the soil moisture of the northern high-latitude continents in winter (Figure 5.7, 5.8 a, b, c):

This increase is due to some or all of the following factors:- enhanced winter precipitation, more precipitation falling as rain rather than snow, more snowmelt and the relatively low rate of increase of potential evaporation with temperature found at lower winter temperatures. Note only a few models (Table 3.2a, entries 11, 12, 16, 20) allow in any way for the effects of groundwater becoming frozen.

2. Most models produce an enhanced large-scale drying of the Earth's surface in northern mid-latitude during northern summer (Figure 5.7, 5.8 d, e, f):

Although there is good agreement among the most recent simulations in this respect, confidence in the reliability of the simulated changes is low in view of the simplified representation of the land surface. In the three high resolution models, the reduction in soil moisture averaged over 35 to 55°N ranges from 17 to 23%. The reasons for this drying are discussed in more detail below. In three of the simulations that do not produce such a drying (Table 3.2a; entries 6, 7, 16), the soil in the control simulation is excessively dry. In the other two (Table 3.2a, entries 11, 12), it has been argued that the simulated changes are equivalent to increased frequency of drought in these latitudes (Rind et al., 1989a).

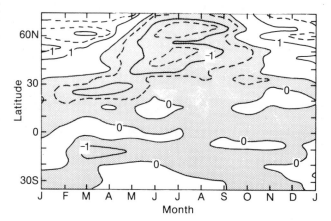

Figure 5.7: Time-latitude diagram of changes in soil moisture due to doubling CO_2. Contours every cm, areas of decrease stippled. (Manabe and Wetherald, 1987).

(a) DJF 2 X CO2 - 1 X CO2 SOIL MOISTURE: CCC

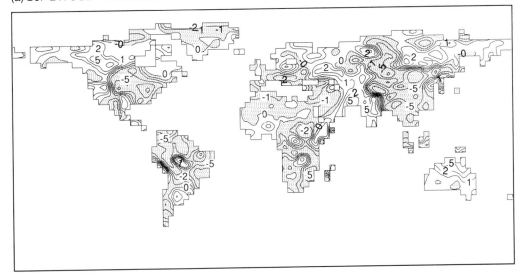

(b) DJF 2 X CO2 - 1 X CO2 SOIL MOISTURE: GFHI

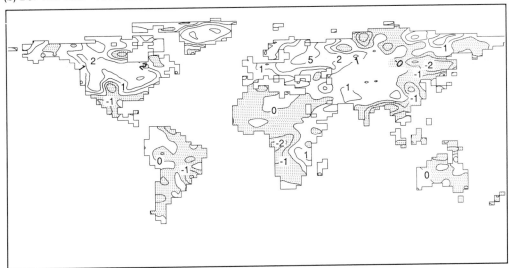

(c) DJF 2 X CO2 - 1 X CO2 SOIL MOISTURE: UKHI

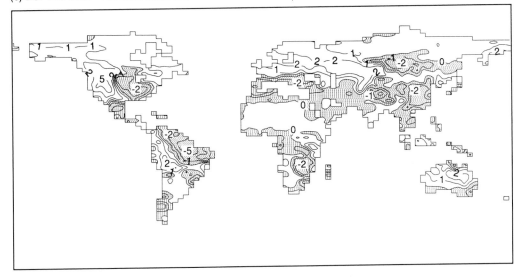

Figure 5.8: Change in soil moisture (smoothed 10-year means) due to doubling CO_2, for months December-January-February, as simulated by three high resolution models: (a) CCC, (b) GFHI, and (c) UKHI. Note that (a) has a geographically variable soil capacity whereas the other two models have the same capacity everywhere. Contours at ± 0, 1, 2, 5 cm, areas of decrease stippled. Also shown in the colour section.

(d) JJA 2 X CO2 - 1 X CO2 SOIL MOISTURE: CCC

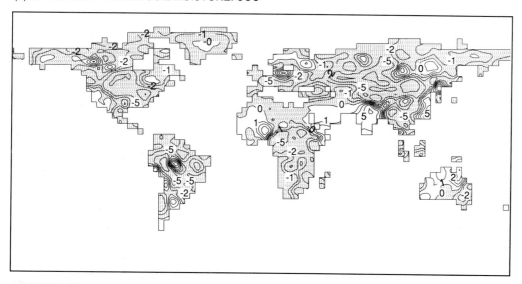

(e) JJA 2 X CO2 - 1 X CO2 SOIL MOISTURE: GFHI

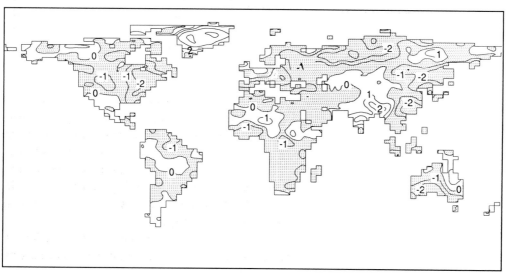

(f) JJA 2 X CO2 - 1 X CO2 SOIL MOISTURE: UKHI

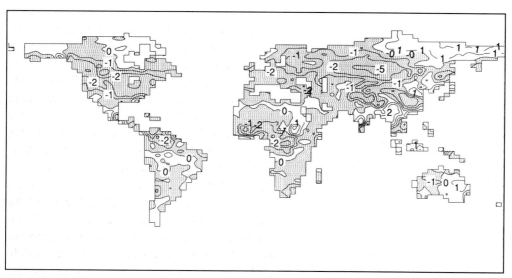

Figure 5.8 continued: Change in soil moisture (smoothed 10-year means) due to doubling CO_2, for months June-July-August, as simulated by three high resolution models: (d) CCC, (e) GFHI, and (f) UKHI. Note that (d) has a geographically variable soil capacity whereas the other two models have the same capacity everywhere. Contours at ± 0, 1, 2, 5 cm, areas of decrease stippled. Also shown in the colour section.

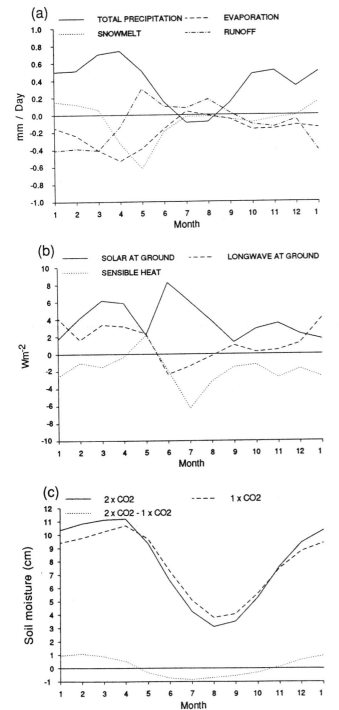

Figure 5.9: Changes in area means due to doubling CO_2 averaged over land between 35 to 55°N. **(a)** Water budget (mm day^{-1}) **(b)** Heat budget (Wm^{-2}) **(c)** Soil moisture (1xCO_2 and 2xCO_2) (from the study by Manabe and Wetherald, 1987).

Since drying of the northern mid-latitude continents in summer could have significant impacts, these changes warrant a close examination of the physical processes responsible, and the fidelity of their representation in models needs to be considered carefully. Hence the water and energy budgets in this region have been analysed in some detail. In the control simulation, all models produce a maximum in soil moisture in winter and spring, and a rapid drying to a minimum in summer (Figure 5.9c). With doubled CO_2, enhanced winter precipitation (and snow melt) (Figure 5.9a) produce higher soil moisture levels into early spring (Figure 5.9c). In the warmer climate, snow melt and the summer drying begin earlier, reducing the soil moisture levels in summer relative to the present climate (Figures 5.8d, e, f, 5.9c). The drying in the 2xCO_2 simulation is also more rapid due to the higher temperatures and in some regions is reinforced by reduced precipitation. Reductions in surface moisture may lead to a drying of the boundary layer, reduced low cloud and hence further warming and drying of the surface (Manabe and Wetherald, 1987) (see also Figure 5.5).

In most models, the soil over much of mid-latitudes (35 to 55°N) is close to saturation in spring in both the 1xCO_2 and 2xCO_2 simulations, so that on enhancing CO_2, the summer drying starts earlier but from the same level (for example, Figure 5.10a). In a minority of models (Meehl and Washington, 1988,1989; Mitchell and Warrilow, 1987) the soil in the 1xCO_2 formulation is not close to saturation, and the enhanced winter precipitation in 2xCO_2 simulation is stored in the soil. Hence, although the summer drying in the 2xCO_2 experiments starts earlier, it starts from a higher level than in the control simulation, and may not become drier before next winter season (Figure 5.10b). Even in these models, the surface becomes drier in the southern mid-latitudes in the 2xCO_2 simulations.

From the experiments carried out to date, the following factors appear to contribute to the simulated summer drying in mid-latitudes.

i)   The soil is close to saturation in late winter (spring in higher latitudes) in the control simulation, so that increased precipitation in the anomaly simulation is run off and is not stored in the soil.

ii) The greater the seasonal variation of soil moisture in the control simulation, the greater the change due to an earlier start to the drying season (for example Figure 5.9c). Of course, if the soil moisture content in the simulation of present climate is very small in summer, it cannot decrease much. A comparison of model and field data over the Soviet Union (Vinnikov and Yeserkepova, 1989) indicates that the simulated soil moisture levels in some models are much too low in summer. The simulated reduction in soil moisture due to doubling CO_2 in such models would then be less than if higher, more realistic levels of soil moisture were present in the control simulation.

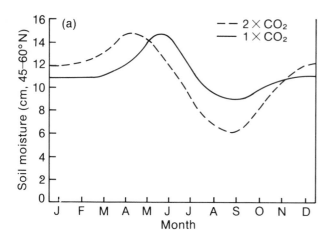

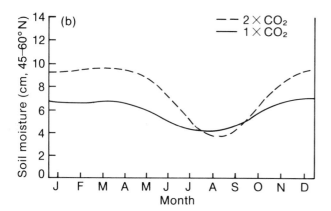

Figure 5.10: Seasonal cycle of soil moisture for normal and doubled CO_2 concentrations, averaged over land, 45-60°N. **(a)** With standard treatment of runoff. **(b)** With snowmelt run off over frozen ground (from Mitchell and Warrilow, 1987).

iii)   In higher latitudes, snow melts earlier. Hence accurate simulation of snow cover is important.

iv)   The changes in soil moisture can be amplified by feedbacks involving changes in cloud.

v)   Enhanced summer drying in mid-latitudes may occur even in models which produce enhanced precipitation. Of course, the drying is more pronounced in those models (and regions) in which precipitation is reduced in summer.

The simulated changes of soil moisture in the tropics vary from model to model, being more directly related to changes in precipitation.

5.2.2.4 Sea-ice changes.

In simulations with enhanced CO_2, both the extent and thickness of sea ice are significantly reduced. In some summer simulations, sea ice is completely removed in the Arctic (Wilson and Mitchell, 1987a; Boer, 1989, personal communication) and around Antarctica (Wilson and Mitchell, 1987a). In other models there are large reductions

in the extent of sea ice, but some cover remains in the Arctic and around Antarctica in summer (Noda and Tokioka, 1989; Meehl and Washington, 1989). Finally, in some models (Wetherald, 1989, pers comm) the extent of sea-ice change is less, but the thickness is reduced by up to a factor of two.

The factors contributing to the differences between models include differences in the sea-ice extent and depth in the control simulation (Spelman and Manabe, 1984), differences in the treatment of sea-ice albedo (for example, Washington and Meehl, 1986), and the inclusion of corrective heat-flux under sea ice in some models (for example Manabe and Wetherald, 1989, personal communication; Boer, 1989, personal communication) and not others.

On the basis of current simulations, it is not possible to make reliable quantitative estimates of the changes in the sea ice extent and depth. It should be noted that the models considered here neglect ice dynamics, leads, salinity effects, and changes in ocean circulation.

5.2.2.5 Changes in mean sea-level pressure

Except in areas close to the equator, sea-level pressure (SLP) changes give an indication of changes in the low-level circulation, including the strength and intensity of the mean surface winds. The changes in SLP have been assessed using the limited number of results available (Table 3.2a, entries 7, 13, 15, 20-22), though in most cases, information on the statistical significance of the changes was not provided.

1. Throughout the year there is a weakening of the north-south pressure gradient in the southern hemisphere extratropics (for example, Figure 5.11 over page), implying a weakening of the mid-latitude westerlies:
Both the subtropical anticyclones and the Antarctic circumpolar low pressure trough diminish in intensity (Figure 5.11 over page). This is presumably due to the relatively strong warming over sea-ice around Antarctica reducing the equator-to-pole temperature gradient (for example, Figure 5.2). Note that at higher levels of the troposphere, the equator-to-pole temperature gradient is increased (for example, Figure 5.2) and may be sufficient to produce **stronger** westerly flow at upper levels (Mitchell and Wilson, 1987a), and that coupled models do not produce a large warming around Antarctica (see Section 6)

2. In December, January and February most models produce higher pressure off Newfoundland, consistent with an eastward shift of the Iceland low, and a general decrease over eastern Siberia, apparently due to a weakening of the Siberian anticyclone (Figure 5.11a).

3. In June, July and August SLP decreases over Eurasia intensifying the monsoon low, and there are increases over

(a) DJF 2 X CO2 - 1 X CO2 MEAN SEA LEVEL PRESSURE: CCC

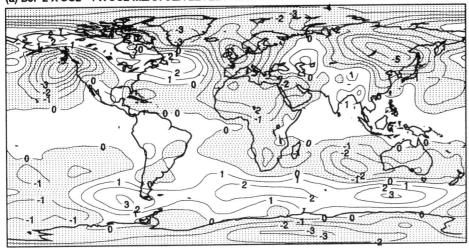

(b) JJA 2 X CO2 - 1 X CO2 MEAN SEA LEVEL PRESSURE: CCC

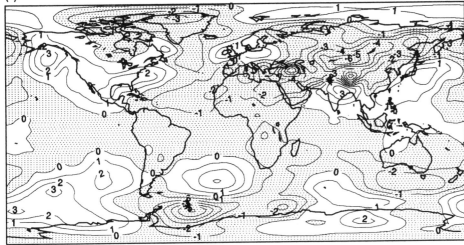

Figure 5.11: Changes in mean sea level pressure due to doubling CO_2 as simulated by a Canadian Climate Centre model (Boer, private communication, 1989). Contours every 1 mb, areas of decrease are stippled. **(a)** December, January and February, **(b)** June, July and August

India, implying a northward shift of the monsoon trough at those longitudes (Figure 5.11b). There is also a weakening of the Azores anticyclone:

Although these features are common to most of the models considered, there are also large differences in the location of individual features from model to model. Hence changes in low level circulation at a particular location are uncertain. The interannual variability of SLP is large, particularly in the extra-tropics in winter. This, as well as differences in model formulation, contributes to inter-model differences.

5.2.2.6 Deep ocean circulation changes.
There have been few equilibrium CO_2 experiments using atmospheric models coupled to a full dynamical model of the ocean, and most so far (Spelman and Manabe, 1984;

Manabe and Bryan, 1985) have assumed a simplified geometry and neglected the seasonal variation of insolation. Two main features have emerged.

1. The large warming of the ocean in high latitudes is propagated downwards to the ocean floor, where it spreads to all latitudes. The warming of the deep oceans is consistent with palaeo-oceanographic data for a warmer climate (the Cretaceous, about 70 million years ago).

2. There is a weakening of the mean meridional oceanic circulation of the Atlantic on doubling CO_2 from present to higher levels (Manabe et al., 1990). This is also seen in the transient response experiments (Section 6).

5.3 Equilibrium Changes in Variability due to Doubling CO$_2$

Changes in the variability and frequency of extreme events may have more impact than changes in mean climate. Changes in the frequency of extreme events may occur in a number of ways. For example, the shape of the frequency distribution may not be altered, but the mean may change, leading to a sharp increase in the frequency of extreme events at one end of the frequency distribution, and a decrease at the other end (Mearns et al., 1984,1990; Parry and Carter, 1985; Wigley, 1985; Pittock, 1989) (Figure 5.12a). Thus the general warming due to increases in greenhouse gases will undoubtedly lead to more frequent "hot" days and fewer "cold" days. Conversely, the mean may be unchanged, but the shape of the frequency distribution may alter. For example, the standard deviation (spread) of the frequency distribution may increase producing more extreme events at both ends of the distribution (Figure 5.12b). Of course, the mean and the spread of the frequency distribution may change simultaneously.

In the first example, the changes in the frequency of extreme events can be calculated by simply shifting the currently observed (or simulated control) frequency distribution by the mean change (Figure 5.12a). Here, the second type of change is considered in more detail. The standard deviation of the frequency distribution is used as a measure of variability, and both the standard deviation of daily values about the monthly mean and the standard deviation of the interannual variation of monthly means are considered, as appropriate. For day-to-day variability, results from only two models were available (Table 3.2a, entries 20, 22), whereas for interannual variability, results

from five models were used (Table 3.2a, entries 7, 11, 13, 15, 20).

Several factors must be borne in mind when deriving changes in standard deviations from numerical simulations. First, a much longer simulation is required to establish that changes in standard deviations, as opposed to changes in means, are statistically significant. Most studies to date use periods of 15 years or less, which is barely sufficient to establish that the simulated changes in standard deviation are unlikely to have occurred by chance. Second, although the models exhibit a measure of agreement on the larger scales, there is much less agreement on the regional scale, especially in the case of precipitation. Hence, only the changes in standard deviations on the larger scales will be considered. Third, information on changes in variability is not as readily available as information on changes in means. Thus the conclusions below are based on a much smaller number of simulations than is generally the case in Section 5.2. Finally, the coupled atmosphere mixed layer models used to derive the results in this section do not reproduce the El Niño - Southern Oscillation phenomena which are the main source of interannual variability in the tropics (Section 7.9.1).

5.3.1 Temperature

5.3.1.1 Day-to-day variability

Results from only two models were available globally. There are reductions in the standard deviation over and around the winter sea-ice margins - notably over Hudson Bay, the Norwegian Sea and the Sea of Okhotsk in January and around Antarctica in July. The decrease in these regions may be attributed to the reduction in the north-south temperature gradient (Figure 5.4), and to a reduction

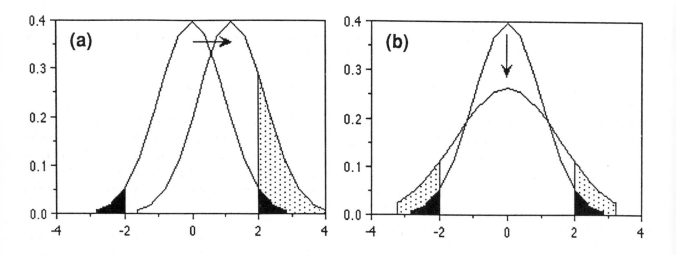

Figure 5.12: Schematic changes in frequency distributions. The solid area marks the 5% extremes of a normal distribution. The shaded area represents an increase in the number of extreme events outside the unperturbed 5% limits. **(a)** Increase in mean only. Note the large increase in the number of extreme events to the right of the original distribution, **(b)** No change in mean but a 50% increase in standard deviation. Note the increase in extreme events on both sides of the original distribution.

in the frequency or intensity of atmospheric disturbances (see Section 5.3.3). There are also reductions over much of the northern hemisphere continents in January, although both models produce increases over parts of Canada and Siberia. A third study (Rind et al., 1989b) reports decreases over the United States though generally they are not statistically significant.

5.3.1.2 Interannual variability
Apart from a general reduction in the vicinity of the winter sea-ice margins, no meaningful patterns of change could be distinguished (Five models were considered, Table 3.2a, entries 7, 11, 13, 15, 20).

5.3.1.3 Diurnal range of temperature
There is no compelling evidence for a general reduction in the amplitude of the diurnal cycle.

The increases in CO_2 and other "greenhouse" gases increase the downward longwave flux at the surface (Section 2). Both the upward flux of longwave radiation and evaporative cooling increase non-linearly with surface temperature. One would expect the increased downward radiation to produce a larger warming at night (when the surface temperature and hence the rate of increase of radiative and evaporative cooling with increase in temperature is smaller) than during the day, and hence a reduction in diurnal range.

Only a few models contain a diurnal cycle (see Table 3.2a). Boer (personal communication) reports a small reduction (0.28°C) in the globally averaged diurnal range of temperature (Table 3.2a, entry 20). In another study (Rind et al., 1989b) the range usually decreased over the United States, especially in summer. However, Cao (personal communication) found that increases in the diurnal range of temperature were evident over much of the northern mid-latitude continents, especially in spring and autumn (Table 3.2a, entry 15) although the global annual mean was reduced by 0.17°C. The amplitude of the diurnal cycle may also be reduced by increases in cloud cover or ground wetness, or altered by changes in the latitude of the snowline. As these quantities (and changes in these quantities) vary greatly from model to model, a reduction in the diurnal cycle seems far from certain.

5.3.2 Precipitation
Precipitation exhibits much more temporal and spatial variability than temperature. As a result, the simulation of the mean (and variability) of precipitation for present day climate is less reliable than for temperature, particularly in low resolution models, and it is only possible to make weak statements concerning changes in variability.

1.There is some indication that variability (interannual standard deviation) increases where mean precipitation

increases and vice-versa (Wilson and Mitchell, 1987b; Rind et al., 1989b), though this is not always the case.
For example, in one study (Rind et al., 1989b) this tendency in interannual variability was found at 60-70% of the grid-points considered. In another study (Wilson and Mitchell, 1987b) the summer rainfall over southern Europe decreased and the maximum number of consecutive days without rainfall increased substantially.

2. There is a consistent increase in the frequency of convective (sub grid-scale) precipitation, usually at the expense of precipitation from the larger scale (resolved) vertical motions (Noda and Tokioka, 1989; Hansen et al., 1989; Mitchell, pers. comm.).

In one study (Noda and Tokioka, 1989) the area of precipitation over the globe decreased even though global mean precipitation increased. There is a tendency for convective motions to penetrate higher (Mitchell and Ingram, 1989; Wetherald and Manabe, 1988) and perhaps over greater depth (Hansen et al., 1989) in a warmer climate. These changes imply an increase in the more intense local rain storms, and hence in run off, at the expense of the gentler but more persistent rainfall events associated with larger scale disturbances. Note that not all models include the diurnal cycle which has a strong modulating influence on convection.

The tendency for local convective instability to increase is likely to be independent of the particular model used as, in a warmer climate, the radiative cooling of the atmosphere and the radiative heating of the surface both increase (Mitchell et al., 1987). These changes must be balanced by the enhanced vertical transport of heat from the surface. Furthermore, given the non-linear increase in potential evaporation with increase in temperature, the increase in vertical heat transport is more likely to be achieved through latent heat rather than by sensible heat, and hence accompanied by a marked increase in convective rainfall.

5.3.3 Winds and Disturbances
Current climate models, particularly those at lower resolution, have limited success in simulating storm tracks and low frequency variability, and do not resolve smaller scale disturbances such as hurricanes explicitly (Sections 4.2.4, 4.6). Hence results from current models at best only give an indication of the likely changes in winds and disturbances.

1. There is some indication of a general reduction in day-to-day and interannual variability in the mid-latitude storm tracks in winter, though the patterns of change vary from model to model:
Here, the standard deviation of variations in mean sea-level pressure (SLP) has been used as an indication of the

frequency and intensity of disturbances. A reduction in mid-latitude synoptic variability might be expected as a result of the reduction in the equator-to-pole temperature gradient at low levels (for example, Figure 5.2). Results on changes in day-to-day variability were available from only two models. There was a general reduction in the standard deviation in mid-latitudes in winter though the patterns of change differed considerably. By applying a time filter to the daily variances of 500mb height, one can pick out the mid-latitude storm tracks (Blackmon, 1975). In winter, Siegmund (1990) found a reduction in the intensity of the filtered variances of 500mb height in mid-latitudes and an increase in high latitudes (Table 3.2a, entry 15). In another study (Bates and Meehl, 1986: Table 3.2a, entry 6) a similar reduction in the filtered variance of 500mb heights was reported. All these changes indicate a decrease in the intensity or frequency (or both) of disturbances resolved on the model grid (typically greater than about 1,000 km) but do not allow one to conclude the same for smaller-scale synoptic disturbances. One study (Bates and Meehl, 1986) reports a reduction in "blocking" (defined as areas of high pressure anomaly which persist for more than seven days) in the southern hemisphere, and changes in the positions but not the intensity of blocking in the Northern Hemisphere though no information was provided on the statistical significance of the results.

In the five models considered, there was a general reduction in the standard deviation of interannual variations in monthly mean SLP. However, the patterns varied considerably from model to model, so no other meaningful conclusions could be drawn.

2. There is some evidence from model simulations and empirical considerations that the frequency per year, intensity and area of occurrence of tropical disturbances may increase, though it is not yet compelling:

It has been observed that tropical storms (hurricanes, typhoons or cyclones) form only where the sea surface temperatures (SSTs) are 27°C or greater. This might lead one to expect a more widespread occurrence of tropical storms in a warmer climate. A recent theoretical model of tropical storms suggests that the maximum possible intensity would increase, with an enhancement of destructive power (as measured by the square of the wind speed) of 40% for an increase of 3°C in SST (Emanuel, 1987). However, Emanuel (1987) did note that very few tropical storms in the present climate actually attained the maximum intensity predicted by his analysis. In a complementary study, Merrill (1988) discussed the environmental influences on hurricane intensification. In agreement with Emanuel, Merrill concluded that the maximum intensity of a tropical storm is bounded above by a monotonically increasing function of sea surface temperature (SST). By compositing intensifying versus non-intensifying systems over a six-year period for the North Atlantic, Merrill was able to identify a number of environmental factors which could inhibit the further deepening of a tropical storm, even if the SSTs are favourable. The non-intensifying composite storms displayed stronger vertical wind shears and uni-directional flow over and near the storm centre than intensifying storms. Gray (1979) identified the need for weak vertical wind-shear over and near the storm centre and enhanced low-level cyclonic vorticity and mid-tropospheric humidity as factors favouring intensification of a tropical cyclone. There is no guarantee that criteria such as the lower bound of SST of 27°C would remain constant with changes in climate. There is little agreement in the simulated changes in tropical circulation due to doubling CO_2 in current climate models (as shown by the differing patterns of changes in tropical precipitation). Furthermore, the models considered in this section ignore changes in ocean circulation which form part of the El Niño phenomenon, and lead to the associated anomalies in SST and atmospheric circulation which have a profound influence on the present distribution and frequency of tropical storms.

High resolution atmospheric models used for weather forecasting show considerable success at predicting the development and track of tropical cyclones (Dell 'Osso and Bengtsson, 1985; Krishnamurti et al., 1989; Morris, 1989) although the horizontal resolution used (~100km) is inadequate to resolve their detailed structure. Krishnamurti et al. (1989) found that the quality of the forecasts decreased as horizontal resolution was decreased, but even so, the simulated maximum wind intensity decreased little until much coarser (above 400 km) resolution was reached. At the lower resolution used in climate studies (250 km or greater) one can choose objective criteria (for example, a warm core and low-level vorticity and surface pressure depression greater than specified limits) to select appropriate cyclones and compare their seasonal and geographical distributions with those of observed tropical storms. In both respects, the simulated storms resemble those observed over most oceans (Manabe et al., 1970; Bengtsson et al., 1985; Broccoli and Manabe, 1990 (Table 3.2a, entry 21)). Thus, although global models cannot resolve hurricanes explicitly, they give a surprisingly good indication of the regions of potential hurricane formation. In contrast to empirical methods, the criteria chosen are not obviously dependent on the present climate.

Using models with prescribed cloudiness, Broccoli and Manabe (1990) found an increase of 20% in the number of storm days (a combined measure of the number and duration of storms) on doubling CO_2. This is attributed to enhanced evaporation leading to increased moisture convergence and latent heat release which is converted to locally transient kinetic energy (stronger winds). In contrast, in an experiment in which cloudiness was allowed to change, the number of storm days decreased by 10 to 15% even though the increase in evaporation was even greater in this experiment. The increases in local energy generation and conversion were smaller, and the associated winds weakened slightly. The reason for this discrepancy has not been found, nor has the role of cloud feedback in these results been identified.

A preliminary experiment with a model which resolves hurricanes (Yamasaki, personal communication) showed an increase in the number, and a decrease in the intensity of tropical disturbances when sea-surface temperatures were increased, but the simulation was very short.

In summary, the maximum intensity of tropical storms may increase, but the distribution and frequency of occurrence will depend on the detailed changes in aspects of circulation in the tropics which are probably not yet adequately simulated by climate models.

5.4 Regional Changes - Estimates for 2030 (assuming IPCC "Business-as-Usual" Scenario)

5.4.1 Introduction

In order to assess the impacts of future changes in climate, one needs to know the changes and rates of change in climate on a regional scale (i.e., areas of order 1000 km square or so). Results from current equilibrium experiments often differ regarding regional variations in the changes. Furthermore, few time-dependent simulations have been carried out (Section 6), none correspond exactly to the IPCC Scenarios and all use low horizontal resolution. Nevertheless, one of the briefs of Working Group I was to provide estimates of changes in 5 selected regions.

In order to provide these regional estimates it has been necessary to make certain assumptions and approximations (Section 5.4.3).

The main conclusions of this section (see Table 5.1 - next page) are:

1. The regional changes in temperature may vary substantially from the global mean, and the magnitudes of regional changes in precipitation and soil moisture are typically 10 to 20% at 2030 under the IPCC "Business-as-Usual" Scenario.
2. Although there is still substantial disagreement in some regions between the models considered, the agreement is better than in earlier studies (e.g., Schlesinger and Mitchell, 1987).

5.4.2 Limitations Of Simulated Regional Changes

Although there is agreement between models on the qualitative nature of the large-scale changes in temperature and to a lesser extent precipitation, there is much less agreement when one considers variations in the changes on a regional (sub-continental) scale, i.e., areas of order $1,000,000 \text{ km}^2$. For example, it is likely that increases in greenhouse gases will increase precipitation near 60 degrees of latitude north, but there is little agreement between models on the variation of the increases with longitude. The horizontal resolution of most models used until now (typically 250-700km) is inadequate to produce an accurate representation of many of the regional features of climate, especially precipitation, which is strongly influenced by topography. The parameterization of processes not explicitly resolvable on the model grid also leads to errors at regional scales. The models in Table 3.2a do not allow for changes or interannual variations in oceanic heat transport.

The nature of inter-model discrepancies in these studies is illustrated by considering the changes averaged over several regions of about $4,000,000 \text{ km}^2$. The regions are chosen so as to represent a range of climates. Different models perform well in different regions. Inconsistencies in the changes produced by different models may be resolved to some extent by selecting those models giving the more realistic simulations of present climate. Such critical evaluations at regional level will best be done by the potential users, and revised as improved model simulations become available. Confidence in any one prediction of spatial *variations* in changes at a regional scale must presently be regarded as low.

An estimate of the changes in temperature, precipitation and soil moisture averaged over the 5 regions selected by IPCC is given in Section 5.4.4. The results are based on the high resolution studies (Table 3.2a, entries 20-22) since in general these produce a better simulation of present day climate (see Section 4). Results from five low resolution models (Grotch 1988, 1989, personal communication) (Table 3.2a, entries 3, 7, 11 ,13 ,15) were also considered. There may be considerable variations within the regions, and in the changes produced by the different models within the regions.

5.4.3 Assumptions Made In Deriving Estimates For 2030

The following assumptions have been made:

i)   **The concentrations of greenhouse gases increase as in the IPCC "Business-as-Usual" Scenario.** This assumes only modest increases in efficiency and gives an effective doubling of CO_2 by about 2020, and an effective quadrupling by about 2080. Reference will also be made to IPCC Scenario B which assumes large efficiency increases and substantial emission controls which delay an effective doubling of CO_2 to about 2040.

ii) **The "best guess" of the *magnitude* of the global mean equilibrium increase in surface temperature due to doubling CO_2 (the climate sensitivity) is**

Table 5.1 *Estimates of changes in areal means of surface air temperature and precipitation over selected regions, from pre-industrial times to 2030, assuming the IPCC "Business-as-Usual" Scenario. These are based on three high resolution equilibrium studies which are considered to give the most reliable regional patterns, but scaling the simulated values to correspond to a global mean warming of 1.8 ˚C, the warming at 2030 assuming the IPCC "best guess" sensitivity of 2.5 ˚C and allowing for the thermal inertia of the oceans. The range of values arises from the use of three different models. For a sensitivity of 1.5 ˚C, the values below should be reduced by 30%; for a sensitivity of 4.5 ˚C they should be increased by 50%. Confidence in these estimates is low, particularly for precipitation and soil moisture. Note that there are considerable variations in the changes within some of these regions.*

REGION	MODEL	TEMPERATURE (˚C)		PRECIPITATION (% change)		SOIL MOISTURE (% change)	
		DJF	JJA	DJF	JJA	DJF	JJA
1 Central North America (35-50˚N,	1	4	2	0	-5	-10	-15
	2	2	2	15	-5	15	-15
80-105˚W)	3	4	3	10	-10	-10	-20
2 South East Asia (5-30˚N,	1	1	1	-5	5	0	5
	2	2	1	0	10	-5	10
70-105˚E)	3	2	2	15	15	0	5
3 Sahel	1	2	2	-10	5	0	-5
(10-20˚N,	2	2	1	-5	5	5	0
20W-40˚E)	3	1	2	0	0	10	-10
4 Southern Europe (35-50˚N,	1	2	2	5	-15	0	-15
	2	2	2	10	-5	5	-15
10W-45˚E)	3	2	3	0	-15	-5	-25
5 Australia	1	1	2	15	0	45	5
(12-45˚S,	2	2	2	5	0	-5	-10
110-155˚E)	3	2	2	10	0	5	0

The numbers 1, 2 and 3 in the third column correspond to the models under entries 20, 21 and 22 respectively in
   Table 3.2a.

2.5˚C. This estimate is based on evidence from both models *and observations* (Section 5.2.1).

iii)   **The most reliable estimate of the regional *patterns* of change is given by the high resolution models.** (Table 3.2a, entries 20-22). These models in general produce a better simulation of present climate than those run at lower resolution (Section 5.4.1) and give results which are more consistent than those from earlier low resolution studies (see for example, Schlesinger and Mitchell, 1987, and Section 5.4.4). Note that although other models give a mean warming which is closer to the "best guess" (for

example, Table 3.2a, entries 17-19) they have a coarser resolution which degrades their simulation of regional climate. Hence the *patterns* of change have been derived from the high resolution models, even though they give a warming which is larger than the "best guess" of 2.5˚C.

iv)   **The patterns of equilibrium and transient climate change are similar.** As stated in Section 5.4.1, the few time dependent simulations that have been run do not use the IPCC Emission Scenarios and so cannot be used directly, and have been run at low horizontal resolution, degrading their capability to simulate regional changes. Recent results from the coupled ocean-atmosphere models (Section 6) indicate that the reduction of the warming due to oceanic thermal inertia is particularly pronounced in the circumpolar ocean of the southern ocean and the northern North Atlantic where deep vertical mixing of water occurs. Elsewhere, the reduction is much smaller and the time-dependent response is similar to the equilibrium response (Section 6, Figure 6.5c). The distribution of the changes in the hydrological cycle was also similar to that at equilibrium, but reduced in magnitude.

v) **The regional changes in temperature, precipitation and soil moisture are proportional to the global mean changes in surface temperatures.** This will be approximately valid except possibly in regions where the changes are associated with a shift in the position of steep gradients, for example where the snowline retreats, or on the edge of a rainbelt which is displaced. In general, this assumption is likely to be less valid for precipitation and soil moisture than for temperature. In the experiment described in detail in Section 6 (Stouffer et al., 1989), the mean temperature response north of 30°S is about 15% higher than the global mean response: this enhancement is omitted in the regional estimates given below.

vi) **The changes in global mean temperature can be derived from a simple diffusion-upwelling box model.** For the Business-as-Usual Scenario, this gives a warming of 1.3 to 2.6°C from pre-industrial times to present, with a "best guess" of 1.8°C (Section 6.6.2). For Scenario B, these estimates should be reduced by about 15%.

Although it is hard to justify some of these assumptions on riguorous scientific grounds, the errors involved are substantially smaller than the uncertainties arising from the threefold range of climate sensitivity. On the basis of the above assumptions, the estimates of regional change have been obtained by scaling the results from the high resolution models by a factor of 1.8/ΔTs where ΔTs is the climate sensitivity of the model involved.

5.4.4 Estimates Of Regional Change; Pre-industrial to 2030 (IPCC "Business-as-Usual" Scenario)

The reader should be aware of the limited ability of current climate models to simulate regional climate change and assumptions made in deriving the regional estimates (Sections 5.4.2 and 5.4.3 respectively). The range of values indicates the range of uncertainty in regional changes arising from using three different models with a similar global sensitivity. The results assume a global mean warming of 1.8°C at 2030, consistent with a global mean sensitivity of 2.5°C (Section 6.6.2). IPCC Scenario B gives results which are about 15% lower. For a sensitivity of 1.5°C, the estimates below should be reduced by 30%, for a sensitivity of 4.5°C, they should be increased by 50%. In general, **confidence in these estimates is low**, especially for the changes in precipitation and soil moisture. The regions are shown in Figure 5.13 and the estimates from the three individual models are given in Table 5.1.

Central North America (35-50°N, 85-105°W)

The warming varies from 2 to 4°C in winter and 2 to 3°C in summer. Precipitation increases range from 0 to 15% in winter whereas there are deceases of 5 to 10% in summer. Soil moisture decreases in summer by 15 to 20% of the present value.

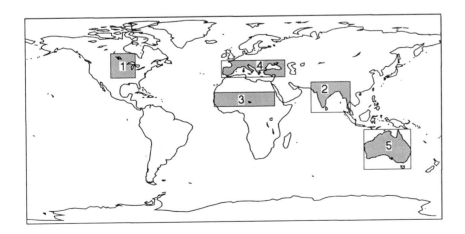

Figure 5.13: IPCC regions for which area means are given in Section 5.4.4 and Table 5.1.

South East Asia (5-30°N, 70-105°E)

The warming varies from 1 to 2°C throughout the year. Precipitation changes little in winter and generally increases throughout the region by 5 to 15% in summer. Summer soil moisture increases by 5 to 10%.

Sahel (10-20°N, 20°W-40°E)

The warming ranges from 1 to 2°C. Area mean precipitation increases and area mean soil moisture decreases marginally in summer. However there are areas of both increase and decrease in both parameters throughout the region which differ from model to model.

Southern Europe (35-50°N, 10°W-45°E)

The warming is about 2°C in winter and varies from 2 to 3°C in summer. There is some indication of increased precipitation in winter, but summer precipitation decreases by 5 to 15%, and summer soil moisture by 15 to 25%.

Australia (10-45°S, 110-155°E)

The warming ranges from 1 to 2°C in summer and is about 2°C in winter. Summer precipitation increases by around 10%, but the models do not produce consistent estimates of the changes in soil moisture. The area averages hide large variations at the sub-continental level.

Many of the differences in these results can be attributed to differences in model resolution, neglect or otherwise of ocean heat transport, and differences in the number of physical processes included and the way they are represented.

5.5 Empirical Climate Forecasting

5.5.1 Introduction

In the light of the poor reliability of regional climate simulations using general circulation models, various authors have suggested the use of data from past climates as indicators of regional climatic relationships for projections of future climate (for example, Flohn, 1977; Budyko et al., 1978; Budyko, 1980; Kellogg and Schware, 1981; Budyko and Izrael, 1987; Budyko et al., 1987). A brief description of the method is given in Section 3.4.1.

The mid-Holocene (5-6 kbp), the Last Interglacial (Eemian or Mikiluno, 125-130 kbp) and the Pliocene (3-4 mbp) have been used as analogues for future climates. January, July and mean annual temperatures, and mean annual precipitation were reconstructed for each of the above three epochs (see Section 7.2.2). Estimates of the mean temperatures over the Northern Hemisphere exceed the present temperature by approximately 1, 2 and 3-4°C during the mid-Holocene, Eemian and Pliocene respectively. These periods were chosen as analogues of future climate for 2000, 2025 and 2050 respectively.

5.5.2 Results

5.5.2.1 Temperature

Winter-time temperature changes in the low and middle latitude zones are quite small for areas dominated by marine climates. Winter cold is, however, less severe in the interior regions of the continents in middle and high latitudes. Summer warming is greater mainly in high latitudes. In some low latitude continental regions there are some areas of cooling due to increasing evaporation resulting from increased precipitation over these regions.

5.5.2.2 Precipitation

The influence of global warming on annual precipitation over the continents appears to be more complicated than for air temperature. During the mid-Holocene, precipitation was greater than at present over most of the northern continents although there were decreases in some regions of the European territory of the Soviet Union, as well as in some central regions of the United States (Figure 7.4b). Reconstructions of the Pliocene climate indicate that precipitation increased over all land areas for which data are available, particularly in a number of areas that are now deserts (Figure 7.2b). For this epoch, the mean latitudinal increase in annual precipitation over the continents of the Northern Hemisphere seems to show little dependence on latitude, averaging approximately 20 cm yr^{-1}. The tentative results for the Eemian, for which data are less complete, indicate that precipitation considerably exceeded the modern value in all regions for which data exist. As discussed in Section 7 the data used in this study have various limitations, and it is possible that the need for datable material to survive has introduced a bias against finding evidence of aridity.

5.5.3 Assessment Of Empirical Forecasts

1. For a climate situation in the past to be a detailed analogue of the likely climate in the next century with increased greenhouse gas concentration, it is necessary for the forcing factors (e.g., greenhouse gases, orbital variations) and the boundary conditions (e.g., ice coverage, topography, etc.) to be similar:

The change in forcing during the mid-Holocene and Eemian was very different to that due to doubling CO_2. During both these periods, CO_2 concentrations were smaller than present, being close to the pre-industrial level (Barnola et al., 1987). The orbital perturbations increase the annual mean radiative heating in high latitudes (up to 5 Wm^{-2} during the mid-Holocene) and reduce it in the tropics (1 Wm^{-2} during the mid-Holocene). The radiative forcing due to doubling CO_2 increases everywhere, from about 2.5 Wm^{-2} in high latitudes to 5 Wm^{-2} in the tropics (Mitchell, 1990). The changes in orbital perturbations produce seasonal anomalies of up to 40 Wm^{-2} at certain latitudes (Berger, 1979) whereas the CO_2 forcing is relatively constant throughout the year. Thus the mid-Holocene and Eemian cannot be considered as reliable analogues for a climate with increased concentrations of greenhouse gases.

The changes in forcing during the Pliocene are less well known. Carbon dioxide levels may have been higher than present, but whether or not they were as high as double present concentrations is disputed (Section 7.2.2.1). Other factors, such as a lower Himalayan massif and an open Isthmus of Panama (which would have profoundly affected the circulation of the North Atlantic) are likely to have altered the climate in those regions. The geographical distribution of data for the Pliocene are limited and there are difficulties in establishing that data from different sites are synchronous (Section 7.2.2). In view of all these factors, it is at best unclear that the reconstructed patterns of climate change during the Pliocene can be regarded as analogues of warming due to increases in greenhouse gases.

2. Because many aspects of climate change respond to these factors and conditions in a non-linear way, direct comparisons with climate situations for which these conditions do not apply cannot be easily interpreted:

The analogue method is based on the assumptions that the patterns of climate change are relatively insensitive to the different changes in forcing factors leading to warming. Recent numerical studies of the equilibrium response to increased CO_2 give a consistent picture of continental scale changes so one can compare the large-scale features from these simulations with those deduced from the palaeo-analogue approach. The main discrepancies are:

i) *The palaeo-climatic data suggest a cooling over large areas of the tropics whereas CO_2 simulations produce a substantial warming.*

   A cooling is consistent with the reduction in insolation in the tropics during the mid-Holocene and Eemian, and is also reproduced in numerical simulations in which the orbital perturbations have been imposed (for example, Kutzbach and Guetter, 1986; Mitchell et al., 1988). As noted above increases in CO_2 produce a radiative warming of the tropics, whereas the relevant changes in orbital properties produce a radiative cooling. Thus on both simple physical grounds and on the basis of model simulations, the palaeo-climatic reconstructions are probably misleading in this respect.

ii) *The palaeo-climatic data suggest that precipitation increases markedly in much of the arid subtropics of the Northern Hemisphere (for example, COHMAP members, 1988; Section 7.2.2.2, 7.2.2.3), whereas recent numerical simulations with enhanced CO_2 indicate little change in these regions.*

   Thus, numerical simulations of the mid-Holocene and Eemian produce increases in precipitation in much of the arid subtropics because the enhanced summer insolation intensifies the summer monsoon circulations. Again, on simple physical grounds and on the basis of model simulations (for example, Kutzbach and Guetter, 1986; Mitchell et al., 1988), it appears that the changes in precipitation in the arid subtropics during these epochs are due to orbital changes.

iii) *The palaeo-climatic data suggest that the warming (in the Northern Hemisphere) in summer would be greatest in high latitudes, whereas in model simulations with increased CO_2 (this section) or orbital perturbations (for example, Kutzbach and Guetter, 1986; Mitchell et al., 1988), the warming is small in high latitudes in summer.*

   The simulated changes may be in error (though there is a plausible physical explanation) or the palaeo-climatic data have been miscompiled or misinterpreted.

From the above, it seems likely that changes in orbital parameters alone can account for much of the changes from present climate found in the mid-Holocene and the Eemian, that some of the large scale effects of the orbital perturbations differ from those expected with an increase in trace gases, and therefore that a necessary condition for these periods to be considered as analogues for future climate change is that the effects of orbital variations should be subtracted out. At present, there is no way of doing this apart from using simulated changes.

In conclusion, the palaeo-analogue approach is unable to give reliable estimates of the equilibrium climatic effect of increases in greenhouse gases as suitable analogues are not available, and it is not possible to allow for the deficiencies in the analogues which are available. Nevertheless, information on past climates will provide useful data against which to test the performance of climate models when run with appropriate forcing and boundary conditions (See Section 4.10). It should be noted that, from the point of view of understanding and testing climate mechanisms and models, palaeo-climatic data on cool epochs may be just as useful as data on warm epochs. Special attention should be paid to times of relatively rapid climatic change when time-dependent effects and ecosystem responses may more closely resemble those to be expected in the coming century.

5.6 The Climatic Effect of Vegetation Changes

5.6.1 Introduction

In addition to the climatic impacts of increasing greenhouse gases, alteration of vegetation cover by man can modify the climate. For small areas this may result in only local impacts, but for large areas it may result in important regional climate change, and may impinge upon regions remote from the area of change.

The vegetative cover (or lack of it) strongly controls the amount of solar radiative heating absorbed by the land surface by varying the albedo (reflectivity). Heat absorbed by the surface, in addition to heating the soil, provides energy for evaporation and for heating the atmosphere directly (sensible heat). Thus, changes in albedo can strongly affect evaporation and atmospheric heating and so influence the hydrological cycle and atmospheric circulation. Other aspects of vegetation cover, such as aerodynamic roughness, stomatal resistance, canopy moisture capacity and rooting depth can affect the partitioning of incoming solar radiation between evaporation and sensible heat.

There are three climatic regions where vegetation changes may have significant impacts on climate: tropical forests, semi-arid and savannah and boreal forests. The first has received considerable attention and is covered in more detail below and in Section 10. Model studies of degradation of vegetation in the Sahel region of Africa, particularly with regard to changes in albedo and soil moisture availability, have shown that rainfall can be reduced over a wide part of the region (Rowntree and Sangster, 1986). Removal of boreal forests has been shown to delay spring snowmelt slightly by increasing albedo (Thomas, 1987).

5.6.2 Global Mean Effects

1. The net effect of deforestation on global mean climate is likely to be small although the regional impacts may be profound:

The conversion of forests to grassland is increasing in the tropics. The current rate of deforestation is estimated to be 0.1×10^6 km^2 yr^{-1} (the total area of tropical forest is about 9×10^6 km^2). Associated with the clearing is a substantial release of CO_2 to the atmosphere (Section 1). The replacement of forest by grassland also increases the reflection of solar radiation to space which tends to cool the climate; but this effect is at present small compared with the warming effect of the accompanying increased CO_2 (see Section 1, 2.2.2, Section 2). The net effect of deforestation is therefore to warm climate. The removal of all the tropical forests could warm the climate by about 0.3°C [3]. Alternatively, if 10% of the Earth's land surface

[3] *Assuming that 13×10^6km^2 of forest is removed and releases 12.5Gt C/10^6km^2 (based on Bolin et al.,1986) with half the resulting CO_2 remaining in the atmosphere. Also, it is assumed that deforestation increases the surface albedo by 5%, and that only 50% of the insolation at the top of the atmosphere reaches the surface. The climate sensitivity is taken to be 3°C for a doubling of CO_2.*

were afforested in addition to the present cover, a global cooling of 0.2 to 0.4°C would be expected.

5.6.3 Regional Effects: Deforestation Of Amazonia

One of the best studied examples of deforestation is the Amazon Basin. Besides changing net carbon storage in Amazonia, deforestation is affecting the regional energy and water balance. A number of modelling studies have concentrated on the climatic impact that might arise from complete deforestation of South America and, in particular, Amazonia (Henderson-Sellers and Gornitz, 1984; Wilson, 1984; Dickinson and Henderson-Sellers, 1988; Lean and Warrilow, 1989; Nobre et al., 1990). The Amazon Basin contains about half of the world's tropical rainforests and plays a significant role in the climate of that region. It is estimated that approximately half of the local rainfall is derived from local evaporation (Salati et al., 1978). The remainder is derived from moisture advected from the surrounding oceans. A major modification of the forest cover could therefore have a significant climatic impact. Reduced evaporation and a general reduction in rainfall, although by variable amounts, was found in most experiments.

1. Total deforestation of the Amazon basin could reduce rainfall locally by 20%

The studies by Lean and Warrilow (1989) and Nobre et al. (1990) show reductions of about 20% in rainfall in simulations in which vegetation parameters for forest were replaced by those for grassland (Figure 5.14). Lean and Warrilow showed that albedo and roughness changes contributed almost equally to the rainfall reduction, although more recent work suggests that the contribution from roughness may have been slightly overestimated.

Nobre et al. suggest that the switch to a more seasonal rainfall regime, which they obtained, would prevent forest recovery. A recent experiment (Lean and Rowntree, 1989, personal communication) considered the impact of setting vegetation cover to desert over South America north of 30°S. Albedos similar to those of the most reflective parts of the Sahara were used. Annual rainfall was reduced by 70%. The seasonal change of rainfall (Figure 5.15) became typical of that observed in semi-arid regions such as the Sahel. This would permit the growth of some rainy season vegetation and thus a desert would be unlikely to be maintained over the whole region. However, the results suggest that a widespread deforestation of the South American tropics could lead to an irreversible decline in rainfall and vegetative cover over at least part of the region.

5.7 Uncertainties

Here we summarize the major uncertainties in model predictions:

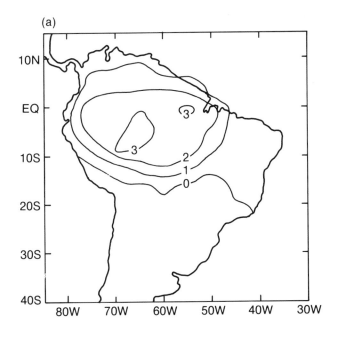

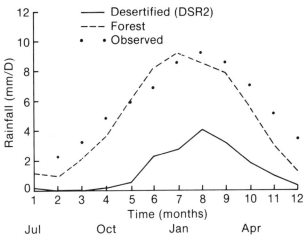

Figure 5.15: Rainfall over South America (2.5 to 30°S, mm day⁻¹). Dashed line: simulated, forested surface, Solid line: simulated, desert surface. Dots, observed. (Lean and Rowntree, personal communication, 1989)

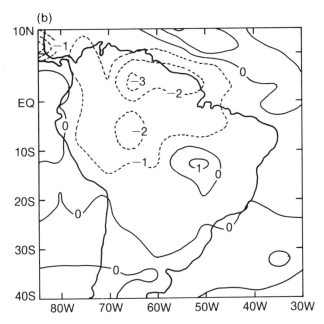

Figure 5.14: Changes in annual means due to deforestation of northern South America (from Nobre et al.,1990): **(a)** Surface temperature (contours every 1°C), **(b)** Precipitation (contours every 1 mm day⁻¹, negative contours are dashed)

One of the largest sources of uncertainty in the simulation of equilibrium climate change lies in the prediction of clouds. It has been shown that clouds can produce either a positive or negative feedback, depending on the model and parameterization of cloud used (Cess et al., 1989; Mitchell et al., 1989) giving an uncertainty of a factor of two or more in the equilibrium warming. Earlier schemes base cloud cover on relative humidity and

prescribed radiative properties: later models use schemes which explicitly represent cloud water and allow cloud radiative properties to vary. The latter are more detailed but not necessarily more accurate as more parameters have to be specified. The radiative effect of clouds depends on cloud height, thickness and fractional cover; on cloud water content and cloud droplet size distribution (and in the case of ice clouds, the size distribution, shape and orientation of particles) (see Section 3.3.4). Thus there is a need to understand both the microphysics of cloud, and their relation to the larger-scale cloud properties. This will require further satellite observations (for example Barkstrom et al., 1986) and carefully designed field studies (for example Raschke, 1988; Cox et al., 1987). In particular, there is a need to refine our knowledge of ice clouds and their radiative properties.

Another large uncertainty lies in the representation of convection in large-scale models. Again, the more detailed (though not necessarily more accurate) parameterizations produce different results from the simpler schemes, including a much greater warming in the tropics. It is less obvious how to reduce this uncertainty, though it may be that a comparison of the observed and simulated response to past anomalies in tropical SSTs may help to eliminate the more unrealistic schemes.

Thirdly, the changes in ground wetness and surface temperature have been shown to be highly sensitive to the treatment of the land surface. In addition, the effects of vegetation and changes in vegetation are ignored in the models used in Table 3.2a. Again process studies, along with satellite measurements (World Meteorological

Organization, 1985, 1987) are needed to guide the development of surface parameterizations and their validations.

Finally the oceans and sea-ice constitute a major source of uncertainty about which more is said in Section 6. Here it has been shown that the distribution of sea-ice and changes in sea-ice extent have a dominant influence on local temperature change, especially in winter. Most of the models considered here ignore salinity effects and possible changes in ocean heat and ice transport: some ignore ocean transport altogether. The inclusion of a more complete representation of the ocean may modify the simulated changes in sea-ice described here, and changes in ocean circulation could produce pronounced local anomalies in SST particularly in the neighbourhood of the major current systems or the main areas of deep water formation, with profound effects on the local climate.

References

Barkstrom, B.R. and ERBE Science team, 1986: First data from the Earth Radiation Budget Experiment (ERBE). *Bull. Am. Met. Soc.*, **67**, 818-824.

Barnola, J.M., D.Raynaud, Y.S. Korotkevich and C.Lorius, 1987: Vostok ice core provides 160,000-year record of atmospheric CO_2. *Nature*, **329**, 408-413.

Bates, G.A. and G.A.Meehl, 1986: The effect of CO_2 concentration on the frequency of blocking in a general circulation model coupled to a simple mixed layer ocean model. *Mon.Wea.Rev.*, **114**, 687-701.

Bengtsson, L., H.Bottger and M.Kanamitsu, 1982: Simulation of hurricane type vortices in a general circulation model. *Tellus*, **34**, 440-457.

Berger, A., 1979: Insolation signatures of Quaternary climatic changes. *Il Nuovo Cimento*, **20(1)**, 63-87.

Blackmon, M., 1975: A climatological spectral study of 500mb geopotential height of the northern hemisphere. *J.Atmos Sci.*, **37**, 1607-1623.

Bolin , B., B.R. Doos, J. Jager and R.A. Warrick (eds), 1986:*"The Greenhouse Effect, Climate Change and Ecosystems*. SCOPE 13, Wiley, New York.

Broccoli, A. and S. Manabe, 1990: Will global warming increase the frequency and intensity of tropical cyclones? (submitted for publication).

Budyko, M.I. et al., 1978: "Forthcoming climate change", *Izvestia AN* USSR.

Budyko, M.I., 1980: "Climate of the past and future", *Gidrometeorizdat*, 352pp.

Budyko, M.I., A.B.Ronov and A.L.Yanshin, 1987: History of the Earth's atmosphere (English translation) Springer-Verlag, Berlin, 139pp.

Budyko, M.I. and Y.Izrael, 1987: "Anthropogenic climate change", *Gidrometeorizdat*, 406pp.

Cess, R.D. and G.L. Potter, 1988: A Methodology for understanding and Intercomparing atmospheric climate feedback processes in general circulation models. *J. Geophys. Res.*, **93**, 8305-8314.

Cess, R.D., G.L. Potter, J.P. Blanchet, G.J. Boer, S.J. Ghan, J.T. Kiehl, H. Le Treut, Z.X. Li, X.Z. Liang, J.F.B. Mitchell, J-J. Morcrette, D.A. Randall, M.R. Riches, E. Roeckner, U. Schlese, A. Slingo, K.E. Taylor, W.M. Washington, R.T. Wetherald, I. Yagai, 1989: Interpretation of cloud-climate feedback as produced by 14 atmospheric general circulation models. *Science*, **245**, 513-516.

Cox, S.K., D.S. McDougall, D.A. Randall and R.A.Schiffer, 1987: FIRE (First ISCCP Regional Experiment). *Bull. Am. Met. Soc.*, **68**, 114-118.

COHMAP Members, 1988: Climatic changes of the last 18,000 years: Observations and model simulations. *Science*, **241**, 1043-1052.

Cunnington, W.M. and J.F.B. Mitchell, 1990: On the dependence of climate sensitivity on convective parametrization. *Climate Dynamics, 4*, 85-93.

Dell'Osso, L. and L.Bengtsson, 1985: Prediction of a typhoon using a finemesh NWP model. *Tellus*, **37A**, 97-105.

Dickinson, R.E., 1986: How will climate change? In *The Greenhouse Effect, Climate Change and Ecosystems*, eds. B. Bolin, B.R. Doos, J. Jager and R.A. Warrick, SCOPE 13, Wiley, New York, 206-270.

Dickinson, R.E. and A.Henderson-Sellers , 1988: Modelling tropical deforestation: a study of GCM land-surface parametrizations. *Quart.J.R.Met.Soc.*, **114**, 439-462.

Emanuel, K.A., 1987: The dependence of hurricane intensity on climate. *Nature*, **326**, 483-485.

Fels, S.B., J.D. Mahlman, M.D. Schwarzkopf and R.W. Sinclair 1980: Stratospheric sensitivity to perturbations in ozone and carbon dioxide: radiative and dynamical response. *J. Atmos. Sci.*, **37**, 2266-2297.

Flohn, H., 1977: Climate and energy: A Scenario to the 21st Century. *Climatic Change*, **1**, 5-20.

Gray, W.M., 1979: Hurricanes: their formation, structure and likely role in the tropical circulation. In *Meteorology over the tropical oceans* (ed. D.B. Shaw) Roy. Meteor. Soc., 155-218.

Grotch, S. L., 1988: Regional intercomparisons of general circulation model predictions and historical climate data. U.S. Dept. of Energy, Washington D.C. Report DOE/NBB-0084 (TR041).

Hansen, J., A.Lacis, D. Rind, L. Russell, P. Stone, I. Fung, R. Ruedy and J. Lerner, 1984: Climate Sensitivity Analysis of Feedback Mechanisms. In *Climate Processes and Climate Sensitivity* (ed. J. Hansen and T. Takahashi) Geophysical Monograph 29, 130-163. American Geophysical Union, Washington DC.

Hansen, J., D. Rind, A. Delgenio, A. Lacis, S. Lebedeff, M. Prather, R. Ruedy and T. Karl 1989: Regional greenhouse climate effects. In *Coping with climate change*, Proceedings of Second North American Conference on Preparing for climatic change. Dec. 6-8, 1988. Washington DC, pp 696.

Henderson-Sellers, A. and V.Gornitz, 1984: Possible climatic impacts of land cover transformations, with particular emphasis on tropical deforestation. *Climatic Change*, **6**, 231-258.

Ingram, W.J., C.A. Wilson and J.F.B. Mitchell, 1989: Modelling climate change: an assessment of sea-ice and surface albedo feedbacks. *J. Geophys Res.*, **94**, 8609-8622.

Kellogg, W.W. and R.Schware, 1981: *Climate Change and Society*. Westview Press, Boulder, Colorado.

Krishnamurti, T.N., D. Oosterhof and N.Dignon, 1989: Hurricane prediction with a high resolution global model. *Mon.Wea.Rev.*, **117**, 631-669.

Kutzbach, J.E., and P.J. Guetter, 1986: The influence of changing orbital parameters and surface boundary conditions on climate simulations for the past 18,000 years, *J.Atmos. Sci .*, **43**, 1726-1759.

Lean, J., and D.A. Warrilow, 1989: Simulation of the regional impact of Amazon deforestation, *Nature*, **342**, 411-413.

Manabe, S. and R.J. Stouffer, 1980: Sensitivity of a global climate model to an increase in the CO_2 concentration in the atmosphere. *J. Geophys. Res.*, **85**, 5529-5554.

Manabe, S., K.Bryan and M.D. Spelman, 1990: Transient response of a global ocean-atmosphere model to a doubling of atmospheric carbon dioxide. *J. Phys. Oceanogr.* (To appear, April).

Manabe, S. and K. Bryan, 1985: CO_2 - Induced change in a coupled Ocean-Atmosphere Model and its Palaeo-climatic Implications. *J. Geophys. Res.*, **90**, C11, 11,689-11,707.

Manabe, S. and R.T. Wetherald, 1987: Large scale changes of soil wetness induced by an increase in atmospheric carbon dioxide. *J. Atmos. Sci.*, **44**, 1211-1235.

Manabe, S., J.L.Holloway Jr. and H.M.Stone, 1970: Tropical circulation in a time integration of a global model of the atmosphere. *J.Atmos.Sci.*, **21**, 580-613.

Mearns, L.O., R.W.Katz and S.H. Schneider, 1984: Extreme high temperature events: changes in their probabilities with changes in mean temperature. *J. Clim. App. Meteorol.*, **23**, 1601-13.

Mearns, L.O., S.H. Schneider, S.L. Thompson and L.R. McDaniel, 1990: Analysis of climatic variability in general circulation models: comparison with observations and changes in variability in $2 \times CO_2$ experiments. (submitted for publication)

Meehl, G.A. and W.M. Washington, 1988: A comparison of soil moisture sensitivity in two global climate models. *J. Atmos. Sci.*, **45**, 1476-1492.

Meehl, G.A. and W.M. Washington, 1989: CO_2 Climate Sensitivity and snow-sea-ice albedo parametrization in an Atmospheric GCM coupled to a Mixed-Layer Ocean Model. *Climatic Change* (to appear).

Merrill, R.T., 1988: Environmental influences on hurricane intensification. *J.Atmos.Sci.*, **45**, 1678-1687.

Mintz, Y., 1984: The sensitivity of numerically simulated climates to land-surface boundary conditions. In *The Global Climate* (ed. J.T. Houghton), 79-106, Cambridge University Press.

Mitchell, J.F.B., 1989: The "Greenhouse effect" and climate change. *Reviews of Geophysics* , **27**, 115-139.

Mitchell, J.F.B., 1990: Greenhouse warming: is the Holocene a good analogue? *J. of Climate* (to appear).

Mitchell, J.F.B. and W.J. Ingram, 1990: On CO_2 and climate. Mechanisms of changes in cloud. *J.of Climate* (submitted).

Mitchell, J.F.B., N.S. Grahame, and K.J. Needham, 1988: Climate Simulations for 9000 years before present: Seasonal variations and the effect of the Laurentide Ice Sheet, *J. Geophys. Res.*, **93**, 8283-8303.

Mitchell, J.F.B., C.A. Wilson and W.M. Cunnington, 1987: On CO_2 climate sensitivity and model dependence of results. *Quart.J.R. Meteorol. Soc.*, **113**, 293-322.

Mitchell, J.F.B. and D.A. Warrilow, 1987: Summer dryness in northern mid-latitudes due to increased CO_2. *Nature*, **330**, 238-240.

Mitchell, J.F.B., C.A. Senior and W.J. Ingram, 1989: CO_2 and climate: a missing feedback? *Nature*, **341**, 132-4.

Morris, R.M., 1989: Forecasting the evolution of tropical storms over the South Indian Oceans and South Pacific during the 1988-9 season using the United Kingdom Meteorological Office operational global model. (unpublished manuscript).

Nobre, C.,J.Shukla, and P.Sellers, 1990: Amazon Deforestation and Climate Change. *Science*, **247**, 1322-5.

Noda, A. and T. Tokioka, 1989: The effect of doubling the CO_2 concentration on convective and non-convective precipitation in a general circulation model coupled with a simple mixed layer ocean. *J. Met. Soc. Japan*, **67**, 1055-67.

Parry, M.L. and T.R. Carter, 1985: The effect of climatic variation on agricultural risk. *Climatic Change*, **7**, 95-110.

Pittock, A.B., 1989: The greenhouse effect, regional climate change and Australian agriculture. *Proc. 5th Agronomy Conf.*, Austn. Soc. Agron., Perth, 289-303.

Rashke, E., 1988: The International Satellite Cloud Climatology Project (ISCCP), and its European regional experiment ICE (International Cirrus Experiment). *Atmos. Res.*, **21**, 191-201.

Raval, A . and V. Ramanathan, 1989: Observational determination of the greenhouse effect. *Nature*, **342**, 758-761.

Rind, D., R. Goldberg, J. Hansen, C. Rosensweig and R. Ruedy, 1989a; Potential evapotranspiration and the likelihood of future drought. (Submitted to *J. Geophys. Res.*)

Rind, D., R. Goldberg and R. Ruedy, 1989b: Change in climate variability in the 21st century. *Climatic Change*, **14**, 5-37.

Roeckner, E., U. Schlese, J. Biercamp and P. Loewe, 1987: Cloud optical depth feedbacks and climate modelling. *Nature*, **329, 138-139**.

Rowntree, P.R. and A.B.Sangster, 1986: Remote sensing needs identified in climate model experiments with hydrological and albedo changes in the Sahel. Proc ISLSCP Conference, Rome, European Space Agency ESA SP-248, pp 175-183.

Salati, E., Marques,J. and L.C.B.Molion, 1978: Origem e Distribuicao das Chuvas na Amazonia. *Interciencia*, **3**, 200-206.

Schlesinger, M.E., 1985: Feedback analysis of results from energy balance and radiative-convective models. In *Projecting the climatic effects of increasing atmosphere carbon dioxide*, ed.M.C. MacCracken and F.M. Luther, pp 280-319, US Department of Energy, Washington DC, 1985. (Available as NTIS, DOE ER-0237 from Nat.Tech. Inf. Serv. Springfield Va.)

Schlesinger, M.E. and J.F.B. Mitchell, 1985: Model Projection of Equilibrium Climatic Response to Increased CO_2 Concentration. In *Projecting the climatic effects of increasing atmosphere carbon dioxide*, ed.M.C. MacCracken and F.M. Luther, pp 280-319, US Department of Energy, Washington DC, 1985. (Available as NTIS, DOE ER-0237 from Nat.Tech. Inf. Serv. Springfield Va.).

Schlesinger, M.E. and J.F.B. Mitchell, 1987: Climate model simulations of the equilibrium climatic response to increased carbon dioxide. *Reviews of Geophysics*, **25**, 760-798.

Siegmund, P., 1990: The effect of doubling CO_2 on the storm tracks in the climate of a GCM. K.N.M.I. report No. 90-01.

Smith, R.N.B., 1990: A scheme for predicting layer clouds and their water content in a general circulation model. *Quart. J.R. Meteorol. Soc.*, **116**, 435-460.

Somerville, R.C.J. and L.A. Remer, 1984: Cloud Optical Thickness Feedbacks in the CO_2 Climate Problem. *J. Geophys Res.*, **89**, 9668-9672.

Somerville, R.C.J. and S. Iacobellis, 1988: Air-sea Interactions and Cirrus Cloud feedbacks on climate. In Proceedings of the 7th American Meteorological Society Conference on Ocean-atmosphere Interaction, 53-55.

Spelman, M.J. and S. Manabe, 1984: Influence of Oceanic Heat Transport upon the Sensitivity of a model climate. *J. Geophys. Res.*, **89**, 571-586.

Stouffer, R.J., S. Manabe and K.Bryan, 1989: Interhemispheric asymmetry in climate response to a gradual increase of CO_2. *Nature*, **342**, 660-662.

Thomas, G., 1987: Land-use change and climate. An investigation of climate modelling and climate monitoring techniques. PhD thesis, University of Liverpool, UK.

UNEP, 1989: Report on Scientific Assessment of Stratospheric Ozone.

U.S. National Academy of Sciences, 1979: Carbon Dioxide and Climate: A Scientific Assessment. Washington D.C. pp22.

Vinnikov, K.Ya., and I.B.Eserkepova, 1989: Empirical data simulations of soil moisture regime. *Meteorologia i Gidrologia*, **11**, 64-72.

Wang, W.-C., G.-Y. Shi and J.T. Kiehl 1990: Incorporation of the thermal radiative effect of CH_4, N_2O, CF_2Cl_2 and $CFCl_3$ into the Community Climate Model. *J.Geophys. Res.* (to appear)

Washington, W.M. and G.A. Meehl, 1984: Seasonal Cycle Experiment on the Climate Sensitivity due to a doubling of CO_2 with an Atmospheric General Circulation Model coupled to a simple mixed layer ocean model. *J. Geophys. Res.*, **89**, 9475-9503.

Washington, W.M. and G.A. Meehl, 1986: General Circulation model CO_2 sensitivity experiments: Snow-ice albedo parameterizations and globally averaged surface air temperature. *Climatic Change*, **8**, 231-241.

Wetherald, R.T. and S.Manabe, 1975: The effect of changing the solar constant on the climate of a general circulation model. *J. Atmos. Sci.*, **32**, 2044 -2059.

Wetherald, R.T. and S. Manabe, 1988: Cloud Feedback Processes in a General Circulation Model. *J. Atmos. Sci.*, **45**, 1397-1415.

Wigley, T.M.L., 1985: Climatology: impact of extreme events. *Nature*, **316**, 106-7.

Wigley, T.M.L. and M.E. Schlesinger, 1985: Analytical solution for the effect of increasing CO_2 on global mean temperature. *Nature*, **315**, 649-652.

Wilson, C.A., and J.F.B. Mitchell, 1987a: A doubled CO_2 Climate Sensitivity experiment with a GCM including a simple ocean. *J. Geophys Res.*, **92**, 13315-13343.

Wilson, C.A. and J.F.B. Mitchell, 1987b: Simulated CO_2 induced climate change over western Europe. *Climatic Change*, **10**, 11-42.

Wilson, M.F., 1984: The construction and use of land surface information in a general circulation model. PhD thesis, University of Liverpool, UK.

World Meteorological Organisation, 1985: Development of the implementation plan for the International Satellite Land Surface Climatology Project (ISLSCP) Phase I (eds H.J.Bolle and S.I.Rasool) WCP-94, Geneva.

World Meteorological Organisation, 1987: Report of the second meeting of the scientific steering group on land surface processes and climate (Toulouse, France, 21-26 May, 1986). WCP-126, Geneva.

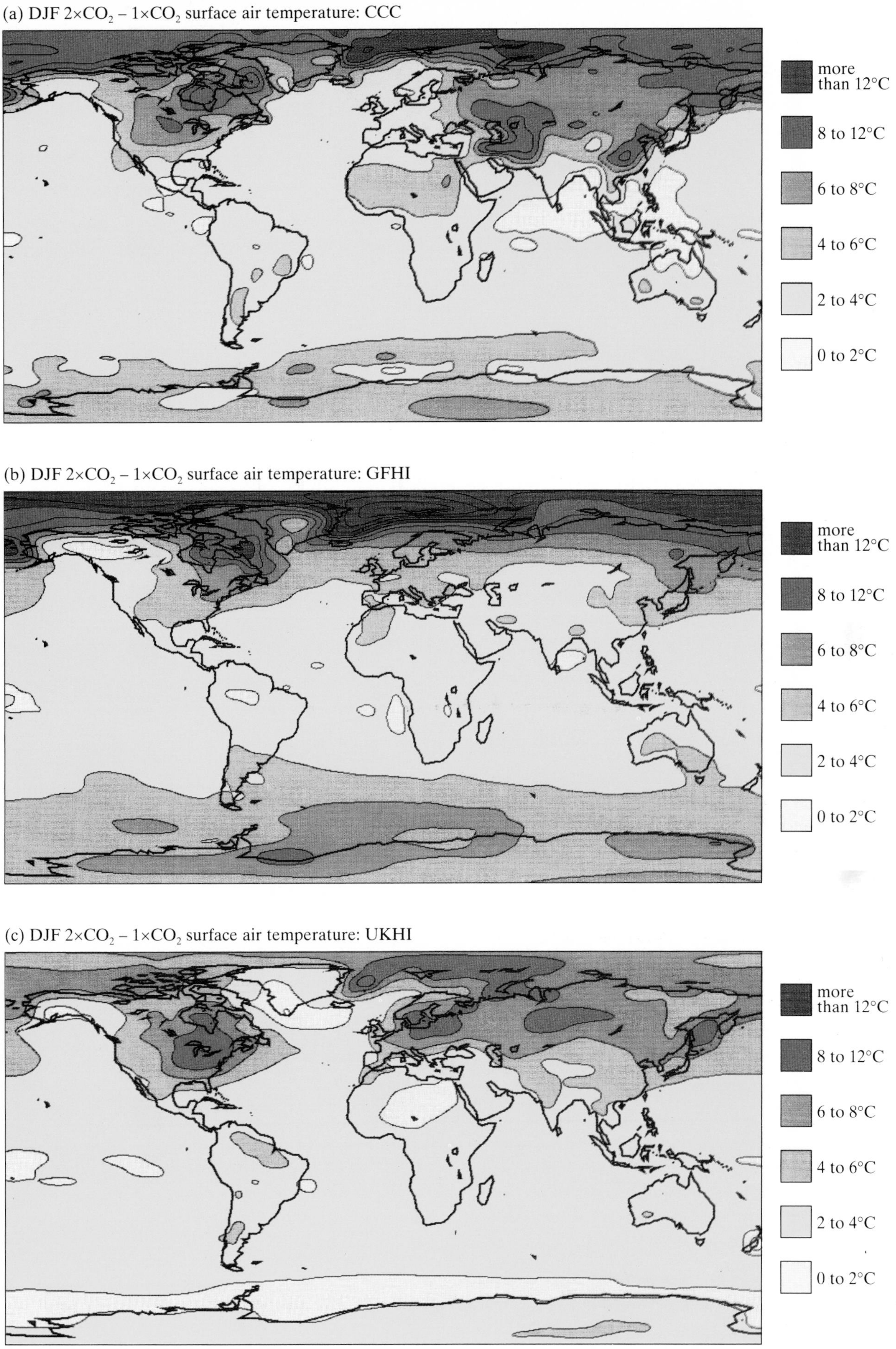

(a) DJF 2×CO$_2$ – 1×CO$_2$ surface air temperature: CCC

	more than 12°C
	8 to 12°C
	6 to 8°C
	4 to 6°C
	2 to 4°C
	0 to 2°C

(b) DJF 2×CO$_2$ – 1×CO$_2$ surface air temperature: GFHI

	more than 12°C
	8 to 12°C
	6 to 8°C
	4 to 6°C
	2 to 4°C
	0 to 2°C

(c) DJF 2×CO$_2$ – 1×CO$_2$ surface air temperature: UKHI

	more than 12°C
	8 to 12°C
	6 to 8°C
	4 to 6°C
	2 to 4°C
	0 to 2°C

Figure 5.4: Change in surface air temperature (ten year means) due to doubling CO$_2$, for months December-January-February, as simulated by three high resolution models: (a) CCC: Canadian Climate Centre (Boer, pers. comm., 1989), (b) GFHI: Geophysical Fluids Dynamics Laboratory (Manabe and Wetherald, pers. comm., 1990), and (c) UKHI: United Kingdom Meteorological Office (Mitchell and Senior, pers. comm., 1990). See legend for contour details.

(d) JJA 2×CO$_2$ – 1×CO$_2$ surface air temperature: CCC

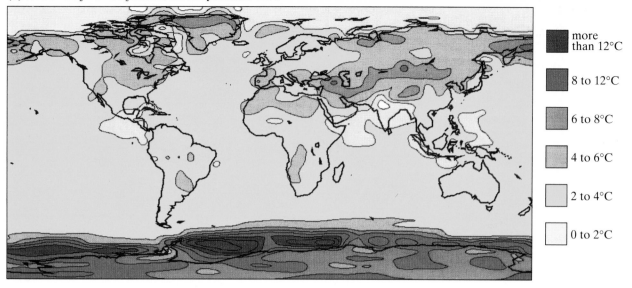

(e) JJA 2×CO$_2$ – 1×CO$_2$ surface air temperature: GFHI

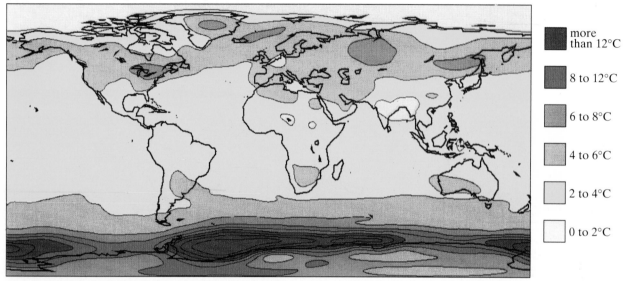

(f) JJA 2×CO$_2$ – 1×CO$_2$ surface air temperature: UKHI

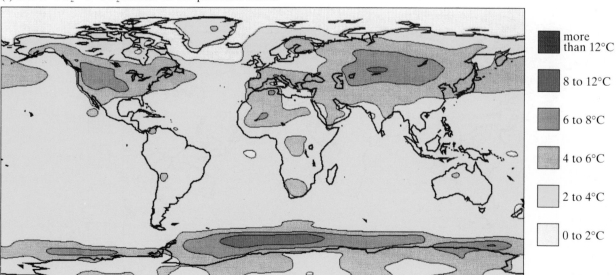

Figure 5.4 continued: Change in surface air temperature (ten year means) due to doubling CO$_2$, for months June-July-August, as simulated by three high resolution models: (d) CCC, (e) GFHI, and (f) UKHI. See legend for contour details.

(a) DJF 2×CO$_2$ – 1×CO$_2$ precipitation: CCC

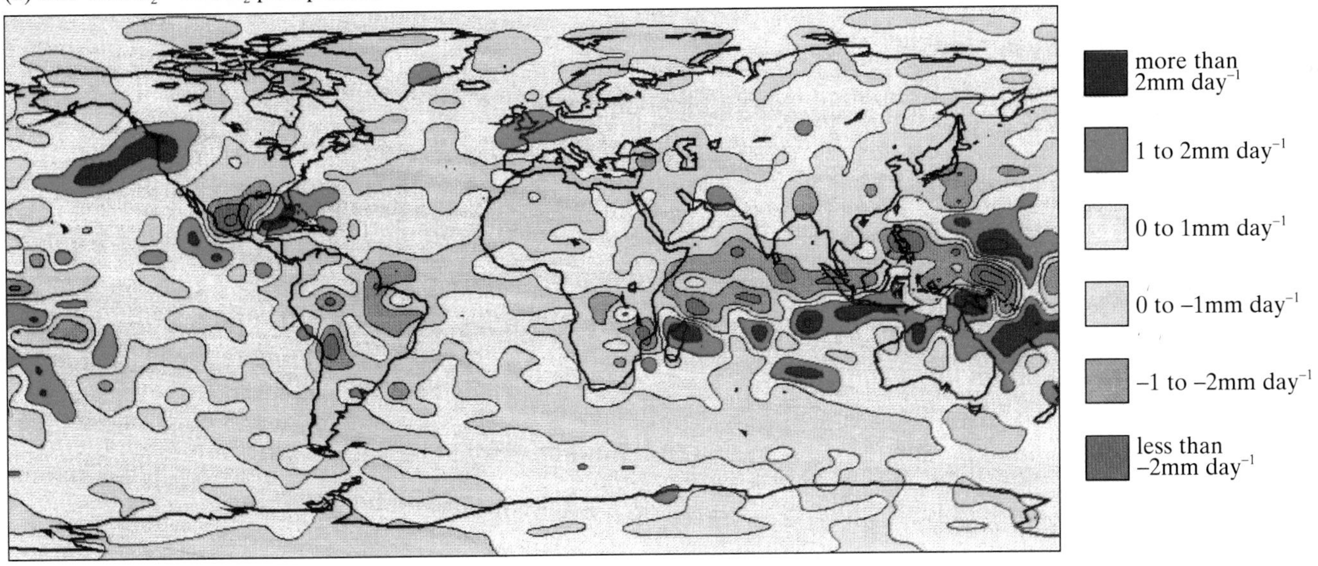

more than 2mm day^{-1}

1 to 2mm day^{-1}

0 to 1mm day^{-1}

0 to –1mm day^{-1}

–1 to –2mm day^{-1}

less than –2mm day^{-1}

(b) DJF 2×CO$_2$ – 1×CO$_2$ precipitation: GFHI

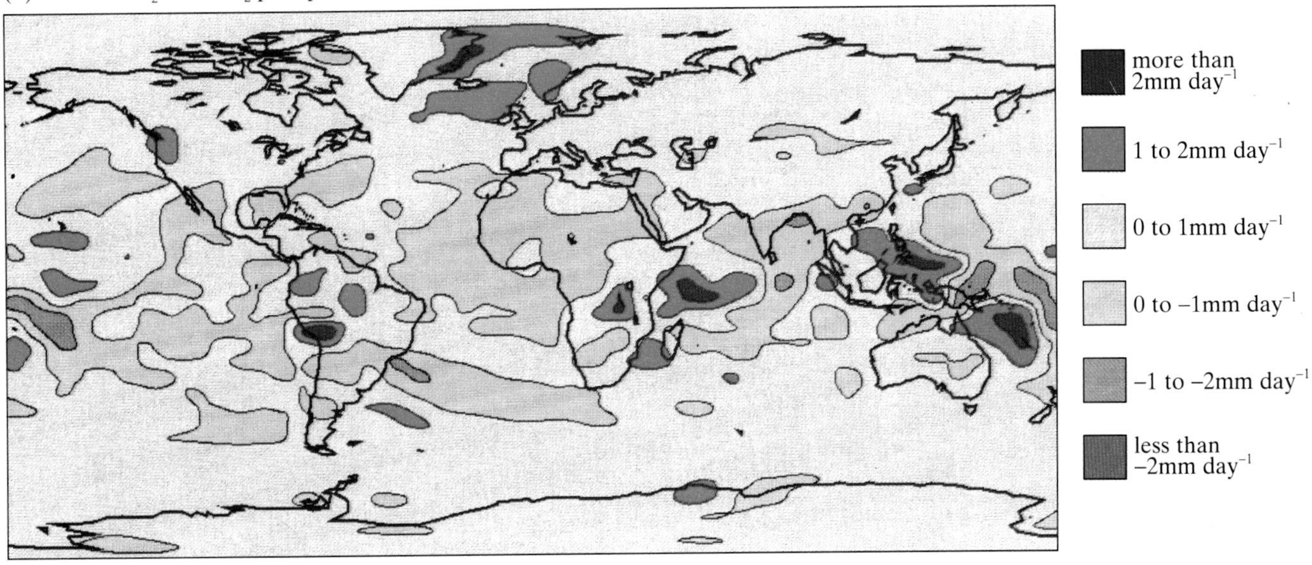

more than 2mm day^{-1}

1 to 2mm day^{-1}

0 to 1mm day^{-1}

0 to –1mm day^{-1}

–1 to –2mm day^{-1}

less than –2mm day^{-1}

(c) DJF 2×CO$_2$ – 1×CO$_2$ precipitation: UKHI

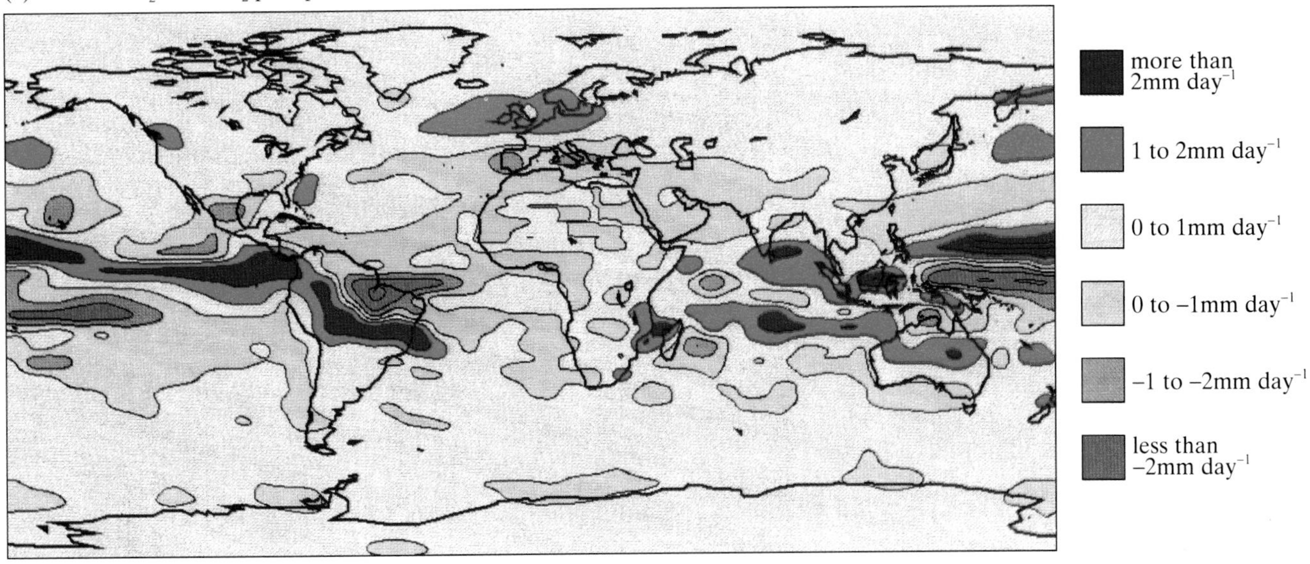

more than 2mm day^{-1}

1 to 2mm day^{-1}

0 to 1mm day^{-1}

0 to –1mm day^{-1}

–1 to –2mm day^{-1}

less than –2mm day^{-1}

Figure 5.6: Change in precipitation (smoothed 10-year means) due to doubling CO$_2$, for months December-January-February, as simulated by three high resolution models: (a) CCC, (b) GFHI, and (c) UKHI. See legend for contour details.

(d) JJA 2×CO$_2$ – 1×CO$_2$ precipitation: CCC

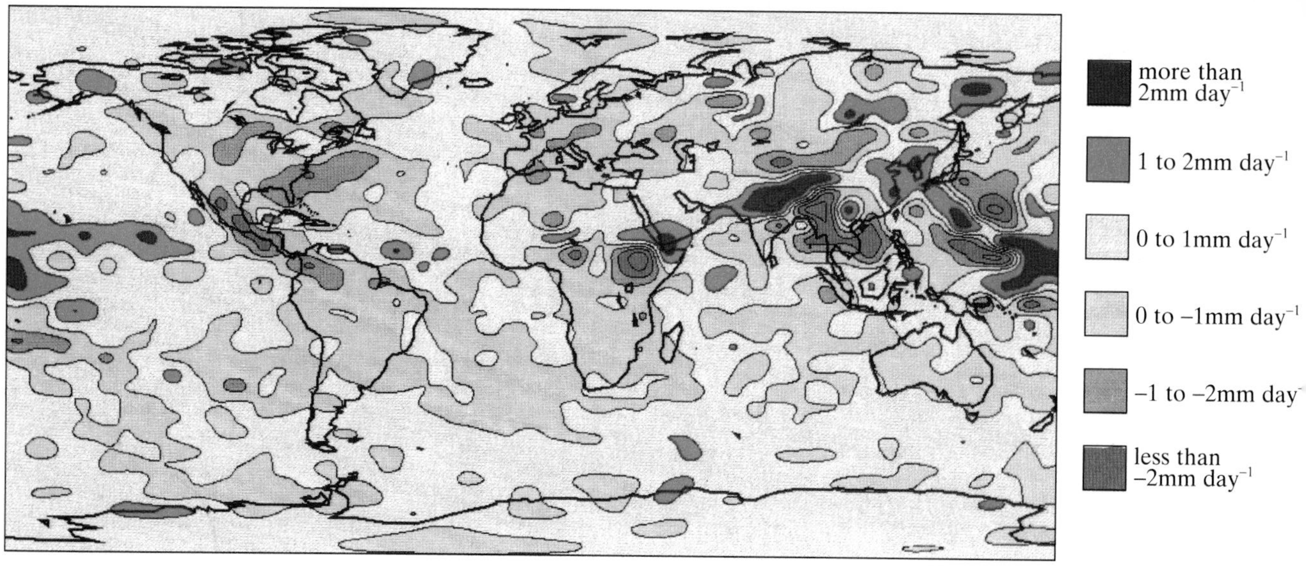

more than 2mm day^{-1}

1 to 2mm day^{-1}

0 to 1mm day^{-1}

0 to −1mm day^{-1}

−1 to −2mm day^{-1}

less than −2mm day^{-1}

(e) JJA 2×CO$_2$ – 1×CO$_2$ precipitation: GFHI

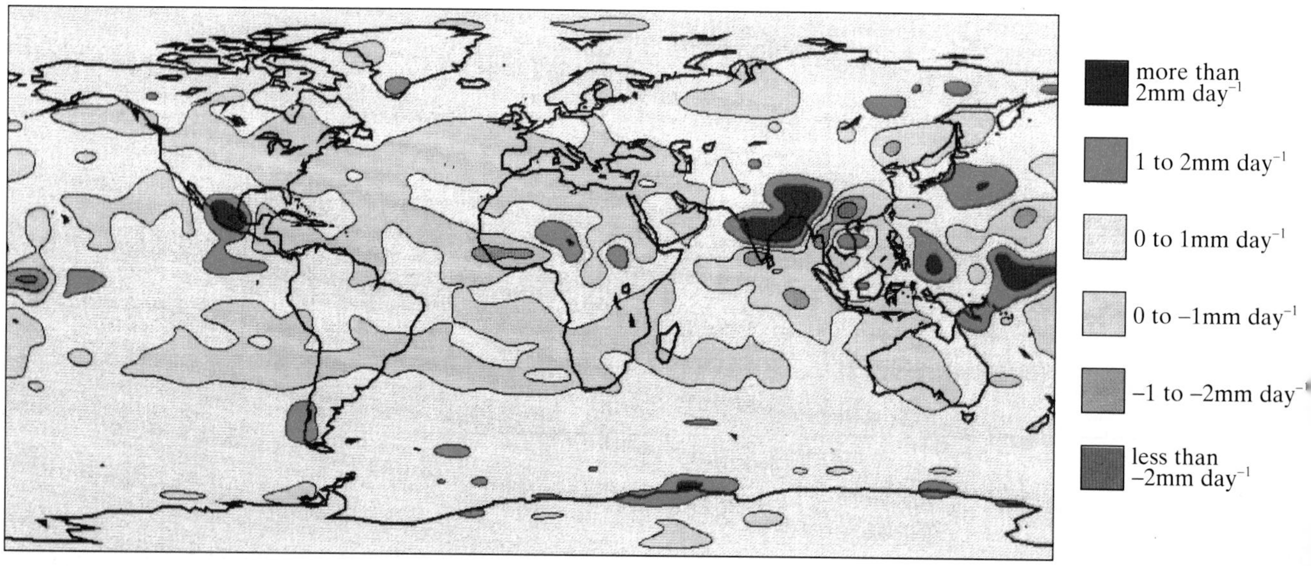

more than 2mm day^{-1}

1 to 2mm day^{-1}

0 to 1mm day^{-1}

0 to −1mm day^{-1}

−1 to −2mm day^{-1}

less than −2mm day^{-1}

(f) JJA 2×CO$_2$ – 1×CO$_2$ precipitation: UKHI

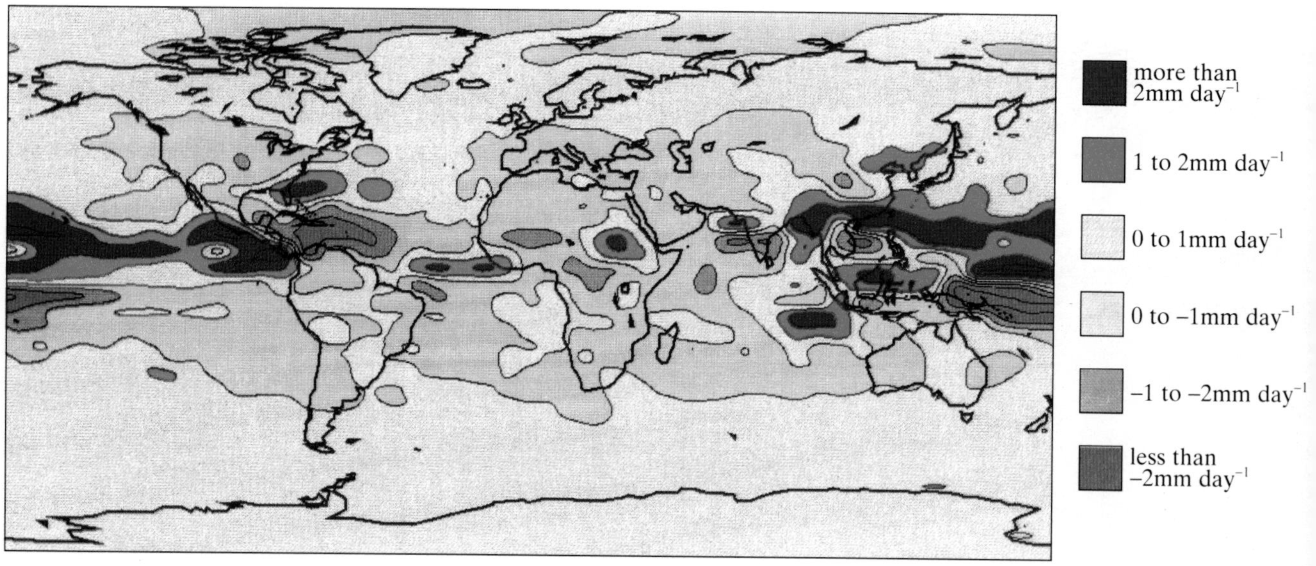

more than 2mm day^{-1}

1 to 2mm day^{-1}

0 to 1mm day^{-1}

0 to −1mm day^{-1}

−1 to −2mm day^{-1}

less than −2mm day^{-1}

Figure 5.6 continued: Change in precipitation (smoothed 10-year means) due to doubling CO$_2$, for months June-July-August, as simulated by three high resolution models: (d) CCC, (e) GFHI, and (f) UKHI. See legend for contour details.

(a) DJF 2×CO$_2$ – 1×CO$_2$ soil moisture: CCC

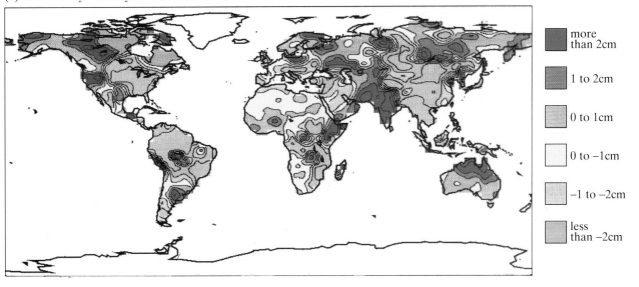

(b) DJF 2×CO$_2$ – 1×CO$_2$ soil moisture: GFHI

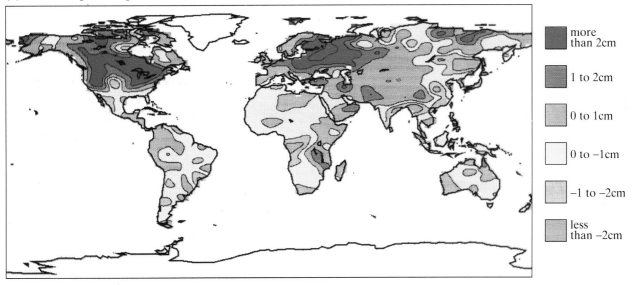

(c) DJF 2×CO$_2$ – 1×CO$_2$ soil moisture: UKHI

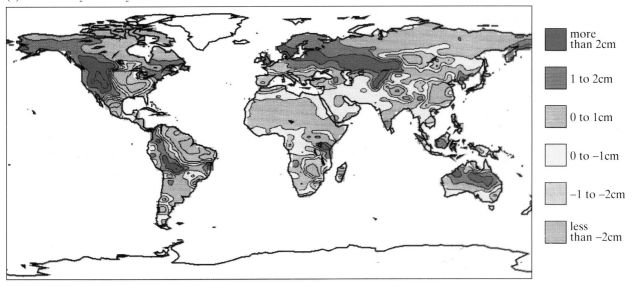

Figure 5.8: Change in soil moisture (smoothed 10-year means) due to doubling CO$_2$, for months December-January-February, as simulated by three high resolution models: (a) CCC, (b) GFHI, and (c) UKHI. Note that (a) has a geographically variable soil capacity whereas the other two models have the same capacity everywhere. See legend for contour details.

(d) JJA 2×CO$_2$ – 1×CO$_2$ soil moisture: CCC

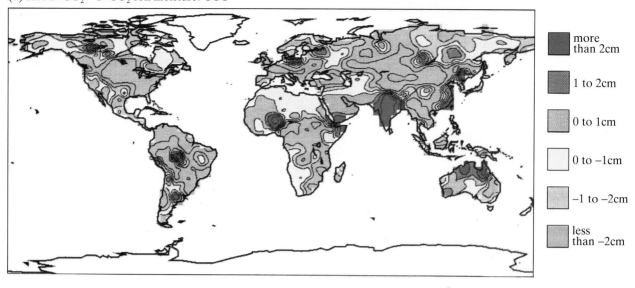

(e) JJA 2×CO$_2$ – 1×CO$_2$ soil moisture: GFHI

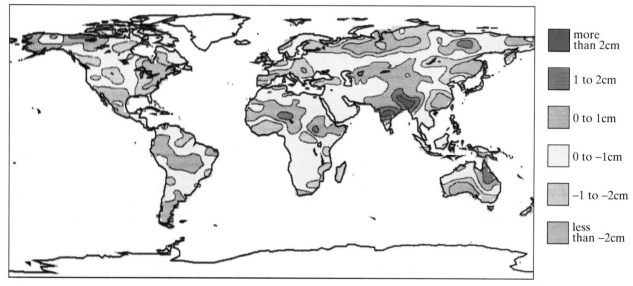

(f) JJA 2×CO$_2$ – 1×CO$_2$ soil moisture: UKHI

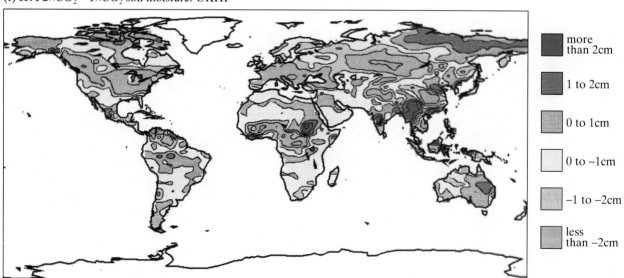

Figure 5.8 continued: Change in soil moisture (smoothed 10-year means) due to doubling CO$_2$, for months June-July-August, as simulated by three high resolution models: (d) CCC, (e) GFHI, and (f) UKHI. Note that (d) has a geographically variable soil capacity whereas the other two models have the same capacity everywhere. See legend for contour details.

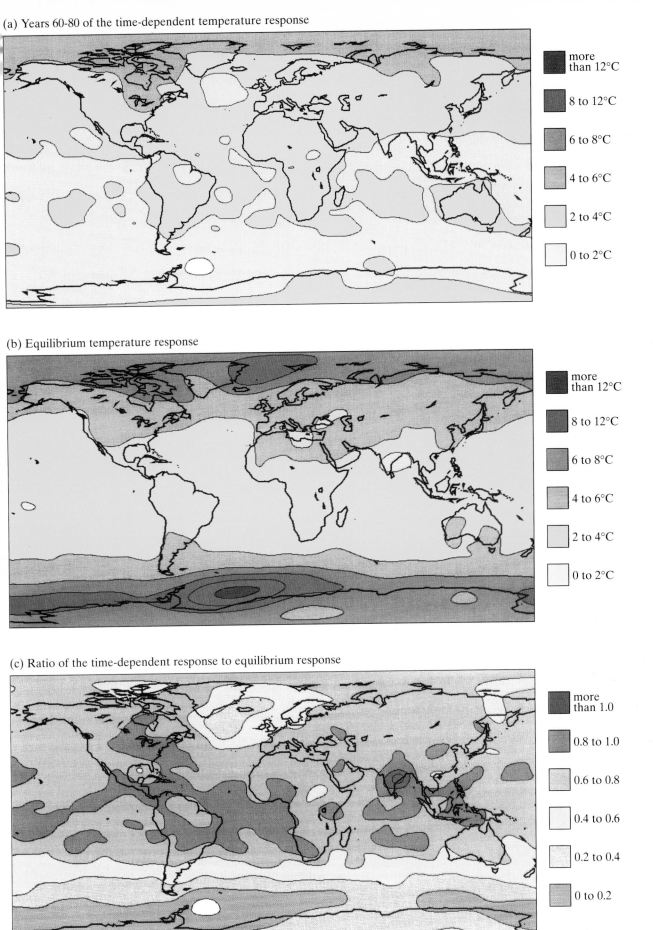

(a) Years 60-80 of the time-dependent temperature response

| more than 12°C |
| 8 to 12°C |
| 6 to 8°C |
| 4 to 6°C |
| 2 to 4°C |
| 0 to 2°C |

(b) Equilibrium temperature response

| more than 12°C |
| 8 to 12°C |
| 6 to 8°C |
| 4 to 6°C |
| 2 to 4°C |
| 0 to 2°C |

(c) Ratio of the time-dependent response to equilibrium response

| more than 1.0 |
| 0.8 to 1.0 |
| 0.6 to 0.8 |
| 0.4 to 0.6 |
| 0.2 to 0.4 |
| 0 to 0.2 |

Figure 6.5: (a) The time-dependent response of surface air temperature (°C) in the coupled ocean-atmosphere model to a 1% yr^{-1} increase of atmospheric CO_2. The difference between the 1% yr^{-1} perturbation run and years 60-80 of the control run when the atmospheric CO_2 concentration approximately doubles is shown. (b) The equilibrium response of surface air temperature (°C) in the atmosphere-mixed layer ocean model to a doubling of atmospheric CO_2. (c) The ratio of the time-dependent to equilibrium responses shown above. From Manabe, pers. comm. (1990). See legends for contour details.

(a) 1950 -1959 surface temperature anomalies

■	more than 0.75°C
	0.5 to 0.75°C
	0.25 to 0.5°C
	0 to 0.25°C
	0 to −0.25°C
	−0.25 to −0.5°C
■	less than −0.5°C

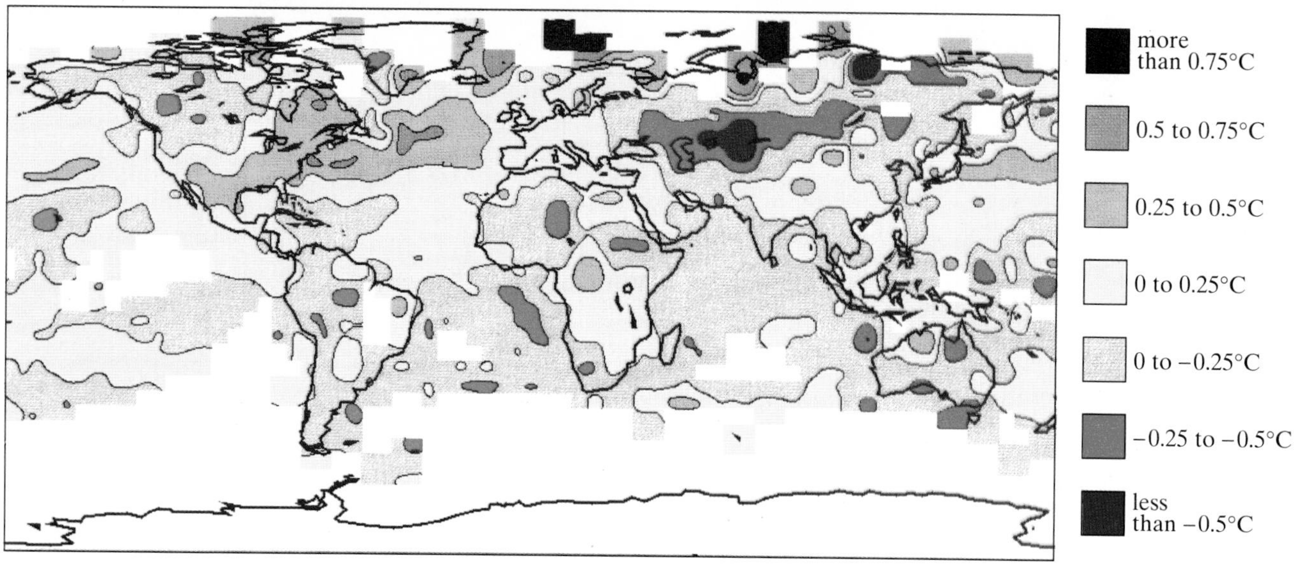

(b) 1967-1976 surface temperature anomalies

■	more than 0.75°C
	0.5 to 0.75°C
	0.25 to 0.5°C
	0 to 0.25°C
	0 to −0.25°C
	−0.25 to −0.5°C
■	less than −0.5°C

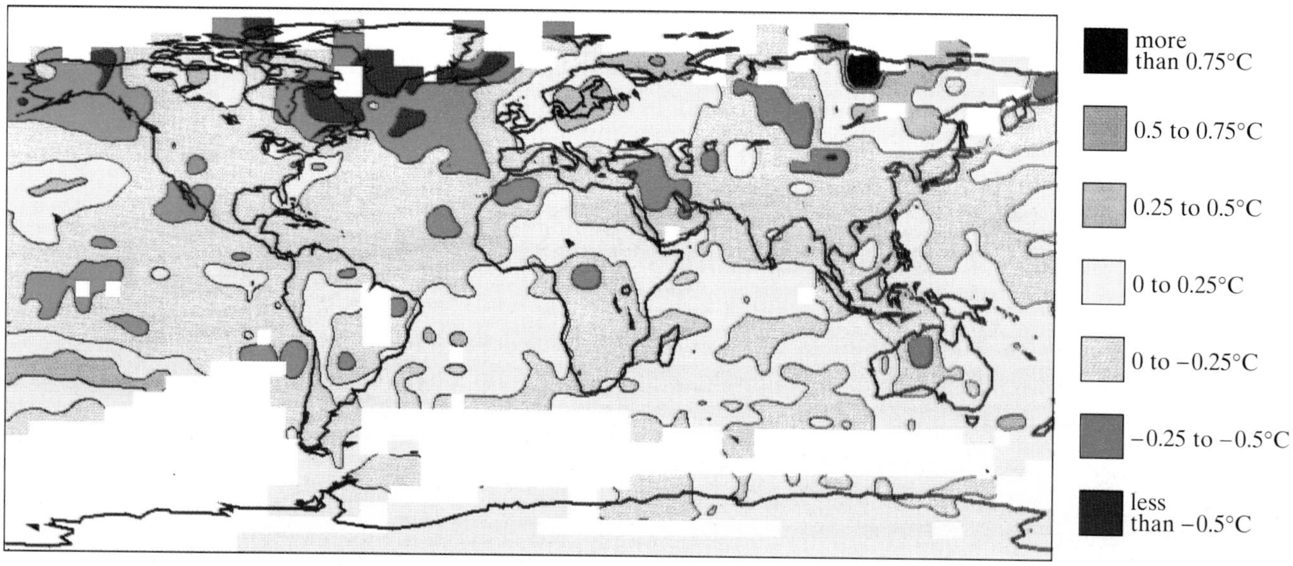

(c) 1980 -1989 surface temperature anomalies

■	more than 0.75°C
	0.5 to 0.75°C
	0.25 to 0.5°C
	0 to 0.25°C
	0 to −0.25°C
	−0.25 to −0.5°C
■	less than −0.5°C

Figure 7.13: Decadal surface temperature anomalies, relative to 1951-80. (a) 1950-1959, (b) 1967-76, (c) 1980-89. Land air temperatures from P.D. Jones and sea surface temperatures from the United Kingdom Meteorological Office. See legend for contour details. White areas show where there are insufficient data.

6

Time-Dependent Greenhouse-Gas-Induced Climate Change

F.P. BRETHERTON, K. BRYAN, J.D. WOODS

Contributors:
J. Hansen; M. Hoffert; X. Jiang; S. Manabe; G. Meehl; S.C.B. Raper; D. Rind; M. Schlesinger; R. Stouffer; T. Volk; T.M.L. Wigley.

CONTENTS

EXECUTIVE SUMMARY

The slowly changing response of climate to a gradual increase in greenhouse gas concentrations can only be modelled rigourously using a coupled ocean-atmosphere general circulation model with full ocean dynamics. This has now been done by a small number of researchers using coarse resolution models out to 100 years. Their results show that:

a) For a steadily increasing forcing, the global rise in temperature is an approximately constant fraction of the equilibrium rise corresponding to the instantaneous forcing for a time that is earlier by a fixed offset. For an atmospheric model with temperature sensitivity 4°C for a doubling of CO_2, this fraction is approximately 66% with an offset of 11 years. In rough terms, the response is about 60% of the current equilibrium value. Extrapolation using robust scaling principles indicates that for a sensitivity of 1.5°C the corresponding values are 85%, 6 years and 80% respectively.

b) The regional patterns of temperature and precipitation change generally resemble those of an equilibrium simulation for an atmospheric model, though uniformly reduced in magnitude. Exceptional regions are around Antarctica and in the northern North Atlantic, where the warming is much less.

c) These results are generally consistent with our understanding of the present circulation in the ocean, as evidenced by geochemical and other tracers. However, available computer power is still a serious limitation on model capability, and existing observational data are inadequate to resolve basic issues about the relative roles of various mixing processes, thus affecting the confidence level that can be applied to these simulations.

d) The conclusions about the global mean can be extended, though with some loss of rigour, by using a simple energy-balance climate model with an upwelling-diffusion model of the ocean, similar to that used to simulate CO_2 uptake. It is inferred that, if a steadily increasing greenhouse forcing were abruptly stabilized to a constant value thereafter, the global temperature would continue to rise at about the same rate for some 10-20 years, following which it would increase much more slowly, approaching the equilibrium value only over many centuries.

e) Based on the IPCC Business-as-Usual scenarios, the energy-balance upwelling diffusion model with best judgement parameters yields estimates of global warming from pre-industrial times (taken to be 1765) to the year 2030 between 1.3°C and 2.8°C, with a best estimate of 2.0°C. This corresponds to a predicted rise from 1990 of 0.7-1.5°C with a best estimate of 1.1°C.

Temperature rise from pre-industrial times to the year 2070 is estimated to be between 2.2°C and 4.8°C with a best estimate of 3.3°C. This corresponds to a predicted rise from 1990 of 1.6°C to 3.5°C, with a best estimate of 2.4°C.

6.1 Introduction

6.1.1 Why Coupled Ocean-Atmosphere Models ?

The responses discussed in Section 5 are for a radiative forcing that is constant for the few years required for the atmosphere and the surface of the ocean to achieve a new equilibrium following an abrupt change. Though the atmospheric models are detailed and highly developed, the treatment of the ocean is quite primitive. However, when greenhouse gas concentrations are changing continuously, the thermal capacity of the oceans will delay and effectively reduce the observed climatic response. At a given time, the realized global average temperature will reflect only part of the equilibrium change for the corresponding instantaneous value of the forcing. Of the remainder, part is delayed by storage in the stably stratified layers of the upper ocean and is realized within a few decades or perhaps a century, but another part is effectively invisible for many centuries or longer, until the heating of the deep ocean begins to influence surface temperature. In addition, the ocean currents can redistribute the greenhouse warming spatially, leading to regional modifications of the equilibrium computations. Furthermore, even without changes in the radiative forcing, interactions between the ocean and atmosphere can cause interannual and inter-decadal fluctuations that can mask longer term climate change for a while.

To estimate these effects, and to make reliable predictions of climate change under realistic scenarios of increasing forcing, coupled atmosphere-ocean general circulation models (GCMs) are essential. Such models should be designed to simulate the time and space dependence of the basic atmospheric and oceanic variables, and the physical processes that control them, with enough fidelity and resolution to define regional changes over many decades in the context of year-to-year variability. In addition, the reasons for differences between the results of different models should be understood.

6.1.2 Types of Ocean Models

The ocean circulation is much less well observed than the atmosphere, and there is less confidence in the capability of models to simulate the controlling processes. As a result, there are several conceptually different types of ocean model in use for studies of greenhouse warming.

The simplest representation considered here regards the ocean as a body with heat capacity modulated by downward diffusion below an upper mixed layer and a horizontal heat flux divergence within the mixed layer. These vary with position, but are prescribed with values that result, in association with a particular atmospheric model running under present climatic conditions, in simulations which fit observations for the annual mean surface temperature, and the annual cycle about that mean.

In such a "no surprises" ocean (Hansen et al., 1988) the horizontal currents do not contribute to modifications of climate change.

A more faithful representation is to treat the additional heat associated with forcing by changing greenhouse gases as a passive tracer which is advected by three-dimensional currents and mixed by specified diffusion coefficients intended to represent the sub-grid scale processes. These currents and mixing coefficients may be obtained from a GCM simulating the present climate and ocean circulation, including the buoyancy field, in a dynamically consistent manner. Using appropriate sources, the distributions of transient and other tracers such as temperature-salinity relationships, ^{14}C, tritium and CFCs are then inferred as a separate, computationally relatively inexpensive, step and compared with observations. Poorly known parameters such as the horizontal and vertical diffusion coefficients are typically adjusted to improve the fit. In box-diffusion models, which generally have a much coarser resolution, the currents and mixing are inferred directly from tracer distributions or are chosen to represent the aggregated effect of transports within a more detailed GCM. Though potential temperature (used to measure the heat content per unit volume) affects the buoyancy of sea-water and hence is a dynamically active variable, a number of studies (e.g., Bryan et al., 1984) have demonstrated that small, thermally driven, perturbations in a GCM do in fact behave in the aggregate very much as a passive tracer. This approach is useful for predicting small changes in climate from the present, in which the ocean currents and mixing co-efficients themselves are assumed not to vary in a significant manner. There are at present no clear guidelines as to to what is significant for this purpose.

A complete representation requires the full power of a high resolution ocean GCM, with appropriate boundary conditions at the ocean surface involving the wind stress, net heat flux and net freshwater flux obtained from an atmospheric model as a function of time in exchange for a simulation of the ocean surface temperature. The explicit simulation of mixing by mesoscale eddies is feasible and highly desirable (Semtner and Chervin, 1988), but it requires high spatial resolution, and so far the computer capacity required for 100 year simulations on a coupled global eddy resolving ocean-atmosphere GCM has not been available. Sea-ice dynamics are needed as well as thermodynamics, which is highly parameterized in existing models. For a fully credible climate prediction, what is required is such a complete, dynamically consistent representation, thoroughly tested against observations.

General circulation models of the coupled ocean atmosphere system have been under development for many years, but they have been restricted to coarse resolution models in which the mixing coefficients in the ocean must be prescribed *ad hoc*, and other unphysical devices are needed to match the ocean and atmospheric components. Despite these limitations, such models are giving results that seem consistent with our present understanding of the broad features of the ocean circulation, and provide an important tool for extending the conclusions of the equilibrium atmospheric climate models to time dependent situations, at least while the ocean circulation does not vary greatly from the present.

Because running coupled ocean-atmosphere GCMs is expensive and time consuming, many of our conclusions about global trends in future climates are based upon simplified models, in which parts of the system are replaced by highly aggregated constructs in which key formulae are inferred from observations or from other, non-interactive, models. An energy-balance atmospheric model coupled to a one dimensional upwelling-diffusion model of the ocean provides a useful conceptual framework, using a tracer representation to aid the interpretation of the results of the GCMs, as well as a powerful tool for quickly exploring future scenarios of climate change.

6.1.3 Major Sources of Uncertainty

An unresolved question related to the coarse resolution of general circulation models is the extent to which the details of the mixing processes and ocean currents may affect the storage of heat on different time scales and hence the fraction of the equilibrium global temperature rise that is realized only after several decades, as opposed to more quickly or much more slowly. Indeed, this issue rests in turn on questions whether the principal control mechanisms governing the sub-grid scale mixing are correctly incorporated. For a more detailed discussion see Section 4.8.

Paralleling these uncertainties are serious limitations on the observational data base to which all these models are compared, and from which the present rates of circulation are inferred. These give rise to conceptual differences of opinion among oceanographers about how the circulation actually functions.

Existing observations of the large scale distribution of temperature, salinity and other geochemical tracers such as tritium and ^{14}C do indicate that near surface water sinks deep into the water column to below the main thermocline, primarily in restricted regions in high latitudes in the North Atlantic and around the Antarctic continent. Associated with these downwelling regions, but not not necessarily co-located, are highly localized patches of intermittent deep convection, or turbulent overturning of a water column that is gravitationally unstable. Though controlled to a significant extent by salinity variations rather than by temperature, this deep convection can transfer heat vertically very rapidly.

However, much less clear is the return path of deep water to the ocean surface through the gravitationally stable thermocline which covers most of the ocean. It is disputed whether the most important process is nearly horizontal motion bulk motion in sloping isopycnal surfaces of constant potential density, ventilating the thermocline laterally. In this view, significant mixing across isopyncal surfaces occurs only where the latter intersect with the well mixed layer just below the ocean-atmosphere interface, which is stirred from above by the wind and by surface heat and water fluxes (Woods, 1984). Another view, still held by some oceanographers, is that the dominant mechanism is externally driven *in situ* mixing in a gravitationally stable environment and can be described by a local bulk diffusivity. To obtain the observations necessary to describe more accurately the real ocean circulation, and to improve our ability to model it for climate purposes, the World Ocean Circulation Experiment is currently underway (see Section 11).

6.2 Expectations Based on Equilibrium Simulations

Besides the different types of model, it is important also to distinguish the different experiments that have been done with them.

With the exception of the few time-dependent simulations described in Sections 6.3 - 6.6, perceptions of the geographical patterns of CO_2-induced climate change have been shaped mainly by a generation of atmospheric GCMs coupled to simple mixed layer or slab ocean models (see review by Schlesinger and Mitchell, 1987). With these specified-depth mixed-layer models having no computed ocean heat transport, CO_2 was instantaneously doubled and the models run to equilibrium. Averages taken at the end of the simulations were used to infer the geographical patterns of CO_2-induced climate change (Section 5).

Generally, the models agreed among themselves in a qualitative sense. Surface air temperature increase was greatest in late autumn at high latitudes in both hemispheres, particularly over regions covered by sea-ice. This was associated with a combination of snow/sea-ice albedo feedback and reduced sea-ice thermal inertia. Soil moisture changes showed a tendency for drying of mid-continental regions in summer, but the magnitude and even the sign of the change was not uniform among the models (see also Section 5.2.2.3). This inconsistency is caused by a number of factors. Some had to do with how soil moisture amounts were computed in the control simulations (Meehl and Washington, 1988), and some had to do with the method of simulating the land surface (Rind et al., 1989). Also, all models showed a strong cooling in the lower stratosphere due to the radiative effects of the increased carbon dioxide.

Recently, a new generation of coupled models has been run with atmospheric GCMs coupled to coarse-grid, dynamical ocean GCMs. These models include realistic geography, but the coarse grid of the ocean part (about 500 km by 500 km) necessitates the parameterization of mesoscale ocean eddies through the use of horizontal heat diffusion. This and other limitations involved with such an ocean model are associated with a number of systematic errors in the simulation (e.g., Meehl, 1989). However, the ability to include an explicitly computed ocean heat transport provides an opportunity to study, for the first time, the ocean's dynamical effects on the geographical patterns of CO_2-induced climate change.

6.3 Expectations Based on Transient Simulations

The first simulations with these global, coupled GCMs applied to the CO_2 problem used the same methodology as that employed in the earlier simple mixed-layer models. That is, CO_2 was doubled instantaneously and the model run for some time-period to document the climate changes. It has been suggested, however, that because of the long thermal response time of the full ocean, and the fact that the warming penetrates downward from the ocean surface into its interior, the traditional concept of a sensitivity experiment to determine a new equilibrium may be less useful with such a coupled system (Schlesinger and Jiang, 1988).

Schlesinger et al. (1985) ran a two-level atmospheric model coupled to a 6-layer ocean GCM for 20 years after instantaneously doubling CO_2. They noted that the model could not have attained an equilibrium in that period, and went on to document changes in climate at the end of the experiment. Washington and Meehl (1989) performed a similar experiment over a 30-year period with instantaneously doubled CO_2 in their global spectral atmospheric GCM coupled to a coarse-grid ocean GCM. Manabe et al. (1990) also used a global spectral atmospheric GCM coupled to a coarse-grid ocean model in an instantaneous CO_2 doubling experiment for a 60-year period.

These model simulations have been referred to as transient experiments in the sense that the time evolution of the whole climate system for a prescribed "switch-on" instantaneous CO_2 doubling could be examined in a meaningful way.

In some respects, all the switch-on coupled GCM experiments agree with the earlier mixed-layer results. In the Northern Hemisphere, warming is larger at higher latitudes, and there is some evidence, though again mixed, of drying in the mid-continental regions in summer. Manabe et al. (1990) also obtained a wetter soil in middle latitudes in winter. In the summer, however, Manabe et al. (1990) found large areas in the middle latitudes where the soil was drier. However, sector-configuration simulations (Bryan et al., 1988) with a coupled GCM first suggested a major difference in the patterns of climate change compared with the earlier mixed-layer model experiments. Around Antarctica, a relative warming minimum, at times even a slight cooling, was evident in these simulations.

6.4 Expectations Based on Time-Dependent Simulations

The term time-dependent in the present context is taken to mean a model simulation with gradually increasing amounts of greenhouse gases. This is what is happening in the real climate system, and such simulations provide us with the first indication of the climate-change signals we may expect in the near future.

To date, three such simulations have been published. One has been performed with an atmospheric model coupled to a simple ocean with fixed horizontal heat transport (Hansen et al., 1988), and the other two have used atmospheric models coupled to coarse-grid, dynamical ocean models, that is, atmosphere-ocean GCMs (Washington and Meehl, 1989; Stouffer et al., 1989). Other studies using coupled ocean-atmosphere GCMs are in progress at the UK Meteorological Office (Hadley Centre) and the Max Plank Institute für Meteorologie, Hamburg.

6.4.1 Changes In Surface Air Temperature

Hansen et al. (1988) performed several simulations with CO_2 and other greenhouse gases increasing at various rates, aimed at assessing the detectability of a warming trend above the inherent variability of a coupled ocean-atmosphere system. The ocean model was a "no surprises" ocean as described in Section 6.1.2, which simulates the spatially varying heat capacity typical of present climate, but precludes feedback to climate change from the ocean currents. At any given time the simulated warming was largest in the continental interior of Asia and at the high latitudes of both hemispheres, though it was first unambiguous in the tropics where the interannual variability is least. Contrasting with some other simulations, regional patterns of climate anomalies also had a tendency to show greater warming in the central and southeast U.S., and less warming in the western U.S.. The Antarctic also warmed about as much as corresponding northern high latitudes, a result that may be sensitive to the assumptions about ocean heat transport in this model.

Washington and Meehl (1989) specified a 1% per year linear increase of CO_2 in their coupled atmosphere-ocean GCM over a 30-year period and documented changes in the ocean-atmosphere system. For this period, there was a tendency for the land areas to warm faster than the oceans, and for the warming to be larger in the surface layer of the ocean than below. Significant areas of ocean surface temperature increase tended to occur between 50°S and 30°N (Figure 6.1). Warming was smaller and less significant around Antarctica. In the high latitudes of the Northern Hemisphere there was no zonally consistent warming pattern, in contrast to the earlier mixed-layer experiments. In fact, there was a cooling in the North Atlantic and Northwest Pacific for the particular five-year period shown in Figure 6.1. Washington and Meehl (1989) show that this cooling was a consequence of alterations in atmospheric and ocean circulation involving changes in precipitation and a weakening of the oceanic thermohaline circulation. However, there was a large inter-annual variability at high latitudes in the model, as occurs in nature, and Washington and Meehl pointed out that the pattern for this five-year period was indicative only of coupled anomalies that can occur in the system. Nevertheless, similar patterns of observed climate anomalies have been documented for temperature trends over the past 20-year period in the Northern Hemisphere (Karoly, 1989; Jones et al., 1988).

Stouffer et al. (1989) performed a time-dependent experiment with CO_2 increasing 1% per year (compounded), and documented geographical patterns of temperature differences for years 61-70. Stouffer et al.(1989) show the continents warming faster than the oceans, and a significant warming minimum near 60°S around Antarctica, as was seen in earlier sector experiments (Bryan et al., 1988). As was also seen in the Washington and Meehl results, there was not a uniform pattern of warming at all longitudes at high latitudes in the Northern Hemisphere. A minimum of warming occurred in the northwestern North Atlantic in association with deep overturning of the ocean. Though the greatest warming occurred at high latitudes of the Northern Hemisphere, the greater variability there resulted in the warming being unambiguously apparent first in the subtropical ocean regions.

The main similarities in the geographical patterns of CO_2-induced temperature change among these three time-dependent experiments are:

(a) ΔT_{991}, DJF, transient minus control, (yr 26—30)

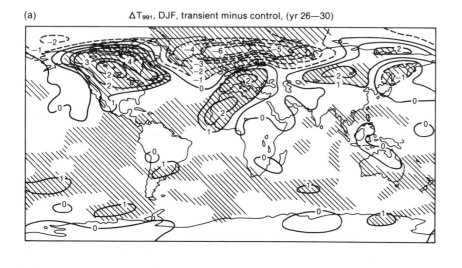

(b) ΔT_{991}, JJA, transient minus control, (yr 26—30)

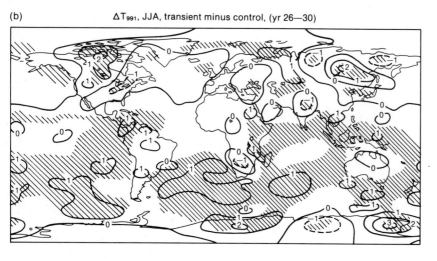

Figure 6.1: Geographical distributions of the surface temperature difference, transient minus control, of years 26-30 for: (a) DJF and (b) JJA (°C, lowest model layer). Differences significant at 5% level are hatched. Adapted from Washington and Meehl (1989).

1) the warming at any given time is less than the corresponding equilibrium value for that instantaneous forcing;

2) the areas of warming are generally greater at high latitudes in the Northern Hemisphere than at low latitudes, but are not zonally uniform in the earlier stages of the time-dependent experiments; and

3) because of natural variability, statistically significant warming is most evident over the subtropical oceans.

The differences between the time-dependent experiment using specified ocean heat transports (Hansen et al., 1988) and the two time-dependent experiments with dynamical ocean models (Washington and Meehl, 1989; Stouffer et al., 1989) are:

1) a warming minimum (or slight cooling) around Antarctica in the models with a dynamical ocean precludes the establishment of the large, positive ice-albedo feedback that contributes to extensive southern high-latitude warming in the mixed-layer models; and

2) a warming minimum in the northern North Atlantic.

These local minima appear to be due to exchange of the surface and deep layers of the ocean associated with upwelling as well as downwelling, or with convective overturning. The downwelling of surface water in the North Atlantic appears to be susceptible to changes of atmospheric circulation and precipitation, and the attendant weakening of the oceanic thermohaline circulation.

6.4.2 Changes In Soil Moisture

As indicated in Section 4.5, large-scale precipitation patterns are very sensitive to patterns of sea-surface temperature anomalies. Rind et al. (1989) link the occurrence of droughts with the climate changes in the time-dependent simulations of Hansen et al. (1988), and predict increased droughts by the 1990s. Washington and Meehl (1989) found in their time-dependent experiment that the soil in mid-latitude continents was wetter in winter and had small changes of both signs in summer. These time-dependent results are consistent with the results from their respective equilibrium climate change simulations (Section 5). A full analysis of the experiment described in Stouffer et al. (1989) is not yet available, but preliminary indications (Manabe, 1990) are that they are similarly consistent, though there may be some small changes in global scale patterns.

6.5 An Illustrative Example

In this section, we illustrate the promise and limitations of interactive ocean-atmosphere models with more details of one of the integrations described in Section 6.4, drawing on

Stouffer et al. (1989) and additional material supplied by Manabe (1990).

6.5.1 The Experiment

Three 100-year simulations are compared with different radiative forcing, each starting from the same, balanced initial state. The concentration of atmospheric carbon dioxide is kept constant in a control run. In two complementary perturbation runs the concentration of atmospheric carbon dioxide is increased or decreased by 1% a year (compounded), implying a doubling or halving after 70 years. This rate of increase roughly corresponds, in terms of CO_2 equivalent units, to the present rate of increase of forcing by all the greenhouse gases. Since greenhouse warming is proportional to the logarithm of carbon dioxide concentration (see Section 2, Table 2.2.4.1), an exponential increase gives a linear increase in radiative forcing.

To reach the initial balanced state the atmospheric model is forced to a steady state with the annual mean and seasonal variation of sea surface temperature given by climatological data. Using the seasonally varying winds from the atmospheric model, the ocean is then forced to a balanced state with the sea surface temperature and sea surface salinity specified from climatological data. For models perfectly representing the present climate and assuming the climatological data are accurate, the fluxes of heat and moisture should agree exactly. In practice the models are less than perfect and the heat and moisture flux fields at the ocean surface have a substantial mismatch.

To compensate for this mismatch, when the two models are coupled an *ad hoc* flux adjustment is added to the atmospheric heat and moisture fluxes. This flux adjustment, which is a fixed function of position and season, is precisely the correction required so that, as long as the radiative forcing of the atmospheric model remains the same, the coupled model will remain balanced and fluctuate around a mean state that includes the observed sea surface temperature and sea surface salinity. When the radiative balance of the atmospheric model is perturbed the coupled model is free to seek a new equilibrium, because the *ad hoc* flux adjustments remain as specified and provide no constraint to damp out departures from the present climate.

The flux adjustments in this treatment are nonphysical and disconcertingly large (Manabe, 1990), but are simply a symptom of the inadequacies in the separate models and of a mismatch between them. Unfortunately, existing measurements of ocean surface fluxes are quite inadequate to determine the precise causes. This device, or its equivalent (e.g., Sausen et al., 1988), is the price that must be paid for a controlled simulation of perturbations from a realistic present-day, ocean-atmosphere climate. Though varying through the annual cycle, the pattern of adjustment is the

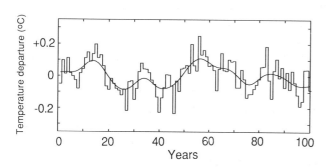

Figure 6.2: The temporal variation of the deviation of global mean surface air temperature (°C) of the coupled ocean-atmosphere model from its long term average. From Manabe (1990), personal communication.

same in all three simulations. Its direct effects thus disappear when differences are considered, and the conclusions from such experiments should be reasonable provided the differences remain small. However, when the simulated ocean circulation or atmospheric state differs greatly from that presently observed, indirect effects are likely to be substantial and too much credence should not be attached to the results. There are currently no quantitative criteria for what differences should be regarded as small for this purpose.

6.5.2 Results
Before examining the changes due to greenhouse forcing, it is instructive to note the random fluctuations in global mean surface air temperature within the control run itself (Figure 6.2). These imply a standard deviation in decadal averages of about ±0.08°C, which, though an accurate representation of the model climatology, appears to be somewhat less than what has been observed during the past 100 years (see Section 8). The regional manifestation shown in Figure 6.3 illustrates the uncertainty that is inherent in estimates of time dependent regional climate change over 20 year periods.

Figure 6.4 shows the difference in 10-year, global average surface air temperature, between the +1% perturbation run and the long term average of the control run, increasing approximately linearly as a function of time. After 70 years, the instantaneous temperature is only 58% of the equilibrium value (4°C, see Wetherald and Manabe, 1988) appropriate to the radiative forcing at that time. This result compares reasonably well with the estimate of 55% obtained by running the box diffusion model of Section 6.6 with similar scenario of radiative forcing and a climate sensitivity of 4°C for a doubling of CO_2 (see Figure 6.7 later). However, a result of both simulations is that the response to a linear increase in forcing with time is also, after a brief initial phase, very close to linear in time. The time dependent response is thus at all times proportional to the instantaneous forcing, but at a reduced magnitude compared to the equilibrium. Close examination of Figures 6.4 and 6.7 show that a straight line fit to the response intersects the time axis at ten years. Since lag in the response increases with time, it is not a useful parameter to describe results. As an alternative, the response is described in terms of a fraction of the equilibrium forcing which corresponds to a time with fixed offset to ten years earlier. Thus in the case of Figure 6.4, the fractional response

STANDARD DEVIATION OF 20 YEAR MEAN CONTROL RUN TEMPERATURE

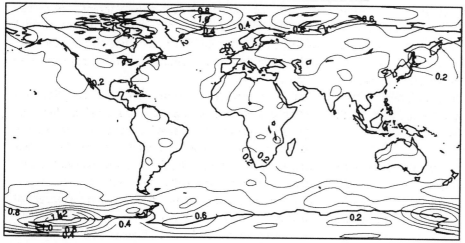

Figure 6.3: The geographic distribution of the standard deviation of 20 year mean surface air temperatures in the control run. From Manabe (1990), personal communication.

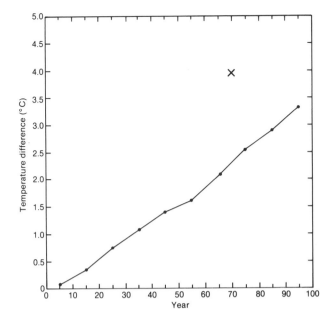

Figure 6.4: The temporal variation of the difference in globally averaged, decadal mean surface air temperature (˚C) between the perturbation run (with 1% /year increase of atmospheric CO_2) and the control run of the coupled ocean-atmosphere model. For comparison, the equilibrium response of global mean surface air temperature of the atmosphere-mixed layer ocean model to the doubling of atmospheric CO_2 is also indicated by x-symbol at 70th year when the gradually increasing CO_2 doubles.

6.5.3 Discussion

A qualitative explanation for these model results seems to lie in strong vertical pathways between the surface and intermediate to deep water in the northern North Atlantic and in the Southern Ocean. In this coupled ocean-atmosphere model, for which the radiative transfer to space is relatively inefficient, downward transfer of additional heat through these pathways short circuits up to 40% of the global greenhouse gas forcing to the deep water, where it mixes into a large volume causing a small local temperature rise. Thence it is carried away by deep currents, and is sequestered for many centuries. As discussed in Section 6.6 below, for a less sensitive atmospheric model, the fraction short circuited would be smaller. Over time-scales of 10-20 years, the remainder of the greenhouse warming brings the upper few hundred meters of most of the world ocean (the seasonal boundary layer and upper part of permanent thermocline) to approximate local equilibrium. The pattern of temperature change resembles that of the equilibrium calculation, though with a response commensurate with the reduced effective global forcing, because non-local processes in the atmosphere dominate inter-regional heat transfers in the surface layers of the ocean.

However, in exceptional regions an additional effect is operating. In the northwestern North Atlantic localized deep convection, which is a realistic feature of the model influenced by salinity contrasts, causes a very efficient heat transfer which, every winter, effectively pins the nearby ocean surface temperature to that of the deep water below. Because ocean currents and atmospheric transports act to smooth out the effects, the surface temperature rise of the whole nearby region is greatly reduced. In the Southern Hemisphere, geochemical tracer studies using the same global ocean model (Toggweiler et al., 1989) show that the most important vertical pathway is associated with the very large, deeply penetrating, wind induced downwelling just north of the Antarctic Circumpolar Current, and compensating upwelling of cold deep water between there and the Antarctic continent. The effect is likewise to reduce the regional temperature rise from what would otherwise be the global response.

Thus, the results described above of coupling this particular ocean and atmosphere GCM are all qualitatively explicable in terms of additional heat being advected as a passive tracer by the simulated present day ocean circulation in a manner similar to inert transient tracers such as tritium, and CFC's. Indeed, the results are consistent with a very simple globally averaged model of the ocean, which involves only a single well-mixed surface layer providing a lag of about 10 years, and a deep layer below of effectively infinite heat capacity, though this model is clearly not unique. However, this explanation is not universally accepted within the oceanographic

would be 58%, with a slope of 68%, and an offset of about 10 years. A lag of about this magnitude was also noted by Washington and Meehl (1989) for their switch-on CO_2 experiment (see their Figure 4).

On a regional scale, a 20-year average centred on 70 years (Figure 6.5(a)) is sufficient to determine general features of climate change that are significant against the background of natural variability (Figure 6.3). These features may be compared to the corresponding instantaneous equilibrium (Figure 6.5(b)). As shown in Figure 6.5(c), which illustrates the ratio of the values in Figures 6.5(a) and 6.5(b), the response is, in general, a relatively constant fraction, 60% - 80%, of the equilibrium. Major exceptions to this general picture are the northern North Atlantic, and the entire Southern Ocean between 40°S and 60°S, where change is largely suppressed.

For the -1% perturbation run, the changes in temperature patterns from the control closely mirror those from the +1% run, for the first 70 years at least, but are opposite sign. This supports the concept that the departure from the present climate can be described as a small perturbation, and is not inconsistent with the interpretation of additional heat in the ocean behaving like a passive tracer.

(a) YEARS 60-80 OF TIME-DEPENDENT TEMPERATURE RESPONSE

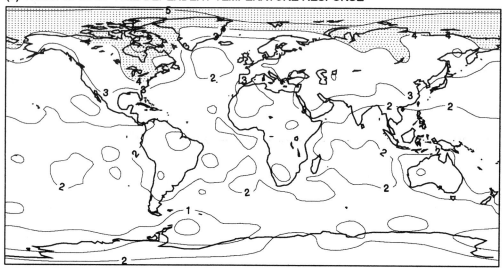

(b) EQUILIBRIUM TEMPERATURE RESPONSE

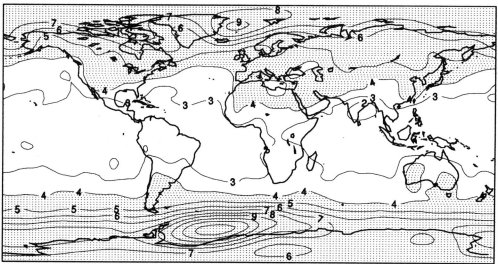

(c) RATIO OF TIME-DEPENDENT RESPONSE TO EQUILIBRIUM RESPONSE

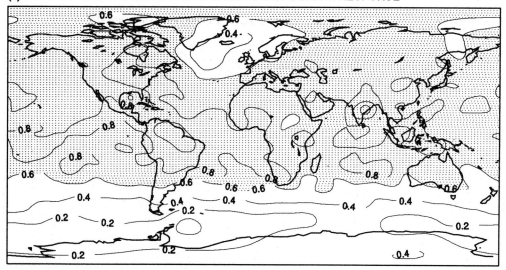

Figure 6.5: (a) The time-dependent response of surface air temperature (°C) in the coupled ocean-atmosphere model to a 1% yr⁻¹ increase of atmospheric CO_2. The difference between the 1% yr⁻¹ perturbation run and years 60-80 of the control run when the atmospheric CO_2 concentration approximately doubles.is shown (b) The equilibrium response of surface air temperature (°C) in the atmosphere-mixed-layer ocean model to a doubling of atmospheric CO_2. (c) The ratio of the time-dependent to equilibrium responses shown above. From Manabe (1990) pers. comm. Also shown in the colour section.

community as an adequate representation of the controlling
processes in the real ocean, reflecting the unresolved
questions described in Section 6.1.3 about heat storage and
the return path of deep water to the surface. Also, from
available analyses of the results of Stouffer et al. (1989) it
is not possible to determine what fraction of the heat
storage is in truly deep water with time scales of return to
the surface of many centuries, as opposed to being in
intermediate water with return times to the surface that
could be shorter (see Section 6.6.3). Thus extrapolation to
other cases should be treated with caution. In particular, the
apparent agreement with the results of the upwelling
diffusion model may prove to be illusory, particularly if
consideration is given to a very different forcing.

6.5.4 Changes In Ocean Circulation
Examination of the long term changes simulated in the
model shows some other trends with potentially important
consequences. Under the influence of increasing surface
temperature and precipitation, the vertical circulation and
overturning in the North Atlantic are becoming systematically weaker (Figure 6.6). That this is not an acc-
idental artefact is confirmed by the minus 1% experiment,
in which the radiative forcing becomes steadily more
negative and this overturning circulation strengthens
significantly. The same does not occur in the Antarctic,
where the controls on exchanges with the sub-surface
waters are different. The indications are, that if the plus 1%
experiment were continued to perhaps 150 years, the
downwelling and deep convection in the North Atlantic
might cease altogether, with climate there and in Western
Europe entering a new regime about which it would be
premature to speculate. There might also be a significant
effect on the carbon cycle and global atmospheric CO_2
levels (Section 1.2.7.1).

Though the potential for substantial climate change is
implicit in such changes in ocean circulation regime,
simulations from present coupled ocean-atmosphere
models must be used with considerable caution in making
such predictions. Both the parameterization of eddy mixing
processes in these models, and the flux adjustment at the
ocean-atmosphere interface, have been selected for the
present ocean circulation, and cannot be expected to
function reliably under drastically different circumstances.

6.6 Projections of Global Mean Change
It is possible to use an energy-balance atmospheric model
coupled to an upwelling-diffusion model of the ocean to
estimate changes in the global-mean surface air temperature induced by different scenarios of radiative forcing
and to help interpret the results from GCMs. Within the
limitations of the tracer representation, it summarizes in
terms of a few parameters the basic results of more
complex simulations of the ocean circulation in time
dependent climate change, and enables rapid extrapolation
to other cases. As in the case of ocean GCMs, the
parameters have been selected to fit geochemical tracer and
water mass data and therefore reflect the present state of
the world ocean. Therefore, the same caveats must be
applied to extrapolating the results of the upwelling
diffusion models to very different climatic regimes.

6.6.1 An Upwelling Diffusion Model
Such a simple climate/ocean model was proposed by
Hoffert et al. (1980) and has since been used in several
studies of the time-dependent response of the climate
system to greenhouse-gas-induced radiative forcing [see,
for example, Harvey and Schneider (1985); Wigley and
Raper (1987), and the review papers by Hoffert and

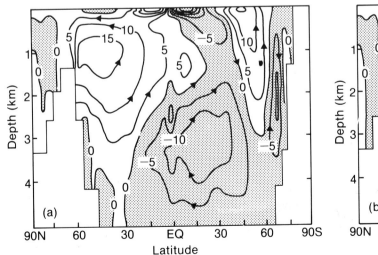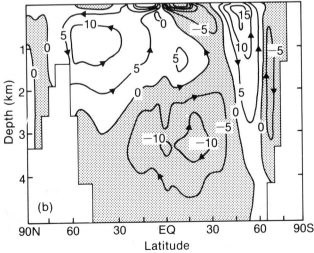

Figure 6.6: The streamfunction describing the vertical circulation in the Atlantic after 100 years: (a) control; (b) increasing forcing. Units are $10^6 m^3 s^{-1}$. From Stouffer et al. (1989).

Flannery (1985) and Schlesinger (1989)]. This simple climate model determines the global-mean surface temperature of the atmosphere and the temperature of the ocean as a function of depth from the surface to the ocean floor. It is assumed that the atmosphere mixes heat efficiently between latitudes, so that a single temperature rise ΔT characterizes the surface of the globe, and that the incremental radiation to space associated with the response to greenhouse gas forcing is proportional to ΔT. The model ocean is subdivided vertically into layers, with the uppermost being the mixed layer. Also, the ocean is subdivided horizontally into a small polar region where water downwells and bottom water is formed, and a much larger nonpolar region where there is a slow uniform vertical upwelling. In the nonpolar region heat is transported upwards toward the surface by the water upwelling there and downwards by physical processes whose bulk effects are treated as an equivalent diffusion. Besides by radiation to space, heat is also removed from the mixed layer in the nonpolar region by a transport to the polar region and downwelling toward the bottom, this heat being ultimately transported upward from the ocean floor in the nonpolar region.

In the simple climate/ocean model, five principal quantities must be specified:

1) the temperature sensitivity of the climate system, ΔT_{2x}, characterized by the equilibrium warming induced by a CO_2 doubling;

2) the vertical profile of the vertical velocity of the ocean in the non-polar region, w;

3) the vertical profile of thermal diffusivity in the ocean, k, by which the vertical transfer of heat by physical processes other than large-scale vertical motion is represented;

4) the depth of the well-mixed, upper layer of the ocean, h; and

5) the change in downwelled sea surface temperature in the polar region relative to that in the nonpolar region, π.

For the following simulations the parameters are those selected by Hoffert et al., 1980, in their original presentation of the model. Globally averaged upwelling, w, outside of water mass source regions is taken as 4 m/yr, which is the equivalent of $42 \times 10^6 m^3 s^{-1}$ of deep and intermediate water formation from all sources. Based upon an e-folding scale depth of the averaged thermocline of 500m (Levitus, 1982), the corresponding k is 0.63 cm^2 s^{-1}. h is taken to be 70 m, the approximate globally averaged depth of the mixed layer (Manabe and Stouffer, 1980). Lastly, two values are considered for π, namely, 1 and 0, the former based on the assumption that the additional heat in surface water that is advected into high latitudes and downwells in regions of deep water formation is

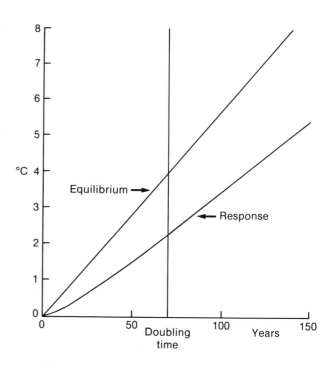

Figure 6.7: The change in surface temperature for a linear increase in greenhouse gas forcing, with an equivalent CO_2 doubling time of 70 years. The simulations were performed with an energy balance/upwelling diffusion ocean model with $\Delta T_{2x} = 4^oC$, an upwelling velocity w of 4 m y^{-1}, a mixed layer depth h of 70 m, a vertical diffusivity k of 0.66 cm^2 s^{-1}, and a π parameter of 1.

transported down rather than rejected to the atmosphere, and the latter on the alternate assumption that the polar ocean temperature remains at the freezing temperature for sea water and therefore does not change. For the latter case to be applicable, the atmosphere would have to accept the additional heat, presumably meaning that the surrounding ocean surface temperature would have to be relatively substantially warmer than elsewhere at that latitude.

Selecting $\pi = 1$, these values of the parameters are used for best judgement estimates of global warming in Sections 8 and 9. The choice is somewhat arbitrary, but the impact of uncertainty must be judged against the sensitivity of the conclusions to their values.

6.6.2 *Model Results*
Figure 6.7 shows the simulated increase in global mean temperature for the radiative forcing function used by Stouffer et al. (1989), with the appropriate atmospheric sensitivity ΔT_{2x} equal to 4°C for a doubling of CO_2 and a π factor of 1. Since the ocean parameters used were chosen independently as standard for best estimate simulations in Sections 7 and 8, the general correspondence with Figure 6.4 provides some encouragement that this simplified

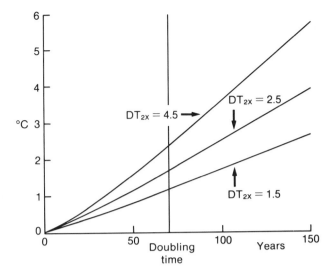

Figure 6.8: As for Figure 6.7 for a π parameter of 1 but for various values of atmospheric climate sensitivity ΔT_{2x}.

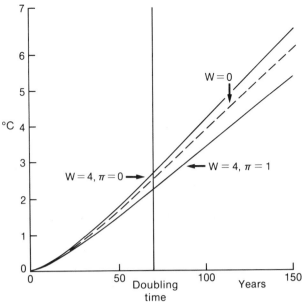

Figure 6.9: As for Figure 6.7 but for w = 4, π = 1 (corresponding to 'Response' curve in Figure 6.7); w = 4, π = 0; and w = 0 (dashed curve). Figure 6.7 shows the equilibrium case.

model is consistent with the ocean GCM. Note, however, the slight upward curvature of the response in Figure 6.7, due to intermediate time scales of 20-100 years associated with heat diffusion or ventilation in the thermocline. A tangent line fit at the 70 year mark could be described as a fractional response that is approximately 55% of the instantaneous forcing, with a slope of 66% superimposed on an offset of about 11 years. For the sense in which the terms percentage response and lag are used here see Section 6.5.2.

Figure 6.8 shows the simulated increase in global mean temperature for the same radiative forcing, but with atmospheric models of differing climate sensitivity. For a sensitivity of 1.5°C for a doubling of CO_2 the response fraction defined by the tangent line at 70 years is approximately 77% of the instantaneous forcing, with a slope of 85% superimposed on an offset of 6 years, whereas for 4.5°C the corresponding values are 52%, 63% and 12 years.

Figure 6.9 compares the response for the standard parameter values with those for a π factor of zero, and for a purely diffusive model with the same diffusivity.

Varying *h* between 50 and 120 m makes very little difference to changes over several decades. The effect of varying *k* between 0.5 and 2.0 cm² s⁻¹ with historical forcing has been discussed by Wigley and Raper (1990). If k/w is held constant, the realized warming varies over this range by about 18% for ΔT_{2x} = 4.5°C, and by 8% for ΔT_{2x} = 1.5°C.

Figure 6.10 shows the effect of terminating the increase of forcing after 70 years. The response with π = 1 continues to grow to a value corresponding to an offset of some 10-

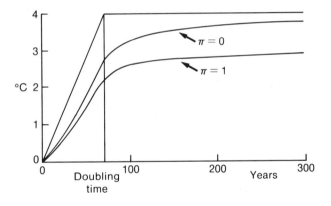

Figure 6.10: As for Figure 6.7 but for a forcing rising linearly with time until an equivalent CO_2 doubling after 70 years, followed by a constant forcing.

20 years, then rises very much more slowly to come to true equilibrium only after many centuries.

Also shown (Figure 6.11) are projections of future climate change using radiative forcing from IPCC Business-as-Usual and B-D emission scenarios, for values of the climate sensitivity ΔT_{2x} equal to 1.5, 2.5 and 4.5°C. These scenarios are discussed in the Annex. Assuming *k* = 0.63, π = 1, and w = 4 ms⁻¹, the realised warming is 1.3, 1.8 and 2.6°C (above pre-industrial temperatures) under the Business-as-Usual Scenario. For Scenario B, these estimates should be reduced by about 15%.

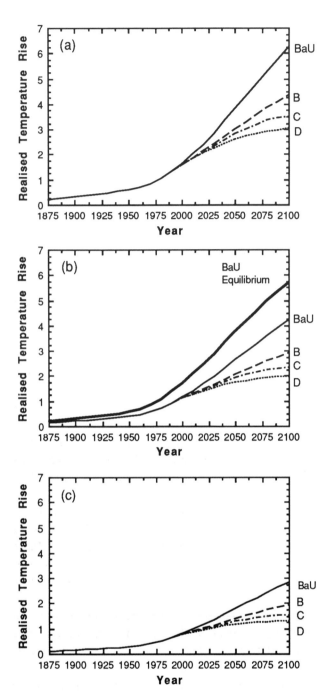

Figure 6.11: The contribution of the change in greenhouse-gas concentrations to the change in global-mean surface air temperature (°C) during 1875 to 1985, together with projections from 1985 to 2100 for IPCC Scenarios BaU-D. The temperature rise is from 1765 (pre-industrial).The simulation was performed with an energy-balance climate/upwelling-diffusion ocean model with an upwelling velocity w of 4 m y^{-1}, mixed layer depth h of 70 m, vertical diffusivity k of 0.66 cm^2 s^{-1}, and a π parameter of 1. The three diagrams are for ΔT_{2x}-values of (a) 4.5°C, (b) 2.5°C and (c) 1.5°C. The equilibrium temperature is also shown for the BaU emissions and 2.5°C climate sensitivity.

6.6.3 Discussion

The model used for these projections is highly simplified, and somewhat different choices could be made of the parameters while still retaining consistency with the observed average thermocline depth and accepted rates of global deep water formation. However, the impact of these residual uncertainties on the time-dependent global mean climate response is relatively small compared to that associated with the cloud-radiation feedback (Figure 6.8). Similar models have also been widely used in interpreting the observed distribution of geochemical tracers in the ocean and for modelling the uptake of CO_2. As discussed in Section 6.1, the detailed physical basis for a model of this type may be questioned, but it yields global average results that are apparently not inconsistent with simulations using more detailed coupled ocean-atmosphere GCMs (Bryan et al., 1984; Schlesinger et al., 1985). Given the limited speed of supercomputers available in 1990, it remains the only tool available for exploring time-dependent solutions for a wide range of forcing scenarios. Thus it is important to appreciate the basic reasons the model gives the results it does.

The increased percentage response for low climate sensitivities shown in Figure 6.8 is a broadly applicable consequence of the distribution of the prescribed forcing into the parallel processes of radiation to space and increasing storage in the ocean. Treating heat in the ocean as a passive tracer, for each process the heat flux at any given time is proportional to the realized temperature rise. However, for radiation the constant of proportionality is inversely proportional to the climate sensitivity, whereas for storage the constant is independent of it. Since both fluxes are positive, and must add to a fixed value, the forcing, decreasing the climate sensitivity will increase the percentage response (though the realized temperature rise will increase). This conclusion, though not the storage proportionality constant, is unaffected by changing the formulation of the storage mechanism in the ocean, provided only it can be modelled by a passive tracer.

The difference between each curve in Figure 6.9 and that for the equilibrium value at the corresponding time is proportional to the rate of increase of heat stored in the oceans, and the area between the curves to the cumulative storage. Compared to pure diffusion ($w = 0$), this storage is increased by upwelling, provided the additional heat that is added, as water rising through the thermocline is brought to ever higher temperatures at the ocean surface, is then retained in the ocean ($\pi = 1$). If, on the other hand, that heat is lost to the atmosphere before the surface water in high latitudes downwells ($\pi = 0$), then the ocean heat storage is reduced, presumably because, once heated, a given parcel of water can retain that heat only for a finite time as the entire volume of the thermocline is recycled into the deep ocean in 150 years or so. After correcting for different surface temperatures, the difference between these two curves measures for the case $\pi = 1$ the effective storage in the deep ocean below the thermocline, where the re-circulation time to the surface is many centuries. The difference between $\pi = 0$ and the equilibrium, on the other hand, measures the retention in the surface layers and the thermocline. Though the latter area is substantially larger than the former, it does not mean that all the heat concerned is immediately available to sustain a surface temperature rise.

Figure 6.10 shows that, within this model with $\pi = 1$, if the increase in forcing ceases abruptly after 70 years, only 40% of the then unrealized global surface temperature increase is realized within the next 100 years, and most of that occurs in the first 20 years. If there is no heat storage in the deep ocean this percentage is substantially higher.

It is thus apparent that the interpretation given in Section 6.5.3 of the results of Stouffer et al. (1989) in terms of a two-layer box model is not unique. However, it is re-assuring that broadly similar results emerge from an upwelling-diffusion model. Nevertheless, it is clear that further coupled GCM simulations, different analyses of experiments already completed, and, above all, more definitive observations, will be necessary to resolve these issues.

6.7 Conclusions

Coupled ocean-atmosphere general circulation models, though still of coarse resolution and subject to technical problems such as the flux adjustment, are providing useful insights into the expected climate response due to a time-dependent radiative forcing. However, only a very few simulations have been completed at this time, and to explore the range of scenarios necessary for this assessment highly simplified upwelling-diffusion models of the ocean must be used instead.

In response to a forcing that is steadily increasing with time, the simulated global rise of temperature in both types of model is approximately a constant fraction of the equilibrium rise corresponding to the forcing at an earlier time. For an atmospheric model with a temperature sensitivity of 4°C for a doubling of CO_2, this constant is approximately 66% with an offset of 11 years. For a sensitivity of 1.5°C, the values are about 85% and 6 years respectively. Changing the parameters in the upwelling diffusion model within ranges supposedly consistent with the global distribution of geophysical tracers can change the response fraction by up to 20% for the most sensitive atmospheric models but by only 10% for the least sensitive.

Indications from the upwelling-diffusion model are that, if after a steady rise the forcing were to be held steady, the response would continue to increase at about the same rate for 10-20 years, but thereafter would increase only much more slowly, taking several centuries to achieve equilibrium. This conclusion depends mostly on the assumption that $\pi = 1$, i.e., greenhouse warming in surface waters is transported downwards in high latitudes by downwelling or exchange with deep water, rather than being rejected to the atmosphere, and is thereafter sequestered for many centuries. Heat storage in the thermocline affects the surface temperature on all time-scales, to some extent even a century or more later.

There is no conclusive analysis of the relative role of different water masses in ocean heat storage for the GCM of Stouffer et al. (1990), but there are significant transfers of heat into volumes of intermediate and deep water from which the return time to the surface is many centuries. Further analysis is required of the implications for tracer distributions of the existing simulations of the circulation in the control run, for comparison with the upwelling diffusion model. Likewise, further simulations of the fully interactive coupled GCM with different forcing functions would strengthen confidence in the use of the simplified tracer representation for studies of near term climate change.

A sudden change of forcing induces transient contrasts in surface temperature between land and ocean areas, affecting the distribution of precipitation. Nevertheless, the regional response pattern for both temperature and precipitation of coupled ocean-atmosphere GCMs for a steadily increasing forcing generally resembles that for an equilibrium simulation, except uniformly reduced in magnitude. However, both such models with an active ocean show an anomalously large reduction in the rise of surface temperature around Antarctica, and one shows a similar reduction in the northern North Atlantic. These anomalous regions are associated with rapid vertical exchanges within the ocean, due to convective overturning or wind-driven upwelling-downwelling.

Coupled ocean-atmosphere GCMs demonstrate an inherent interannual variability, with a significant fraction on decadal and longer timescales. It is not clear how realistic current simulations of the statistics of this variability really are. At a given time, warming due to greenhouse gas forcing is largest in high latitudes of the northern hemisphere, but because the natural variability is greater there, the warming first becomes clearly apparent in the tropics.

The major sources of uncertainty in these conclusions arise from inadequate observations to document how the present ocean circulation really functions. Even fully interactive GCMs need mixing parameters that must be adjusted *ad hoc* to fit the real ocean, and only when the processes controlling such mixing are fully understood and reflected in the models can there be confidence in climate simulations under conditions substantially different from the present day. There is an urgent need to establish an operational system to collect oceanographic observations routinely at sites all around the world ocean,

Present coarse resolution coupled ocean-atmosphere general circulation models are yielding results that are broadly consistent with existing understanding of the general circulation of the ocean and of the atmospheric climate system. With increasing computer power and improved data and understanding based upon planned ocean observation programs, it may well be possible within the next decade to resolve the most serious of the remaining technical issues and achieve more realistic simulations of the time-dependent coupled ocean-atmosphere system responding to a variety of greenhouse forcing scenarios. Meanwhile, useful estimates of global warming under a variety of different forcing scenarios may be made using highly simplified upwelling-diffusion models of the ocean, and with tracer simulations using GCM reconstructions of the existing ocean circulation.

References

Bryan, K., F. G. Komro, S. Manabe and M. J. Spelman, 1982: Transient climate response to increasing atmospheric carbon dioxide. *Science*, **215**, 56-58.

Bryan, K., F. G. Komro, and C. Rooth, 1984: The ocean's transient response to global surface anomalies. In *Climate Processes and Climate Sensitivity*, Geophys. Monogr. Ser., **29** , J. E. Hansen and T. Takahashi (eds.), Amer. Geophys. Union, Washington, D.C., 29-38.

Bryan, K., S. Manabe, and M. J. Spelman, 1988: Interhemispheric asymmetry in the transient response of a coupled ocean-atmosphere model to a CO_2 forcing. *J. Phys. Oceanogr.*, **18**, 851-867.

Harvey, L.D.D. and S.H. Schneider, 1985: Transient climate response to external forcing on 10^0 - 10^4 year time scales. *J. Geophys. Res.*, **90**, 2191-2222.

Hansen, J., I. Fung, A. Lacis, D. Rind, S. Lebedeff, R. Ruedy and G. Russell, 1988: Global climate changes as forecast by the Goddard Institute for Space Sciences three dimensional model. *J. Geophys. Res.*, **93**, 9341-9364.

Hoffert, M. I., and B. F. Flannery, 1985: Model projections of the time-dependent response to increasing carbon dioxide. In *Projecting the Climatic Effects of Increasing Carbon Dioxide*, DOE/ER-0237, edited by M. C. MacCracken and F. M. Luther, United Stated Department of Energy, Washington, DC, pp. 149-190.

Hoffert, M. I., A. J. Callegari, and C.-T. Hsieh, 1980: The role of deep sea heat storage in the secular response to climatic forcing. *J. Geophys. Res.*, **85** (C11), 6667-6679.

Jones, P.D., T. M. L. Wigley, C. K. Folland, and D. E. Parker, 1988: Spatial patterns in recent worldwide temperature trends. *Climate Monitor*, **16**, 175-186

Karoly, D., 1989: Northern hemisphere temperature trends: A possible greenhouse gas effect ? *Geophys. Res. Lett.*, **16**, 465-468.

Levitus, S., 1982: Climatological Atlas of the World Ocean, NOAA Prof. Papers, **13**, U.S. Dept. of Commerce, Washington D.C., 17311.

Manabe, S., and R. J. Stouffer, 1980: Sensitivity of a global climate model to an increase of CO_2 concentration in the atmosphere. *J. Geophys. Res.*, **85**, 5529-5554.

Manabe, S., K. Bryan, and M. J. Spelman, 1990: Transient response of a global ocean-atmosphere model to a doubling of atmospheric carbon dioxide. *J. Phys. Oceanogr.*, To be published.

Manabe, S., 1990: Private communication.

Meehl, G. A., and W. M. Washington, 1988: A comparison of soil-moisture sensitivity in two global climate models. *J. Atmos. Sci.*, **45**, 1476-1492.

Meehl, G. A., 1989: The coupled ocean-atmosphere modeling problem in the tropical Pacific and Asian monsoon regions. *J. Clim.*, **2**, 1146-1163.

Rind, D., R. Goldberg, J. Hansen, C. Rosenzweig, and R. Ruedy, 1989: Potential evapotranspiration and the likelihood of future drought. *J. Geophys. Res.*, in press.

Sausen, R., K. Barthel, and K. Hasselmann, 1988: Coupled ocean-atmosphere models with flux correction. *Clim Dyns*, **2**, 145-163.

Semtner, A.J. and R.M. Chervin, 1988: A simulation of the global ocean circulation with resolved eddies. *J. Geophys. Res.*, **93**, 15502-15522.

Schlesinger, M. E., 1989: Model projections of the climate changes induced by increased atmospheric CO_2. In *Climate and the Geo-Sciences: A Challenge for Science and Society in the 21st Century*, A Berger, S. Schneider, and J. C. Duplessy, Eds., Kluwer Academic Publishers, Dordrecht, 375-415.

Schlesinger, M. E. and X. Jiang, 1988: The transport of CO_2-induced warming into the ocean: An analysis of simulations by the OSU coupled atmosphere-ocean general circulation model. *Clim. Dyn.*, **3**, 1-17.

Schlesinger, M. E. and J. F. B. Mitchell, 1987: Climate model simulations of the equilibrium climatic response to increased carbon dioxide. *Rev. Geophys.*, **25**, 760-798.

Schlesinger, M. E., W. L. Gates and Y. -J. Han, 1985: The role of the ocean in CO_2-induced climate change: Preliminary results from the OSU coupled atmosphere-ocean general circulation model. In *Coupled Ocean-Atmosphere Models*, J. C. J. Nihoul (ed), Elsevier Oceanography Series, **40**, 447-478.

Stouffer, R. J., S. Manabe and K. Bryan, 1989: Interhemispheric asymmetry in climate response to a gradual increase of atmospheric carbon dioxide. *Nature*, **342**, 660-662.

Toggweiler, J. R., K. Dixon and K. Bryan, 1988: Simulations of Radiocarbon in a Coarse-Resolution, World Ocean Model. II: Distributions of Bomb-produced Carbon-14. *J.Geophys Res.*, **94(C10)**, 8243-8264.

Washington, W. M., and G. A. Meehl, 1989: Climate sensitivity due to increased CO_2: Experiments with a coupled atmosphere and ocean general circulation model. *Clim. Dyn.*, **4**, 1-38.

Wetherald, R.T. and S. Manabe, 1988: Cloud feedback processes in general circulation models. *J. Atmos. Sci.*, **45(8)**, 1397-1415.

Wigley, T. M. L., and S. C. B.Raper, 1987: Thermal expansion of sea water associated with global warming. *Nature*, **330**, 127-131

Wigley, T. M. L., and S.C.B. Raper, 1989: Future changes in global-mean temperature and thermal-expansion-related sea level rise. In *"Climate and Sea Level Change: Observations, Projections and Implications"*, (eds R. A. Warrick and T. M. L. Wigley) To be published.

Woods, J.D., 1984: Physics of thermocline ventilation. In: *Coupled ocean-atmosphere circulation models.* Ed. J.C.J. Nihoul, Elsevier.

7

Observed Climate Variations and Change

C.K. FOLLAND, T.R. KARL, K.YA. VINNIKOV

Contributors:
J.K. Angell; P. Arkin; R.G. Barry; R. Bradley; D.L. Cadet; M. Chelliah; M. Coughlan; B. Dahlstrom; H.F. Diaz; H Flohn; C. Fu; P. Groisman; A. Gruber; S. Hastenrath; A. Henderson-Sellers; K. Higuchi; P.D. Jones; J. Knox; G. Kukla; S. Levitus; X. Lin; N. Nicholls; B.S. Nyenzi; J.S. Oguntoyinbo; G.B. Pant; D.E. Parker; B. Pittock; R. Reynolds; C.F. Ropelewski; C.D. Schönwiese; B. Sevruk; A. Solow; K.E. Trenberth; P. Wadhams; W.C. Wang; S. Woodruff; T. Yasunari; Z. Zeng; and X. Zhou.

CONTENTS

EXECUTIVE SUMMARY

***** There has been a real, but irregular, increase of global surface temperature since the late nineteenth century.

***** There has been a marked, but irregular, recession of the majority of mountain glaciers over the same period.

***** Precipitation has varied greatly in sub-Saharan Africa on time scales of decades.

*** Precipitation has progressively increased in the Soviet Union over the last century.

*** A steady increase of cloudiness of a few percent has been observed since 1950 over the USA.

* A larger, more sudden, but less certain increase of cloudiness has been observed over Australia.

Observational and palaeo-climatic evidence indicates that the Earth's climate has varied in the past on time scales ranging from many millions of years down to a few years. Over the last two million years, glacial-interglacial cycles have occurred on a time scale of 100,000 years, with large changes in ice volume and sea level. During this time, average global surface temperatures appear to have varied by about 5-7°C. Since the end of the last ice age, about 10,000 BP, globally averaged surface temperatures have fluctuated over a range of up to 2°C on time scales of centuries or more. Such fluctuations include the Holocene Optimum around 5,000-6,000 years ago, the shorter Medieval Warm Period around 1000 AD (which may not have been global) and the Little Ice Age which ended only in the middle to late nineteenth century. Details are often poorly known because palaeo-climatic data are frequently sparse.

The instrumental record of surface temperatures over the land and oceans remains sparse until after the middle of the nineteenth century. It is common, therefore, to emphasize trends in the global instrumental record from the late nineteenth century. The record suggests a global (combined land and ocean) average warming of 0.45±0.15°C since the late nineteenth century, with an estimated small (less than 0.05°C) exaggeration due to urbanisation in the land component. The greater part of the global temperature increase was measured prior to the mid-1940s. Global warming is indicated by three independent data sets: air temperatures over land, air temperatures over the ocean, and sea surface temperatures. The latter two data sets show only a small lag compared with land temperatures. A marked retreat of mountain glaciers in all parts of the world since the end of the nineteenth century provides further evidence of warming.

The temperature record of the last 100 years shows significant differences in behaviour between the Northern and Southern Hemispheres. A cooling of the Northern Hemisphere occurred between the 1940s and the early 1970s, while Southern Hemisphere temperatures remained nearly constant from the 1940s to about 1970. Since 1970 in the Southern Hemisphere and 1975 in the Northern Hemisphere, a more general warming has been observed, concentrated into the period 1975-1982, with little global warming between 1982 and 1989. However, changes of surface temperature in different regions of the two hemispheres have shown considerable contrasts for periods as long as decades throughout the last century, notably in the Northern Hemisphere.

Over periods as short as a few years, fluctuations of global or hemispheric temperatures of a few tenths of a degree are common. Some of these are related to the El Niño-Southern Oscillation phenomenon in the tropical Pacific. Evidence is also emerging of decadal time scale variability of ocean circulation and deep ocean heat content that is likely to be an important factor in climate change.

It is not yet possible to deduce changes in precipitation on global or even hemispheric scales. Some regions have, however, experienced real changes over the past few decades. A large decline in summer seasonal rainfall has been observed in sub-Saharan Africa since the 1950s but precipitation appears to have increased progressively over the Soviet Union during the last century.

Reliable records of sea-ice and snow are too short to discern long-term changes. Systematic changes in the number and intensity of tropical cyclones are not apparent, though fluctuations may occur on decadal time scales. There is no evidence yet of global scale changes in the frequency of extreme temperatures. Increases in cloud cover have been reported from the oceans and some land areas. Uncertainties in these records are mostly too large to allow firm conclusions to be drawn. Some of the changes are artificial, but increases of cloudiness over the USA and Australia over the last forty years may be real.

We conclude that despite great limitations in the quantity and quality of the available historical temperature data, the evidence points consistently to a real but irregular warming over the last century. A global warming of larger size has almost certainly occurred at least once since the end of the last glaciation without any appreciable increase in greenhouse gases. Because we do not understand the reasons for these past warming events it is not yet possible to attribute a specific proportion of the recent, smaller, warming to an increase of greenhouse gases.

7.1 Introduction

This Section focuses on changes and variations in the modern climate record. To gain a longer term perspective and to provide a background to the discussion of the palaeo-analogue forecasting technique in Section 3, variations in palaeo-climate are also described. Analyses of the climate record can provide important information about natural climate variations and variability. A major difficulty in using observed records to make deductions about changes resulting from recent increases in greenhouse gases (Sections 1 and 2) is the existence of natural climatic forcing factors that may add to, or subtract from, such changes. Unforced internal variability of the climate system will also occur, further obscuring any signal induced by greenhouse gases.

Observing the weather, and converting weather data to information about climate and climate change, is a very complex endeavour. Virtually all our information about modern climate has been derived from measurements which were designed to monitor weather rather than climate change. Even greater difficulties arise with the proxy data (natural records of climate-sensitive phenomena, mainly pollen remains, lake varves and ocean sediments, insect and animal remains, glacier termini) which must be used to deduce the characteristics of climate before the modern instrumental period began. So special attention is given to a critical discussion of the quality of the data on climate change and variability and our confidence in making deductions from these data. Note that we have not made much use of several kinds of proxy data, for example tree ring data, that can provide information on climate change over the last millennium. We recognise that these data have an increasing potential; however their indications are not yet sufficiently easy to assess nor sufficiently integrated with indications from other data to be used in this report.

A brief discussion of the basic concepts of climate, climate change, climate trends etc, together with references to material containing more precise definitions of terms, is found in the Introduction at the beginning of this Report.

7.2 Palaeo-Climatic Variations and Change

7.2.1 Climate Of The Past 5,000,000 Years

Climate varies naturally on all time scales from hundreds of millions of years to a few years. Prominent in recent Earth's history have been the 100,000 year Pleistocene glacial-interglacial cycles when climate was mostly cooler than at present (Imbrie and Imbrie, 1979). This period began about 2,000,000 years before the present time (BP) and was preceded by a warmer epoch having only limited glaciation, mainly over Antarctica, called the Pliocene. Global surface temperatures have typically varied by 5-7°C

through the Pleistocene ice age cycles, with large changes in ice volume and sea level, and temperature variations as great as 10-15°C in some middle and high latitude regions of the Northern Hemisphere. Since the beginning of the current interglacial epoch about 10,000 BP, global temperatures have fluctuated within a much smaller range. Some fluctuations have nevertheless lasted several centuries, including the Little Ice Age which ended in the nineteenth century and which was global in extent.

Proxy data clearly indicate that the Earth emerged from the last ice age 10,000 to 15,000 BP (Figure 7.1). During this glacial period, continental-size ice sheets covered much of North America and Scandinavia, and world sea level was about 120m below present values. An important cause of the recurring glaciations is believed to be variations in seasonal radiation receipts in the Northern Hemisphere. These variations are due to small changes in the distance of the Earth from the sun in given seasons, and slow changes in the angle of the tilt of the Earth's axis which affects the amplitude of the seasonal insolation. These "Milankovitch" orbital effects (Berger, 1980) appear to be correlated with the glacial-interglacial cycle since glacials arise when solar radiation is least in the extratropical Northern Hemisphere summer.

Variations in carbon dioxide and methane in ice age cycles are also very important factors; they served to modify and perhaps amplify the other forcing effects (see Section 1). However, there is evidence that rapid changes in climate have occurred on time scales of about a century which cannot be directly related to orbital forcing or to changes in atmospheric composition. The most dramatic of these events was the Younger Dryas cold episode which involved an abrupt reversal of the general warming trend in progress around 10,500 BP as the last episode of continental glaciation came to a close. The Younger Dryas was an event of global significance; it was clearly observed in New Zealand (Salinger, 1989) though its influence may not have extended to all parts of the globe (Rind et al., 1986). There is, as yet, no consensus on the reasons for this climatic reversal, which lasted about 500 years and ended very suddenly. However, because the signal was strongest around the North Atlantic Ocean, suggestions have been made that the climatic reversal had its physical origin in large changes in the sea surface temperature (SST) of the North Atlantic Ocean. One possibility is that the cooling may have resulted from reduced deep water production in the North Atlantic following large-scale melting of the Laurentide Ice sheet and the resulting influx of huge amounts of low density freshwater into the northern North Atlantic ocean (Broecker et al., 1985). Consequential changes in the global oceanic circulation may have occurred (Street-Perrott and Perrott, 1990) which may have involved variations in the strength of the thermohaline circulation in the Atlantic. This closed oceanic circulation involves northward flow of water near the ocean surface, sinking in the sub-Arctic and a return flow at depth. The relevance of the Younger Dryas to today's conditions is that it is possible that changes in the thermohaline circulation of a qualitatively similar character might occur quite quickly during a warming of the climate induced by greenhouse gases. A possible trigger might be an increase of precipitation over the extratropical North Atlantic (Broecker, 1987), though the changes in ocean circulation are most likely to be considerably smaller than in the Younger Dryas. Section 6 gives further details.

The period since the end of the last glaciation has been characterized by small changes in global average temperature with a range of probably less than 2°C (Figure 7.1), though it is still not clear whether all the fluctuations indicated were truly global. However, large regional

changes in hydrological conditions have occurred, particularly in the tropics. Wetter conditions in the Sahara from 12,000 to 4,000 years BP enabled cultural groups to survive by hunting and fishing in what are today almost the most arid regions on Earth. During this time Lake Chad expanded to become as large as the Caspian Sea is today (several hundred thousand km^2, Grove and Warren, 1968). Drier conditions became established after 4,000 BP and many former lake basins became completely dry (Street-Perrot and Harrison, 1985). Pollen sequences from lake beds of northwest India suggest that periods with subdued monsoon activity existed during the recent glacial maximum (Singh et al., 1974), but the epoch 8,000 to 2,500 BP experienced a humid climate with frequent floods.

There is growing evidence that worldwide temperatures were higher than at present during the mid-Holocene (especially 5,000-6,000 BP), at least in summer, though carbon dioxide levels appear to have been quite similar to those of the pre-industrial era at this time (Section 1). Thus parts of western Europe, China, Japan, the eastern USA were a few degrees warmer in July during the mid-Holocene than in recent decades (Yoshino and Urushibara, 1978; Webb et al., 1987; Huntley and Prentice, 1988; Zhang and Wang, 1990). Parts of Australasia and Chile were also warmer. The late tenth to early thirteenth centuries (about AD 950-1250) appear to have been exceptionally warm in western Europe, Iceland and Greenland (Alexandre 1987; Lamb, 1988). This period is known as the Medieval Climatic Optimum. China was, however, cold at this time (mainly in winter) but South Japan was warm (Yoshino, 1978). This period of widespread warmth is notable in that there is no evidence that it was accompanied by an increase of greenhouse gases.

Cooler episodes have been associated with glacial advances in alpine regions of the world; such "neo-glacial" episodes have been increasingly common in the last few thousand years. Of particular interest is the most recent cold event, the "Little Ice Age", which resulted in extensive glacial advances in almost all alpine regions of the world between 150 and 450 years ago (Grove, 1988) so that glaciers were more extensive 100-200 years ago than now nearly everywhere (Figure 7.2). Although not a period of continuously cold climate, the Little Ice Age was probably the coolest and most globally-extensive cool period since the Younger Dryas. In a few regions, alpine glaciers advanced down-valley even further than during the last glaciation (for example, Miller, 1976). Some have argued that an increase in explosive volcanism was responsible for the coolness (for example Hammer, 1977; Porter, 1986); others claim a connection between glacier advances and reductions in solar activity (Wigley and Kelly, 1989) such as the Maunder and Sporer solar activity minima (Eddy, 1976), but see also Pittock (1983). At present, there is no

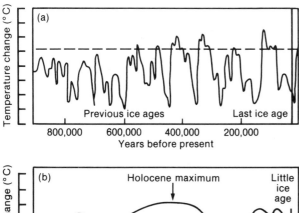

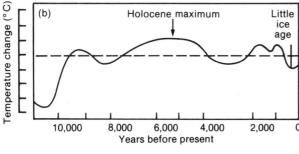

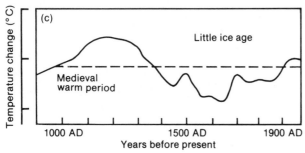

Figure 7.1: Schematic diagrams of global temperature variations since the Pleistocene on three time-scales: (a) the last million years, (b) the last ten thousand years, and (c) the last thousand years. The dotted line nominally represents conditions near the beginning of the twentieth century.

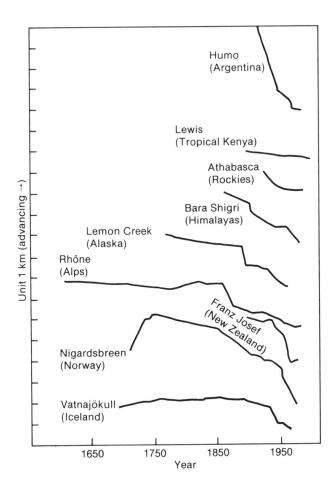

Figure 7.2: Worldwide glacier termini fluctuations over the last three centuries (after Grove, 1988, and other sources).

agreed explanation for these recurrent cooler episodes. The Little Ice Age came to an end only in the nineteenth century. Thus some of the global warming since 1850 could be a recovery from the Little Ice Age rather than a direct result of human activities. So it is important to recognise that natural variations of climate are appreciable and will modulate any future changes induced by man.

7.2.2 Palaeo-Climate Analogues from Three Warm Epochs

Three periods from the past have been suggested by Budyko and Izrael (1987) as analogues of a future warm climate. For the second and third periods listed below, however, it can be argued that the changed seasonal distribution of incoming solar radiation existing at those times may not necessarily have produced the same climate as would result from a globally-averaged increase in greenhouse gases.

1) The climate optimum of the Pliocene (about 3,300,000 to 4,300,000 years BP),

2) The Eemian interglacial optimum (125,000 to 130,000 years BP),

3) The mid-Holocene (5,000 to 6,000 years BP).

Note that the word "optimum" is used here for convenience and is taken to imply a warm climate. However such a climate may not be "optimal" in all senses.

7.2.2.1 Pliocene climatic optimum (about 3,300,000 to 4,300,000 BP)

Reconstructions of summer and winter mean temperatures and total annual precipitation have been made for this

(a)

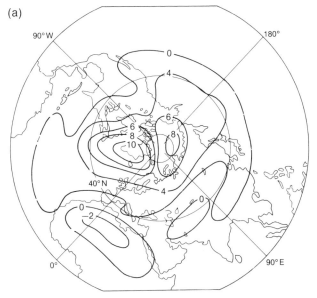

(b)

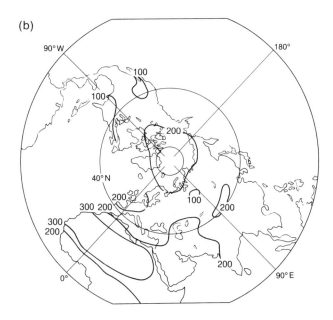

Figure 7.3: (a) Departures of summer air temperature (°C) from modern values for the Pliocene climatic optimum (4.3 to 3.3 million years BP) (from Budyko and Izrael, 1987).

(b) Departures of annual precipitation (mm) from modern values for the Pliocene climatic optimum (from Budyko and Izrael, 1987; Peshy and Velichko, 1990).

period by scientists in the USSR. Many types of proxy data were used to develop temperature and precipitation patterns over the land masses of the Northern Hemisphere (Budyko and Izrael, 1987). Over the oceans, the main sources of information were cores drilled in the bed of the deep ocean by the American Deep-sea Ocean Core Drilling Project. Some of these reconstructions are shown in Figure 7.3a and b.

Figure 7.3a suggests that mid-latitude Northern Hemisphere summer temperatures averaged about 3-4°C higher than present-day values. Atmospheric concentrations of carbon dioxide are estimated by Budyko et al. (1985) to have been near 600 ppm, i.e., twice as large as immediately pre-industrial values. However Berner et al. (1983) show lower carbon dioxide concentrations. So there is some doubt about the extent to which atmospheric carbon dioxide concentrations were higher than present values during the Pliocene. Figure 7.3b is a partial reconstruction of Northern Hemisphere annual precipitation; this was generally greater during the Pliocene. Of special interest is increased annual precipitation in the arid regions of Middle Asia and Northern Africa where temperatures were lower than at present in summer.

Uncertainties associated with the interpretation of these reconstructions are considerable and include:

1) Imprecise dating of the records, especially those from the continents (uncertainties of 100,000 years or more);

2) Differences from the present day surface geography, including changes in topography; thus Tibet was at least 1000m lower than now and the Greenland ice sheet may have been much smaller;

3) The ecology of life on Earth from which many of the proxy data are derived was significantly different.

See also Sections 5.5.3 and 4.10.

7.2.2.2 Eemian interglacial optimum (125,000-130,000 years BP)

Palaeo-botanic, oxygen-isotope and other geological data show that the climates of the warmest parts of some of the Pleistocene interglacials were considerably warmer (1 to 2°C) than the modern climate. They have been considered as analogues of future climate (Budyko and Izrael, 1987; Zubakov and Borzenkova, 1990). Atmospheric carbon dioxide reached about 300 ppm during the Eemian optimum (Section 1) but a more important cause of the warmth may have been that the eccentricity of the Earth's orbit around the sun was about twice the modern value, giving markedly more radiation in the northern hemisphere summer. The last interglacial optimum (125,000-130,000 years BP) has sufficient information (Velichko et al., 1982, 1983, 1984 and CLIMAP, 1984) to allow quantitative reconstructions to be made of annual and seasonal air

Figure 7.4: Departures of summer air temperature (°C) from modern values for the Eemian interglacial (Velichko et al., 1982, 1983, 1984).

temperature and annual precipitation for part of the Northern Hemisphere. For the Northern Hemisphere as a whole, mean annual surface air temperature was about 2°C above its immediately pre-industrial value. Figure 7.4 shows differences of summer air temperature, largest (by 4-8°C) in northern Siberia, Canada and Greenland. Over most of the USSR and Western Europe north of 50-60°N, temperatures were about 1-3°C warmer than present. South of these areas, temperatures were similar to those of today, and precipitation was substantially larger over most parts of the continents of the Northern Hemisphere. In individual regions of Western Europe, the north of Eurasia and Soviet Central Asia and Kazakhstan, annual precipitation has been estimated to have been 30-50% higher than modern values.

It is difficult to assess quantitatively the uncertainties associated with these climate reconstructions. The problems include:

1) Variations between the timing of the deduced thermal maximum in different records;

2) The difficulties of obtaining proxy data in arid areas;

3) The absence of data from North America and many other continental regions in both Hemispheres.

7.2.2.3 Climate of the Holocene Optimum (5,000-6,000 years BP)

The Early and Middle Holocene was characterized by a relatively warm climate with summer temperatures in high northern latitudes about 3-4°C above modern values. Between 9,000 and 5,000 years BP, there were several short-lived warm epochs, the last of which, the mid Holocene optimum, lasted from about 6,200 to 5,300 years BP (Varushchenko et al., 1980). Each warm epoch was accompanied by increased precipitation and higher lake levels in subtropical and high latitudes (Singh et al., 1974; Swain et al., 1983). However, the level of such mid-latitude lakes as the Caspian Sea, Lake Geneva and the Great Basin lakes in the USA was lowered (COHMAP, 1988; Borzenkova and Zubakov, 1984).

Figures 7.5a and b show maps of summer surface air temperature (as departures from immediately pre-industrial values) and annual precipitation for the mid-Holocene optimum in the Northern Hemisphere. This epoch is

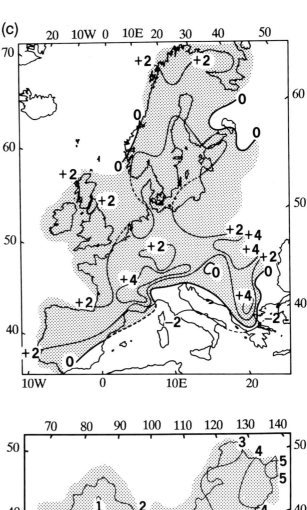

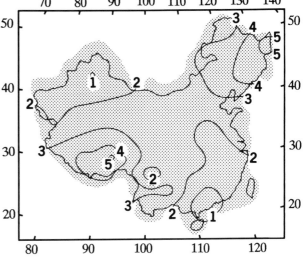

Figure 7.5 (continued): (c) Summer temperatures (relative to the mid-twentieth century) in Europe and China between 5,000 and 6,000 BP (after Huntley and Prentice, 1988; Wang et al., personal communication).

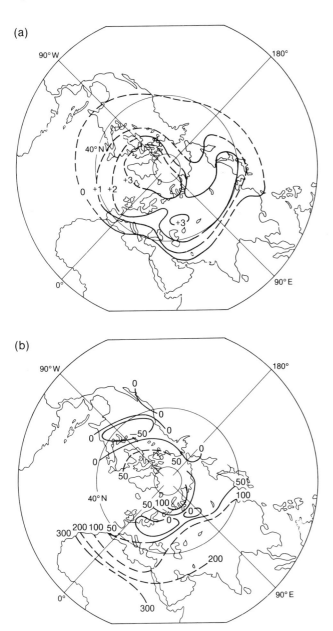

Figure 7.5: Departures of: (a) summer temperature (°C), (b) annual precipitation (mm), from modern values for the Holocene climatic optimum (Borzenkova and Zubakov, 1984; Budyko and Izrael, 1987).

sometimes used as an analogue of expected early-21st century climate. The greatest relative warmth in summer (up to 4°C) was in high latitudes north of 70°N (Lozhkin and Vazhenin, 1987). In middle latitudes, summer temperatures were only 1-2°C higher and further south summer temperatures were often lower than today, for example in Soviet Central Asia, the Sahara, and Arabia. These areas also had increased annual precipitation. Annual precipitation was about 50-100 mm higher than at present in the Northern regions of Eurasia and Canada but in central regions of Western Europe and in southern regions of the European USSR and West Siberia there were small decreases of annual precipitation. The largest decrease in annual precipitation took place in the USA, especially in central and eastern regions (COHMAP, 1988).

The above reconstructions are rather uncertain; thus Figure 7.5a disagrees with reconstructions of temperature over north east Canada given by Bartlein and Webb (1985). However the accuracy of reconstructions is increasing as more detailed information for individual regions in both hemispheres becomes available. For instance, the CLIMANZ project has given quantitative estimates of Holocene temperature and precipitation in areas from New Guinea to Antarctica for selected times (Chappell and Grindrod, 1983). Detailed mid-Holocene reconstructions of summer temperature in Europe and China are shown in Figure 7.5c.

7.3 The Modern Instrumental Record

The clearest signal of an enhanced greenhouse effect in the climate system, as indicated by atmosphere/ocean general circulation models, would be a widespread, substantial increase of near-surface temperatures. This section gives special attention to variations and changes of land surface air temperatures (typically measured at about two metres above the ground surface) and sea surface temperatures (SSTs) since the mid-nineteenth century. Although earlier temperature, precipitation, and surface pressure data are available (Lamb, 1977), spatial coverage is very poor. We focus on changes over the globe and over the individual hemispheres but considerable detail on regional space scales is also given.

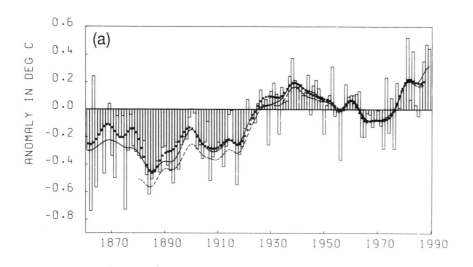

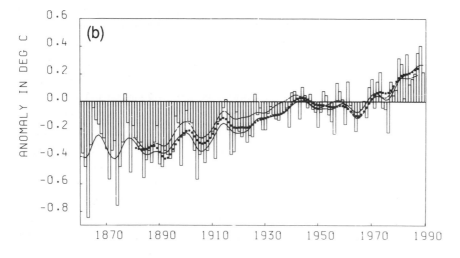

Figure 7.6: Land air temperatures, expressed as anomalies relative to 1951-80. Annual values from P.D. Jones. Smoothed curves of values from P.D. Jones (1861-1989) (solid lines), Hansen and Lebedeff (1880-1987) (dashed lines), and Vinnikov et al. (1861-1987 NH and 1881-1987 SH) (dots). (a) Northern Hemisphere, (b) Southern Hemisphere.

7.4 Surface Temperature Variations and Change

7.4.1 Hemispheric and Global

7.4.1.1 Land

Three research groups (Jones et al., 1986a,b; Jones, 1988; Hansen and Lebedeff, 1987, 1988; and Vinnikov et al., 1987, 1990) have produced similar analyses of hemispheric land surface air temperature variations (Figure 7.6) from broadly the same data. All three analyses indicate that during the last decade globally-averaged land temperatures have been higher than in any decade in the past 100 to 140 years. (The smoothed lines in Figure 7.6, as for all the longer time series shown in this Section, are produced by a low pass binomial filter with 21 terms operating on the annual data. The filter passes fluctuations having a period of 20 years or more almost unattenuated).

Figure 7.6 shows that temperature increased from the relatively cool late nineteenth century to the relatively warm 1980s, but the pattern of change differed between the two hemispheres. In the Northern Hemisphere the temperature changes over land are irregular and an abrupt warming of about 0.3°C appears to have occurred during the early 1920s. This climatic discontinuity has been pointed out by Ellsaesser et al. (1986) in their interpretation of the thermometric record. Northern Hemisphere temperatures prior to the climatic discontinuity in the 1920s could be interpreted as varying about a stationary mean climate as shown by the smoothed curve. The nearest approach to a monotonic trend in the Northern Hemisphere time series is the decrease of temperature from the late 1930s to the mid-1960s of about 0.2°C. The most recent warming has been dominated by a relatively sudden increase of nearly 0.3°C over less than ten years before 1982. Of course, it is possible to fit a monotonic trend line to the entire time series; such a trend fitted to the current version of the Jones (1988) data gives a rate of warming of 0.53°C/100 years when the trend is calculated from 1881 to 1989 or the reduced, if less reliable, value of 0.45°C/100 years if it is calculated from 1861. Clearly, this is a gross oversimplification of the observed variations, even though the computed linear trends are highly significant in a statistical sense.

The data for the Southern Hemisphere include the Antarctic land mass, since 1957, except for the data of Vinnikov et al. (1987, 1990). Like the Northern Hemisphere, the climate appears stationary throughout the latter half of the nineteenth century and into the early part of the twentieth century. Subsequently, there is an upward trend in the data until the late 1930s, but in the next three decades the mean temperature remains essentially stationary again. A fairly steady increase of temperature resumes before 1970, though it may have slowed recently. Linear trends for the Southern Hemisphere are 0.52°C/100 years from 1881 to 1989, but somewhat less, and less reliable, at 0.45°C/100 years for the period 1861-1989.

The interpretation of the rise in temperature shown in Figure 7.6 is a key issue for global warming, so the accuracy of these data needs careful consideration. A number of problems may have affected the record, discussed in turn below:

1) Spatial coverage of the data is incomplete and varies greatly;
2) Changes have occurred in observing schedules and practices;
3) Changes have occurred in the exposures of thermometers;
4) Stations have changed their locations;
5) Changes in the environment, especially urbanisation, have taken place around many stations.

Land areas with sufficient data to estimate seasonal anomalies of temperature in the 1860s and 1980s are shown (with ocean areas) in Figure 7.7. Decades between these times have an intermediate coverage. There are obvious gaps and changes in coverage. Prior to 1957, data for Antarctica are absent while some other parts of the global land mass lack data as late as the 1920s, for example many parts of Africa, parts of China, the Russian and Canadian Arctic, and the tropics of South America. In the 1860s coverage is sparsest; thus Africa has little or no data and much of North America is not covered. The effect of this drastically changing spatial coverage on hemispheric temperature variations has been tested by Jones et al. (1986a, b) who find that sparse spatial coverage exaggerates the variability of the annual averages. The reduction in variability of the Northern Hemisphere annual time series after about 1880 (Figure 7.6) is attributed to this effect. Remarkably, their analysis using a "frozen grid" experiment (see Section 7.4.1.3 for a detailed discussion for the combined land and ocean data) suggests that changes of station density since 1900 have had relatively little impact on estimates of hemispheric land temperature anomalies. However, prior to 1900, the decadal uncertainty could be up to 0.1°C. This is quite small relative to the overall change. Thus varying data coverage does not seem to have had a serious impact on the magnitude of the perceived warming over land over the last 125 years.

Another potential bias arises from changes in observation schedules. Even today, there is no international standard for the calculation of mean daily temperature. Thus each country calculates mean daily temperature by a method of its choice, such as the average of the maximum and the minimum, or some combination of hourly readings weighted according to a fixed formula. As long as each country continues the same practice, the **shape** of the temperature record is unaffected. Unfortunately, few countries have maintained the same practice over the past century; biases have therefore been introduced into the climate record, some of which have been corrected for in

(a) 1861-1870 PERCENTAGE COVERAGE OF OBSERVATIONS
CONTOURS AT 10%, 50%, 90%

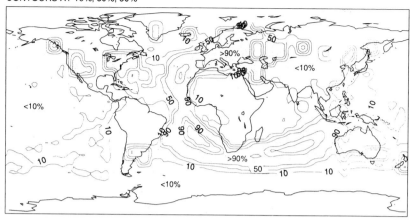

(b) 1980-1989 PERCENTAGE COVERAGE OF OBSERVATIONS
CONTOURS AT 10%, 50%, 90%

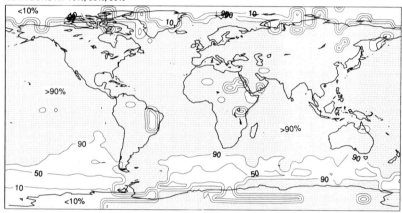

Figure 7.7: Coverage of land surface air (P.D. Jones) and sea surface (UK Meteorological Office) temperature data. Isopleths are percentage of seasons with one or more month's data in 5° x 5° boxes in a given decade. Contours drawn for 90%, 50% and 10% coverage. (a) 1861-70, (b) 1980-89.

existing global data sets, but some have not. These biases can be significant; in the USA a systematic change in observing times has led to a nominal 0.2°C decrease of temperature in the climate record since the 1930s (Karl et al., 1986). The effects of changing observation time have only been partly allowed for in the USA temperature data used in analyses presented here. So an artificial component of **cooling** of rather less than 0.2°C may exist in the USA part of the temperature analyses for this reason, offsetting the warming effects of increasing urbanisation in that country. Artificial changes of temperature of either sign may exist in other parts of the world due to changes in observation time but have not been investigated.

Substantial systematic changes in the exposure of thermometers have occurred. Because thermo-meters can be affected by the direct rays of the sun, reflected solar radiation, extraneous heat sources and precipitation, there has been a continuous effort to improve their exposures

over the last 150 years. Additional biases must accompany these changes in the thermometric record. Since many of the changes in exposure took place during the nineteenth and early twentieth centuries, that part of the record is most likely to be affected. Recently, Parker (1990) has reviewed the earlier thermometer exposures, and how they evolved, in many different countries. The effects of exposure changes vary regionally (by country) and seasonally. Thus tropical temperatures prior to the late 1920s appear to be too high because of the placement of thermometers in cages situated in open sheds. There is also evidence that for the mid-latitudes prior to about 1880 summer temperatures may be too high and winter temperatures too low due to the use of poorly screened exposures. This includes the widespread practice of exposing thermometers on the north walls of buildings. These effects have not yet been accounted for in existing analyses (see Section 7.4.2.2).

Changes in station environment can seriously affect temperature records (Salinger, 1981). Over the years, stations often have minor (usually under 10 km) relocations and some stations have been moved from rooftop to ground level. Even today, international practice allows for a variation of thermometer heights above ground from 1.25 to 2 metres. Because large vertical temperature gradients exist near the ground, such changes could seriously affect thermometer records. When relocations occur in a random manner, they do not have a serious impact on hemispheric or global temperature anomalies, though they impair our ability to develop information about much smaller scale temperature variations. A bias on the large scale can emerge when the character of the changes is not random. An example is the systematic relocations of some observing stations from inside cities in many countries to more rural airport locations that occurred several decades ago. Because of the heat island effect within cities, such moves tend to introduce artificial cooling into the climate record. Jones et al. (1986a, b) attempt in some detail to adjust for station relocations when these appear to have introduced a significant bias in the data but Hansen and Lebedeff (1988) do not, believing that such station moves cancel out over large time and space averages. Vinnikov et al. (1990) do adjust for some of these moves. There are several possible correction procedures that have been, or could be, applied to the Jones (1988) data set (Bradley and Jones, 1985; Karl and Williams, 1987). All depend on denser networks of stations than are usually available except in the USA, Europe, the Western Soviet Union and a few other areas.

Of the above problems, increasing urbanisation around fixed stations is the most serious source of systematic error for hemispheric land temperature time series that has so far been identified. A number of researchers have tried to ascertain the impact of urbanisation on the temperature record. Hansen and Lebedeff (1987) found that when they removed all stations having a population in 1970 of greater than 100,000, the trend of temperature was reduced by 0.1°C over 100 years. They speculated that perhaps an additional 0.1°C of bias might remain due to increases in urbanisation around stations in smaller cities and towns. Jones et al. (1989) estimate that the effect of urbanisation in their quality-controlled data is no more than 0.1°C over the past 100 years. This conclusion is based on a comparison of their data with a dense network of mostly rural stations over the USA. Groisman and Koknaeva (1990) compare the data from Vinnikov et al. (1990) with the rural American data set and with rural stations in the Soviet Union and find very small warm relative biases of less than 0.05°C per 100 years. In the USA, Karl et al. (1988) find that increases due to urbanisation can be significant (0.1°C), even when urban areas have populations as low as 10,000. Other areas of the globe are now being studied. Preliminary results indicate that the effects of urbanisation are highly regional and time dependent. Changes in urban warming in China (Wang et al., 1990) appear to be quite large over the past decade, but in Australia they are rather less than is observed in the USA (Coughlan et al., 1990). Recently, Jones and co-workers (paper in preparation) have compared trends derived from their quality-controlled data, and those of Vinnikov et al. (1990), with specially selected data from more rural stations in the USSR, eastern China, and Australia. When compared with trends from the more rural stations, only small (positive) differences of temperature trend exist in the data used in Jones (1988) and Vinnikov et al. (1990) in Australia and the USSR (of magnitude less than 0.05°C/100 years). In eastern China, the data used by Vinnikov et al. (1990) and Jones (1988) give **smaller** warming trends than those derived from the more rural stations. This is an unexpected result. It suggests that either (1) the more rural set is sometimes affected by urbanisation or, (2) other changes in station characteristics over-compensate for urban warming bias. Thus it is known that the effects of biases due to increased urbanisation in the Hansen and Lebedeff (1987) and the Vinnikov et al. (1990) data sets are partly offset by the artificial cooling introduced by the movement of stations from city centres to more rural airport locations during the 1940s to 1960s (Karl and Jones, 1990). Despite this, some of these new rural airport locations may have suffered recently from increasing urbanisation.

In light of this evidence, the estimate provided by Jones et al. (1989) of a maximum overall warming bias in all three land data sets due to urbanisation of 0.1°C/100 years, or less, is plausible but not conclusive.

7.4.1.2 Sea

The oceans comprise about 61% of the Northern Hemisphere and 81% of the Southern Hemisphere. Obviously, a compilation of global temperature variations must include ocean temperatures. Farmer et al. (1989) and Bottomley et al. (1990) have each created historical analyses of global ocean SSTs which are derived mostly from observations taken by commercial ships. These data are supplemented by weather ship data and, in recent years, by an increasing number of drifting and moored buoys. The Farmer et al. (1989) analyses are derived from a collection of about 80 million observations assembled in the Comprehensive Ocean-Atmosphere Data Set (COADS) in the USA (Woodruff et al., 1987). The data set used by Bottomley et al. (1990) is based on a slightly smaller collection of over 60 million observations assembled by the United Kingdom Meteorological Office. Most, but not all, of the observations in the latter are contained in the COADS data set.

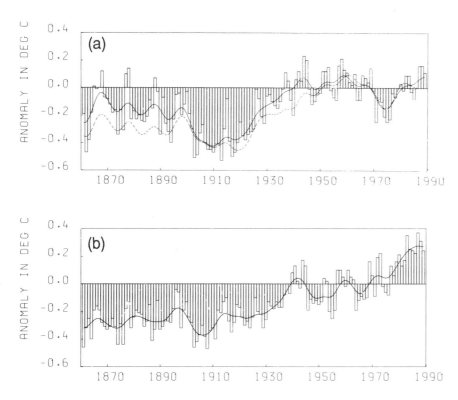

Figure 7.8: Sea surface temperature anomalies 1861-1989, relative to 1951-1980. Annual values (bars) and solid curves from UK Meteorological Office data. Dashed curves from Farmer et al. (1989). (a) Northern Hemisphere, (b) Southern Hemisphere.

Long-term variations of SSTs over the two hemispheres, shown in Figure 7.8, have been, in general, similar to their land counterparts. The increase in temperature has not been continuous. There is evidence for a fairly rapid cooling in SST of about 0.1 to 0.2°C at the beginning of the twentieth century in the Northern Hemisphere. This is believed to be real because night marine air temperatures show a slightly larger cooling. The cooling strongly affected the North Atlantic, especially after 1903, and is discussed at length by Helland-Hansen and Nansen (1920). The cool period was terminated by a rapid rise in temperature starting after 1920. This resembled the sudden warming of land temperatures, but lagged it by several years. Subsequent cooling from the late 1950s to about 1975 lagged that over land by about five years, and was followed by renewed warming with almost no lag compared with land data. Overall warming of the Northern Hemisphere oceans since the late nineteenth century appears to have been slightly smaller than that of the land (Figure 7.8a), and may not have exceeded 0.3°C.

In the Southern Hemisphere ocean (remembering that the Southern Ocean has always been poorly measured) there appear to have been two distinct stable climatic periods, the first lasting until the late 1920s, the second lasting from the mid-1940s until the early 1960s. Since the middle 1970s, SSTs in the Southern Hemisphere have continued to rise to their highest levels of record. Overall

warming has certainly exceeded 0.3°C since the nineteenth century, but has probably been less than 0.5°C (Figure 7.8b), and has been slightly less than the warming of the land. However, if the increases of temperature are measured from the time of their minimum values around 1910, the warming of the oceans has been slightly larger than that of the land. Despite data gaps over the Southern Ocean, the global mean ocean temperature variations (Figure 7.9) tend to take on the characteristics of the Southern Hemisphere because a larger area of ocean is often sampled in the Southern Hemisphere than the Northern Hemisphere. Overall warming in the global oceans between the late nineteenth century and the latter half of the twentieth century appears to have been about 0.4°C.

Significant differences between the two SST data sets presented in Figure 7.8 result mainly from differing assumptions concerning the correction of SST data for instrumental biases. The biases arose chiefly from changes in the method of sampling the sea water for temperature measurement. Several different types of bucket have been used for sampling made, for example, of wood, canvas, rubber or metal, but the largest bias arose from an apparently rather sudden transition from various un-insulated buckets to ship engine intake tubes in World War II. A complex correction procedure developed by Folland and Parker (1989) and Bottomley et al. (1990), which

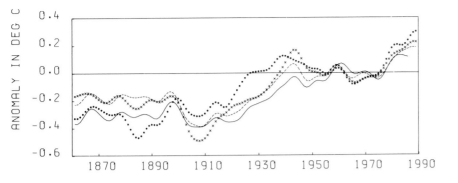

Figure 7.9: Global sea surface, night marine air (crosses) and land air temperature anomalies (dots) 1861-1989, relative to 1951-80. Sea surface temperatures are values from Farmer et al. (1989) (solid line) and the UK Meteorological Office (dashed line). Night marine air temperatures from UK Meteorological Office. Land air temperatures are equally weighted averages of data from Jones, Hansen and Lebedeff, and Vinnikov et al.

creates geographically varying corrections, has been also been used in nearly the same form by Farmer et al. Differences in the two data sets remain, however, primarily because of different assumptions about the mix of wooden versus canvas buckets used during the nineteenth century. Despite recommendations by Maury (1858) to use wooden buckets with the thermometer inserted for four to five minutes, such buckets may have been much less used in practice (Toynbee, 1874; correspondence with the Danish Meteorological Service, 1989) possibly because of damage iron-banded wooden buckets could inflict upon the hulls of ships. Some differences also result, even as recently as the 1970s, because the data are not always derived from identical sources (Woodruff, 1990).

No corrections have been applied to the SST data from 1942 to date. Despite published discussions about the differences between "bucket" and engine intake SST data in this period (for example James and Fox, 1972), there are several reasons why it is believed that no further corrections, with one reservation noted below, are needed. Firstly the anomalies in Figure 7.8 are calculated from the mean conditions in 1951-1980. So only **relative** changes in the mix of data since 1942 are important. Secondly many of the modern "buckets" are insulated (Folland and Parker, 1990) so that they cool much less than canvas buckets. A comparison of about two million bucket and four million engine intake data for 1975-1981 (Bottomley et al., 1990) reveals a global mean difference of only 0.08°C, the engine intake data being the warmer. Thus a substantial change in the mix of data types (currently about 25-30% buckets) must occur before an appreciable artificial change will occur in Figure 7.8. This conclusion is strongly supported by the great similarity between time series of globally-averaged anomalies of colocated SST and night marine air temperature data from 1955 to date (not shown). Less

perfect agreement between 1946 and the early 1950s (SST colder) suggests that uninsulated bucket SST data may have been more numerous then than in 1951-80, yielding an overall cold bias of up to 0.1°C on a global average.

Marine air temperatures are a valuable test of the accuracy of SSTs after the early 1890s. Biases of day-time marine air temperatures are so numerous and difficult to overcome that only night-time marine air temperatures have been used. The biases arise during the day because overheating of the thermometers and screens by solar insolation has changed as ships have changed their physical characteristics (Folland et al., 1984). On the other hand appreciable biases of night-time data are currently believed to be confined to the nineteenth century and much of the Second World War. Night marine air temperatures have been found to be much too high relative to SST, or to modern values, in certain regions and seasons before 1894 (Bottomley et al., 1990). These values were corrected using SSTs, but subsequently (except in 1941-1945) night marine air temperature data constitute independent evidence everywhere, although corrections are also made for the increasing heights of ship decks (Bottomley et al., 1990). Figure 7.9 indicates that multi-decadal global variations of corrected night-time marine air temperature have been quite similar to those of SST. To provide a complete picture, Figure 7.9 shows the Farmer et al. and UK Meteorological Office global SST curves separately along with a global land air temperature series created by averaging the series of Jones, Hansen and Lebedeff and Vinnikov et al. Both SST and night marine air temperature data appear to lag the land data by at least five years during the period of warming from 1920 to the 1940s. However some of the apparent warmth of the land at this time may be erroneous due to the use of open shed screens in the tropics (Section 7.4.1.1).

The above results differ appreciably in the nineteenth century from those published by Oort et al. (1987) who followed the much less detailed correction procedure of Folland et al. (1984) to adjust the COADS SST and **all hours** marine air temperature data sets. Newell et al. (1989) also present an analysis, quite similar to that of the above authors, based on a UK Meteorological Office data set that was current in early 1988. All these authors obtain higher values of global SST and marine air temperature in the middle to late nineteenth century, typically by about 0.1°C and 0.15°C respectively, than are indicated in this report. It is our best judgement that the more recent analyses represent a real improvement, but the discrepancies highlight the uncertainties in the interpretation of early marine temperature records. Yamamoto et al. (1990a) have tried to quantify changing biases in the COADS **all hours** marine air temperature data using a mixture of weather ship air temperature data from the 1940s to 1970s and selected land air temperature data, mainly in three tropical coastal regions, to calculate time varying corrections. Based on these corrections, Yamamoto et al. (1990b) calculate a global air temperature anomaly curve for 1901-1986 of similar overall character to the night marine air temperature curve in Figure 7.9, but with typically 0.15°C warmer anomalies in the early part of the twentieth century, and typically 0.1°C cooler anomalies, in the warm period around 1940-1950. Recent data are similar. It could be argued that the corrections of Yamamoto et al. may be influenced by biases in the land data, including warm biases arising from the use of tropical open sheds earlier this century. Warm biases may also exist in some ocean weather ship day-time air temperature data (Folland, 1971). Although we believe that the night marine air temperature analysis in Figure 7.9 minimises the known sources of error, the work of Yamamoto et al. underlines the level of uncertainty that exists in trends derived from marine air temperature data.

7.4.1.3 Land and sea combined

Combined land and sea surface temperatures show a significant increase of temperature from the late nineteenth to the late twentieth century (Figures 7.10a to c). These data are an average of two data sets: a combination of the Jones land data and the Farmer et al. SST data, and a combination of the Jones land data and the UK Meteorological Office SST data. Note that the relative contributions of land and sea to the combined data have varied according to changing data availability over land and ocean (bottom of Figure 7.10c). Over the globe, the combined data gives an increase of temperature of 0.45°C between the average for the two decades 1881-1900 and the decade 1980-89. The comparable increase for the Northern Hemisphere is 0.42°C and for the Southern Hemisphere 0.48°C. A similar calculation for the changes

of temperature between 1861-1880 and 1980-89 gives 0.45°C, 0.38°C and 0.53°C respectively. A linear trend fitted between 1890 and 1989 gives values of 0.50°C/100 years (globe), 0.47°C/100 years (Northern Hemisphere) and 0.53°C/100 years (Southern Hemisphere); a linear trend fitted between 1870 and 1989 gives the reduced values of 0.41°C, 0.39°C and 0.43°C/100 years respectively.

Apparent decadal rates of change of smoothed global combined temperature have varied from an increase of 0.21°C between 1975 and 1985 (largely between 1975 and 1981), to a decrease of 0.19°C between 1898 and 1908 (though data coverage was quite poor around 1900). Surprisingly, the maximum magnitudes of decadal change (warming or cooling) over land and ocean (SST) have been quite similar (Figure 7.9) at about 0.25°C. Smoothed night global marine air temperature showed the largest apparent change around 1900, with a maximum cooling of 0.32°C between 1898 and 1908, though this value is very uncertain.

Combined land and ocean temperature has increased rather differently in the Northern than in the Southern Hemisphere (Figure 7.10). A rapid increase in Northern Hemisphere temperature during the 1920s and into the 1930s contrasts with a more gradual increase in the Southern Hemisphere. Both hemispheres had relatively stable temperatures from the 1940s to the 1970s, though with some evidence of cooling in the Northern Hemisphere. Since the 1960s in the Southern Hemisphere, but after 1975 in the Northern Hemisphere temperatures have risen, with the overall rise being more pronounced in the Southern Hemisphere. Only a small overall rise was observed between 1982 and 1989.

An important problem concerns the varying spatial coverage of the combined marine and land observations. Figure 7.7 indicates that this has been far from uniform in time or space, and even today coverage is not comprehensive. Ships have followed preferred navigational routes, and large areas of the ocean have been inadequately sampled. The effect that this may have on global estimates of SSTs has been tested in "frozen grid" analyses (Bottomley et al., 1990) and in eigenvector analyses (Folland and Colman, 1988). In the frozen grid analyses, global and hemispheric time series were recalculated using data from 5°x5° boxes having data in nominated earlier decades, for example 1861-1870. Remarkably, the small coverage of this period (Figure 7.7a) appears surprisingly adequate to estimate long term trends, probably because the data are distributed widely in both hemispheres throughout the last 125 years. An eigenvector analysis of combined land and ocean data (Colman, personal communication) isolates an underlying signal of century time scale climate change which is surprisingly uniform geographically, and very like Figure 7.10c, even though gross regional changes vary because of other factors. Figure 7.10d shows the

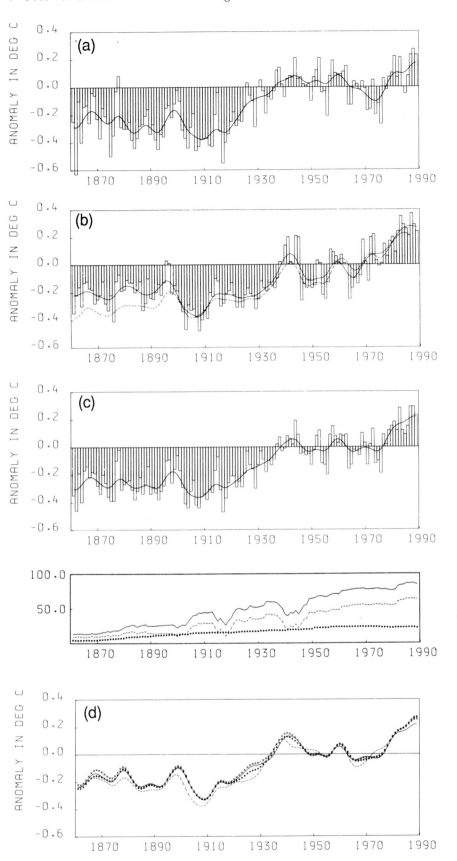

Figure 7.10: Combined land air and sea surface temperatures, 1861-1989, relative to 1951-1980. Land air temperatures from P.D. Jones and sea surface temperatures from the UK Meteorological Office and Farmer et al. (1989). Sea surface temperature component is the average of the two data sets. **(a)** Northern Hemisphere, **(b)** Southern Hemisphere, **(c)** Globe. Percentage coverage of the data is shown for Figure 7.10c, expressed as a percentage of total global surface area for land (dotted line) ocean (dashed line) separately, and for combined data (solid line) plotted annually. 100% coverage would imply that all 5° x 5° boxes had data in two or three months in each season of the year. **(d)** "Frozen grid" analyses for 1861-1989 for the globe, using land data as above and UK Meteorological Office SST data: 1861-70 coverage (dashed), 1901-10 coverage (dotted), 1921-30 coverage (crosses) and all data (solid line).

results of a frozen grid analysis applied to a combination of the Jones land data and the UK Meteorological Office SST data. Frozen grids were defined for 1861-1870, 1901-1910, 1921-30 and global series using these were compared with that incorporating all data (also shown in Figure 7.9). Varying the data coverage has only a small effect on trends; confining the data to the 1861-70 grid augments the temperature increase since the late nineteenth century by about 0.05°C at most. However, omission of much of the Southern Ocean in these tests, as well as air temperatures over the Arctic ocean north of 80°N, is a cause for concern. So the uncertainties in trends due to varying data coverage may be underestimated by Figure 7.10d.

In models forced with enhanced greenhouse gases (Sections 5 and 6), warming over the land is substantially greater than that over the ocean, so that the steadier and larger warming of the Southern Hemisphere in recent decades is not predicted. The latter may be part of a global-scale natural fluctuation of the ocean circulation (Street-Perrott and Perrott, 1990). Comparing Northern Hemisphere land and ocean, Figure 7.11 shows that the land is now relatively warmer than the oceans by the about the same amount as in the period of global warming during the first half of this century. Southern Hemisphere land (not shown) shows no recent warming relative to the oceans.

In summary, the overall increase of temperature since the nineteenth century can be estimated as follows. It is probable that a small residual positive bias remains in the land surface temperatures due to increasing urbanisation (Section 7.4.1.1). However, the contribution of the urbanisation bias is at least halved in recent decades in the combined ocean and land data, so it is unlikely to exceed 0.05°C. From Figure 7.10d, allowing for areas never adequately sampled, we estimate that varying data coverage produces an uncertainty in trend of at least ±0.05°C. We also recognise the existence of several other sources of bias, highlighted by some disagreements between individual analyses, but are uncertain as to their true sign. Therefore our best estimates of the lower and upper limits of global warming between the late nineteenth century and the 1980s are about 0.3°C and not more than 0.6°C respectively, slightly less than most previous estimates.

7.4.2 Regional, Seasonal, and Diurnal Space and Time Scales

We show that regional, including zonally-averaged, climate variations often do not match those of the globe as a whole. Some apparent differences between large-scale variations of temperature in different seasons are shown. Finally we discuss variations of maximum and minimum daily temperatures over the relatively restricted areas so far analysed.

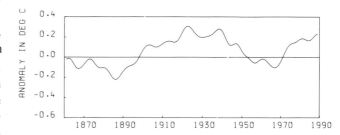

Figure 7.11: Differences between land air and sea surface temperature anomalies, relative to 1951-80, for the Northern Hemisphere, 1861-1989. Land air temperatures from P.D. Jones. Sea surface temperatures are averages of UK Meteorological Office and Farmer et al. (1989) values.

7.4.2.1 Land and Sea

Regional time series suffer from many near-random errors that can nearly cancel in analyses of global and hemispheric temperatures (Kleshchenko et al., 1988). The 20° latitude x 60° longitude areas analysed in Figure 7.12a have been chosen to be large enough to minimise random errors and yet be small enough to capture the individual character of regional temperature changes. Figure 7.12a demonstrates considerable regional variability in temperature trends which nevertheless evolve coherently between adjacent regions. Particularly striking is the peak warmth in the north east Atlantic and Scandinavian regions around 1940-45 followed by a sharp cooling, and the strong warming in the South Atlantic and much of the Indian Ocean since about 1965.

Figure 7.12b shows zonally-averaged land air temperature and SST anomalies using the same data as in Figure 7.12a. Almost uniformly cooler conditions in the nineteenth century are clearly seen in all zones, extending into the early twentieth century. Warming around 1920-1940 occurs in most zones, except perhaps over the northern part of the Southern Ocean, with a strong warming, exceeding 0.8°C, occurring to the north of 60°N over this period. Note that the polar cap (north of 80°N) has insufficient data for analysis and insufficient data exist to calculate representative zonal means south of 40°S until after 1950. The cooling after 1950 was mainly confined to the Northern Hemisphere, though weak cooling is evident in the Southern Hemisphere tropics between about 1940 and the early 1950s. There was renewed warming in most Southern Hemisphere zones before 1970. This warming continued until the early 1980s but then slowed markedly. However very little change of temperature is evident over Antarctica (south of 60°S) since records began there around 1957. Renewed warming is seen in the Northern Hemisphere in all zones after the early 1970s, **including small rises in high latitudes**, a fact hitherto little appreciated, probably because of the marked cooling in the

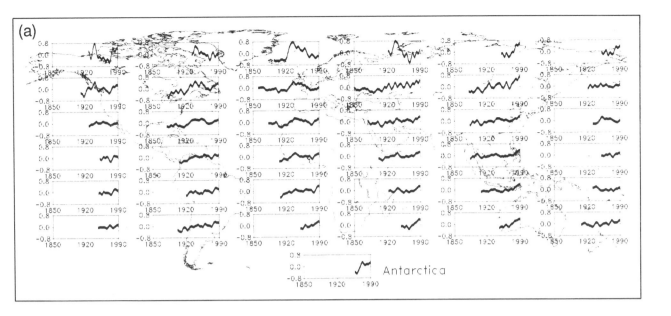

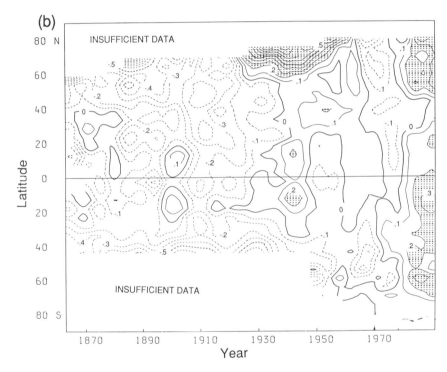

Figure 7.12: (a) Regional surface temperature anomaly variations over 20° latitude x 60° longitude boxes and Antarctica (regions south of 60°S). (b) Zonal averages of combined sea surface and land air temperature data, 1861-1989. Land air temperatures from P.D. Jones and sea surface temperatures from the UK Meteorological Office.

Atlantic/Barents Sea sector in recent decades (Figure 7.12a) which is not seen elsewhere in high latitudes. The temperature curve cited by Lindzen (1990) as showing recent Arctic cooling is in fact one representative of the North Atlantic Arctic sector only (much as Figure 7.12, third curve from the left, northernmost row) and is therefore not properly representative of high latitudes of the Northern Hemisphere as a whole. General circulation models with enhanced concentrations of carbon dioxide tend to show largest increases of annual mean temperature

in Northern Hemisphere polar latitudes. The rate of warming has slowed again in many Northern Hemisphere zones in recent years and, almost simultaneously, cooling in the middle to high latitude Atlantic sector has ceased.

Figure 7.13 shows the pattern of temperature anomalies in 1950-59, 1967-76 and 1980-89. Much of the Southern Hemisphere has warmed steadily since 1950-59, with a few exceptions, for example, parts of Brazil and Antarctica. In the Northern Hemisphere the middle decade of those shown was coolest (see Figures 7.6, 7.8, 7.10). The most

(a) 1950 - 1959 SURFACE TEMPERATURE ANOMALIES

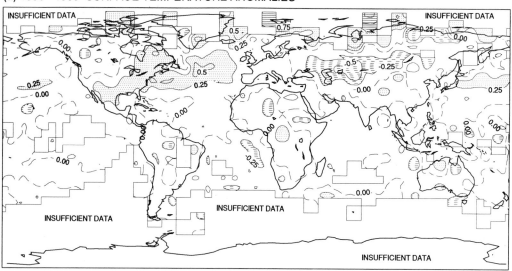

(b) 1967 - 1976 SURFACE TEMPERATURE ANOMALIES

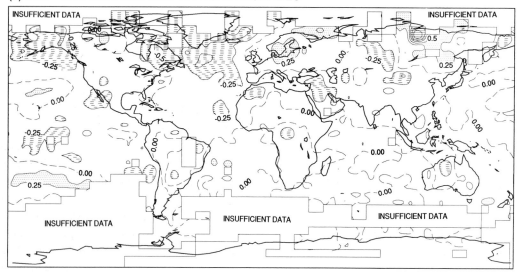

(c) 1980 - 1989 SURFACE TEMPERATURE ANOMALIES

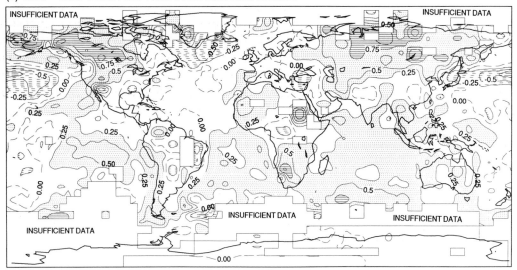

Figure 7.13: Decadal surface temperature anomalies, relative to 1951-80. Isopleths every 0.25°C; dashed isopleths are negative values; dotted positive anomalies 0.25 to 0.5°C; heavy-shaded greater than 0.5°C. Heavily shaded negative values less than -0.25°C. Land air temperatures from P.D. Jones and sea surface temperatures from the UK Meteorological Office. (a) 1950-59, (b) 1967-76, (c) 1980-89. Also shown in the colour section.

consistent recent warming is found in subtropical and tropical regions, especially the Indian Ocean and regions near to, and including, the tropical South Atlantic. By contrast, cooling occurred in parts of the extratropical North Pacific and North Atlantic, especially between 1970 and 1985. Recent warming has also been weak or absent over the Canadian archipelago, the Eastern Soviet Union, and Europe. In Section 7.9 it will be shown that some of these regional temperature variations are linked to regional-scale fluctuations in the circulation of the atmosphere, so it is not surprising that pronounced variations in regional temperature trends occur.

7.4.2.2 Seasonal variations and changes

Figure 7.14 indicates that the increase of land-based temperatures in the Northern Hemisphere since 1975 has largely consisted of an increase between December and May, but with little increase between June and November. In the Southern Hemisphere there is little difference in recent seasonal trends (not shown). Of some concern are substantial differences in seasonal trends before 1900 in the Northern Hemisphere. The relative warmth of summer and coolness of winter at that time reflect considerably greater seasonal differences of the same character in the continental interiors of North America and Asia (not shown). It is not clear whether a decrease in the seasonal cycle of temperature that commenced around 1880 is real; it could be due to changes in the circulation of the atmosphere or it may reflect large, seasonally dependent, biases in some nineteenth century land data. The latter might arise from the progressive changes of thermometer exposure known to have occurred then (Section 7.4.1.1). The Southern Hemisphere (not shown) shows a similar decrease in the seasonal cycle of temperature in the last

part of the nineteenth century, but with less than half the amplitude of that in the Northern Hemisphere.

7.4.2.3 Day-time and Night-time

Because the ocean has a large heat capacity, diurnal temperature variations in the ocean and in the overlying air are considerably muted compared with those over land and, from a climatic point of view, are likely to change little. Over land, diurnal variations are much less restricted so the potential for relative variations in maximum and minimum temperature is much larger. Such relative changes might result from changes in cloudiness, humidity, atmospheric circulation patterns, windiness or even the amount of moisture in the ground. Unfortunately, it is not yet possible to assess variations of maximum and minimum temperature on a hemispheric or global scale. However in the regions discussed below, multi-decadal trends of day-time and night-time temperatures have been studied and do not always appear to be the same.

Figure 7.15a, second panel shows a rise of minimum temperatures (these usually occur around dawn) in the USA. The rise has not been reflected in maximum temperatures (which usually occur during mid-afternoon). (See also Section 7.10.1). Similar behaviour has been found in other parts of North America (Karl et al., 1984). Appreciably different variations of maximum and minimum temperatures on decadal time-scales are also observed at inland stations in Australia (Figure 7.15b). It is unlikely that urban heat islands play a significant role in these variations as the data for both countries have been extensively scrutinized for urban heat island biases. In China (Figure 7.15c), the minimum temperature also appears to have risen more than the maximum. It is uncertain to what extent increases in urbanisation contribute to the changes in China, especially as urban heat island biases tend to be greatest during the night. Over New Zealand, a strong influence of atmospheric circulation variations on variations in daily maxima relative to daily minima has been observed (Salinger, 1981). This is an indication that the above results can only be fully understood when changes in atmospheric circulation over these countries have been studied in some detail.

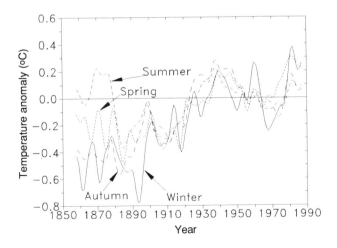

Figure 7.14: Smoothed seasonal land surface air temperature anomalies, relative to 1951-80, for the Northern Hemisphere. Data from P.D. Jones.

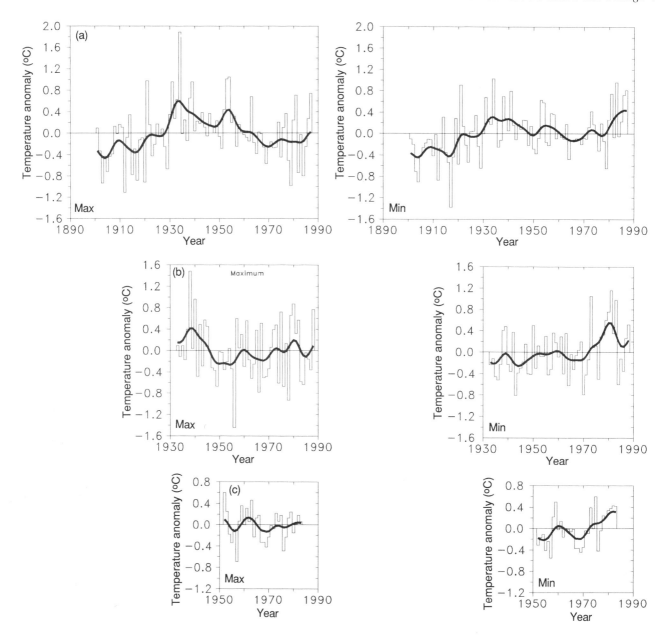

Figure 7.15: Changes of maximum (day-time) and minimum (night-time) temperatures. (a) United States, (b) South-eastern Australia, (c) China.

7.5 Precipitation and Evaporation Variations and Changes

7.5.1 Precipitation Over Land

Several large-scale analyses of precipitation changes over the Northern and Southern Hemisphere land masses have been carried out (Bradley et al., 1987; Diaz et al., 1989; Vinnikov et al., 1990). These have demonstrated that during the last few decades precipitation has tended to increase in the mid-latitudes, but decrease in the Northern Hemisphere subtropics and generally increase throughout the Southern Hemisphere. However, these large-scale features contain considerable spatial variability. Figure 7.16 illustrates this variability for three regions in the Northern Hemisphere and East Africa. Annual precipitation over the Soviet Union displays a remarkably consistent increase over the twentieth century (Figure 7.16a). An apparent increase in precipitation has been found over northern Europe (Schönweise and Birrong, 1990) with a suggestion of a decrease in extreme southern Europe, though these data have not yet been corrected for changing instrumental biases. In the tropics, East African rainfall departures from normal show significant decadal

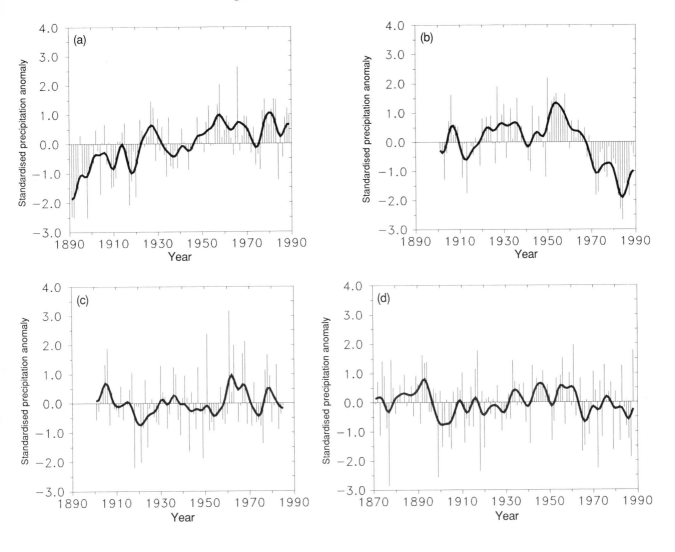

Figure 7.16: Standardised regional annual precipitation anomalies. (a) USSR, (b) Sahel, (c) East Africa, (d) All-India monsoon. Note that Sahel values are annual averages of standardised **station** values restandardised to reflect the standard deviation of these annual averages.

variability, but consistent trends are absent (Figure 7.16c). Summer monsoon rainfall in India also reflects multi-decadal changes in climate (Figure 7.16d), but consistent trends are also absent. The period 1890-1920 was characterised by a high frequency of droughts in India, while 1930-1964 had a much lower frequency. Since 1965 the frequency of droughts has again been higher relative to 1930-1964 (Gadgil, 1988), mostly in the wet areas of north eastern India (Gregory, 1989).

The dramatic drying of sub-Saharan Africa shown in Figure 7.16b deserves special comment. Various explanations have been proposed, reviewed in Druyan (1989); see also Semazzi et al. (1988) and Wolter (1989). The most consistent result of these studies was to show, over the last few decades, a pattern of anomalously high SSTs in the Atlantic south of about 10°N, lower than normal SSTs in the Atlantic to the north of 10°N and higher

SSTs in the tropical Indian Ocean (Figure 7.13c). There has been a distinct weakening of some these patterns recently and a return to near normal rainfall in 1988 and 1989. Such large-scale changes of SST appear to have a major impact on the sub-Saharan atmospheric circulation (Folland et al., 1990; Wolter, 1989). Although SST changes appear to be strongly related to the decreased rainfall since the 1950s, they are probably not the only cause (Nicholson, 1989). Folland et al. (1990) show, however, that at least 60% of the variance of Sahel rainfall between 1901 and 1988 on time scales of one decade and longer is explained by worldwide SST variations. Reductions of rainfall occurred at much the same time immediately south of the Sahel and over much of Ethiopia and the Caribbean.

It is important to consider the accuracy of the precipitation data sets. Precipitation is more difficult to monitor than temperature as it varies much more in time and space. A higher spatial density of data is needed to provide an analysis of variations and trends of comparable accuracy. High density data often reside within national meteorological centres, but there is no regular international exchange. The number of stations required to sample a regional rainfall climate adequately varies with region and an adequate number may not always be available.

A severe problem for analysing multi-decadal variations of precipitation lies in the fact that the efficiency of the collection of precipitation by raingauges varies with gauge siting, construction and climate (Sevruk, 1982; Folland, 1988; Legates and Willmott, 1990). Major influences are the wind speed during rain, the size distribution of precipitation particle sizes, and the exposure of the raingauge site. Fortunately, appropriate climatological averages of the first two, highly variable, quantities can be used to assess usefully their effects over a long enough period (Folland, 1988, Appendix 1). Collection efficiency has tended to increase as operational practices have improved, often in poorly documented ways that may give artificial upward trends in precipitation in some regions. Thus precipitation data are not completely compatible between countries due to the lack of agreed standards. Of particular concern is the measurement of snowfall from conventional gauges where errors of at least 40% in long-term collection efficiency can occur. When precipitation errors are expressed as a percentage of the true rainfall, it is not surprising that they tend to be greatest in high latitude, windy, climates and least in wet equatorial regions.

Vinnikov et al., (1990) have carried out detailed corrections to USSR data for the varying aerodynamic and wetting problems suffered by gauges. These corrections are incorporated in the record shown in Figure 7.16a, though no aerodynamic corrections were thought necessary in summer. In winter the (positive) aerodynamic corrections can be large and vary from 5% to 40% (the latter for snow). Wetting corrections, which are also positive and tend to be largest in summer, varied typically in the range 4% to 10% and were applied after correction for aerodynamic effects. Despite these large biases, comparisons of data sets over the USSR from Bradley et al. (1987), who only partially corrected for biases, and Vinnikov et al., who corrected more extensively, show that most of the important long-term variations are apparent in both data sets (Bradley and Groisman, 1989). Many of the major variations apparent in precipitation records are evident in hydrological data, such as the rise in the levels of the North American Great Lakes, Great Salt Lake and the Caspian Sea during the early 1980s, and the severe desiccation of the Sahel. Nevertheless the lack of bias corrections in most rainfall data outside the USSR is a severe impediment to quantitative assessments of rainfall trends.

7.5.2 *Rainfall Over The Oceans*

Quantitative estimates of precipitation over the oceans are limited to the tropics where they are still very approximate. The mean temperature of the upper surfaces of convective clouds, deduced from satellite measurements of outgoing long-wave thermal radiation (OLR), are used to estimate mean rainfall over periods of days upwards. The colder the clouds, the less is OLR and the heavier the rainfall (Section 4 gives references). Nitta and Yamada (1989) found a significant downward trend in OLR averaged over the global equatorial belt 10°N to 10°S between 1974 and 1987, implying an increase of equatorial rainfall over that time. Arkin and Chelliah (1990) have investigated Nitta and Yamada's results for this Report. They find that inhomogeneities in the OLR data are sufficiently serious to cast doubt on Nitta and Yamada's conclusions. However the latter's claim that equatorial SST has risen over this period seems justified (Flohn and Kapala, 1989, and Figures 7.12 and 7.13). This trend is likely to result in increased deep convection and more rainfall there (Gadgil et al., 1984; Graham and Barnett, 1987).

Section 7.5 has shown that some regional-scale rainfall trends have occurred over land. However, much more attention needs to be paid to data quality and to improving data coverage before more comprehensive conclusions can be drawn about precipitation variations over the global land surface. Precipitation cannot yet be measured with sufficient accuracy over the oceans to reliably estimate trends, even though quite modest changes in SST in the tropics could give rise to important changes in the distribution of tropical rainfall (see also Section 7.9.1).

7.5.3 *Evaporation from the Ocean Surface*

It is difficult to estimate trends in evaporation from the oceans. An increase is, however, expected as a result of an increase in greenhouse gases (Section 5). The most important problem concerns the reliability of measurements of wind speed that are an essential component of evaporation estimates. Oceanic wind speeds have apparently increased in recent decades. However, Cardone et al. (1990) have demonstrated that much of this increase can be explained by changes in the methods of estimating wind speed from the state of the sea surface, and changes in the heights of anemometers used to measure wind speed on ships. Until these problems are substantially reduced, it is considered that estimates of trends in evaporation are unlikely to be reliable.

7.6 Tropospheric Variations and Change

7.6.1 *Temperature*

Tropospheric and stratospheric temperatures are central to the problem of "greenhouse" warming because general circulation models (Section 5) predict that temperature change with enhanced concentrations of greenhouse gases will have a characteristic profile in these layers, with more warming in the mid-troposphere than at the surface over many parts of the globe, and cooling in much of the stratosphere. One of the "fingerprint" techniques (Section 8) for detecting anthropogenic climate change depends in part on an ability to discriminate between tropospheric

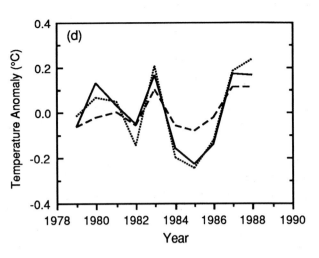

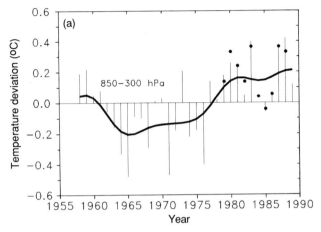

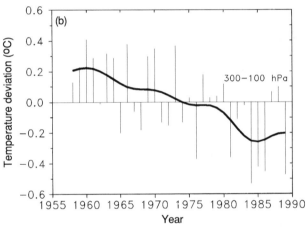

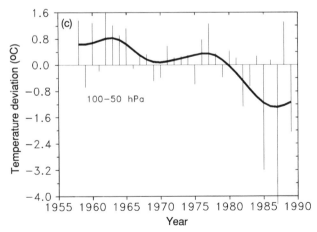

Figure 7.17 (continued): (d) Annual global anomalies for 1979-1988: tropospheric satellite temperatures from Spencer and Christy (1990) (solid line); 850-300mb radiosonde temperatures based on Angell (1988) (dots); combined land and sea surface temperatures as in Figure 7.10 (dashed line). All anomalies are referred to the average of their respective data sets for **1979-1988**.

Figure 7.17: Temperature anomalies in the troposphere and lower stratosphere 1958-1989, based on Angell (1988).
(a) Annual global values for 850-300mb. Dots are values from Spencer and Christy (1990). (b) 300-100mb. (c) Annual values for Antarctic (60°S-90°S) for 100-50mb.

warming and stratospheric cooling (Barnett and Schlesinger, 1987). Observational studies of variations in recent temperature changes with height have been made by numerous authors, for example Parker (1985), Barnett (1986), Sellers and Liu (1988) and Karoly (1989). Layer mean temperatures from a set of 63 radiosonde stations covering the globe have been derived by Angell (1988). Most stations have operated continuously only since about 1958 (the International Geophysical Year). The network is zonally well distributed, but about 60% of the stations are in the Northern Hemisphere and only 40% in the Southern Hemisphere. Layer mean temperatures from this network have been integrated for the globe. Figure 7.17a shows that, over the globe as a whole, mid-tropospheric (850-300 mb) temperatures increased by about 0.4°C between the late 1960s and mid-1980s, with much of the rise concentrated between 1975 and the early 1980s as at the surface. Zonal average anomalies for 850-300mb (not shown) indicate that the largest changes occurred in the zone 10°S to 60°S followed by the equatorial region (10°N to 10°S), with little trend north of 60°N or south of 60°S. This finding is in good agreement with surface data (Figures 7.12a and b).

In the upper troposphere (300-100 mb), Figure 7.17b shows that there has been a rather steady decline in temperature since the late 1950s and early 1960s, in general **disagreement** with model simulations that show warming at these levels when the concentration of greenhouse gases is increased (Section 5). The greatest change in temperature has been in the lower stratosphere (100-50 mb) where the decrease after 1980 is much beyond the variability of the previous decades. It is mostly attributed to changes over and around Antarctica (Figure 7.17c) where the cooling since 1973 has reached nearly 10°C in austral spring and 2°C in summer (Angell, 1988) but with small values of cooling in other seasons. A small amount of lower stratospheric cooling has been observed elsewhere in the Southern Hemisphere, mainly in the tropics, and also in the equatorial belt (10°N to 10°S). The abrupt decrease over Antarctica in spring may at least be partly related to the formation of the "ozone hole".

Temperatures derived from radiosondes are subject to instrumental biases. These biases have not been assessed in the data used by Angell (1988), although there have been many changes in radiosonde instrumentation over the last 31 years. In 1984-85, international radiosonde comparisons were carried out (Nash and Schmidlin, 1987). Systematic differences between various types of radiosonde were determined for a series of flights which penetrated the tropopause. The estimated heights of the 100mb surface generally differed by up to 10-20 geopotential metres which is equivalent to average differences of 0.25°C in the layer from the surface to 100 mb.

7.6.2 *Comparisons of Recent Tropospheric and Surface Temperature Data*

A measure of the robustness of the tropospheric data derived by Angell (1988), at least in recent years, can be obtained by comparing his 850-300mb data with ten years of independent satellite measurements analysed by Spencer and Christy (1990) for 1979-1988. Spencer and Christy have used the average of measurements from microwave sounding units (MSU) aboard two USA National Oceanographic and Atmospheric Administration (NOAA) TIROS-N series of satellites to derive global temperatures in the mid-troposphere. Although surface and mid-tropospheric data are likely to show rather different changes in their values over individual regions, better, but not perfect, coupling is expected when the data are averaged over the globe as a whole. Figure 7.17d compares the annual global combined land air temperature and SST data used in Figure 7.10 with annual values of these two tropospheric data sets for the period 1979-1988; in each case the 10 annual anomalies are calculated from their respective 1979-1988 averages. The agreement between the three data sets is surprisingly good, despite recent suggestions that it is poor. Thus the correlations and root mean squared differences between the surface and MSU data are 0.85 and 0.08°C respectively, while the correlation between the surface and the radiosonde data is 0.91. The correlation between the two tropospheric data sets is, as expected, slightly higher at 0.96 with a root mean squared difference of 0.02°C. The latter represents excellent agreement given the relatively sparse network of radiosondes. Note that annual values in both tropospheric data sets have nearly twice the variability of the surface values, as measured by their standard deviation. This partly explains why the root mean square difference between the MSU and surface data is appreciably larger than that between the two tropospheric data sets, despite the high correlation. This is, arguably, an indication of genuine climatological differences between the interannual variability of mid-troposphere temperatures and those of the surface. All three data sets show a small positive trend over the period 1979-1988, varying from 0.04°C/decade for the MSU data to 0.13°C/decade for the surface data. These trends are not significantly different over this short period and again reflect surprisingly great agreement. Further discussion of these results is given in Jones and Wigley (1990).

7.6.3 *Moisture*

Water vapour is the most abundant greenhouse gas, and its increases are expected to augment the warming due to increases of other greenhouse gases by about 50%. Trenberth et al. (1987) estimate that doubling carbon dioxide concentrations would increase the global concentration of water vapour by about 20%, and Hansen et al. (1984) estimate a 33% increase.

There is evidence that global water vapour has been a few percent greater during the 1980s than during the 1970s (Elliott et al., 1990). Hense et al. (1988), and Flohn et al. (1990) find a 20% increase in water vapour content in the mid-troposphere over the equatorial Pacific from 1965-1986 with at least a 10% rise between the surface and the 300mb level. Despite great uncertainties in these data some increase seems to have taken place. Because of numerous changes in radiosondes a global assessment of variations prior to 1973 is difficult and trends after 1973 have an uncertain accuracy. See also Section 8.

7.7 Sub-Surface Ocean Temperature and Salinity Variations

The sub-surface ocean data base is now just becoming sufficient for climate change studies in the North Atlantic and North Pacific basins to be carried out. A few, long, local time series of sub-surface measurements exist, sufficient to alert the scientific community to emerging evidence of decadal-scale temperature variability in the Atlantic Ocean. Beginning about 1968, a fresh, cold water mass with its origins in the Arctic Ocean appears to have circulated around the sub-Arctic gyre of the North Atlantic Ocean. This event has been described by Dickson et al. (1988) as the "Great Salinity Anomaly". Some of this cold, fresh water penetrated to the deep waters of the North Atlantic (Brewer et al., 1983). The marked cool anomalies in the North Atlantic SST shown in Figure 7.13 for 1967-76 partly reflect this event.

Recently, Levitus (1989a, b, c, d) has carried out a major study of changes of sub-surface temperature and salinity of the North Atlantic Ocean between 1955-59 and 1970-74. 1955-59 was near the end of a very warm period of North Atlantic surface waters, but by 1970-74 the subsequent cool period was well developed (Figure 7.13). Cooler water extended from near the sea surface to 1400m depth in the subtropical gyre (30-50°N). Beneath the subtropical gyre, a warming occurred between the two periods. North of this gyre there was an increase in the temperature and salinity of the western sub-arctic gyre. The density changes associated with these changes in temperature and salinity indicate that the transport of the Gulf Stream may have decreased between the two periods. Temperature difference fields along 24.5°N and 36.5°N presented by Roemmich and Wunsch (1984), based on data gathered during 1981 and the late 1950s, are consistent with these ideas.

Antonov (1990) has carried out a complementary study for the North Atlantic and North Pacific using subsurface temperature data held in the USSR and SST data from the UK Meteorological Office. He finds that zonal averages of temperature changes between 1957 and 1981 show statistically significant cooling in the upper layers and a warming below 600m when averaged over the North Atlantic as a whole. This agrees well with Levitus' results for the North Atlantic. Basin mean temperature changes (1957 to 1981) for the North Atlantic and North Pacific, as computed by Antonov, are shown in Figure 7.18.

The reasons for some of these changes are partially understood. For example, the cooling of the upper 1400m

of the subtropical gyre was due to an upward displacement of cooler, fresher water. Why this displacement occurred is not definitely known, but most probably is related to changes in the large-scale wind field over the North Atlantic. Of particular importance is the temperature increase of approximately 0.1°C over, on average, a thousand metre thick layer in the deep North Atlantic because it represents a relatively large heat storage. Even the upper few metres of the ocean can store as much heat as the entire overlying atmospheric column of air. Scientists have long recognized (Rossby, 1959) that the ocean could act to store large amounts of heat, through small temperature changes in its sub-surface layers, for hundreds or thousands of years. When this heat returns to the atmosphere/cryosphere system it could also significantly affect climate. Section 6 gives more details.

The magnitude and extent of the observed changes in the temperature and salinity of the deep North Atlantic are thus large enough that they cannot be neglected in future theories of climate change.

7.8 Variations and Changes in the Cryosphere

Snow, ice, and glacial extent are key variables in the global climate system. They can influence the global heat budget through regulation of the exchange of heat, moisture, and momentum between the ocean, land, and atmosphere. Accurate information on cryospheric changes is essential for full understanding of the climate system. Cryospheric data are also integrators of the variations of several variables such as temperature, sunshine amount and precipitation, and for sea-ice, changes in wind stress. Therefore caution must be exercised when interpreting a cryospheric change. Variations in the Greenland and Antarctic ice sheets are discussed in Section 9.

7.8.1 Snow Cover
Surface-based observations of snow cover are sufficiently dense for regional climate studies of the low lying areas of the Northern Hemisphere mid-latitudes. Unfortunately, a hemisphere-wide data set of mid-latitude snow cover observations has not yet been assembled (Barry and Armstrong, 1987). In fact sustained high-quality measurements are generally incomplete (Karl et al., 1989). Since 1966 Northern Hemisphere snow cover maps have been produced operationally on a weekly basis using satellite imagery by NOAA. The NOAA data contain snow/no-snow information for 7921 grid boxes covering the globe and were judged by Scialdone and Robock (1987) as the best of four data sets which they compared. Deficiencies have been noted by Wiesnet et al. (1987) such as: until 1975 the charts did not consistently include Himalayan snow cover; there were occasional extensions of the southern edge of the snow cover beyond observed

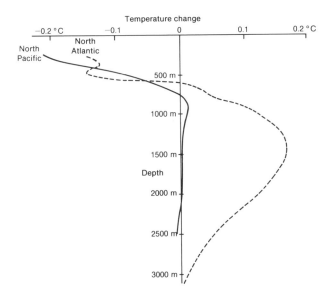

Figure 7.18: Sub-surface ocean temperature changes at depth between 1957 and 1981 in the North Atlantic and North Pacific. Adapted from Antonov (1990).

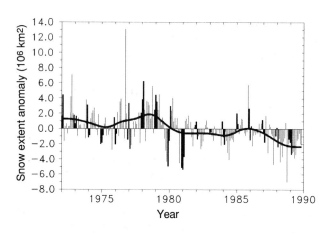

Figure 7.19: Northern Hemisphere snow extent anomalies. Data from NOAA (USA).

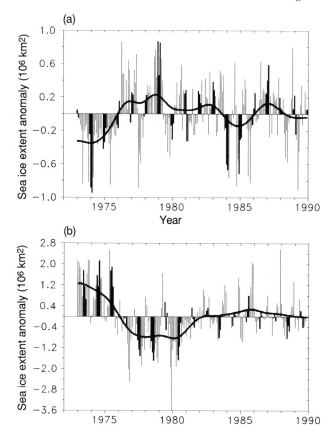

Figure 7.20: (a) Northern Hemisphere, and (b) Southern Hemisphere sea-ice extent anomalies. Data from NOAA (USA).

surface limits; the seasonal variation of sunlight limits polar coverage in the visible wavelengths; and scattered mountain snows are omitted because of the coarse grid resolution. Data are believed to be usable from 1972 with caution, but are better from 1975 onwards.

Consistent with the surface and tropospheric temperature measurements is the rapid decrease in snow cover extent around 1980 (Figure 7.19). This decrease is largest during the transition seasons. Robinson and Dewey (1990) note that the reduction in snow cover extent during the 1980s is largest in Eurasia where they calculate decreases during autumn and spring of about 13% and 9% respectively relative to the 1970s.

7.8.2 Sea-ice Extent and Thickness

There has been considerable interest in the temporal variability of global sea-ice in both the Arctic and Antarctic (for example, Walsh and Sater, 1981; Sturman and Anderson, 1985). This interest has been increased by general circulation model results suggesting that greenhouse warming may be largest at high latitudes in the Northern Hemisphere. It must be recognized, though, that sea-ice is strongly influenced by surface winds and ocean currents so that the consequences of global warming for changes in sea-ice extent and thickness are unlikely to be straightforward.

Sea-ice limits have long been observed by ships, and harbour logs often contain reported dates of the appearance and disappearance of harbour and coastal ice. These observations present many problems of interpretation (Barry, 1986) though they are thought to be more reliable after about 1950. Changes and fluctuations in Arctic sea-ice extent have been analysed by Mysak and Manak (1989); they find no long term trends in sea-ice extent between 1953 and 1984 in a number of Arctic ocean regions but substantial decadal time scale variability was

evident in the Atlantic sector. These variations were found to be consistent with the development, movement and decay of the "Great Salinity Anomaly" noted in Section 7.7.

Sea-ice conditions are now reported regularly in marine synoptic observations, as well as by special reconnaissance flights, and coastal radar. Especially importantly, satellite observations have been used to map sea-ice extent routinely since the early 1970s. The American Navy Joint Ice Center has produced weekly charts which have been digitised by NOAA. These data are summarized in Figure 7.20 which is based on analyses carried out on a 1° latitude x 2.5° longitude grid. Sea-ice is defined to be present when its concentration exceeds 10% (Ropelewski, 1983). Since about 1976 the areal extent of sea-ice in the Northern Hemisphere has varied about a constant climatological level but in 1972-1975 sea-ice extent was significantly less. In the Southern Hemisphere since about 1981, sea-ice extent has also varied about a constant level. Between 1973 and 1980 there were periods of several years when Southern Hemisphere sea-ice extent was either appreciably more than or less than that typical in the 1980s.

Gloersen and Campbell (1988) have analysed the Scanning Multi-channel (dual polarization) Microwave Radiometer data from the Nimbus 7 satellite from 1978-1987. They find little change in total global ice area, but a significant decrease of open water within the ice. Their time series is short, and it is uncertain whether the decrease is real.

Sea-ice thickness is an important parameter but it is much more difficult to measure than sea-ice extent. The heat flux from the underlying ocean into the atmosphere depends on sea-ice thickness. Trends in thickness over the Arctic Ocean as a whole could be a sensitive indicator of global warming. The only practical method of making extensive measurements is by upward-looking sonar from submarines. Apart from a very recent deployment of moorings, data gathering has been carried out on voyages by military submarines. In the past, repeated tracks carried out in summer have either found no change in mean thickness (Wadhams, 1989) or variations that can be ascribed to interannual variability in summer ice limits and ice concentration (McLaren, 1989). Recently, however, Wadhams (1990) found a 15% or larger decrease in mean sea-ice thickness between October 1976 and May 1987 over a large region north of Greenland. Lack of a continuous set of observations makes it impossible to assess whether the change is part of a long term trend. In the Antarctic no measurements of thickness variability exist and so far only one geographically extensive set of sea-ice thickness data is available (Wadhams et al., 1987).

7.8.3 Land Ice (Mountain Glaciers)

Measurements of glacial ice volume and mass balance are more informative about climatic change than those of the extent of glacial ice, but they are considerably scarcer. Ice volume can be determined from transects of bedrock and ice surface elevation using airborne radio-echo sounding measurements. Mass balance studies performed by measuring winter accumulation and summer ablation are slow and approximate, though widely used. Section 9 discusses changes in the Greenland and Antarctic ice-caps so attention is confined here to mountain glaciers.

A substantial, but not continuous, recession of mountain glaciers has taken place almost everywhere since the latter half of the nineteenth century (Grove, 1988). This conclusion is based on a combination of mass balance analyses and changes in glacial terminus positions, mostly the latter. The recession is shown in Figure 7.2; evidence for glacial retreat is found in the Alps, Scandinavia, Iceland, the Canadian Rockies, Alaska, Central Asia, the Himalayas, on the Equator, in tropical South America, New Guinea, New Zealand, Patagonia, the sub-Antarctic islands and the Antarctic Peninsula (Grove, 1988). The rate of recession appears to have been generally largest between about 1920 and 1960.

Glacial advance and retreat is influenced by temperature, precipitation, and cloudiness. For example, at a given latitude glaciers tend to extend to lower altitudes in wetter, cloudier, maritime regions with cooler summers than in continental regions. The complex relation between glaciers and climate makes their ubiquitous recession since the nineteenth century remarkable; temperature changes appear to be the only plausible common factor (Oerlemans, 1988). The response time of a glacier to changes in environmental conditions varies with its size so that the larger the glacier, the slower is the response (Haeberli et al., 1989). In recent decades glacial recession has slowed in some regions. Makarevich and Rototaeva (1986) show that between 1955 and 1980 about 27% of 104 North American glaciers were advancing and 53% were retreating, whereas over Asia only about 5% of nearly 350 glaciers were advancing. Wood (1988) found that from 1960 to 1980 the number of retreating glaciers decreased. This may be related to the relatively cool period in the Northern Hemisphere over much of this time (Figure 7.10). However, Patzelt (1989) finds that the proportion of retreating Alpine glaciers has increased sharply since the early 1980s so that retreat has dominated since 1985 in this region. A similar analysis for other mountain regions after 1980 is not yet available.

7.8.4 Permafrost

Permafrost may occur where the mean annual air temperatures are less than -1°C and is generally continuous where mean annual temperature is less than -7°C. The vertical profile of temperature measurements in permafrost that is obtained by drilling boreholes can indicate integrated changes of temperature over decades and longer. However, interpretation of the profiles requires knowledge of the ground conditions as well as natural or human-induced changes in vegetation cover. Lachenbruch and Marshall (1986) provide evidence that a 2 to 4°C warming has taken place in the coastal plain of Alaska at the permafrost surface over the last 75 to 100 years, but much of this rise is probably associated with warming prior to the 1930s. Since the 1930s, there is little evidence for sustained warming in the Alaskan Arctic (see Figure 7.12a and Michaels, 1990). A fuller understanding of the relationship between permafrost and temperature requires better information on changes in snow cover, seasonal variations of ground temperature, and the impact of the inevitable disturbances associated with the act of drilling the bore holes (Barry, 1988).

7.9 Variations and Changes in Atmospheric Circulation

The atmospheric circulation is the main control behind regional changes in wind, temperature, precipitation, soil moisture and other climatic variables. Variations in many of these factors are quite strongly related through large-scale features of the atmospheric circulation, as well as through interactions involving the land and ocean surfaces. One goal of research into regional changes of atmospheric circulation is to show that the changes of temperature, rainfall and other climatic variables are consistent with the changes in frequency of various types of weather pattern.

Climates at the same latitude vary considerably around the globe, while variations in regional temperatures that occur on decadal time scales are far from uniform but form distinctive large-scale patterns as indicated in Figure 7.13. The spatial scale of these climatic patterns is partly governed by the regional scales of atmospheric circulation patterns and of their variations. Changes in weather patterns may involve changes in the quasi-stationary atmospheric long waves in the extratropics or in monsoonal circulations (van Loon and Williams, 1976). Both phenomena have a scale of several thousand kilometres. Their large-scale features are related to the fixed spatial patterns of land and sea, topography, sea temperature patterns, and the seasonal cycle of solar heating.

Persistent large-scale atmospheric patterns tend to be wavelike so that regional changes of atmospheric heating, if powerful and persistent enough, can give rise to a sequence of remote atmospheric disturbances. Thus a number of well-separated areas of anomalous temperature and precipitation of opposite character may be produced. The best known examples are, in part, related to the large changes in SSTs that accompany the El Niño-Southern Oscillation (ENSO), whereby changes in the atmosphere over the tropical Pacific often associated with the SST changes there are linked to atmospheric circulation changes in higher latitudes (Wallace and Gutzler, 1981). The 1988 North American drought has been claimed to be partly a response to persistent positive tropical SST anomalies located to the west of Mexico and to the north of the cold La Niña SST anomalies existing at that time (Trenberth et al., 1988). Such localised SST anomalies may themselves have a much larger scale "cause" (Namias, 1989).

An emerging topic concerns observational evidence that the 11 year solar cycle and the stratospheric quasi-biennial oscillation (QBO) of wind direction near the equator are linked to changes in tropospheric circulation in the Northern Hemisphere (van Loon and Labitzke, 1988). Coherent variations in tropospheric circulation are claimed to occur over each 11 year solar cycle in certain regions, but their character depends crucially on the phase (easterly or westerly) of the QBO. No mechanism has been proposed for this effect, and the data on the QBO cover only about 3.5 solar cycles, so that the reality of the effect is very uncertain. However, Barnston and Livesey (1989), in a careful study, find evidence for statistically significant influences of these factors on atmospheric circulation patterns in the Northern Hemisphere extratropics in winter.

Many previous, largely unsubstantiated, claims of links between the 11 year, and other solar cycles, and climate are reviewed in Pittock (1983). Section 2 discusses current thinking about the possible magnitude of the physical forcing of global climate by solar radiation changes in some detail.

Several examples are now given of links between changes in atmospheric circulation over the last century and regional-scale variations or trends of temperature.

7.9.1 El Niño-Southern Oscillation (ENSO) Influences

ENSO is the most prominent known source of interannual variability in weather and climate around the world, though not all areas are affected. The Southern Oscillation (SO) component of ENSO is an atmospheric pattern that extends over most of the global tropics. It principally involves a seesaw in atmospheric mass between regions near Indonesia and a tropical and sub-tropical south east Pacific Ocean region centred near Easter Island. The influence of ENSO sometimes extends to higher latitudes (see Section 7.9.3). The El Niño component of ENSO is an anomalous warming of the eastern and central tropical Pacific Ocean. In major "Warm Events", warming extends over much of the tropical Pacific and becomes clearly linked to the atmospheric SO pattern. An opposite phase of "Cold Events" with opposite patterns of the SO is sometimes referred to as "La Niña", ENSO events occur every 3 to 10 years and have far-reaching climatic and economic influences around the world (Figure 7.21a, adapted from Ropelewski and Halpert, 1987). Places especially affected include the tropical central and East Pacific islands, the coast of north Peru, eastern Australia, New Zealand (Salinger, 1981), Indonesia, India (Parthasarathy and Pant, 1985) and parts of eastern (Ogallo, 1989), and southern Africa (van Heerden et al., 1988). A fuller description of ENSO can be found in Rasmusson and Carpenter (1982) and Zebiak and Cane (1987). Over India, the occurrence of ENSO and that of many droughts (see Section 7.5.1) is strikingly coincident. Droughts tend to be much more frequent in the first year of an ENSO event, though, intriguingly, this is often before the ENSO event has fully developed. However not all Indian droughts are associated with ENSO.

While ENSO is a natural part of the Earth's climate, a major issue concerns whether the intensity or frequency of ENSO events might change as a result of global warming. Until recently, the models used to examine the climatic consequences of enhanced greenhouse forcing had such simplified oceans that ENSOs could not be simulated. Some models now simulate ENSO-like, but not entirely realistic, SST variations (Section 4); unfortunately, long term variations in ENSO cannot be studied yet using models. The observational record reveals that ENSO events have changed in frequency and intensity in the past. The

(a)

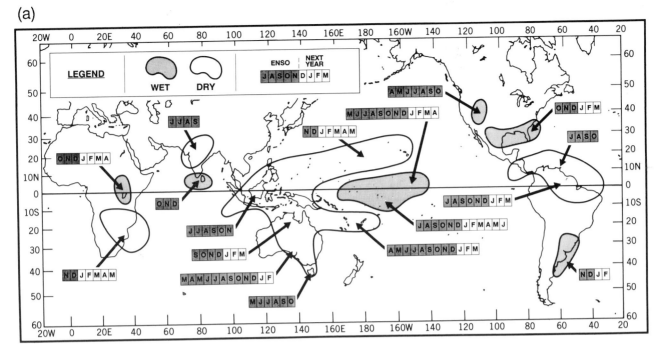

(b)

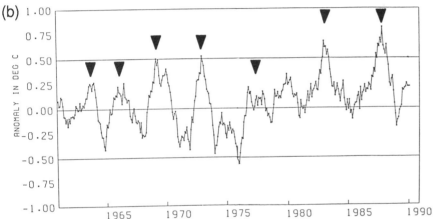

Figure 7.21: (a) Schematic diagram of areas and times of the year with a consistent ENSO precipitation signal (adapted from Ropelewski and Halpert, 1987). (b) Monthly tropical sea surface and land air temperature anomalies 1961-1989; land data from P.D. Jones and sea surface temperature data from the UK Meteorological Office. Tropics extend from 20°N to 20°S. Arrows mark maximum ENSO warmth in the tropics.

strong SO fluctuations from 1880 to 1920 led to the discovery and naming of the SO (Walker and Bliss, 1932) and strong SO events are clearly evident in recent decades. A much quieter period occurred from the late 1920s to about 1950, with the exception of a very strong multi-year ENSO in 1939-42 (Trenberth and Shea, 1987; Cooper et al., 1989). Quinn et al. (1987) (covering the past 450 years) and Ropelewski and Jones (1987) have documented historical ENSO events as seen on the northwest coast of South America. Therefore, the potential exists for a longer palaeo-record based on river deposits, ice cores, coral growth rings and tree rings.

During ENSO events, the heat stored in the warm tropical western Pacific is transferred directly or indirectly to many other parts of the tropical oceans. There is a greater than normal loss of heat by the tropical oceans, resulting in a short period warming of many, though not all, parts of the global atmosphere (Pan and Oort, 1983). Consequently, warm individual years in the record of global temperatures (Figure 7.10) are often associated with El Niños. Maxima in global temperatures tend to occur about three to six months after the peak warmth of the El Niño (Pan and Oort, 1983). Figure 7.21b shows monthly anomalies of combined land surface air temperatures and SST for the global tropics from 1961-1989. The strong, coherent, warming influence of the 1972-73, 1982-83 and 1986-88 ENSO events on the record of tropical temperature is very clear, as is the cold influence of the strong La Niña episodes of 1974-75 and 1988-89.

From an inter-decadal perspective, ENSO is a substantial source of climatic noise which can dominate the tropical temperature record. It is possible to remove ENSO signals statistically (for example Jones, 1989) to give a smoother global temperature curve; in this way 20 to 30% of decadal and shorter time scale variance is removed. The warming of the globe in the last 15 years then is reduced by about 0.1°C, i.e., by about one half. However other temperature signals might also be partly removed at the same time, for example those relating to the effects of volcanic eruptions, though some ability to separate these signals has recently been shown (Mass and Portman, 1989) (see also Section 2). As it is unclear whether ENSO might change with and contribute directly to long-term global warming, it seems preferable to retain ENSO variability as an integral part of the global climate record.

7.9.2 The North Atlantic

The early twentieth century cooling of the Northern Hemisphere oceans (Figure 7.8) was accompanied by a period of intensified westerlies in the extratropical Northern Hemisphere, especially in the Atlantic sector, that affected most of the year. An extensive discussion is given in Lamb (1977). The global warming which took place in the 1920s and 1930s (Figure 7.10) was largest in the extratropical North Atlantic and in the Arctic (Figure 7.12), and coincided with the latter part of the period of intense westerlies. The westerly epoch is regarded as finishing around 1938 (Makrogiannis et al., 1982). The effects of the enhanced westerlies on surface climate were clearest in winter when there was an absence of very cold outbreaks over Europe and winters were persistently mild. Rogers (1985) noted that the best correlation between temperatures in Europe and wind direction is with the westerly component, largely reflecting whether or not the encroaching air masses have had an oceanic moderating influence imposed on them. Figure 7.22 shows an index of westerly flow expressed as the difference in atmospheric pressure measured at mean sea level between the Azores and Iceland in winter. High index epochs have stronger or more frequent westerly flow across the extratropical North Atlantic. Also shown is the air temperature anomaly over northern Europe and the adjacent seas from 45°N-60°N and 5°W-35°E in winter. The inter-decadal variations of the pressure index are strikingly large, with weakest flow centred around the late 1960s (less westerlies) and a return to a stronger westerlies recently (Flohn et al., 1990). Wallen (1986) notes that the period of intense westerly flow also affected summer temperatures in western Europe, making them generally cooler between the late 1890s and about 1920, giving striking decreases in the differences between July and January temperatures. This suggests that at least a small part of the decrease in the annual range of

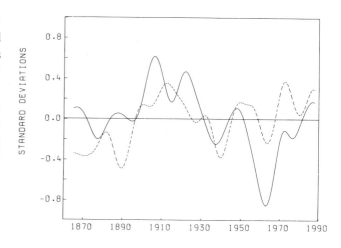

Figure 7.22: Smoothed standardised indices in winter (using binomial filter with 21 terms applied three times) of difference in atmospheric pressure at mean sea level between Ponta Delgada, Azores and Stykkisholmur, Iceland (solid line) and surface temperature for 45°-60°N, 5°W-35°E, based on land air (P.D. Jones) and night marine air temperature (UK Meteorological Office). December 1866-February 1867 to December 1989-February 1990.

Northern Hemisphere land surface air temperature seen at this time in Figure 7.14 may be real.

Variations in the westerly index of Figure 7.22 are associated with changes in the depth of the Iceland low pressure centre both near the surface and in the troposphere (van Loon and Rogers, 1978). European temperatures well downstream are quite strongly related to this Atlantic pressure index, especially during winter, being warmest when the index is largest, i.e., pressure over Iceland is lowest relative to the Azores. Relative to the level of the westerly pressure index, there is a long-term increase of European winter surface air temperature in Figure 7.22, much as noted by Moses et al. (1987) for a small set of stations in western Europe. Above normal winter temperatures in Europe tend to go hand-in-hand with below normal temperatures in Greenland and the Canadian Arctic where there are increased northerlies as a result of the deeper Iceland Low. Stronger westerlies over the Atlantic, do not, therefore, account for the Arctic warming of the 1920s and 1930s on their own: in fact they preceded it by 20 years. Iceland (Einarsson, 1984) and Spitzbergen began to warm after about 1917 (Lamb, 1977), whereas Greenland and Northern Canada did not warm until the mid-1920s (Rogers, 1985). When both Greenland and Northern Europe had above normal temperature, especially in the winter half year, the atmospheric circulation was more zonal around most of the Arctic, not just in the Atlantic sector. This was associated with increased cyclonic activity over the whole Arctic basin which increased the frequency of zonal flows over the higher latitudes of the continents. Note that the character of the warming experienced in the higher latitude Northern Hemisphere in the the 1920s and 1930s differs from that of the mid-1970s to early 1980s (Figure 7.10) when the North Atlantic and Arctic stayed cool, or in parts, cooled further.

Inter-decadal changes in the west African monsoon circulation which have particularly affected Sub-Saharan African rainfall (Figure 7.16) were introduced in Section 7.5.1. The main change in atmospheric circulation has been in the convergence of winds into sub-Saharan Africa in summer from the north and the south (Newell and Kidson, 1984); less intense convergence gives less rainfall (Folland et al., 1990). Drier years are also often accompanied by a slightly more southerly position of the main wind convergence (rain bearing) zone. The North Atlantic subtropical high pressure belt also tended to extend further southward and eastward during the summer in the dry Sahel decades (Wolter and Hastenrath, 1989).

7.9.3 The North Pacific

Circulation changes in the North Pacific have recently been considerable and have been linked with regional temperature changes. Figure 7.23 shows a time series of mean sea level pressure for the five winter months November to March, averaged over most of the extra-tropical North Pacific for 1946-1988 (Trenberth, 1990). This index is closely related to changes in the intensity of the Aleutian low pressure centre. It is also quite strongly linked to a pattern of atmospheric circulation variability known as the Pacific-North American (PNA) pattern (Wallace and Gutzler, 1981) which is mostly confined to the North Pacific and to extratropical North America. All five winter months showed a much deeper Aleutian Low in the period 1977 to 1988, with reduced pressure over nearly all the extratropical North Pacific north of about 32°N (Flohn et al., 1990). The change in pressure in Figure 7.23 appears to have been unusually abrupt; other examples of such "climatic discontinuities" have been analysed (Zhang et al., 1983), though discontinuities can sometimes be artifacts of the statistical analysis of irregular time series. The stronger Aleutian Low resulted in warmer, moister air being carried into Alaska while much colder air moved south over the North Pacific. These changes account for the large Pacific temperature anomalies for 1980-89 shown in Figure 7.13, which are even clearer for the decade 1977-86 (not shown). This decade had a positive anomaly, (relative to 1951-80), of over 1.5°C in Alaska and negative anomaly of more than 0.75°C in the central and western North Pacific.

The above changes are likely to have been related to conditions in the equatorial Pacific. 1977-1987 was a period when much of the tropical Pacific and tropical Indian Oceans had persistently above normal SSTs (Nitta

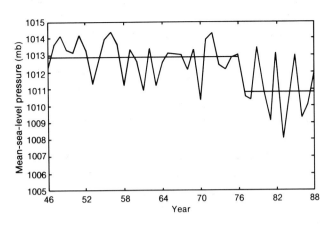

Figure 7.23: Time series of mean North Pacific sea level pressures averaged over 27.5 to 72.5°N, 147.5°E to 122.5°W for November through March. Means for 1946-76 and 1977-87 are indicated.

and Yamada, 1989, and Figure 7.12a). Very strong El Niño events and a lack of cold tropical La Niña events in the period 1977-1987 (Figure 7.21b) contributed to this situation.

7.9.4 Southern Hemisphere

In Antarctica strong surface temperature inversions form in winter, but elsewhere in the Southern Hemisphere maritime influences dominate. The SO has a pronounced influence on precipitation over Australia (Pittock, 1975) and also affects New Zealand temperatures and precipitation (Gordon, 1986). However, the best documented regional circulation-temperature relationship in the Southern Hemisphere is that between an index of the meridional (southerly and northerly) wind (Trenberth, 1976) and New Zealand temperature. The index is calculated by subtracting sea level pressure values measured at Hobart (Tasmania) from those at Chatham Island (east of New Zealand). A tendency for more northerly mean flow across New Zealand (Hobart pressure relatively low), especially from about 1952 to 1971, has been related to generally warmer conditions in New Zealand after 1950. However a return to more southerly (colder) flow after 1971 is not strongly reflected in New Zealand temperatures, so the recent warmth may be related to the general increase in temperature in much of the Southern Hemisphere (Figures 7.10b, 7.12 and 7.13). This finding indicates that regional temperature changes due to a future greenhouse warming are likely to result from an interplay between large scale warming and changes in local weather patterns.

7.10 Cloudiness

Clouds modify both the shortwave (solar) and longwave (terrestrial) radiation, the former by reflection and the latter by absorption. Therefore they may cause a net warming or cooling of global temperature, depending on their type, distribution, coverage, and radiative characteristics (Sommerville and Remer, 1984; Cess and Potter, 1987). Ramanathan et al. (1989) show that with today's distribution and composition of clouds, their overall effect is to cool the Earth (Section 3.3.4). Changes in cloudiness are therefore likely to play a significant role in climate change. Furthermore, local and regional climate variations can be strongly influenced by the amount of low, middle, and high clouds.

Observations of cloudiness can be made from the Earth's surface by trained observers from land stations or ocean vessels, or by automated systems. Above the Earth's surface, aircraft or space platforms are used (Rossow, 1989; McGuffie and Henderson-Sellers, 1989a). Surprisingly, surface-based observations of cloudiness give closely similar results to space-based observations. Careful and detailed intercomparisons, undertaken as a preliminary part of the International Satellite Cloud Climatology Project (ISCCP) by Sze et al. (1986), have demonstrated conclusively that surface and space-based observations are highly correlated. Space-based observations of cloudiness from major international programs, such as ISCCP, are not yet available for periods sufficiently long to detect long-term changes.

7.10.1 Cloudiness Over Land

Henderson-Sellers (1986, 1989) and McGuffie and Henderson-Sellers (1989b) have analyzed changes in total cloud cover over Europe and North America during the twentieth century. It was found that annual mean cloudiness increased over both continents. Preliminary analyses for Australia and the Indian sub-continent also give increases in cloudiness. The increases are substantial: 7% of initial cloudiness/50 years over India, 6%/80 years over Europe, 8%/80 years for Australia, and about 10%/90 years for North America. These changes may partly result from alterations in surface-based cloud observing practices and in the subsequent processing of cloud data. This may be especially true of the large increase in cloudiness apparently observed in many areas in the 1940s and 1950s. At this time (about 1949 or later, depending on the country) the synoptic meteorological code, from which many of these observations are derived, generally underwent a major change but not in the USA, USSR and Canada. Observers began recording cloud cover in "oktas" (eighths) instead of in tenths. When skies were partly cloudy, it is possible that some observers who had been used to making observations in the decimal system converted decimal observations of cloud cover erroneously to the same number of oktas, thereby overestimating the cloud cover.

Recently, Karl and Steurer (1990) have compared daytime cloudiness statistics over the USA with data from automated sunshine recorders. They indicate that there was a much larger increase of annual cloud cover during the 1940s than can be accounted for by the small observed decrease in the percentage of possible sunshine. The large increase of cloudiness may be attributed to the inclusion of the obscuring effects of smoke, haze, dust, and fog in cloud cover reports from the 1940s onward (there being no change in the recording practice from tenths to oktas in the USA). The increase in cloudiness after 1950 may be real because an increase is consistent with changes in the temperature and precipitation records in the USA, including the decreased diurnal temperature range seen in Figure 7.15.

Observed land-based changes in cloudiness are difficult to assess. Nonetheless, total cloud amount appears to have increased in some continental regions, a possibility supported by noticeable reductions in the diurnal range of temperature in some of these regions. Elsewhere the cloudiness record cannot be interpreted reliably.

7.10.2 Cloudiness Over The Oceans

Ocean-based observations of cloud cover since 1930 have been compiled by Warren et al. (1988). The data are derived from maritime synoptic weather observations. Their number varies between 100,000 and 2,000,000 each year, increasing with time, and the geographic coverage also changes. The data indicate that an increase in marine cloudiness, exceeding one percent in total sky covered on a global basis, took place from the 1940s to the 1950s. This increase is not reflected in the proportion of observations having a clear sky or a complete overcast. The largest increases were in stratocumulus clouds in Northern Hemisphere mid-latitudes and in cumulonimbus in the tropics. Since 1930, mean cloudiness has increased by 3-4 percent of the total area of sky in the Northern Hemisphere and by about half of this value in the Southern Hemisphere. Fixed ocean weather ships, placed after 1945 in the North Atlantic and North Pacific with well trained observers, showed no trends in cloudiness between the 1940s and 1950s when other ship data from nearby locations showed relatively large increases, changes of the same sign as those in available land records (Section 7.10.1). It is clearly not possible to be confident that average global cloudiness has really increased.

7.11 Changes of climate Variability and Climatic Extremes

Aspects of climate variability include those associated with day-to-day changes, inter-seasonal and interannual variations, and the spatial variability associated with horizontal gradients. A pervasive problem for assessments of changes in the temporal variability of climate is the establishment of a reference level about which to calculate that variability. For example, the results of an investigation to determine whether the variability of monthly mean values was changing could give different results depending on how the average, or baseline, climate was calculated. If the climate was changing, calculations of variability would depend on whether a fixed baseline was used or whether it was allowed to change smoothly or discontinuously. Additionally, attempts to identify whether the number of extremes is decreasing or increasing may be critically dependent on the definition of the threshold value above which an extreme is defined. An attempt was made for this Report to estimate changes in the number of extremes of monthly average temperature values for a globally distributed set of about 150 stations, all having at least 60 years of record. The results were found to be very sensitive to the threshold chosen to define the extreme, and to the baseline climatology selected, so that no firm deductions about changes in extremes or variability were possible.

Interest in climatic variability often includes that on daily time-scales: a common question concerns whether daily temperatures above or below given threshold values, for example frosts, are changing in their frequency. Although considerable local climatic information exists on daily time-scales in any national meteorological data centre, such information covering the globe as a whole is not readily available at any one centre. Thus many of the data needed for a comprehensive assessment of changes in variability still need to be assembled and a scientifically sound method of analysis needs to be developed. Nevertheless a few comments can be made about variability which are discussed below.

7.11.1 Temperature
Several researchers have assessed relationships between anomalously cold or warm seasons and daily temperature variability in the USA. Brinkmann (1983) and Agee (1982) both find reduced day-to-day variability in anomalously warm winters, though Brinkmann (1983) provides evidence of enhanced day-to-day variability during anomalously warm summers. It is likely, however, that these relat-

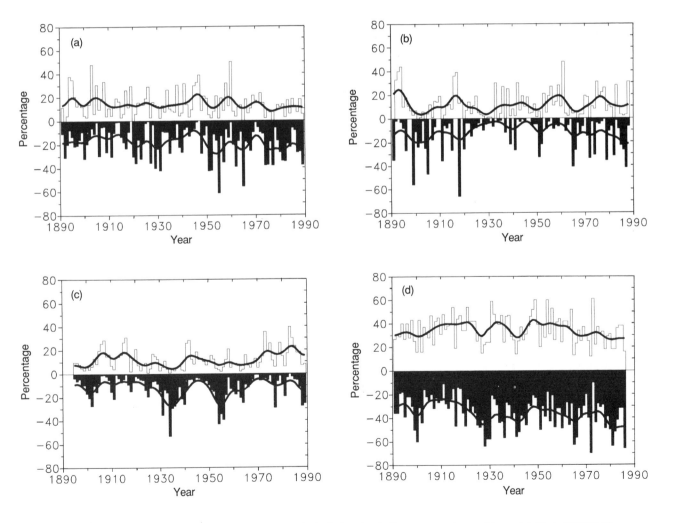

Figure 7.24: Regional drought and moisture index series. Unshaded region indicates excessive moisture. (a) Part of USSR (from Meshcherskaya and Blazhevich, 1977), (b) India, (c) USA, (d) China.

ionships are highly sensitive to the choice of region. There appears to be little relation between interannual variability and the relative warmth or coldness of decadal averages. Although Diaz and Quayle (1980) found a tendency for increased variability in the USA during the relatively warm years of the mid-twentieth century (1921-1955), Karl (1988) found evidence for sustained episodes (decades) of very high and low interannual variability with little change in baseline climate. Furthermore, Balling et al. (1990) found no relationship between mean and extreme values of temperature in the desert southwest of the USA.

7.11.2 Droughts and Floods

An important question concerns variations in areas affected by severely wet ("flood") or drought conditions. However, drought and moisture indices calculated for Australia (not shown), parts of the Soviet Union, India, the USA, and China (Figure 7.24, previous page) do not show systematic long-term changes. Although this does not represent anything like a global picture, it would be difficult to envisage a worldwide systematic change in variability without any of these diverse regions participating. It is noteworthy that the extended period of drought in the Sahel (Figure 7.16) between 1968 and 1987 exhibited a **decreased interannual** variability of rainfall compared with the previous 40 years even though the number of stations used remained nearly the same.

7.11.3 Tropical Cyclones

Tropical cyclones derive their energy mainly from the latent heat contained in the water vapour evaporated from the oceans. As a general rule, for tropical cyclones to be sustained, SSTs must be at or above 26°C to 27°C at the present time. Such values are confined to the tropics, as well as some subtropical regions in summer and autumn. The high temperatures must extend through a sufficient depth of ocean that the wind and wave action of the storm itself does not prematurely dissipate its energy source. For a tropical cyclone to develop, its parent disturbance must be about 7° of latitude or more from the equator. Many other influences on tropical cyclones exist which are only partly understood. Thus ENSO modulates the frequency of tropical storms in some regions, for example over the north-west Pacific, mainly south of Japan (Li, 1985; Yoshino, 1990), East China (Fu and Ye, 1988) and in the central and southwest Pacific (Revell and Gaulter, 1986). The reader is referred to Nicholls (1984), Gray (1984), Emanuel (1987) and Raper (1990) for more detail.

Have tropical cyclone frequencies or their intensities increased as the globe has warmed over this century? Current evidence does not support this idea, perhaps because the warming is not yet large enough to make its impact felt. In the North Indian Ocean the frequency of tropical storms has noticeably decreased since 1970 (Figure

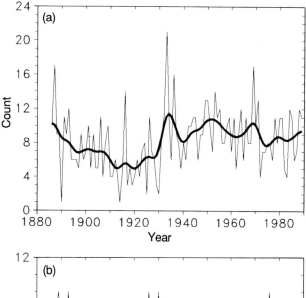

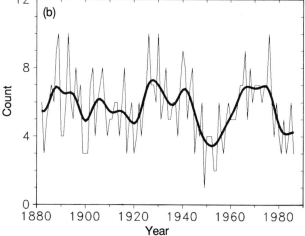

Figure 7.25: Estimated number of tropical cyclones in (a) Atlantic, and (b) North Indian Oceans over the last century. Data in (b) is less reliable before about 1950.

7.25) while SSTs have risen here since 1970, probably more than in any other region (Figure 7.13). See also Raper (1990). There is little trend in the Atlantic, though pronounced decadal variability is evident over the last century. There have been increases in the recorded frequency of tropical cyclones in the eastern North Pacific, the southwest Indian Ocean, and the Australian region since the late 1950s (not shown). However, these increases are thought to be predominantly artificial and to result from the introduction of better monitoring procedures. Relatively good records of wind speed available from the North Atlantic and western Pacific oceans do not suggest that there has been a change toward more **intense** storms either.

7.11.4 Temporales of Central America

Temporales are cyclonic tropical weather systems that affect the Pacific side of Central America and originate in the Pacific Inter-tropical Convergence Zone. Very heavy rainfall totals over several days occur with these systems, but unlike hurricanes, their winds are usually weak. Their atmospheric structure is also quite different from that of hurricanes as they possess a cold mid-tropospheric core. Temporales typically last several days, are slow moving and cause damaging floods and landslides in the mountainous regions of Central America. Records of temporales are available since the 1950s. They were markedly more frequent in the earlier than in the later part of this period. Thus there was an average of 2.4 temporales per year in 1952-1961 (Hastenrath, 1988); only 1959 had no temporales. In 1964-1983 the average reduced to 0.45 temporales per year and in 12 of these years there were none.

When the evidence in Section 7.11 is taken together, we conclude that there is no evidence of an increasing incidence of extreme events over the last few decades. Indeed some of the evidence points to recent decreases, for example in cyclones over the North Indian Ocean and temporales over Central America.

7.12 Conclusions

The most important finding is a warming of the globe since the late nineteenth century of $0.45 \pm 0.15°C$, supported by a worldwide recession of mountain glaciers. A quite similar warming has occurred over both land and oceans. This conclusion is based on an analysis of new evidence since previous assessments (SCOPE 29, 1986) and represents a small reduction in previous best estimates of global temperature change. The most important diagnosis that could **not** be made concerns temperature variations over the Southern Ocean. Recent transient model results (Section 6) indicate that this region may be resistant to long term temperature change. A data set of blended satellite and ship SST data is now becoming available and may soon provide an initial estimate of recent Southern Ocean temperature changes.

Precipitation changes have occurred over some large land regions in the past century but the data sets are so poor that only changes of large size can be monitored with any confidence.

Some substantial regional atmospheric circulation variations have occurred over the last century, notably over the Atlantic and Europe. Regional variations in temperature trends have also been quite substantial. This indicates that, in future, regional climatic changes may sometimes be quite diverse.

Natural climate variations have occurred since the end of the last glaciation. The Little Ice Age, in particular, involved global climate changes of comparable magnitude to the warming of the last century. It is possible that some of the warming since the nineteenth century may reflect the cessation of Little Ice Age conditions. The rather rapid changes in global temperature seen around 1920-1940 are very likely to have had a mainly natural origin. Thus a better understanding of past variations is essential if we are to estimate reliably the extent to which the warming over the last century, and future warming, is the result of an increase of greenhouse gases.

References

Agee, E.M., 1982: A diagnosis of twentieth century temperature records at West Lafayette, *Indiana. Clim. Change*, **4**, 399-418.

Alexandre, P., 1987: Le Climat en Europe au Moyen Age. Contribution a l'histoire des variations climatiques de 1000 a 1425, d'apres les sources narratives de l'Europe Occidentale. Ec. Hautes Etud. Sci. Soc.: *Rech. Hist. Sci. Soc.*, **24**, Paris, 827pp.

Angell, J.K., 1988: Variations and trends in tropospheric and stratospheric global temperatures, 1958-87. *J. Clim.*, **1**, 1296-1313.

Antonov, J.I., 1990: Recent climatic changes of vertical thermal structure of the North Atlantic and North Pacific Oceans. *Meteorol. i Gidrolog.*, **4**, (in press).

Arkin, P.A., and M. Chelliah, 1990: An assessment of variations of outgoing longwave radiation over the tropics, 1974-1987. Draft Report to IPCC WG1, Section 7.

Balling, R.C., Jr, J.A. Skindlow and D.H. Phillips, 1990: The impact of increasing summer mean temperatures on extreme maximum and minimum temperatures at Phoenix, Arizona. *J. Clim.*, (in press).

Barnett, T.P., 1986: Detection of changes in the global tropospheric temperature field induced by greenhouse gases. *J. Geophys. Res.*, **91**, 6659-6667.

Barnett, T.P., and M.E. Schlesinger, 1987: Detecting changes in global climate induced by greenhouse gases. *J. Geophys. Res.*, **92**, 14772-14780.

Barnston, A.G., and R.E. Livesey, 1989: A closer look at the effect of the 11-year solar cycle and the quasi-biennial oscillation on Northern Hemisphere 700mb height and extratropical North American surface temperature. *J. Clim.*, **2**, 1295-1313.

Barry, R.G., 1988: Permafrost data and information: status and needs. In: K. Senneset, (Ed.), *Permafrost*. Fifth International Conf. Proc., Vol. 1, Tapir Publishers, Trondheim, pp119-122.

Barry, R.G., 1986: The sea ice data base. In: Untersteiner (Ed.), *The Geophysics of Sea Ice*. Plenum Press, New York, 1099-1134.

Barry, R.G., and R.L. Armstrong, 1987: Snow cover data management. The role of WDC A for Glaciology. *Hydrol. Sci. J.*, **32**, 281-295.

Bartlein, P.J., and T. Webb III, 1985: Mean July temperatures at 6000BP in eastern North America: regression equations for estimates from fossil pollen data. In: *Climatic Change in Canada*, 5, C.R. Harrington (Ed.). Syllogeus 55, National Museum of Canada, Ottawa, 301-342.

Berger, A., 1980: The Milankovitch astronomical theory of palaeoclimates: a modern review. *Vistas in Astronomy*, **24**, 103-122.

Berner, R.A., A.C. Lasaga and R.M. Garrels, 1983: The carbonate-silicate geochemical cycle and its effect on carbon dioxide over the past 110 million years. *Am. J. Sc.*, **283**, 7, 641-683.

Borzenkova, I.I., and V.A. Zubakov, 1984: Climatic Optimum of the Holocene as a model of the global climate at the beginning of the 21st century. *Meteorol. i Gidrolog.*, **N8**, 69-77. Also Sov. Met. Hydr., 8, 52-58.

Bottomley, M., C.K. Folland, J. Hsiung, R.E. Newell and D.E. Parker, 1990: Global Ocean Surface Temperature Atlas (GOSTA). Joint Meteorological Office/Massachusetts Institute of Technology Project. Project supported by US Dept of Energy, US National Science Foundation and US Office of Naval Research. Publication funded by UK Depts of Energy and Environment. 20+iv pp and 313 Plates. HMSO, London.

Bradley, R.S., and P.D. Jones, 1985: Data bases for isolating the effects of increasing carbon dioxide concentration. In: *Detecting the Climatic Effects of increasing carbon dioxide*, M.C. MacCracken and F.M. Luther (Eds). USA Dept of Energy, Carbon Dioxide Research Division, pp29-53.

Bradley, R.S., H.F. Diaz, J.K. Eischeid, P.D. Jones, P.M. Kelly and C.M. Goodess, 1987: Precipitation fluctuations over Northern Hemisphere land areas since the mid-19th century. *Science*, **237**, 171-175.

Bradley, R.S., and P. Ya. Groisman, 1989: Continental scale precipitation variations in the 20th century. In: *Proc. Int. Conf. Precip. Measurements*, WMO, Geneva (in press).

Brewer, P.G., W.S. Broecker, W.J. Jenkins, P.B. Rhines, C.G. Rooth, J.H. Swift, T. Takahashi and R.T. Williams, 1983: A climatic freshening of the deep Atlantic (north of 50°N) over the past 20 years. *Science*, **222**, 1237-1239.

Brinkmann, W.A.R., 1983: Variability of temperature in Wisconsin. *Mon. Weath. Rev.*, **111**, 172-180.

Broecker, W.S., 1987: Unpleasant surprises in the greenhouse. *Nature*, **328**, 123-126.

Broecker, W.S., D. Peteet and D. Rind, 1985: Does the ocean-atmosphere system have more than one stable mode of operation? *Nature*, **315**, 21-26.

Budyko, M., A.B. Ronov and A.L. Yanshin, 1985: *The history of the Earth's atmosphere*. Leningrad, Gidrometeoizdat, 209pp (English trans.: Springer-Verlag, 1987, 139pp).

Budyko, M., and Yu. A. Izrael (Eds), 1987: *Anthropogenic Climatic Changes*. L. Gidrometeoizdat, 404pp.

Cardone, V.J., J.G. Greenwood and M.A. Cane, 1990: On trends in historical marine wind data. *J. Clim.*, **3**, 113-127.

Cess, R.D., and G.L. Potter, 1987: Exploratory studies of cloud radiative forcing with a general circulation model. *Tellus*, **39A**, 460-473.

Chappell, J.M.A., and Grindrod, A., (Eds), 1983: Proceedings of the first CLIMANZ conference, Howman's Gap, 1981. Dept of Biogeography and Geomorphology, Australian National University, Canberra, 2 vols.

CLIMAP Project Members, 1984: The last interglacial ocean. *Quatern. Res.*, **21**, 2, 123-224.

COHMAP Members, 1988: Climate changes of the last 18,000 years: observations and model simulations. *Science*, **241**, 1043-1052.

Cooper, N.S., K.D.B. Whysall and G.R. Bigg, 1989: Recent decadal climate variations in the tropical Pacific. *J. Climatol.*, **9**, 221-242.

Coughlan, M.J., R. Tapp and W.R. Kininmonth, 1990: Trends in Australian temperature records. Draft contribution to IPCC WGI, Section 7.

Diaz, H.F., and R.G. Quayle, 1980: The climate of the United States since 1985: spatial and temporal changes. *Mon. Weath. Rev.*, **108**, 249-266.

Diaz, H.F., R.S. Bradley and J.K. Eischeid, 1989: Precipitation fluctuations over global land areas since the late 1800s. *J. Geophys. Res.*, **94**, 1195-1210.

Dickson, R.R., J. Meincke, S.A. Malmberg and A.J. Lee, 1988: The great salinity anomaly in the northern North Atlantic, 1968-82. *Prog. Oceanogr.*, **20**, 103-151.

Druyan, L.M., 1989: Advances in the study of Sub-saharan drought. *Int. J. Climatol.*, **9**, 77-90.

Eddy, J., 1976: The Maunder Minimum. *Science*, **192**, 1189-1202.

Einarsson, M.A., 1984: Climate of Iceland. In: Chapter 7 of *Climate of the Oceans*, *World Survey of Climatology*, 15, H van Loon, (Ed.), pp673-697.

Ellsaesser, H.W., M.C. MacCracken, J.J. Walton, and S.L. Grotch, 1986: Global climatic trends as revealed by the recorded data. *Rev. Geophys.*, **24**, 745-792.

Elliott, W.P., M.E. Smith and J.K. Angell, 1990: On monitoring tropospheric water vapour changes using radiosonde data. Proc. Workshop on Greenhouse Gas Induced Climate Change, Amherst, MA, 8-12 May 1989, Elsevier , (in press).

Emanuel, K.A., 1987: The dependence of hurricane intensity on climate. *Nature*, **326**, 483-485.

Farmer, G., T.M.L. Wigley, P.D. Jones and M. Salmon, 1989: Documenting and explaining recent global-mean temperature changes. Climatic Research Unit, Norwich, Final Report to NERC, UK, Contract GR3/6565.

Flohn, H., and A. Kapala, 1989: Changes in tropical sea-air interaction processes over a 30-year period. *Nature*, **338**, 244-246.

Flohn, H., A. Kapala, H.R. Knoche and H. Machel, 1990: Recent changes of the tropical water and energy budget and of midlatitude circulations. Submitted to *Climate Dynamics*.

Folland, C.K., 1971: Day-time temperature measurements on weather ship "Weather Reporter". *Met. Mag.*, **100**, 6-14.

Folland, C.K., 1988: Numerical models of the raingauge exposure problem, field experiments and an improved collector design. *Q. J. Roy. Met. Soc.*, **114**, 1485-1516.

Folland, C.K., and A.W. Colman, 1988: An interim analysis of the leading covariance eigenvectors of worldwide sea surface temperature anomalies for 1951-80. LRFC 20, available from National Meteorological Library, Meteorological Office, Bracknell, UK.

Folland, C.K., D.E. Parker and F.E. Kates, 1984: Worldwide marine temperature fluctuations 1856-1981. *Nature*, **310**, 670-673.

Folland, C.K., and D.E. Parker, 1989: Observed variations of sea surface temperature. NATO Advanced Research Workshop on Climate-Ocean Interaction, Oxford, UK, 26-30 Sept 1988. Kluwer Academic Press, pp 31-52.

Folland, C.K., J. Owen, M.N. Ward and A. Colman, 1990: Prediction of seasonal rainfall in the Sahel region using empirical and dynamical methods. *J. Forecasting* , (in press).

Fu, C., and D. Ye, 1988: The tropical very low-frequency oscillation on interannual scale. *Adv. Atmos. Sci.*, **5**, 369-388.

Gadgil, S.C., P.V. Joseph and N.V. Joshi, 1984: Ocean-atmosphere coupling over monsoon regions. *Nature*, **312**, 141-143.

Gadgil, S., 1988: Recent advances in monsoon research with particular reference to the Indian Ocean. *Aust. Met. Mag.*, **36**, 193-204.

Gloersen, P., and W.J. Campbell, 1988: Variations in the Arctic, Antarctic, and global sea ice covers during 1979-1987 as observed with the Nimbus 7 Scanning Multichannel Microwave Radiometer. *J. Geophys. Res.*, **93**, 10,666-10,674.

Gordon, N.D., 1986: The Southern Oscillation and New Zealand weather. *Mon. Weath. Rev.*, **114**, 371-387.

Graham, N.E., and T.P. Barnett, 1987: Sea surface temperature, surface wind divergence, and convection over tropical oceans. *Science*, **238**, 657-659.

Gray, W.M., 1984: Atlantic seasonal hurricane frequency: Part 1: El Niño and 30mb quasi-biennial oscillation influences. *Mon. Weath. Rev.*, **112**, 1649-1683.

Gregory, S., 1989: The changing frequency of drought in India, 1871-1985. *J. Geog.*, **155**, 322-334.

Groisman, P.Ya., and V.V. Koknaeva, 1990: The influence of urbanisation on the estimate of the mean air temperature change in the 20th century over the USSR and USA territories obtained by State Hydrological Institute network. Draft contribution to IPCC WGI, Section 7.

Grove, A.T., and A. Warren, 1968: Quaternary landforms and climate on the south side of the Sahara. *J. Geog.*, **134**, 194-208.

Grove, J.M., 1988: *The Little Ice Age*. Methuen, London, 498pp.

Haeberli, W., P. Mulle, P. Alean and H. Bosch, 1989: Glacier changes following the Little Ice Age - a survey of the international data bases and its perspectives. In: J. Oerlemans, (Ed.), *Glacier Fluctuations and Climate Change*, 77-101.

Hammer, C., 1977: Past volcanism revealed by Greenland Ice Sheet impurities. *Nature*, **270**, 482-486.

Hansen, J.E., A. Lacis, D. Rind, G. Russell, P. Stone, I. Fung, R. Rudy and J Lerner, 1984: Climate sensitivity: analysis of feedback mechanisms. In *Climate Processes and Climate Sensitivity*, J.E. Hansen and T. Takahashi (Eds), *Geophys. Monogr.*, **29**, pp viii + 368, Washington.

Hansen, J., and S. Lebedeff, 1987: Global trends of measured surface air temperature. *J. Geophys. Res.*, **92**, 13345-13372.

Hansen, J., and S. Lebedeff, 1988: Global surface temperatures: update through 1987. *Geophys. Res. Letters.*, **15**, 323-326.

Hastenrath, S.L., 1988: *Climate and Circulation in the Tropics.* Second edition Reidel, Dordrecht, 455pp.

Helland-Hansen B., and F. Nansen, 1920: Temperature variations in the North Atlantic Ocean and in the atmosphere: introductory studies on the cause of climatological variations. (Translated from German). *Smithsonian Misc.*, Coll., **70**(4), 408pp.

Henderson-Sellers, A., 1986: Cloud changes in a warmer Europe. *Clim. Change*, **8**, 25-52.

Henderson-Sellers, A., 1989: North American total cloud amount variations this century. *Glob. and Planet. Change*, **1**, 175-194.

Hense, A., P. Krahe and H. Flohn, 1988: Recent fluctuations of tropospheric temperature and water vapour content in the tropics. *J. Meteor. Atmos. Phys.*, **38**, 215-227.

Huntley, B., and I.C. Prentice, 1988: July temperatures in Europe from pollen data, 6000 years before present. *Science*, **241**, 687-690.

Imbrie, J., and K.P. Imbrie, 1979: *Ice Ages - solving the mystery.* Macmillan, London. 224pp.

James, R.W., and P.T. Fox, 1972: Comparative sea-surface temperature measurements. Marine Science Affairs Report No 5, WMO 336, 27pp.

Jones, P.D., 1988: Hemispheric surface air temperature variations: recent trends and an update to 1987. *J. Clim.*, **1**, 654-660.

Jones, P.D., 1989: The influence of ENSO on global temperatures. *Climate Monitor*, **17**, 80-89.

Jones, P.D., P.M. Kelly, G.B. Goodess and T.R. Karl, 1989: The effect of urban warming on the Northern Hemisphere temperature average. *J. Clim.*, **2**, 285-290.

Jones, P.D., S.C.B. Raper, R.S. Bradley, H.F. Diaz, P.M. Kelly and T.M.L. Wigley, 1986a: Northern Hemisphere surface air temperature variations, 1851-1984. *J. Clim. Appl. Met.*, **25**, 161-179.

Jones, P.D., S.C.B. Raper, R.S. Bradley, H.F. Diaz, P.M. Kelly and T.M.L. Wigley, 1986b: Southern Hemisphere surface air temperature variations, 1851-1984. *J. Clim. Appl. Met.*, **25**, 1213-1230.

Jones, P.D., and T.M.L. Wigley, 1990: Satellite data under scrutiny. *Nature*, **344**, 711.

Karl, T.R., G. Kukla and J. Gavin, 1984: Decreasing diurnal temperature range in the United States and Canada from 1941-1980. *J. Clim. Appl. Met.*, **23**, 1489-1504.

Karl, T.R., C.N. Williams Jr and P.J. Young, 1986: A model to estimate the time of observation bias associated with monthly mean maximum, minimum and mean temperatures for the United States. *J. Clim. Appl. Met.*, **25**, 145-160.

Karl, T.R., and C.N. Williams, Jr., 1987: An approach to adjusting climatological time series for discontinuous inhomogeneities. *J. Clim. Appl. Met.*, **26**, 1744-1763.

Karl, T.R., 1988: Multi-year fluctuations of temperature and precipitation: The gray areas of climate change. *Clim. Change*, **12**, 179-197.

Karl, T.R., H.F. Diaz and G. Kukla, 1988: Urbanization: its detection and effect in the United States climate record. *J. Clim.*, **1**, 1099-1123.

Karl, T.R., J.D. Tarpley, R.G. Quayle, H.F. Diaz, D.A. Robinson and R.S. Bradley, 1989: The recent climate record: What it can and cannot tell us. *Rev. Geophys.*, **27**, 405-430.

Karl, T.R., and P.M. Steurer, 1990: Increased cloudiness in the United States during the first half of the twentieth century: Fact or fiction. In Review, *Geophys. Res. Lett.*

Karl, T.R., and P.D. Jones, 1990: Reply to comments on "Urban bias in area-averaged surface temperature trends". *Bull. Amer. Met. Soc.*, **71** , (in press).

Karoly, D.J., 1989: Northern Hemisphere temperature trends: A possible greenhouse gas effect? *Geophys. Res. Lett.*, **16**, 465-468.

Kleshchenko, L.K., V.T. Radiuhin and R.N. Khostova, 1988: On the comparison of Northern Hemisphere grid point mean monthly air temperature archives. Proceedings of VNIIGMI-WDC, 153, (in Russian).

Lachenbruch, A.H., and B. Vaughn Marshall, 1986: Changing climate: geothermal evidence from permafrost in the Alaskan Arctic. *Science*, **234**, 689-696.

Lamb, H.H., 1977: *Climate Present, Past and Future.* Vol. 2, Methuen.

Lamb, H.H., 1988: Climate and life during the Middle Ages, studied especially in the mountains of Europe. In: *Weather, Climate and Human Affairs.* Routledge, London, 40-74.

Legates, D.R., and C.J. Willmott, 1990: Mean sesonal and spatial variability in gauge corrected, global precipitation. *Int. J. Climatol.*, **10**, 111-127.

Levitus, S., 1989a: Interpentadal variability of temperature and salinity at intermediate depths of the North Atlantic Ocean, 1970-74 versus 1955-59. *J. Geophys. Res. Oceans*, **94**, 6091-6131.

Levitus, S., 1989b: Interpentadal variability of salinity in the upper 150m of the North Atlantic Ocean, 1970-74 versus 1955-59. *J. Geophys. Res. Oceans*, **94**, (in press).

Levitus, S., 1989c: Interpentadal variability of temperature and salinity in the deep North Atlantic Ocean, 1970-74 versus 1955-59. *J. Geophys. Res. Oceans*, **94**, 16125-16131.

Levitus, S., 1989d: Interpentadal variability of steric sea level and geopotential thickness of the North Atlantic Ocean, 1970-74 versus 1955-59. *J. Geophys. Res. Oceans*, **94**, (in press).

Li, C.Y., 1985: El Niño and typhoon activities in the western Pacific. *Kexue Tongbao*, **30**, 1087-1089.

Lindzen, R.S., 1990: Some coolness concerning global warming. *Bull. Am. Met. Soc.*, **71**, 288-299.

Lozhkin A.B., and L.H. Vazhenin, 1987: The features of the plant cover development in the Kolyma lowland in the Early Holocene. In: *The Quaternary Period in the North-East of Asia.* Magadan.

Madden R.A., and J. Williams, 1978: The correlation between temperature and precipitation in the United States and Europe. *Mon. Weath. Rev.*, **106**, 142-147.

Makarevitch, K.G., and O.V. Rototaeva, 1986: Present-day fluctuations of mountain glaciers in the Northern Hemisphere. Data of Glaciological Studies, Publ. No. 57, Soviet Geophysical Committee, Academy of Sciences of the USSR, Moscow, 157-163.

Makrogiannis, T.J., A.A. Bloutsos and B.D. Giles, 1982: Zonal index and circulation change in the North Atlantic area, 1873-1972. *J. Climatol.*, **2**, 159-169.

Mass, C.F., and D.A. Portman, 1989: Major volcanic eruptions and climate: a critical evaluation. *J. Clim.*, **2**, 566-593.

Maury, M.F., 1858: Explanations and sailing directions to accompany the wind and current charts. Vol 1, pp383 +51 plates. W.A. Harris, Washington, DC.

McGuffie, K., and A. Henderson-Sellers, 1989a: Almost a century of "imaging" clouds over the whole-sky dome, *Bull. Amer. Met. Soc.*, **70**, 1243-1253.

McGuffie, K., and A. Henderson-Sellers, 1989b: Is Canadian cloudiness increasing?, *Atmos. Ocean*, **26**, 608-633.

Mechcherskaya, A.V., and V.G. Blazhevich, 1977: Catalogues of precipitation and temperature anomalies for basic agricultural regions of the southern part of the European

territory of the USSR, Northern Kazakhstan and West Siberia. *Meteorol. i. Gidrolog.*, **9**, 76-84.

Michaels, P.T., 1990: Regional 500mb heights prior to the radiosonde era. *Theoret. and Appl. Climat.*, (in press)

Miller, G.H., 1976: Anomalous local glacier activity, Baffin Island, Canada: palaeo-climatic implications. *Geology*, **4**, 502-504.

McLaren, A.S., 1989: The under-ice thickness distribution of the Arctic Basin as recorded in 1958 and 1970. *J Geophys. Res.*, **94**, 4971-4983.

Moses, T., G.N. Kiladis, H.F. Diaz and R.G. Barry, 1987: Characteristics and frequency of reversals in mean sea level pressure in the North Atlantic sector and their relationship to long term temperature trends. *J. Climatol.*, **7**, 13-30.

Mysak, L.A., and D.A. Manak, 1989: Arctic sea-ice extent and anomalies, 1953-1984. *Atmosphere-Ocean*, **27**, 376-405.

Namias, J., 1989: Cold winters and hot summers. *Nature*, **338**, 15-16.

Nash, J., and F.J. Schmidlin, 1987: WMO International Radiosonde Comparison (UK 1984, USA 1985). Final Report. WMO Instruments and Observing Methods Report No. 30. WMO/TD No. 195, 103pp.

Newell, R.E., and J.W. Kidson, 1984: African mean wind changes between Sahelian wet and dry periods. *J. Climatol.*, **4**, 27-33.

Newell, N.E., R.E. Newell, J. Hsiung and Z.-X. Wu, 1989: Global marine temperature variation and the solar magnetic cycle. *Geophys. Res. Lett.*, **16**, 311-314.

Nicholls, N., 1984: The Southern Oscillation, sea surface temperature, and interannual fluctuations in Australian tropical cyclone activity. *J. Climatol.*, **4**, 661-670.

Nicholson, S.E., 1989: African drought: characteristics, causal theories and global teleconnections. *Geophys. Monogr.*, **52**, 79-100.

Nitta, T., and S. Yamada, 1989: Recent warming of tropical sea surface temperature and its relationship to the Northern Hemisphere circulation. *J. Met. Soc. Japan*, **67**, 375-383.

Oerlemans, J., 1988: Simulation of historical glacier variations with a simple climate-glacier model. *J. Glaciol.*, **34**, 333-341.

Ogallo, L.J., 1989: The spatial and temporal patterns of the East African seasonal rainfall derived from principal component analysis. *Int. J. Climatol.*, **9**, 145-167.

Oort, A.H., Y.-H Pan, R.W. Reynolds and C.F. Ropelewski, 1987: Historical trends in the surface temperature over the oceans based on the COADS. *Clim. Dynam.*, **2**, 29-38.

Pan, Y.-H., and A.H. Oort, 1983: Global climate variations connected with sea surface temperature anomalies in the eastern equatorial Pacific Ocean for the 1958-1973 period. *Mon. Weath. Rev.*, **111**, 1244-1258.

Parker, D.E., 1985: On the detection of temperature changes induced by increasing atmospheric carbon dioxide. *Q. J. Roy. Met. Soc.*, **111**, 587-601.

Parker, D.E., 1990: Effects of changing exposure of thermometers at land stations. Draft contribution to IPCC WGI, Section 7.

Parthasarathy, B., and G.B. Pant, 1985: Seasonal relationships between Indian summer monsoon rainfall and the Southern Oscillation. *J. Climatol.*, **5**, 369-378.

Patzelt, G., 1989: Die 1980er-Vorstossperiode der Alpengletcher. *Oesterreicher Alpenverein*, **44**, 14-15.

Peshy, A. and A.A. Velichko, 1990: (Eds) Palaeo-climatic and palaeo-environment reconstruction from the late Pleistocene to Holocene. In: *Palaeo-geographic Atlas for the Northern Hemisphere*. Budapest (In press)

Pittock, A.B., 1975: Climatic change and patterns of variation in Australian rainfall. *Search*, **6**, 498-504.

Pittock, A.B., 1983: Solar variability, weather and climate: an update. *Q. J. Roy. Met. Soc.*, **109**, 23-55.

Porter, S.C., 1986: Pattern and forcing of Northern Hemisphere glacier variations during the last millennium. *Quat. Res.*, **26(1)**, 27-48.

Quinn, W.H., V.T. Neal and S.E. Antunez de Mayolo, 1987: El Niño occurrences over the past four and a half centuries. *J. Geophys. Res.*, **92**, 14449-14462.

Ramanathan, V., R.D. Cess, E.F. Harrison, P. Minnis, B.R. Barkstrom, E. Ahmad and D. Hartman, 1989: Cloud-radiative forcing and climate: results from the Earth Radiation Budget Experiment. *Science*, **243**, 57-63.

Raper, S.C.B, 1990: Observational data on the relationships between climatic change and the frequency and magnitude of severe tropical storms. In: *Climate and sea level change: Observations, Projections, Implications*. R.A Warrick and T.M.L. Wigley (Eds), Cambridge University Press (in press).

Rasmusson, E.M., and T.H. Carpenter, 1982: Variations in tropical sea surface temperature and surface wind fields associated with the Southern Oscillation/El Niño. *Mon. Weath. Rev.*, **110**, 354-384.

Revell, C.G., and S.W. Gaulter, 1986: South Pacific tropical cyclones and the Southern Oscillation. *Mon. Weath. Rev.* **114**, 1138-1145.

Rind, D., D. Peteet, W. Broecker, A. McIntyre and W.F. Ruddiman, 1986: The impact of cold North Atlantic sea surface temperatures on climate: implications for the Younger Dryas cooling (11-10K). *Clim. Dynam.*, **1**, 3-33.

Robinson, D.A., and K.F. Dewey, 1990: Recent secular variations in the extent of Northern Hemisphere snow cover. Submitted to *Nature*.

Roemmich, D., and C. Wunsch, 1984: Apparent changes in the climatic state of the deep North Atlantic Ocean. *Nature*, **307**, 447-450.

Rogers, J.C., 1985: Atmospheric circulation changes associated with the warming over the Northern North Atlantic in the 1920s. *J. Clim. Appl. Met.*, **24**, 1303-1310.

Ropelewski, C.F., 1983: Spatial and temporal variations in Antarctic sea ice (1973-1982). *J. Clim. Appl. Met.*, **22**, 470-473.

Ropelewski, C.F., and M.S. Halpert, 1987: Global and regional scale precipitation patterns associated with the El Niño/Southern Oscillation. *Mon. Weath. Rev.*, **115**, 1606-1626.

Ropelewski, C.F., and P.D. Jones, 1987: An extension of the Tahiti-Darwin Southern Oscillation Index. *Mon. Weath. Rev.*, **115**, 2161-2165.

Rossby, C.G., 1959: Current problems in meteorology. In: *The Atmosphere and Sea in Motion*, Rockefeller Institute Press, New York, pp9-50.

Rossow, W.B., 1989: Measuring cloud properties from space: a review. *J. Clim.*, **2**, 201-213.

Salinger, M.J., 1981: New Zealand climate: The instrumental record. Victoria University of Wellington, 327pp.

Salinger, M.J., 1989: New Zealand climate: from ice age to present. Environmental Monitoring in New Zealand, 32-40.

Schönwiese, C.-D., and W. Birrong, 1990: European precipitation trend statistics 1851-1980 including multivariate assessments of the anthropogenic CO_2 signal. *Z. Meteorol.*, **40**, 92-98.

Scialdone, J., and A. Robock, 1987: Comparisons of northern hemisphere snow cover data sets. *J. Clim. Appl. Met.*, **26**, 53-68.

SCOPE 29, 1986: *The greenhouse effect, climatic change and ecosystems*. B. Bolin, B.R. Doos, J. Jager and R.W. Warrick (Eds). Wiley, Chichester.

Sellers, W.D., and W. Liu, 1988: Temperature pátterns and trends in the upper troposphere and lower stratosphere. *J. Clim.*, **1**, 573-581.

Semazzi, F.H.M., V.Mehta and Y.C. Sud, 1988: An investigation of the relationship between Sub-saharan rainfall and global sea surface temperatures. *Atmosphere-Ocean*, **26**, 118-138.

Sevruk, B., 1982: Methods of correcting for systematic error in point precipitation measurement for operational use. *Operational Hydrology Report* No 21, WMO No 589.

Singh, G., R.D.Joshi, S.K. Chopra and A.B. Singh, 1974: Late quaternary history of vegetation and climate of the Rajasthan Desert, India. *Phil. Trans. Roy. Soc. Lond.*, Series B, **267**, 179-195.

Sommerville, R.J., and L.A. Remer, 1984: Cloud optical thickness feedbacks in the CO_2 climate problem, *J. Geophys. Res.*, **89**, 9668-9672.

Spencer, R.W., and J.R. Christy, 1990: Precise monitoring of global temperature trends from satellites. *Science*, **247**, 1558-1562.

Street-Perrott, F.A., and S.P. Harrison, 1985: Lake levels and climate reconstruction. In: *Palaeo-climate Analysis and Modeling*, (Ed.) A. Hecht. J. Wiley, New York, pp291-340.

Street-Perrott, F.A., and R.A. Perrott, 1990: Abrupt climate fluctuations in the tropics: The influence of the Atlantic circulation. *Nature*, **343**, 607-612.

Sturman, A.P., and M.R. Anderson, 1985: A comparison of Antarctic sea ice data sets and inferred trends in ice area. *J. Clim. Appl. Met.*, **24**, 275-280.

Swain, A.M., J.E. Kutzbach and S. Hastenrath, 1983: Estimates of Holocene precipitation for Rajasthan, India, based on pollen and lake level data. *Quatern. Res.*, **19**, 1-17.

Sze, G., F. Drake, M. Desbois and A. Henderson-Sellers, 1986: Total and low cloud amounts over France and Southern Britain in the summer of 1983: comparison of surface-observed and satellite-retrieved values. *Int. J. Rem. Sens.*, **7**, 1031-1050.

Toynbee, H., 1874: Appendix D of Report of Proc. Conf. Maritime Meteorology, London, 1874. HMSO.

Trenberth, K.E., 1976: Fluctuations and trends in indices of the Southern Hemispheric circulation. *Q.J. Roy. Met. Soc.*, **102**, 65-75.

Trenberth, K.E., G.W. Branstator and P.A. Arkin, 1988: Origins of the 1988 North American Drought. *Science*, **242**, 1640-1645.

Trenberth, K.E., and D.J. Shea, 1987: On the evolution of the Southern Oscillation. *Mon. Weath. Rev.*, **115**, 3078-3096.

Trenberth, K.E., J.R. Christy and J.G. Olson, 1987: Global atmospheric mass, surface pressure and water vapour variations. *J. Geophys. Res.*, **92**, 14,815-14,826.

Trenberth, K.E., 1990: Recent observed interdecadal climate changes in the Northern Hemisphere. *Bull. Am. Met. Soc.*, **71**, (in press).

van Heerden, J., D.E. Terblanche and G.C. Schultze, 1988: The Southern Oscillation and South African summer rainfall. *J. Climatol.*, **8**, 577-598.

van Loon, H. and J.C. Rogers, 1978: The seesaw in winter temperature between Greenland and northern Europe. Part 1: General description. *Mon. Weath. Rev.*, **106**, 296-310.

van Loon, H., and J. Williams, 1976: The connection between trends of mean temperature and circulation at the surface. Pt. I: Winter. *Mon. Weath. Rev.*, **104**, 365-380.

van Loon, H., and K. Labitzke, 1988: Association between the 11-year solar cycle, the QBO, and the atmosphere. Part II: Surface and 700mb in the Northern Hemisphere winter. *J. Climate*, **1**, 905-920.

Varushchenko S.I., A.N. Varushchenko and R.K. Klige, 1980: *Changes in the Caspian Sea regime and closed water bodies in palaeotimes.* M. Nauka, 238pp.

Velichko A.A., M.P. Grichuk, E.E. Gurtovaya, E.M. Zelikson and O.K. Borisova, 1982: Palaeo-climatic reconstructions for the optimum of the Mikulino Interglacial in Europe. Izv. Acad. Sci. USSR, *Ser. Geogr.*, **N1**, 15.

Velichko A.A., M.P. Grichuk, E.E. Gurtovaya and E.M. Zelikson, 1983: Palaeo-climate of the USSR territory during the optimum of the last interglacial (Mikulino). Izv. Acad. Sci. USSR, *Ser. Geogr.* **N1**, 5-18.

Velichko A.A., M.S. Barash, M.P. Grichuk, Ye.Ye. Gurtoyova and E.M. Zelikson, 1984: Climate of the Northern Hemisphere in the last interglacial (Mikulino). Izv. Acad. Sci. USSR, *Ser. Geogr.* **N6**, 30-45.

Vinnikov, K.Ya., P. Ya. Groisman, and K.M. Lugina, 1990: The empirical data on modern global climate changes (temperature and precipitation). *J. Clim.*, (in press).

Vinnikov, K.Ya., P.Ya. Groisman, K.M. Lugina, and A.A. Golubev, 1987: Variations in Northern Hemisphere mean surface air temperature over 1881-1985. *Meteorology and Hydrology*, **1**, 45-53 (In Russian).

Wadhams, P., M.A. Lange and S.F. Ackley, 1987: The ice thickness distibution across the Atlantic sector of the Antarctic Ocean in midwinter. *J. Geophys. Res.*, **92**, 14535-14552.

Wadhams, P., 1989: Sea-ice thickness distribution in the Trans Polar Drift Stream. Rapp. P-v Reun. Cons. *Int. Explor. Mer*, **188**, 59-65.

Wadhams, P., 1990: Evidence for thinning of the Arctic sea ice cover north of Greenland. *Nature* , (in press).

Walker, G.T., and E.W. Bliss, 1932; World Weather Vol. *Mem. Roy. Met. Soc.*, **4**, 53-84.

Wallace, J.M., and D.S. Gutzler, 1981: Teleconnections in the geopotential height field during the Northern Hemisphere winter. *Mon. Weath. Rev.*, **109**, 784-812.

Wallen, C.C., 1986: Impact of present century climate fluctuations in the Northern Hemisphere. *Geograf. Ann.*, Ser. A, **68**, 245-278.

Walsh, J.E. and J.E. Sater, 1981: Monthly and seasonal variability in the ocean-ice-atmosphere systems of the North Pacific and the North Atlantic. *J. Geophys. Res.*, **86**, 7425-7445.

Wang, W-C., Z. Zeng, and T.R. Karl, 1990: Urban warming in China. Draft contribution to IPCC WGI, Section 7.

Warren, S.G., C.J. Hahn, J. London, R.M. Chervin and R.L. Jenne, 1988: Global distribution of total cloud cover and cloud type amounts over the ocean. NCAR Tech. Note TN 317+STR, 42pp.

Webb, T. III, P.J. Bartlein and J.E. Kutzbach, 1987: Climatic change in eastern North America during the past 18000 years: comparisons of pollen data with model results. In: *The Geology of North America*, Vol K-3; North America and Adjacent Oceans During the Last Deglaciation (eds. W.F. Ruddiman and H.E. Wright). Geol. Soc. of America, Boulder, pp447-462.

Wiesnet, D.R., C.F. Ropelewski, G.J. Kukla, and D.A. Robinson, 1987: A discussion of the accuracy of NOAA satellite-derived global seasonal snow cover measurements. In: B.E. Goodison, R.G. Barry, and J. Dozier, (Eds), *Large-Scale Effects of Seasonal Snow Cover*, IAHS Publ. 166, pp291-304. IAHS Press, Wallingford, UK.

Wigley, T.M.L., and P.M. Kelly, 1989: Holocene climatic change, 14C wiggles and variations in solar irradiance. *Phil. Trans. Roy. Soc. Lond.*, (in press).

Wolter, K., 1989: Modes of tropical circulation, Southern Oscillation, and Sahel rainfall anomalies. *J. Clim.*, **2**, 149-172.

Wolter, K. and S. Hastenrath, 1989: Annual Cycle and Long-Term Trends of Circulation and Climate Variability over the Tropical Oceans. *J. Clim.*, **2**, 1329-1351.

Wood, F.B., 1988: Global alpine glacier trends, 1960s to 1980s. *Arct. Alp. Res.*, **20**, 404-413.

Woodruff, S.D., R.J. Slutz, R.J. Jenne, and P.M. Steurer, 1987: A Comprehensive Ocean-Atmosphere Data Set. *Bull. Am. Met. Soc.*, **68**, 1239-1250.

Woodruff, S.D., 1990: Preliminary comparison of COADS (US) and MDB (UK) ship reports. Draft contribution to IPCC WG1, Section 7.

Yamamoto, R.M., R.M. Hoshaio and N. Nishi, 1990a: An estimate of systematic errors in air temperature data in COADS. *J. Clim.* , (in press).

Yamamoto, R.M., R.M. Hoshaio, N. Nishi and Y Kakuno, 1990b: Interdecadal warming over the oceans. *J. Clim.* , (in press)

Yoshino, M.M., 1978: Regionality of climatic change in Monsoon Asia. In: *Climatic change and food production.* K. Takahashi and M.M. Yoshino (Eds), Univ Tokyo Press, Tokyo, 331-342.

Yoshino, M.M., 1990: Climate-related impact research activities-Japan. Int. Networkshop, Boulder, Co, March 14-17, 1989, (in press).

Yoshino, M.M., and K. Urushibara, 1978: Palaeoclimate in Japan since the Last Ice Age. *Climatological Notes*, **22**, 1-24.

Zebiak, S.E., and M.A. Cane, 1987: A model El Niño--Southern Oscillation. *Mon. Weath. Rev.*, **115**, 2262-2278.

Zhang, C., J. Yang and Lin, X., 1983: The stages nature of long-term variations of the atmospheric centres of action. *Sci. Atmos. Sinica*, **7**, 364-374.

Zhang, Y., and W.-C. Wang, 1990: The surface temperature in China during the mid-Holocene. In: *Zhu Kezhen Centennial Memorial Collections.* Science Press, Beijing, (in press).

Zubakov, V.A., and I.I. Borzenkova, 1990: *Global Climate during the Cenozoic.* Amsterdam, Elsevier, 453pp.

8

Detection of the Greenhouse Effect in the Observations

T.M.L. WIGLEY, T.P. BARNETT

Contributors:
T.L. Bell; P. Bloomfield; D. Brillinger; W. Degefu; C.K. Folland; S. Gadgil; G.S. Golitsyn; J.E. Hansen; K. Hasselmann; Y. Hayashi; P.D. Jones; D.J. Karoly; R.W. Katz; M.C. MacCracken; R.L. Madden; S. Manabe; J.F.B. Mitchell; A.D. Moura; C. Nobre; L.J. Ogallo; E.O. Oladipo; D.E. Parker; A.B. Pittock; S.C.B. Raper; B.D. Santer; M.E. Schlesinger; C.-D. Schönwiese; C.J.E. Schuurmans; A. Solow; K.E. Trenberth; K.Ya. Vinnikov; W.M. Washington; T. Yasunari; D. Ye; W. Zwiers.

CONTENTS

EXECUTIVE SUMMARY

Global-mean temperature has increased by 0.3-0.6°C over the past 100 years. The magnitude of this warming is broadly consistent with the theoretical predictions of climate models, but it remains to be established that the observed warming (or part of it) can be attributed to the enhanced greenhouse effect. This is the detection issue.

If the sole cause of the warming were the Man-induced greenhouse effect, then the implied climate sensitivity would be near the lower end of the accepted range of model predictions. Natural variability of the climate system could be as large as the changes observed to date, but there are insufficient data to be able to estimate its magnitude or its sign. If a significant fraction of the observed warming were due to natural variability, then the implied climate sensitivity would be even lower than model predictions. However, it is possible that a larger greenhouse warming has been offset partially by natural variability and other factors, in which case the climate sensitivity could be at the high end of model predictions.

Global-mean temperature alone is an inadequate indicator of greenhouse-gas-induced climatic change. Identifying the causes of any global-mean temperature change requires examination of other aspects of the changing climate, particularly its spatial and temporal characteristics. Currently, there is only limited agreement between model predictions and observations. Reasons for this include the fact that climate models are still in an early stage of development, our inadequate knowledge of natural variability and other possible anthropogenic effects on climate, and the scarcity of suitable observational data, particularly long, reliable time series. An equally important problem is that the appropriate experiments, in which a realistic model of the global climate system is forced with the known past history of greenhouse gas concentration changes, have not yet been performed.

Improved prospects for detection require a long-term commitment to comprehensively monitoring the global climate system and potential climate forcing factors, and to reducing model uncertainties. In addition, there is considerable scope for the refinement of the statistical methods used for detection. We therefore recommend that a comprehensive detection strategy be formulated and implemented in order to improve the prospects for detection. This could be facilitated by the setting up of a fully integrated international climate change detection panel to coordinate model experiments and data collection efforts directed towards the detection problem.

Quantitative detection of the enhanced greenhouse effect using objective means is a vital research area, because it is closely linked to the reduction of uncertainties in the magnitude of the effect and will lead to increased confidence in model projections. The fact that we are unable to reliably detect the predicted signals today does not mean that the greenhouse theory is wrong, or that it will not be a serious problem for mankind in the decades ahead.

8.1 Introduction

8.1.1 The Issue

This chapter addresses the question "Have we detected the greenhouse effect?", or, stated more correctly, have we detected changes in climate that can, with high statistical confidence, be attributed to the enhanced greenhouse effect associated with increasing trace gas concentrations? It is important to answer this question, because detecting the enhanced greenhouse effect will provide direct validation of models of the global climate system. Until we can identify aspects of greenhouse-gas-induced changes in the observed climate record with high confidence, there will always be doubts about model validity, and hence about even the most general predictions of future climatic change. Even when detection has occurred, uncertainties regarding the magnitude and spatial details of future changes will still remain.

Previous reviews of the greenhouse problem (N.R.C., 1983; MacCracken and Luther, 1985; Bolin et al., 1986) have also addressed the detection issue. They have concluded that the enhanced greenhouse effect has not yet been detected unequivocally in the observational record. However, they have also noted that the global-mean temperature change over the past 100 years is consistent with the greenhouse hypothesis, and that there is no convincing observational evidence to suggest that the model-based range of possible climate sensitivity [1] values is wrong. The purpose of the present review is to re-evaluate these conclusions in the light of more recent evidence.

8.1.2 The Meaning Of "Detection"

The word "detection" has been used to refer to the identification of a significant change in climate (such as an upward trend in global-mean temperature). However, identifying a change in climate is not enough for us to claim that we have detected the enhanced greenhouse effect, even if statistical methods suggest that the change is statistically significant (i.e., extremely unlikely to have occurred by chance). To claim detection in a useful and practical way, we must not only identify a climatic change, but we must **attribute** at least part of such a change to the enhanced greenhouse effect. It is in this stricter sense that the word "detection" is used here. Detection requires that the observed changes in climate are in accord with detailed model predictions of the enhanced greenhouse effect, demonstrating that we understand the cause or causes of the changes.

[1] *Climate sensitivity is defined here as the equilibrium global-mean temperature change for a CO_2 doubling (ΔT_{2x}). ΔT_{2x} is thought to lie in the range 1.5°C to 4.5°C (see Section 5).*

To illustrate this important difference, consider changes in global-mean temperature. A number of recent analyses have claimed to show a statistically significant warming trend over the past 100 years (Hansen et al., 1988; Tsionis and Elsner, 1989; Wigley and Raper, 1990). But, is this warming trend due to the enhanced greenhouse effect? We have strong evidence that changes of similar magnitude and rate have occurred prior to this century (see Section 7). Since these changes were certainly **not** due to the enhanced greenhouse effect, it might be argued that the most recent changes merely represent a natural, long-time-scale fluctuation.

The detection problem can be conveniently described in terms of the concepts of "signal" and "noise" (Madden and Ramanathan, 1980). Here, the signal is the predicted, time-dependent climate response to the enhanced greenhouse effect. The noise is any climatic variation that is not due to the enhanced greenhouse effect [2]. Detection requires that the observed signal is large relative to the noise. In addition, in order to be able to attribute the detected signal to the enhanced greenhouse effect, it should be one that is specific to this particular cause. Global-mean warming, for example, is not a particularly good signal in this sense, because there are many possible causes of such warming.

8.1.3 Consistency Of The Observed Global-Mean Warming With The Greenhouse Hypothesis

Global-mean temperature has increased by around 0.3-0.6°C over the past 80-100 years (see Section 7). At the same time, greenhouse gas concentrations have increased substantially (Section 1). Is the warming consistent with these increases? To answer this question, we must model the effects of these concentration changes on global-mean temperature and compare the results with the observations. Because of computing constraints and because of the relative inflexibility of coupled ocean-atmosphere GCMs, we cannot use such models for this purpose. Instead, we must use an upwelling-diffusion climate model to account for damping or lag effect of the oceans (see Section 6). The response of such a model is determined mainly by the climate sensitivity (ΔT_{2x}), the magnitude of ocean mixing (specified by a diffusion coefficient, K), and the ratio of the temperature change in the regions of sinking water relative to the global-mean change (π). Uncertainties in these parameters can be accounted for by using a range of values.

[2] *Noise, as used here, includes variations that might be due to other anthropogenic effects (see Section 2), and natural variability. "Natural variability" refers to all natural climatic variations that are unrelated to Man's activities, embracing both the effects of external forcing factors (such as solar activity and volcanic eruptions), and internally-generated variability. Uncertainties in the observations also constitute a form of noise.*

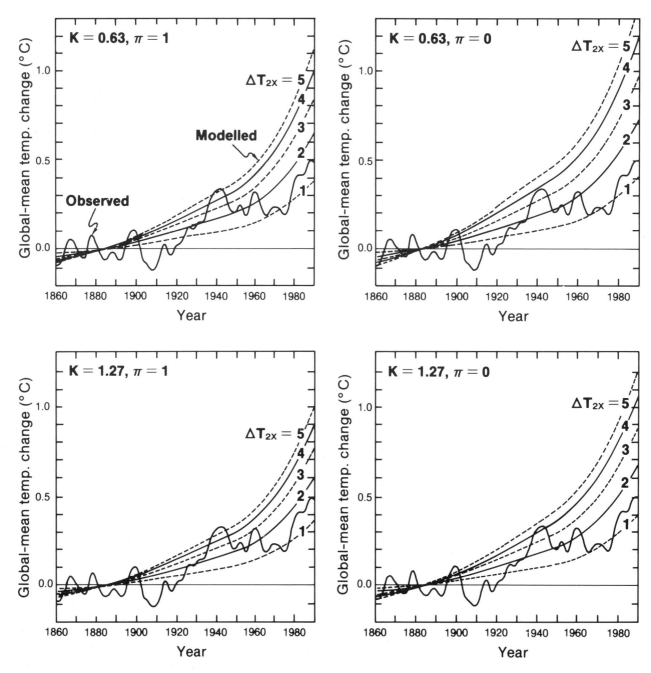

Figure 8.1: Observed global-mean temperature changes (1861-1989) compared with predicted values. The observed changes are as given in Section 7, smoothed to show the decadal and longer time scale trends more clearly. Predictions are based on observed concentration changes and concentration/forcing relationships given in Section 2, and have been calculated using the upwelling-diffusion climate model of Wigley and Raper (1987). To provide a common reference level, modelled and observed data have been adjusted to have zero mean over 1861-1900. To illustrate the sensitivity to model parameters, model results are shown for $\Delta T_{2x} = 1, 2, 3, 4$ and $5°C$ (all panels) and for four K, π combinations. The top left panel uses the values recommended in Section 6 (K = $0.63cm^2sec^{-1}$, $\pi = 1$). Since sensitivity to K is relatively small and sensitivity to π is small for small ΔT_{2x}, the best fit ΔT_{2x} depends little on the choice of K and π.

The model is forced from 1765-1990 using concentration changes and radiative forcing/concentration relationships given in Section 2.

Figure 8.1 compares model predictions for various model parameter values with the observed warming over 1861-1989. The model results are clearly qualitatively consistent with the observations on the century time-scale. Agreement on long time-scales is about all that one might expect. On shorter time-scales, we know that the climate system is subject to internal variability and to a variety of external forcings, which must obscure any response to greenhouse forcing. Although we cannot explain the observed shorter time-scale fluctuations in detail, their magnitude is compatible with our understanding of natural climatic variability. Essentially, they reflect the noise against which the greenhouse signal has to be detected.

While the decadal time-scale noise is clear, there may also be substantial century time-scale noise. This noise makes it difficult to infer a value of the climate sensitivity from Figure 8.1. Internal variability arising from the modulation of random atmospheric disturbances by the ocean (Hasselmann, 1976) may produce warming or cooling trends of up to 0.3°C per century (Wigley and Raper, 1990; see Figure 8.2), while ocean circulation changes and the effects of other external forcing factors such as volcanic eruptions and solar irradiance changes

and/or other anthropogenic factors (see Section 2) could produce trends of similar magnitude. On time-scales of order a decade, some of these (volcanic eruptions, sulphate aerosol derived cloud albedo changes) clearly have a negative forcing effect, while others have uncertain sign. If the net century time-scale effect of all these non-greenhouse factors were close to zero, the climate sensitivity value implied by Figure 8.1 would be in the range 1°C to 2°C. If their combined effect were a warming, then the implied sensitivity would be less than 1°C, while if it were a cooling, the implied sensitivity could be larger than 4°C. The range of uncertainty in the value of the sensitivity becomes even larger if uncertainties in the observed data (Section 7) are accounted for.

From this discussion, one may conclude that an enhanced greenhouse effect could already be present in the climate record, even though it cannot yet be reliably detected above the noise of natural climatic variability. The goal of any detection strategy must be to achieve much more than this. It must seek to establish the credibility of the models within relatively narrow limits and to reduce our uncertainty in the value of the climate sensitivity parameter. In this regard, global-mean temperature alone is an inadequate indicator of greenhouse-gas-induced climatic change. Identifying the causes of any global-mean temperature change requires examination of other aspects of the changing climate, particularly its spatial and temporal characteristics.

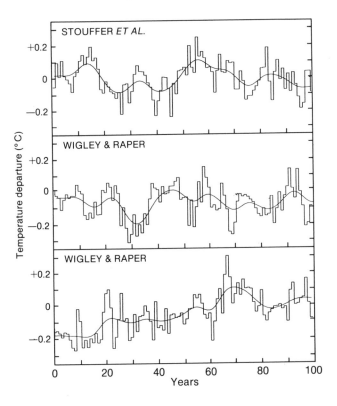

Figure 8.2: Simulated natural variability of global-mean temperature. The upper panel shows results from the 100-year control run with the coupled ocean/atmosphere GCM of Stouffer et al. (1989). These data are also shown in Figure 6.2. The lower two panels are 100-year sections from a 100,000 year simulation using the upwelling-diffusion model employed in Figure 8.1, with the same climate sensitivity as the Stouffer et al. model (ΔT_{2x} = 4°C). The upwelling-diffusion model is forced with random inter-annual radiative changes chosen to match observed inter-annual variations in global-mean temperature (Wigley and Raper, 1990). The consequent low-frequency variability arises due to the modulating effect of oceanic thermal inertia. Most 100-year sections are similar in character to the middle panel, and are qualitatively indistinguishable from the coupled ocean/atmosphere GCM results. However, a significant fraction show century time-scale trends as large or larger than that in the lower panel. Longer GCM simulations may therefore reveal similar century time-scale variability.

8.1.4 Attribution And The Fingerprint Method

Given our rudimentary understanding of the magnitude and causes of low-frequency natural variability, it is virtually impossible to demonstrate a cause-effect relationship with high confidence from studies of a single variable. (However, if the global warming becomes sufficiently large, we will eventually be able to claim detection simply because there will be no other possible explanation). Linking cause and effect is referred to as "attribution". This is the key issue in detection studies - we must be able to attribute the observed changes (or part of them) to the enhanced greenhouse effect. Confidence in the attribution is increased as predictions of changes in various components of the climate system are borne out by the observed data in more and more detail. The method proposed for this purpose is the "fingerprint method"; namely, identification of an observed multivariate signal [3] that has a structure unique to the predicted enhanced greenhouse effect (Madden and Ramanathan, 1980; Baker and Barnett, 1982; MacCracken and Moses, 1982). The

---

[3] *A multivariate signal could be changes in a single climate element (such as temperature) at many places or levels in the atmosphere, or changes in a number of different elements, or changes in different elements at different places.*

current scientific focus in the detection issue is therefore on multivariate or fingerprint analyses. The fingerprint method is essentially a form of model validation, where the perturbation experiment that is being used to test the models is the currently uncontrolled emission of greenhouse gases into the atmosphere. The method is discussed further in Section 8.3. First, however, we consider some of the more general issues of a detection strategy.

8.2 Detection Strategies

8.2.1 *Choosing Detection Variables*

There are many possible climate elements or sets of elements that we could study to try to detect an enhanced greenhouse effect. In choosing the ones to study, the following issues must be considered:

> the strength of the predicted signal and the ease with which it may be distinguished from the noise,
> uncertainties in both the predicted signal and the noise, and
> the availability and quality of suitable observed data.

8.2.1.1 *Signal-to-noise ratios*

The signal-to-noise ratio provides a convenient criterion for ranking different possible detection variables. The stronger the predicted signal relative to the noise, the better the variable will be for detection purposes, all other things being equal. For multivariate signals, those for which the pattern of natural variability is distinctly different from the pattern of the predicted signal will automatically have a high signal-to-noise ratio.

Signal-to-noise ratios have been calculated for a number of individual climate elements from the results of $1 \times CO_2$ and $2 \times CO_2$ equilibrium experiments using atmospheric GCMs coupled to mixed-layer oceans (Barnett and Schlesinger, 1987; Santer et al., 1990; Schlesinger et al., 1990). The highest values were obtained for free troposphere temperatures, near-surface temperatures (including sea-surface temperatures), and lower to middle tropospheric water vapour content (especially in tropical regions). Lowest values were found for mean sea level pressure and precipitation. While these results may be model dependent, they do provide a useful preliminary indicator of the relative values of different elements in the detection context.

Variables with distinctly different signal and noise patterns may be difficult to find (Barnett and Schlesinger, 1987). There are reasons to expect parallels between the signal and the *low-frequency* noise patterns, at least at the zonal and seasonal levels, simply because such characteristics arise through feedback mechanisms that are common to both greenhouse forcing and natural variability.

8.2.1.2 *Signal uncertainties*

Clearly, a variable for which the signal is highly uncertain cannot be a good candidate as a detection variable. Predicted signals depend on the models used to produce them. Model-to-model differences (Section 5) point strongly to large signal uncertainties. Some insights into these uncertainties may also be gained from studies of model results in attempting to simulate the present-day climate (see Section 4). A poor representation of the present climate would indicate greater uncertainty in the predicted signal (e.g., Mitchell et al., 1987). Such uncertainties tend to be largest at the regional scale because the processes that act on these scales are not accurately represented or parameterized in the models. Even if a particular model is able to simulate the present-day climate well, it will still be difficult to estimate how well it can define an enhanced greenhouse signal. Nevertheless, validations of simulations of the present global climate should form at least one of the bases for the selection of detection variables.

A source of uncertainty here is the difference between the results of equilibrium and transient experiments (see Section 6). Studies using coupled ocean-atmosphere GCMs and time-varying CO_2 forcing have shown reduced warming in the areas of deep water formation (i.e., the North Atlantic basin and around Antarctica) compared with equilibrium results (Bryan et al., 1988; Washington and Meehl, 1989; Stouffer et al., 1989). These experiments suggest that the regional patterns of temperature change may be more complex than those predicted by equilibrium simulations. The results of equilibrium experiments must therefore be considered as only a guide to possible signal structure.

The most reliable signals are likely to be those related to the largest spatial scales. Small-scale details may be eliminated by spatial averaging, or, more generally, by using filters that pass only the larger scale (low wave number) components. (Note that some relatively small-scale features may be appropriate for detection purposes, if model confidence is high.) An additional benefit of spatial averaging or filtering is that it results in data compression (i.e., reducing the dimensionality of the detection variable), which facilitates statistical testing. Data compression may also be achieved by using linear combinations of variables (e.g., Bell, 1982, 1986; Karoly, 1987, 1989).

8.2.1.3 *Noise uncertainties*

Since the expected man-made climatic changes occur on decadal and longer time-scales, it is largely the low-frequency characteristics of natural variability that are important in defining the noise. Estimating the magnitude of low-frequency variability presents a major problem because of the shortness and incompleteness of most instrumental records. This problem applies particularly to new satellite-based data sets.

In the absence of long data series, statistical methods may be used to estimate the low frequency variability (Madden and Ramanathan, 1980; Wigley and Jones, 1981), but these methods depend on assumptions which introduce their own uncertainties (Thiébaux and Zwiers, 1984). The difficulty arises because most climatological time-series show considerable persistence, in that successive yearly values are not independent, but often significantly correlated. Serially correlated data show enhanced low-frequency variability which can be difficult to quantify.

As an alternative to statistically-based estimates, model simulations may be used to estimate the low-frequency variability, either for single variables such as global-mean temperature (Robock, 1978; Hansen et al., 1988; Wigley and Raper, 1990) or for the full three-dimensional character of the climate system (using long simulations with coupled ocean-atmosphere GCMs such as that of Stouffer et al., 1989). Internally-generated changes in global-mean temperature based on model simulations are shown in Figure 8.2.

8.2.1.4 Observed data availability

The final, but certainly not the least important factor in choosing detection variables is data availability. This is a severe constraint for at least two reasons, the definition of an evolving signal and the quantification of the low-frequency noise. Both require adequate spatial coverage and long record lengths, commodities that are rarely available. Even for surface variables, global-scale data sets have only recently become available (see Section 7). Useful upper air data extend back only to the 1950s, and extend above 50mb (i.e., into the lower stratosphere) only in recent years. Comprehensive three-dimensional coverage of most variables has become available only recently with the assimilation of satellite data into model-based analysis schemes. Because such data sets are produced for meteorological purposes (e.g., model initialisation), not for climatic purposes such as long-term trend detection, they contain residual inhomogeneities due to changes in instrumentation and frequent changes in the analysis schemes. In short, we have very few adequately observed data variables with which to conduct detection studies. It is important therefore to ensure that existing data series are continued and observational programmes are maintained in ways that ensure the homogeneity of meteorological records.

8.2.2 Univariate Detection Methods

A convenient way to classify detection studies carried out to date is in terms of the number of elements (or variables) considered, i.e., as univariate or multivariate studies. The key characteristic of the former is that the detection variable is a single time series. Almost all published univariate studies have used temperature averaged over a large area as the detection parameter. A central problem in such studies is defining the noise level, i.e., the low-frequency variability (see 8.2.1.3).

There have been a number of published variations on the univariate detection theme. One such has been referred to as the noise reduction method. In this method, the effects of other external forcing factors such as volcanic activity and/or solar irradiance changes or internal factors such as ENSO are removed from the record in some deterministic (i.e., model-based) or statistical way (Hansen et al., 1981; Gilliland, 1982; Vinnikov and Groisman, 1982; Gilliland and Schneider, 1984; Schönwiese, 1990). This method is fraught with uncertainty because the history of past forcings is not well known. There are no direct observations of these forcing factors, and they have been inferred in a variety of different ways leading to a number of different forcing histories (Wigley et al., 1985; Schönwiese, 1990). The noise reduction principle, however, is important. Continued monitoring of any of the factors that might influence global climate in a deterministic way (solar irradiance, stratospheric and tropospheric aerosol concentrations, etc.) can make a significant contribution to facilitating detection in the future.

As noted above in the case of global-mean temperature, univariate detection methods suffer because they consider change in only one aspect of the climate system. Change in a single element could result from a variety of causes, making it difficult to attribute such a change specifically to the enhanced greenhouse effect. Nevertheless, it is useful to review recent changes in a number of variables in the light of current model predictions (see also Wood, 1990).

8.2.3 Evaluation Of Recent Climate Changes

8.2.3.1 Increase of global-mean temperature

The primary response of the climate system to increasing greenhouse gas concentrations is expected to be a global-mean warming of the lower layers of the atmosphere. In Section 8.1.3, the observed global-mean warming of 0.3-0.6°C over the past century or so was compared with model predictions. It was noted that the observed warming is compatible with the enhanced greenhouse hypothesis, but that we could not claim to have detected the greenhouse effect on this basis alone. It was also noted that the directly implied climate sensitivity (i.e., the value of ΔT_{2x}) was at the low end of the expected range, but that the plethora of uncertainties surrounding an empirical estimation of ΔT_{2x} precludes us drawing any firm quantitative conclusions. The observed global warming is far from being a steady, monotonic upward trend, but this does not mean that we should reject the greenhouse hypothesis. Indeed, although our understanding of natural climatic variability is still quite limited, one would certainly expect substantial natural fluctuations to be superimposed on any greenhouse-related warming trend.

8.2.3.2 Enhanced high-latitude warming, particularly in the winter half-year

Most model simulations suggest that the warming north of 50°N in the winter half of the year should be enhanced due to feedback effects associated with sea-ice and snow cover (Manabe and Stouffer, 1980; Robock, 1983; Ingram et al., 1989). In the Southern Hemisphere, results from simulations with atmospheric GCMs coupled to ocean GCMs do not show this enhancement (Bryan et al., 1988; Washington and Meehl, 1989; Stouffer et al., 1989). Figures 8.3 and 8.4 show observed annual and winter temperature changes for various latitude bands. Over the past 100 years, high northern latitudes have warmed slightly more than the global mean, but only in winter and spring. Since the 1920s, however, the annual-mean temperature for the area north of 50°N shows almost no trend, except in recent years. Summer and autumn temperatures have actually cooled since the mid to late

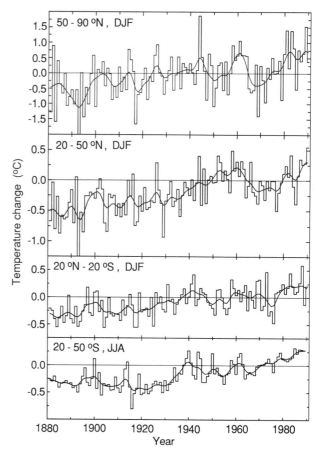

Figure 8.4: Observed variations in winter temperature for various latitude bands (DJF, dated by the January in the Northern Hemisphere and tropics, and JJA in the Southern Hemisphere). The temperatures used are air temperature data over land areas and sea-surface temperature data for the oceans, as described in Section 7. The smooth curves are filtered values designed to show decadal and longer time-scale trends more clearly. Note the compressed scale in the upper panel.

1930s. High-latitude Southern Hemisphere data are inadequate to make any meaningful comparisons.

The observed northern high-latitude winter enhancement is broadly consistent with model predictions. However, some of the latitudinal and seasonal details of observed temperature changes are contrary to equilibrium model predictions. This result has little bearing on the detection issue for two reasons. First, the variability of temperatures in high latitudes is greater than elsewhere and published calculations have shown that this is not an optimum region for signal detection, based on signal-to-noise ratio considerations (Wigley and Jones, 1981). Second, there are still considerable doubts about the regional and seasonal details of the evolving greenhouse signal. Failure to identify a particular spatial pattern of change could be because the signal has not yet been correctly specified, although it is equally likely to be because the noise still dominates.

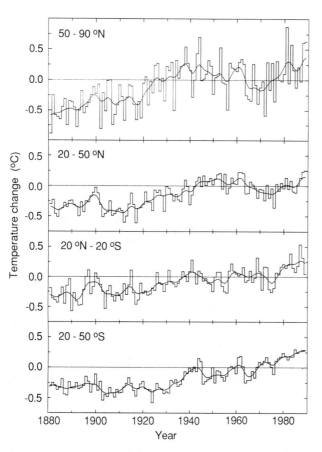

Figure 8.3: Observed variations in annual-mean temperature for various latitude bands. The temperatures used are air temperature data over land areas and sea-surface temperature data for the oceans, as described in Section 7. The smooth curves are filtered values designed to show decadal and longer time-scale trends more clearly.

8.2.3.3 *Tropospheric warming and stratospheric cooling*

All equilibrium model simulations show a warming to near the top of the troposphere (Section 5). Trends near the tropopause and for the lower stratosphere, at least up to 50mb, differ in sign between models. Above 50mb, all models show a cooling. It has been suggested that this contrast in trends between the troposphere and stratosphere might provide a useful detection fingerprint (Epstein, 1982; Parker, 1985; Karoly, 1987, 1989), but this is not necessarily the case for a number of reasons. First, identification of such a signal is hampered because observations above 50mb are of limited duration and generally of poorer quality than those in the troposphere. Second, there are reasons to expect natural variability to show a similar contrast between stratospheric and tropospheric trends (Liu and Schuurmans, 1990).

Stratospheric cooling alone has been suggested as an important detection variable, but its interpretation is difficult because it may be caused by a number of other factors, including volcanic eruptions and ozone depletion. Furthermore, the physics of greenhouse-gas-induced stratospheric cooling is much simpler than that of tropospheric warming. It is quite possible for models to behave correctly in their stratospheric simulations, yet be seriously in error in the lower atmosphere. Validation of the stratospheric component of a model, while of scientific importance, may be of little relevance to the detection of an enhanced greenhouse effect in the troposphere.

Nevertheless, there is broad agreement between the observations and equilibrium model simulations. While the observations (Angell, 1988) show a global-mean cooling trend from 1958 between 100hPa and 300hPa (Section 7, Figure 7.17), which appears to conflict with model results, this cooling is apparent only between 10 and 30°N (where it is not statistically significant) and south of 60°S (where it is associated with the ozone hole). There are no noticeable trends in other regions. Data compiled by Karoly (1987, 1989) show a warming trend since 1964 up to around 200hPa in the Southern Hemisphere, to 100hPa in the Northern Hemisphere to 60°N and a more complex (but largely warming) behaviour north of 60°N. Near the tropopause and in the lower stratosphere, temperatures have cooled since 1964. The main difference between recent observations and model simulations is in the level at which warming reverses to cooling. Although the models show large model-to-model differences, this level is generally lower in the observations. This difference may be associated with poor vertical resolution and the inadequate representation of the tropopause in current climate models.

8.2.3.4 *Global-mean precipitation increase*

Equilibrium experiments with GCMs suggest an increase in global-mean precipitation, as one might expect from the associated increase in atmospheric temperature. However, the spatial details of the changes are highly uncertain (Schlesinger and Mitchell, 1987; and Section 5). Observations from which the long-term change in precipitation can be determined are available only over land areas (see Bradley et al., 1987; Diaz et al., 1989; and Section 7), and there are major data problems in terms of coverage and homogeneity. These difficulties, coupled with the recognized model deficiencies in their simulations of precipitation and the likelihood that the precipitation signal-to-noise ratio is low (see 8.2.1.1), preclude any meaningful comparison.

8.2.3.5 *Sea level rise*

Increasing greenhouse gas concentrations are expected to cause (and have caused) a rise in global-mean sea level, due partly to oceanic thermal expansion and partly to melting of land-based ice masses (see Section 9). Because of the strong dependence of sea level rise on global-mean temperature change, this element, like global-mean precipitation, cannot be considered as an independent variable. Observations show that global-mean sea level has risen over the past 100 years, but the magnitude of the rise is uncertain by a factor of at least two (see Section 9). As far as it can be judged, there has been a positive thermal expansion component of this sea level rise. Observational evidence (e.g., Meier, 1984; Wood, 1988) shows that there has been a general long-term retreat of small glaciers (but with marked regional and shorter time-scale variability) and this process has no doubt contributed to sea level rise. Both thermal expansion and the melting of small glaciers are consistent with global warming, but neither provides any independent information about the cause of the warming.

8.2.3.6 *Tropospheric water vapour increase*

Model predictions show an increase in tropospheric water vapour content in association with increasing atmospheric temperature. This is of considerable importance, since it is responsible for one of the main feedback mechanisms that amplifies the enhanced greenhouse effect (Raval and Ramanathan, 1989). Furthermore, a model-based signal-to-noise ratio analysis (see Section 8.2.1.1) suggests that this may be a good detection variable. However, the brevity of available records and data inhomogeneities preclude any conclusive assessment of trends. The available data have inhomogeneities due to major changes in radiosonde humidity instrumentation. Since the mid 1970s, there has been an apparent upward trend, largest in the tropics (Flohn and Kapala, 1989; Elliott et al., 1990). However, the magnitude of the tropical trend is much larger than any expected greenhouse-related change, and it is likely that natural variability is dominating the record.

8.3 Multivariate or Fingerprint Methods

8.3.1 Conspectus

The fingerprint method, which involves the simultaneous use of more than one time series, is the only way that the attribution problem is likely to be solved expeditiously. In its most general form one might consider the time evolution of a set of three-dimensional spatial fields, and compare model results (i.e., the signal to be detected) with observations. There are, however, many potential difficulties both in applying the method and in interpreting the results, not the least of which is reliably defining the greenhouse-gas signal and showing *a priori* that it is unique.

In studies that have been performed to date, predicted changes in the three-dimensional structure of a single variable (mean values, variances and/or spatial patterns) have been compared with observed changes. The comparison involves the testing of a "null hypothesis", namely that the observed and modelled fields do not differ. Rejection of the null hypothesis can be interpreted in several ways. It could mean that the model pattern was not present in the observations (i.e., in simplistic terms, that there was no enhanced greenhouse effect), or that the signal was obscured by natural variability, or that the prediction was at fault in some way, due either to model errors or because the chosen prediction was inappropriate. We know *a priori* that current models have numerous deficiencies (see Sections 4 and 5), and that, even on a global scale, the predicted signal is probably obscured by noise (Section 8.1.3). Furthermore, most studies to date have only used the results of equilibrium simulations, rather than the more appropriate time-dependent results of coupled ocean-atmosphere GCM experiments [4]. Because of these factors, published work in this area can only be considered as exploratory, directed largely towards testing the methods and investigating potential statistical problems.

8.3.2 Comparing Changes In Means And Variances

Means, time variances and spatial variances of the fields of observed and predicted changes have been compared for a number of variables by Santer et al. (1990). Predicted changes were estimated from the equilibrium $1 \times CO_2$ and $2 \times CO_2$ simulations using the Oregon State University (OSU) atmospheric GCM coupled to a mixed-layer ocean (Schlesinger and Zhao, 1989). In all cases (different variables, different months) the observed and modelled fields were found to be significantly different: i.e., for these

---

[4] *In this regard, the correct experiment, simulation of changes to date in response to observed greenhouse-gas forcings, has not yet been performed. Because a realistic model simulation would generate its own substantial natural variability, a number of such experiments may be required in order to ensure that representative results are obtained.*

tests the null hypothesis of no difference was rejected and the model signal could not be identified in the observations. As noted above this is not an unexpected result.

8.3.3 Pattern Correlation Methods

The basic approach in pattern correlation is to compare the observed and modelled time-averaged patterns of change (or changing observed and modelled patterns) using a correlation coefficient or similar statistic. The word "pattern" is used in a very general sense - it may refer to a two-point pattern involving two time series of the same variable, or to a many-point pattern involving the full three-dimensional spatial fields of more than one variable. In some studies, time-standardized variables have been used. This has the advantage of giving greater emphasis to those spatial regions in which the time variance (i.e., the noise) is smallest.

Four examples of pattern correlation detection studies have appeared in the literature, all involving comparisons of observed and modelled temperature changes (Barnett, 1986; Barnett and Schlesinger, 1987; Barnett, 1990; Santer et al., 1990). Barnett (1986) and Barnett and Schlesinger (1987) used the covariance between the patterns of standardized observed and modelled changes as a test statistic. Equilibrium $1 \times CO_2$ and $2 \times CO_2$ results from the OSU atmospheric GCM coupled to a mixed-layer ocean were employed to generate the multivariate predicted signal. This pattern was then correlated with observed changes relative to a reference year on a year-by-year basis. A significant trend in the correlation would indicate the existence of an increasing expression of the model signal in the observed data, which could be interpreted as detection of an enhanced greenhouse signal. A marginally significant trend was apparent, but this was not judged to be a robust result.

Santer et al. (1990) used the same model data and the spatial correlation coefficient between the time-averaged patterns of observed and predicted change as a detection parameter. The observed changes used were the differences between two decades, 1947-56 and 1977-86. Statistically significant differences between observed and model patterns of temperature change were found in all months but February (for which the amount of common variance was very small, less than 4%).

Barnett (1990) compared observed data with the time-evolving spatial fields from the GISS transient GCM run (Hansen et al., 1988). The model run uses realistic time-dependent forcing beginning in the year 1958, and accounts for the lag effect of oceanic thermal inertia by using a diffusion parameterization of heat transport below the mixed layer. Comparisons were made using spatial correlation coefficients between decadal means of the evolving signal and the equivalent pattern in the observations. There was virtually no similarity between modelled and observed temperature patterns.

The largely negative results obtained in these studies can be interpreted in a variety of ways, as noted in Section 8.3.1. Because of this, failure to detect the model signal in the data cannot be taken as evidence that there is no greenhouse-gas signal in the real world. Future multivariate detection studies should employ coupled ocean-atmosphere GCMs forced with observed greenhouse-gas concentration changes over more than just the past few decades.

8.4 When Will The Greenhouse Effect be Detected ?

The fact that we have not yet detected the enhanced greenhouse effect leads to the question: when is this likely to occur? As noted earlier, detection is not a simple yes/no issue. Rather, it involves the gradual accumulation of evidence in support of model predictions, which, in parallel with improvements in the models themselves, will increase our confidence in them and progressively narrow the uncertainties regarding such key parameters as the climate sensitivity. Uncertainties will always remain. Predicting when a certain confidence level might be reached is as difficult as predicting future climate change - more so, in fact, since it requires, at least, estimates of both the future signal and the future noise level.

Nevertheless, we can provide some information on the time-scale for detection by using the "unprecedented change" concept mentioned briefly in Section 8.1.4. This should provide an upper bound to the time for detection, since more sophisticated methods should produce earlier results. We take a conservative view as a starting point, namely that the magnitude of natural variability is such that all of the warming of the past century could be attributed to this "cause". (Note that this is not the same as denying the existence of an enhanced greenhouse effect. With such a noise level, the past warming could be explained as a 1°C greenhouse effect offset by 0.5°C natural variability.) We then assume, again somewhat arbitrarily, that a further 0.5°C warming (i.e., a total warming of 1°C since the late nineteenth century) is required before we could say, with high confidence, that the only possible explanation would be that the enhanced greenhouse effect was as strong as predicted by climate models. Given the range of uncertainty in future forcing predictions and future model-predicted warming, when would this elevated temperature level be reached?

The answer is given in Figure 8.5. The upper curve shows the global-mean warming for the Business-as-Usual Scenario (see Appendix 1) assuming a set of upwelling-diffusion climate model parameters that maximizes the warming rate (viz., $\Delta T_{2X} = 4.5°C$, $K = 0.63$ cm^2 sec^{-1} and $\pi = 0$). Under these circumstances, "detection" (as defined above) would occur in 12 years. The lower curve shows

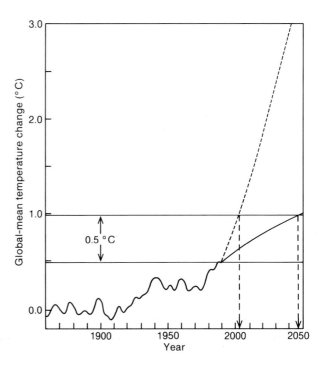

Figure 8.5: Observed global-mean temperature changes (as in Figure 8.1) and extreme predictions of future change. If a further 0.5°C warming were chosen as the threshold for detection of the enhanced greenhouse effect, then this would be reached sometime between 2002 and 2047. In practice, detection should be based on more sophisticated methods which would bring these dates closer to the present.

the global-mean warming for the lowest forcing Scenario ("D" in the Annex) with model parameters chosen to minimize the warming rate (viz., $\Delta T_{2X} = 1.5°C$, $K = 1.27$ cm^2 sec^{-1} and $\pi = 1$). Detection does not occur until 2047.

On the basis of this simple analysis alone we might conclude that detection with high confidence is unlikely to occur before the year 2000. If stringent controls are introduced to reduce future greenhouse gas emissions, and if the climate sensitivity is at the low end of the range of model predictions, then it may be well into the twenty-first century before we can say with high confidence that we have detected the enhanced greenhouse effect.

The time limits inferred from Figure 8.5 are, of course, only a rough guide to the future and they are almost certainly upper bound values. Nevertheless, the time frame for detection is likely to be of order a decade or more. In order to detect the enhanced greenhouse effect within this time frame it is essential to continue the development of models and to ensure that existing observing systems, for both climate variables and potential climate forcing factors, be maintained or improved.

8.5 CONCLUSIONS

Because of the strong theoretical basis for enhanced greenhouse warming, there is considerable concern about the potential climatic effects that may result from increasing greenhouse-gas concentrations. However, because of the many significant uncertainties and inadequacies in the observational climate record, in our knowledge of the causes of natural climatic variability and in current computer models, scientists working in this field cannot at this point in time make the definitive statement: "Yes, we have now seen an enhanced greenhouse effect".

It is accepted that global-mean temperatures have increased over the past 100 years, and are now warmer than at any time in the period of instrumental record. This global warming is consistent with the results of simple model predictions of greenhouse-gas-induced climate change. However, a number of other factors could have contributed to this warming and it is impossible to prove a cause and effect relationship. Furthermore, when other details of the instrumental climate record are compared with model predictions, while there are some areas of agreement, there are many areas of disagreement.

The main reasons for this are:

1) The inherent variability of the climate system appears to be sufficient to obscure any enhanced greenhouse signal to date. Poor quantitative understanding of low-frequency climate variability (particularly on the 10-100 year time scale) leaves open the possibility that the observed warming is largely unrelated to the enhanced greenhouse effect.

2) The lack of reliability of models at the regional spatial scale means that the expected signal is not yet well defined. This precludes any firm conclusions being drawn from multivariate detection studies.

3) The ideal model experiments required to define the signal have not yet been performed. What is required are time-dependent simulations using realistic time-dependent forcing carried out with fully coupled ocean-atmosphere GCMs.

4) Uncertainties in, and the shortness of available instrumental data records mean that the low-frequency characteristics of natural variability are virtually unknown for many climate elements.

Thus, it is not possible at this time to attribute all, or even a large part, of the observed global-mean warming to the enhanced greenhouse effect on the basis of the observational data currently available. Equally, however, we have no observational evidence that conflicts with the model-based estimates of climate sensitivity. Thus, because of model and other uncertainties we cannot preclude the possibility that the enhanced greenhouse effect *has* contributed substantially to past warming, nor even that the greenhouse-gas-induced warming has been greater than that observed, but is partly offset by natural variability and/or other anthropogenic effects.

References

Angell, J.K., 1988: Variations and trends in tropospheric and stratospheric global temperatures, 1958-87. *J. Clim*, **1**, 1296-1313.

Baker, D.J. and Barnett, T.P., 1982: Possibilities of detecting CO_2-induced effects. In, *Proceedings of the Workshop on First Detection of Carbon Dioxide Effects*, DOE/CONF-8106214 (H. Moses and M.C. MacCracken, Coordinators), Office of Energy Research, U.S. Dept. of Energy, Washington, D.C., 301-342.

Barnett, T.P., 1986: Detection of changes in global tropospheric temperature field induced by greenhouse gases. *J. Geophys. Res.*, **91**, 6659-6667.

Barnett, T.P., 1990: An attempt to detect the greenhouse gas signal in a transient GCM simulation. In, *Greenhouse-Gas-Induced Climatic Change: A Critical Appraisal of Simulations and Observations* (M.E. Schlesinger, Ed.), Elsevier Science Publishers, Amsterdam (in press).

Barnett, T.P. and Schlesinger, M.E., 1987: Detecting changes in global climate induced by greenhouse gases. *J. Geophys. Res*, **92**, 14,772-14,780.

Bell, T.L., 1982: Optimal weighting of data to detect climatic change: An application to the carbon dioxide problem. *J. Geophys. Res*, **87**, 11,161-11,170.

Bell, T.L., 1986: Theory of optimal weighting of data to detect climatic change. *J. Atmos. Sci.*, **43**, 1694-1710.

Bolin, B., Döös, B.R., Jäger, J. and Warrick, R.A. (Eds.), 1986: *The Greenhouse Effect, Climatic Change, and Ecosystems*. SCOPE Vol. 29, John Wiley and Sons Ltd., Chichester, 539pp.

Bradley, R.S., Diaz, H.F., Eischeid, J.K., Jones, P.D., Kelly, P.M. and Goodess, C.M., 1987: Precipitation fluctuations over Northern Hemisphere land areas since the mid-19th century. *Science*, **237**, 171-175.

Bryan, K., Manabe, S. and Spelman, M.J., 1988: Interhemispheric asymmetry in the transient response of a coupled ocean-atmosphere model to a CO_2 forcing. *J. Phys. Oceanog.*, **18**, 851-867.

Diaz, H.F., Bradley, R.S. and Eischeid, J.K., 1989: Precipitation fluctuations over global land areas since the late 1800s. *J. Geophys. Res.*, **94**, 1195-1210.

Elliott, W.P., Smith, M.E. and Angell, J.K., 1990: On monitoring tropospheric water vapour changes using radiosonde data. In, *Greenhouse-Gas-Induced Climatic Change: A Critical Appraisal of Simulations and Observations* (M.E. Schlesinger, Ed.), Elsevier Science Publishers, Amsterdam (in press).

Epstein, E.S., 1982: Detecting climate change. *J. App. Met.*, **21**, 1172-1182.

Flohn, H. and Kapala, A., 1989: Changes of tropical sea-air interaction processes over a 30-year period. *Nature*, **338**, 244-246.

Gilliland, R.L, 1982: Solar, volcanic and CO_2 forcing of recent climatic change. *Clim. Change*, **4**, 111-131.

Gilliland, R.L. and Schneider, S.H, 1984: Volcanic, CO_2 and solar forcing of Northern and Southern Hemisphere surface air temperatures. *Nature*, **310**, 38-41.

Hansen, J., Fung, I., Lacis, A., Rind, D., Lebedeff, S., Ruedy, R. and Russell, G., 1988: Global climate changes as forecast by Goddard Institute for Space Studies three-dimensional model. *J. Geophys. Res*, **93**, 9341-9364.

Hansen, J., Johnson, D., Lacis, A., Lebedeff, S., Lee, P. Rind, D. and Russell, G., 1981: Climate impact of increasing atmospheric carbon dioxide. *Science*, **213**, 957-966.

Hasselmann, K., 1976: Stochastic climate models, 1, Theory. *Tellus*, **28**, 473-485.

Ingram, W.J., Wilson, C.A. and Mitchell, J.F.B., 1989: Modeling climate change: an assessment of sea ice and surface albedo feedbacks. *J. Geophys. Res.*, **94**, 8609-8622.

Karoly, D.J., 1987: Southern Hemisphere temperature trends: A possible greenhouse gas effect? *Geophys. Res. Lett.*, **14**, 1139-1141.

Karoly, D.J., 1989: Northern Hemisphere temperature trends: A possible greenhouse gas effect? *Geophys. Res. Lett.*, **16**, 465-468.

Liu, Q. and Schuurmanns, C.J.E., 1990: The correlation of tropospheric and stratospheric temperatures and its effect on the detection of climate changes. *Geophys. Res. Lett.*(in press).

MacCracken, M.C. and Luther, F.M. (Eds.), 1985: *Detecting the Climatic Effects of Increasing Carbon Dioxide*. U.S. Department of Energy, Carbon Dioxide Research Division, Washington, D.C., 198 pp.

MacCracken, M.C. and Moses, H., 1982: The first detection of carbon dioxide effects: Workshop Summary, 8-10 June 1981, Harpers Ferry, West Virginia. *Bull. Am. Met. Soc.*, **63**, 1164-1178.

Madden, R.A. and Ramanathan, V., 1980: Detecting climate change due to increasing carbon dioxide. *Science*, **209**, 763-768.

Manabe, S. and Stouffer, R.J., 1980: Sensitivity of a global climate model to an increase of CO_2 concentration in the atmosphere. *J. Geophys. Res.*, **85**, 5529-5554.

Meier, M.F., 1984: Contribution of small glaciers to global sea level. *Science*, **226**, 1418-1421.

Mitchell, J.F.B., Wilson, C.A. and Cunnington, W.M., 1987: On CO_2 climate sensitivity and model dependence of the results. *Q.J. R. Met. Soc.*, **113**, 293-322.

National Research Council, 1983: *Changing Climate. Report of the Carbon Dioxide Assessment Committee* (W.A. Nierenberg, Committee Chairman). Board on Atmospheric Sciences and Climate, National Academy Press, Washington, D.C., 496.

Parker, D.E., 1985: On the detection of temperature changes induced by increasing atmospheric carbon dioxide. *Q .J. R. Met. Soc.*, **111**, 587-601.

Ramanathan, V., Callis, L., Cess, R., Hansen, J., Isaksen, I., Kuhn, W., Lacis, A., Luther, F., Mahlman, J., Reck, R. and Schlesinger, M., 1987: Climate-chemical interactions and effects of changing atmospheric trace gases. *Rev .Geophys.*, **25**, 1441-1482.

Raval, A. and Ramanathan, V., 1989: Observational determination of the greenhouse effect. *Nature*, **342**, 758-761.

Robock, A., 1978: Internally and externally caused climate change. *J. Atmos. Sci.*, **35**, 1111-1122.

Robock, A., 1983: Ice and snow feedbacks and the latitudinal and seasonal distribution of climate sensitivity. *J. Atmos. Sci.*, , **40**, 986-997.

Santer, B.D., Wigley, T.M.L., Schlesinger, M.E. and Jones, P.D., 1990: Multivariate methods for the detection of greenhouse-gas-induced climate change. In, *Greenhouse-Gas-Induced Climatic Change: A Critical Appraisal of Simulations and Observations* (M.E. Schlesinger, Ed.), Elsevier Science Publishers, Amsterdam (in press).

Schlesinger, M.E., Barnett, T.P. and Jiang, X.-J., 1990: On greenhouse gas signal detection strategies. In, *Greenhouse-Gas-Induced Climatic Change: A Critical Appraisal of Simulations and Observations* (M.E. Schlesinger, Ed.), Elsevier Science Publishers, Amsterdam (in press).

Schlesinger, M.E. and Mitchell, J.F.B., 1987: Climate model simulations of the equilibrium climatic response to increased carbon dioxide. *Rev. Geophys.*, **25**, 760-798.

Schlesinger, M.E. and Zhao, Z.-C., 1989: Seasonal climate changes induced by doubled CO_2 as simulated by the OSU atmospheric GCM mixed-layer ocean model. *J. Clim.*, **2**, 459-495.

Schönwiese, C.-D., 1990: Multivariate statistical assessments of greenhouse-induced climatic change and comparison with the results from general circulation models. In, *Greenhouse-Gas-Induced Climatic Change: A Critical Appraisal of Simulations and Observations* (M.E. Schlesinger, Ed.), Elsevier Science Publishers, Amsterdam (in press).

Stouffer, R.J., Manabe, S. and Bryan, K., 1989: Interhemispheric asymmetry in climate response to a gradual increase of atmospheric CO_2. *Nature* **342**, 660-662.

Thiébaux, H.J. and Zwiers, F.W., 1984: The interpretation and estimation of effective sample size. *J. Clim. & App. Met.*, **23**, 800-811.

Tsionis, A.A. and and Elsner, J.B., 1989: Testing the global warming hypothesis. *Geophys. Res. Lett.*, **16**, 795-797.

Vinnikov, K. Ya. and Groisman, P. Ya., 1982: Empirical study of climate sensitivity. *Isvestiya AS USSR, Atmospheric and Oceanic Physics* , **18(11)**, 1159-1169.

Washington, W.M. and Meehl, G.A., 1989: Climate sensitivity due to increased CO_2: experiments with a coupled atmosphere and ocean general circulation model. *Clim. Dynam.*, **4**, 1-38.

Wigley, T.M.L. and Jones, P.D., 1981: Detecting CO_2-induced climatic change. *Nature*, **292**, 205-208.

Wigley, T.M.L. and Raper, S.C.B., 1987: Thermal expansion of seawater associated with global warming. *Nature*, **330**, 127-131.

Wigley, T.M.L. and Raper, S.C.B., 1990: Natural variability of the climate system and detection of the greenhouse effect. *Nature*, **344**, 324-327.

Wigley, T.M.L., Angell, J.K. and Jones, P.D., 1985: Analysis of the temperature record. In, *Detecting the Climatic Effects of Increasing Carbon Dioxide*, DOE/ER-0235 (M.C. MacCracken and F.M. Luther, Eds.), U.S. Dept. of Energy, Carbon Dioxide Research Division, Washington, D.C., 55-90.

Wood, F.B., 1988: Global alpine glacier trends, 1960s-1980s. *Arctic and Alpine Research,* **20**, 404-413.

Wood, F.B., 1990: Monitoring global climate change: The case of greenhouse warming. *Bull. Am. Met. Soc.*, **71**, 42-52.

9

Sea Level Rise

R. WARRICK, J. OERLEMANS

Contributors:
P. Beaumont; R.J. Braithwaite; D.J. Drewery; V. Gornitz; J.M. Grove; W. Haeberli;
A. Higashi; J.C. Leiva; C.S. Lingle; C. Lorius; S.C.B. Raper; B. Wold;
P.L. Woodworth.

CONTENTS

EXECUTIVE SUMMARY

This Section addresses three questions:

Has global-mean sea level been rising during the last 100 years?

What are the causal factors that could explain a past rise in sea level?

And what increases in sea level can be expected in the future?

Despite numerous problems associated with estimates of globally-coherent, secular changes in sea level based on tide gauge records, we conclude that it is highly likely that sea level has been rising over the last 100 years. There is no new evidence that would alter substantially the conclusions of earlier assessments regarding the rate of change. Our judgement is that:

The average rate of rise over the last 100 years has been 1.0 - 2.0 mm yr^{-1}.

There is no firm evidence of accelerations in sea level rise during this century (although there is some evidence that sea level rose faster in this century compared to the previous two centuries).

As to the possible causes and their specific contributions to past sea level rise, the uncertainties are very large, particularly for Antarctica. However, in general it appears that the observed rise can be explained by thermal expansion of the oceans, and by the increased melting of mountain glaciers and the margin of the Greenland ice sheet. From present data it is impossible to judge whether the Antarctic ice sheet as a whole is currently out of balance and is contributing, either positively or negatively, to changes in sea level.

Future changes in sea level were estimated for each of the IPCC forcing scenarios (using the same simple box model as in Section 6). For each scenario, three projections - best estimate, high and low - were made corresponding to the estimated range of uncertainty in each of the potential contributing factors. It is found that:

For the IPCC Business-as-Usual Scenario at year 2030, global-mean sea level is 8 - 29 cm higher than today, with a best-estimate of 18 cm. At the year 2070, the rise is 21 - 71 cm, with a best-estimate of 44 cm.

Most of the contribution is estimated to derive from thermal expansion of the oceans and the increased melting of mountain glaciers and small ice caps.

On the decadal time scale, the role of the polar ice sheets is expected to be minor, but they contribute substantially to the total uncertainty. Antarctica is expected to contribute negatively to sea level due to increased snow accumulation associated with warming. A rapid disintegration of the West Antarctic Ice Sheet due to global warming is unlikely within the next century.

For the lower forcing scenarios (B,C and D), the sets of sea level rise projections are similar, at least until the mid-21st century. On average these projections are approximately one-third lower than those of the Business-as-Usual Scenario.

Even with substantial decreases in the emissions of the major greenhouse gases, future increases in temperature and, consequently, sea level are unavoidable - a sea level rise "commitment" - due to lags in the climate system.

This present assessment does not foresee a sea level rise of ≥1 metre during the next century. Nonetheless, the implied rate of rise for the best-estimate projection corresponding to the IPCC Business-as-Usual Scenario is about 3-6 times faster than over the last 100 years.

9.1 Sea Level Rise: Introduction

This section is primarily concerned with decade-to-century changes in global-mean sea level, particularly as related to climatic change. First, the evidence for sea level rise during the last 100 years is reviewed as a basis for looking for climate-sea level connections on a decade-to-century timescale. Next, the possible contributing factors - thermal expansion of the oceans and the melting of land ice - to both past and future sea level change are examined. Finally, the issue of future sea level due to global warming is addressed.

9.2 Factors Affecting Sea Level

Changes in sea level occur for many reasons on different time and space scales. Tide gauges measure sea level variations in relation to a fixed benchmark and thus record "relative sea level" change due both to vertical land movements and to real (eustatic) changes in the ocean level. Vertical land movements result from various natural isostatic movements, sedimentation, tectonic processes and even anthropogenic activities (e.g., groundwater and oil extraction). In parts of Scandinavia, for instance, relative sea level is decreasing by as much as 1m per century due to isostatic "rebound" following the last major glaciation. In attempting to identify a globally-coherent, secular trend in MSL, the vertical land movements "contaminate" tide gauge records and have to be removed.

Eustatic sea level is also affected by many factors. Differences in atmospheric pressure, winds, ocean currents and density of seawater all cause spatial and temporal variations in sea level in relation to the geoid (the surface of constant gravitational potential corresponding to the surface which the ocean would assume if ocean temperature and salinity were everywhere 0°C and 35 o/oo, respectively, and surface air pressure was everywhere constant). Changes in the geoid itself, due to re-distribution of mass within the Earth, are irrelevant on the decadal-century timescales under consideration. Over these timescales, the most important climate-related factors are likely to be thermal expansion of the oceans and melting of land ice (but not floating ice shelves or sea ice).

9.3 Has Sea Level Been Rising Over the Last 100 Years?

It is highly likely that global-mean sea level (MSL) has been rising. This is the general conclusion of no fewer than 13 studies of MSL change over various periods during the last 100 years (Table 9.1). The estimates range from about 0.5mm/yr to 3.0mm/yr, with most lying in the range 1.0-2.0mm/yr.

Table 9.1: Estimate of Global Sea-Level Change (updated from Barnett, 1985; Robin, 1986).

Rate (mm/yr)	Comments	References
>0.5	Cryologic estimate	Thorarinsson (1940) †
1.1 ± 0.8	Many stations, 1807-1939	Gutenburg (1941)
1.2 - 1.4	Combined methods	Kuenen (1950)
1.1 ± 0.4	Six stations, 1807-1943	Lisitzin (1958, in Lisitzin 1974)
1.2	Selected stations, 1900-1950	Fairbridge & Krebs (1962)
3.0	Many stations, 1935-1975	Emery et al. (1980)
1.2	Many stations -> regions, 1880-1980	Gornitz et al. (1982)
1.5	Many stations, 1900-1975	Klige (1982)
1.5 ± 0.15 †	Selected stations, 1903-1969	Barnett (1983)
1.4 ± 0.14 †	Many stations -> regions, 1881-1980	Barnett (1984)
2.3 ± 0.23 †	Many stations -> regions, 1930-1980	Barnett (1984)
1.2 ± 0.3 †	130 stations, 1880-1982	Gornitz & Lebedeff (1987)
1.0 ± 0.1 †	130 stations >11 regions, 1880-1982	Gornitz & Lebedeff (1987)
1.15	155 stations, 1880-1986	Barnett (1988)
2.4 ± 0.9 §	40 stations, 1920-1970	Peltier & Tushingham (1989; 1990)
1.7 ± 0.13 §	84 stations, 1900-1980	Trupin and Wahr (1990)

† = Value plus 95% confidence interval

§ = Mean and standard deviation

In addition, several assessments of the likely rate of past sea level rise have been made: 12±5cm since 1900 from the SCOPE 29 assessment (Bolin et al., 1986); 10-25cm since 1900 from the US.DOE assessment (MacCracken and Luther, 1985); and 10-20cm over last 100 years from the PRB assessment (Polar Research Board, 1985). These assessments also include detailed reviews of the literature (Barnett, 1985; Aubrey, 1985; Robin, 1986) Rather than repeat these, we shall focus on the most recent studies and ask whether they provide any new information that would substantially alter previous assessments.

9.3.1 Comparison of Recent Estimates

The analyses by Gornitz and Lebedeff (1987; also see Gornitz, 1990) used tide-gauge data from 130 stations with minimum record length of 20 years to estimate the average rate of sea level change over the period 1880-1982. This analysis differed from previous analyses (Gornitz et al., 1982) by including a more careful correction for vertical land movements using extensive data from ^{14}C dated Holocene sea level indicators (see below). This correction significantly reduced the spread of the trend estimates from the individual stations (Figure 9.1).

Using two different averaging techniques to produce composite global MSL curves (averaging individual stations versus regional trends), the study obtained estimates of 1.2 ± 0.3mm/yr and 1.0 ± 0.1mm/yr respectively. These results do not differ significantly from their previous findings.

The study by Barnett (1988) is an update of previous work (Barnett, 1983, 1984) in which 155 stations are analysed over the period 1880-1986. A rate of 1.15mm/yr is obtained, in close agreement with the rates noted above. However, from a comparison of the composite global sea

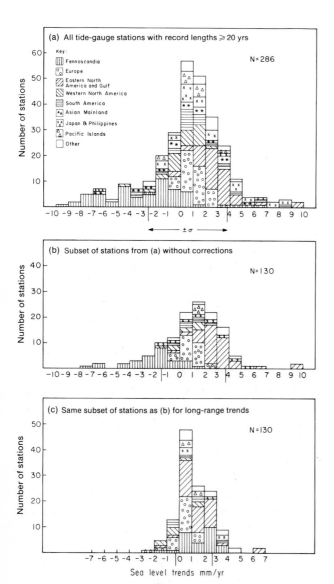

Figure 9.1: Histogram of number of tide-gauge stations vs sea-level trends. Triangle indicates mean rate of sea-level rise; lines indicate +/- sigma. **(a)** All tide-gauge stations with record length > 20 years; raw data. **(b)** Subset of tide-gauge stations; long-range trends included. **(c)** Same subset of stations as (a); long range trends subtracted. From Gorntiz (1990).

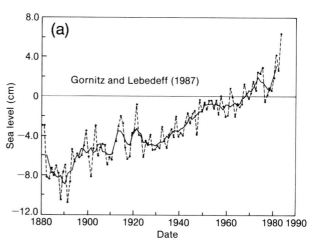

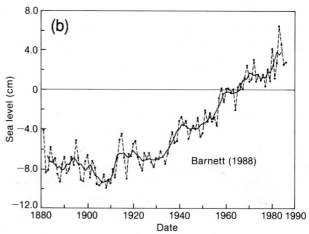

Figure 9.2: Global-mean sea level rise over the last century. The baseline is obtained by setting the average for the period 1951-1970 to zero. The dashed line represents the annual mean, and the solid line the 5-year running mean. **(a)** Gornitz and Lebedeff (1987), **(b)** Barnett (1988).

Table 9.2 *Time-dependency of the Tide Gauge Records (Modified from Peltier & Tushingham, 1990)*

Window Width (yr)	Start Year	End Year	No. of Records Available	LR Est. of SL Rise (mm/yr)	SD of LR Estimate (mm/yr)	EOF Est. of SL Rise (mm/yr)	SD of EOF Est. (mm/yr)
Fixed Window Width							
51	1890	1940	11	1.6	1.5	0.7	0.7
51	1900	1950	20	1.6	0.9	1.2	0.6
51	1910	1960	27	1.8	0.8	1.4	0.6
51	1920	1970	40	2.3	0.8	2.4	0.9
51	1930	1980	33	2.0	1.1	1.5	0.6
Variable Window Width							
71	1900	1970	13	1.9	0.8	1.2	0.5
66	1905	1970	17	1.9	0.8	1.2	0.5
61	1910	1970	24	1.9	0.8	1.5	0.6
56	1915	1970	29	2.1	0.8	1.7	0.7
51	1920	1970	40	2.3	0.8	2.4	0.9
46	1925	1970	52	2.2	1.0	2.3	1.1
41	1930	1970	66	1.9	1.1	1.9	1.1
36	1935	1970	82	1.6	1.5	1.4	1.0

All tide gauge records have been reduced using the standard model.

level curves (Figure 9.2), it is apparent that while Gornitz and Lebedeff's curve appears linear over the entire time period, Barnett's curve suggests a steeper rate of rise over about 1910-1980 - approximately 1.7mm/yr. This is more nearly in line with estimates of Peltier and Tushingham (1990) for the same time period (see Table 9.2).

Peltier and Tushingham (1989, 1990) select a minimum record length of 51 years, correct the data for ongoing glacial isostatic adjustments using a geophysical model, and analyse the corrected data using both linear regression (LR) techniques and empirical orthogonal function (EOF) analyses. From a final total of 40 stations over the time period 1920-1970, they conclude that the global rate of sea level rise is 2.4 ± 0.9mm/yr. This rate is considerably higher than most other estimates noted in Table 9.1. However, the authors caution that the results are sensitive to variations in the analysis procedure. As shown in Table 9.2, variations in either the record length or period of record have large effects on the estimated rate of rise. In fact, for all combinations other than their preferred period 1920-1970 (chosen to maximize the number of stations with a minimum 50-year record length), the estimated rates are lower and, in a number of cases, compare favourably with those of Gornitz and Lebedeff (1987), Barnett (1988) and Trupin and Wahr (1990).

Why the differences? Possible reasons have to do with choice of minimum record length, period of record, number of stations, geographical representation, correction procedures for vertical land movements, and methods of data aggregation and analysis. Unfortunately, these factors are interrelated and not easily isolated from published studies. Nevertheless, it is significant that, despite the differences, both the recent and earlier studies all find a positive trend in global MSL. This seems to be a rather robust finding. There is, however, the possibility that all the studies could be systematically biased.

9.3.2 Possible Sources of Error

There are several potential sources of systematic bias common to all such studies. Firstly, they make use of the same global MSL dataset, that of the Permanent Service for Mean Sea Level (PSMSL), an International Council of Scientific Unions databank located at the Bidston Observatory, U.K. (Pugh et al., 1987). The PSMSL collects data from approximately 1300 stations worldwide. However, only 850 of these are suitable for time series work (the PSMSL "Revised Local Reference" (RLR) dataset), and 420 of these are 20 years or more in length. Tide gauge records contain many signals other than a secular trend. These stem primarily from large interannual meteorological and oceanographic forcings on sea level and, in principle, can be modelled and thereby removed from the record. In practice, the variability is such that accurate trends can be computed only given 15-20 years of data, which significantly reduces the size of the dataset available for analysis.

Secondly, there is an historical geographical bias in the dataset in favour of Northern Europe, North America and Japan. Areas of Africa, Asia, ocean islands and polar regions are sparsely represented. The geographical bias inherent in any global dataset will propagate into all studies. This bias can be reduced (but not eliminated) by treating regional subsets of the dataset as independent information, as has been done in the recent studies described above.

The problem of geographical bias is now being addressed with the establishment of the Global Level of the Sea Surface (GLOSS) global tide gauge network coordinated by the Intergovernmental Oceanographic Commission (IOC) (Pugh, 1990). Most islands involved now have tide gauges and most continental GLOSS stations (other than polar sites) are now operational, but much work remains to improve standards and the reliability of observations.

Finally, perhaps the most important source of error stems from the difficulties involved in removing vertical land movements from the dataset. In addition to the effects noted above, most mid-latitude stations located on continental margins are especially susceptible to effects from sedimentation, groundwater and oil extraction, and tectonic influences and could be undergoing general submergence, which, unless accounted for, could introduce a positive bias into any global MSL secular trend (Pirazzoli et al., 1987). In order to identify a globally-coherent trend that can be linked to changes in global climate, such effects have to be removed. The issue is how to do so.

In the future, the inherent ambiguity between land and ocean level changes in a tide gauge record will be solved by the use of advanced geodetic methods, but such data are not available for present analysis (Carter et al., 1989). In lieu of new geodetic data, one approach adopted by recent analyses has been to model explicitly the expected geology-induced MSL changes at each tide gauge site by the use of ancillary Holocene data (e.g., molluscs, corals, peats. Gornitz et al., 1982; Gornitz and Lebedeff, 1987) or by the use of geodynamic models of the Earth (Peltier and Tushingham, 1989; 1990). The other approach is simply to assemble a sufficiently broad geographical spread of records such that (it is hoped) the net contribution of land movements reduces to zero (Barnett, 1983; 1984; 1988).

These differences in approach probably account substantially for the different results noted in Table 9.1. But it cannot be said with confidence that vertical land movements (or, that is, the failure to account adequately for

them), along with reliance on a single dataset and problems of geographical bias, have not systematically biased all studies in the same direction.

9.3.3 Accelerations in Sea Level Rise

Is there evidence of any "accelerations" (or departures from long-term linear trends) in the rate of sea level rise? From examinations of both composite regional and global curves and individual tide-gauge records, there is no convincing evidence of an acceleration in global sea level rise during the twentieth century. For longer periods, however, there is weak evidence for an acceleration over the last 2-3 centuries.

Long-term analyses are hindered by the scarcity of tide-gauge records longer than 100-120 years. Data are limited to a few stations in Europe and North America. Woodworth (1990) inspected individual tide gauge records in Europe and found that although there is no general evidence for an increasing (or decreasing) rate of MSL change during the past century, a regionally-coherent acceleration of the order of 0.4mm/year per century is apparent over the last 2-3 centuries. This finding is supported by Gornitz and Solow (1989) who find weak evidence for an increase in the trend around 1895. Similar conclusions were reached by Ekman (1988) from an examination of one of the longest tide-gauge records, at Stockholm. Extension of such findings to the global scale, however, should be carried out with caution.

We now turn to the possible contributing factors to see if we can explain the past rise.

9.4 Possible Contributing Factors To Past and Future Sea Level Rise

There are four major climate-related factors that could possibly explain a rise in global MSL on the 100-year time scale. These are:

1) thermal expansion of the oceans;
2) glaciers and small ice caps;
3) the Greenland ice sheet; and
4) the Antarctic ice sheet (including the special case of the West Antarctic ice sheet).

In this section, we examine the sensitivity of each factor to changes in climate (particularly temperature), and estimate its possible contribution to past sea level change. In the subsequent section, attention is then turned to future sea level change.

9.4.1 Thermal Expansion of The Oceans

At constant mass, the volume of the oceans, and thus sea level, will vary with changes in the density of sea-water. Density is inversely related to temperature. Thus, as the oceans warm, density decreases and the oceans expand - a "steric" rise in sea level. Marked regional variations in sea-water density and volume can also result from changes in salinity, but this effect is relatively minor at the global scale.

In order to estimate oceanic expansion (past or future), changes in the interior temperature, salinity and density of the oceans have to be considered, either empirically or by models. Unfortunately, observational data are scant, both in time and space (Barnett, 1985). A few recent analyses have been carried out on the limited time-series data. For instance, Roemmich (1985) examined the 1955-1981 Panuliris series of deep hydrographic stations off Bermuda, and Thomson and Tabata (1987) examined the Station PAPA (northeast Pacific Ocean) steric height anomalies for a similar 27-year record. The latter study found that open ocean steric heights are increasing linearly at 0.93mm/year. However, in this and other studies, the large interannual variability creates too much "noise" to be confident of the estimate derived from such a short time-series. Moreover, the limited geographical coverage makes inference to the global scale problematic. In a few decades, current efforts such as the World Ocean Circulation Experiment (WOCE) will be begin to fill the data gaps and overcome these problems.

An alternative approach could be based on numerical models of the ocean's circulation (Barnett, 1985). Ideally, detailed three-dimensional models could describe the various oceanic mixing processes and could simulate heat transfer and expansion effects throughout the oceans. However, such models are in the early stages of development and applications to problems of global warming and thermal expansion are few in number. A drawback of this sort of model is that the computing time required precludes numerous runs for sensitivity analyses.

Instead, for the present assessment a simple upwelling-diffusion energy-balance climate model is used. Typically, this type of model represents the world's land and oceans by a few "boxes", and complicated processes of oceanic mixing are simplified in one or more parameters (for review see Hoffert and Flannery, 1985). Such a model was used to estimate the transient global warming (see Sections 6 and 8 for other results, and Section 8 for the justification for using this type of model). The inclusion of expansion coefficients in the model (varying with depth and, possibly, latitude) allows the sea level changes to be estimated as well. In order to maintain consistency throughout this assessment, both past and future (see below) thermal expansion effects are also estimated with this modelling technique, bearing in mind that full understanding of the dynamic processes and their effects on the depths and timing of ocean warming will eventually require more physically realistic models.

The model of Wigley and Raper (1989) was forced by past changes in radiative forcing due to increasing atmospheric concentrations of greenhouse gases (see Section 2). The internal model parameters that most affect the output are the diffusivity (K), the sinking water to global mean temperature change ratio (π) and the climate sensitivity (ΔT_{2x}, the global-mean equilibrium temperature change for a CO_2 doubling) (see Section 6). In order to estimate past thermal expansion effects, the parameter values were constrained to maintain consistency with

Table 9.3 *Some physical characteristics of glacier ice on Earth.   Sources: Flint (1971), Radok et al. (1982), Drewry (1983), Haeberli et al. (1988), Ohmura and Reeh (1990).   Estimated accuracy: † =15%, †† = 30%, otherwise better than 10%.*

	Antarctica (grounded ice)	Greenland	Glaciers & small ice caps
Area (10^6 km^2)	11.97	1.68	0.55
Volume (10^6 km^3 ice)	29.33	2.95	0.11 ††
Mean thickness (m)	2,488	1,575	200 ††
Mean elevation (m)	2,000	2,080	-
Equivalent sea level (m)	65	7	0.35 ††
Accumulation (10^{12} kg/yr)	2200 ††	535 †	-
Ablation (10^{12} kg/yr)	< 10 ††	280 ††	-
Calving (10^{12} kg/yr)	2200 ††	255 ††	-
Mean equilibrium - line altitude (m)	-	950 †	0 - 6,300
Mass turnover time (yr)	~15,000	~5,000	50 - 1,000

observed global warming over the same time period (i.e , 0.3 - 0.6°C; see Section 7). For the period 1880-1985, the resultant range of sea level rise due to thermal expansion is about 2-6cm (also see Wigley and Raper, 1987; 1990).

9.4.2 Land Ice

A distinction is made between glaciers and small ice caps, the Greenland ice sheet and the Antarctic ice sheet, since different climatic characteristics and different response times are involved. Table 9.3 lists some of their physical properties.

A large uncertainty exists regarding the volume of glaciers and small ice caps. Although the total area is relatively well-known (Haeberli et al., 1988), the mean thickness is not. Here a value of 200m is adopted, which really is a first-order estimate. Fortunately, on small time scales, (up to several decades), it is the surface area that largely determines the changes in runoff.

Although the positive and negative contributions to the mass budget of the Greenland and Antarctic ice sheets noted in Table 9.3 sum up to zero, it is actually unknown how close the ice sheets are to equilibrium (a more detailed discussion is given below). This introduces the problem of choosing an initial state for model integrations to estimate future sea level. In a previous assessment (Oerlemans, 1989), the year AD 1850 was taken as a starting point, and it was suggested to simply add the "unexplained part" of past sea level rise to the calculated future contributions from thermal expansion and land ice. The unexplained trend can also be associated with long-term changes of the ice sheets, but also to unknown tectonic effects. We return to this particular problem later.

From Table 9.3 a notable difference between the Greenland and Antarctic ice is evident. On Greenland there is significant ablation (melting *and* runoff, evaporation); on Antarctica ablation is a negligible component in the total mass budget. This is also reflected in the altitude of the equilibrium line (= zero annual mass gain). It is instructive to consider this in the light of a generalised mass-balance curve, where mass gain and loss are plotted as a function of *annual* surface air temperature (Figure 9.3). The resulting mass balance is in metres of water equivalent per year. Depending on the annual temperature range, ablation occurs for annual temperatures higher than -15 to -10°C. At the lower reaches, accumulation increases with temperature, reaches a maximum in the vicinity of the freezing point and then decreases.

The net specific balance thus shows two ranges with different behaviour. In range (a) mass balance increases with temperature, in range (b) it decreases with temperature. This makes the response of ice masses to climatic change complicated. Most glaciers and the Greenland ice sheet are mainly, but not entirely, in region (b). The Antarctic ice sheet with its much colder climate is

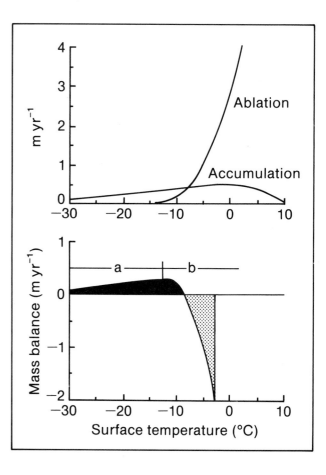

Figure 9.3: Dependence of ablation (evaporation and runoff) and accumulation on annual surface temperature (upper panel). The dependence of net annual balance (lower panel) on temperature changes sign, which complicates the response of glaciers to climate change. The picture is schematic and will change from place to place. In regions with excessive precipitation, the mean temperature at equilibrium is higher.

situated in region (a). In case of a climatic warming, one expects an increasing surface mass balance for the Antarctic ice sheet (contributing to a sea level lowering) and a decreasing mass balance for the other ice bodies (contributing to sea level rise).

9.4.3 Glaciers and Small Ice Caps

The majority of valley glaciers has been retreating over the last hundred years. Although long records of glacier length are only available for some glaciers in the European-North Atlantic region (Figure 9.4), geomorphological investigations have made it clear that the trend of glacier retreat has generally been world-wide since the Little Ice Age (Grove, 1988). Wastage was most pronounced in the middle of the 20th century. Around 1960, many glaciers started to advance. In the 1980s this advance slowed down

During the period 1900-1961, global mean temperature rose by approximately 0.35°C (see Section 7). This yields a sensitivity in terms of sea level rise of 1.3 mm yr^{-1} per degree (with the ocean area equal to 361 million km^2).

This sensitivity value is broadly supported by mass balance studies. In the study noted above Meier found a net glacier *mass balance* of -0.38 ± 0.2m/yr for 25 well studied glaciers, converting to a sensitivity of -1 m yr^{-1} per degree temperature rise. This can be compared to other, more direct estimates. For example, based on an analysis of 29 years of climatological and mass balance data from the Hintereisferner (a well-studied glacier in the Austrian Alps), Greuell (1989) finds a sensitivity of -0.41 m yr^{-1} per degree warming. Kuhn (1990) suggest a global value of the order of -0.5 m yr^{-1} per degree. Sensitivity tests with an energy balance model for a glacier surface, including albedo feedback, yields values ranging from -0.45 m yr^{-1} per degree warming for drier climates to -0.7 m yr^{-1} per degree warming for moist climates (Oerlemans, 1990). Altogether, these studies come up with smaller values than the -1 m yr^{-1} per degree derived from Meier's estimate of change in glacier mass balance combined with a figure for the global temperature change. It is probably the use of this *global* temperature change from which the discrepancy arises. It is known that summer temperature is the important parameter, and it also seems unlikely that the mean change over the glacierized regions can be represented by the global mean temperature. The global sensitivity value inferred by these studies is 1.2± 0.6 mm yr^{-1} per degree warming.

In our judgement glacier shrinkage will continue and accelerate in a warming climate.

9.4.4 *The Greenland Ice Sheet.*

Estimates of the mass budget of the Greenland ice sheet have been hampered by a pronounced lack of data. Accumulation measurements have been done on a few traverses only (see Radok et al., 1982). Systematic ablation measurements in the marginal zone have been carried out in several places, but all located in the southwestern part of Greenland (Braithwaite and Oleson, 1989). The only profile from the ice margins to the region well above equilibrium line is the EGIG profile (Expedition Glaciologique Internationale au Groenland, West Greenland, at about 70°N latitude). Here mass balance and meteorological measurements have been carried out in the summer seasons of 1959 and 1967 (Ambach, 1963, 1979).

Table 9.4 lists estimates of the total mass budget of the Greenland ice sheet as compiled by Robin (1986). The differences appear quite large. The zeros in the column "balance" should not be interpreted as an indication for a balanced state - equilibrium has only been assumed. When going through the original papers it becomes clear that, on the basis of currently available data, there is considerable

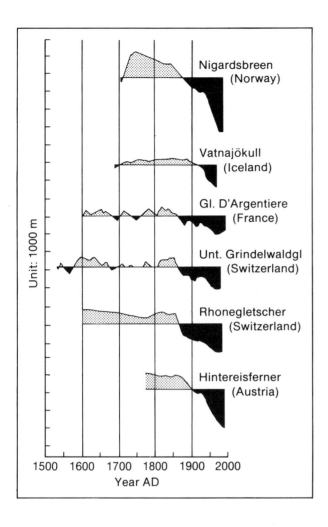

Figure 9.4: Variations of some selected glaciers as measured by their length. Data from Bjornsson (1979); Ostrem et al. (1977); Kasser (1967, 1973); Kasser and Haeberli (1979); Muller (1977); Vivian (1975); Haeberli (1985).

or stopped in several glacier basins (e.g., Haeberli et al., 1989a), but not everywhere. In Scandinavia, for instance, mass balance remained positive on the maritime glaciers and close to zero on the others.

The main published estimate of the contribution of retreating glaciers to past sea level rise is that of Meier (1984). In his analysis, Meier assumed that the magnitude of the long-term changes, for which data are sparse, are proportional to the difference between summer and winter balance, for which data are more abundant. This allowed extrapolation of measurements on a few glaciers to a world-wide scale. Meier estimated that during the period 1900-1961, glacier retreat contributed 2.8 cm, or 0.46 ± 0.26 mm yr^{-1}, to global sea level rise. By using data from three well-documented glaciers in temperate climate regions, Meier also extrapolated the record to encompass 1885-1974. This gives an average rate of sea level rise which is somewhat less than 0.46 mm yr^{-1}.

Table 9.4 *Estimates of the mass budget of the Greenland ice sheet in 10^{12} kg/yr(updated from Robin, 1986)*

Source	Accumulation	Ablation	Calving	Balance
Bader (1961)	+630	-120 to -270	-240	+270 to +120
Benson (1962)	+500	-272	-215	+13
Bauer (1968)	+500	-330	-280	-110
Weidick (1984)	+500	-295	-205	0
Reeh (1985)	+487	-169	-318	0
Ohmura & Reeh (1990)	+535			

Table 9.5 *Estimates of the sensitivity of the Greenland mass balance to climatic change. T = temperature, P = precipitation, C = cloudiness. Expressed in rate of change of global mean sea level (mm/yr).*

Source	T (+1°C)	P (+5%)	C (+5%)	Remarks
Ambach & Kuhn (1989)	+0.31	-0.13		Analysis of EGIG data
Bindschadler (1985)	+0.45			EGIG data/retreating margin
Braithwaite and Olesen (1990)	+0.36 to +0.48			Energy balance calculation
Oerlemans (1990)	+0.37	-0.11	-0.06	Energy balance Model

uncertainty regarding the current state of balance of the ice sheet. An imbalance of up to 30% of the annual mass turnover cannot be excluded.

A few studies have also been undertaken to detect changes in some selected area. Along the EGIG line in central West Greenland, there is some indication of slight thickening in the interior part of the ice sheet. A study along the "Oregon State University line" in South Greenland suggested a close balance between accumulation and ice discharge, at least in the interior part (Kostecka and Whillans, 1988).

On the basis of satellite altimetry, Zwally (1989) found that the mass balance of the southern part of the ice sheet has been positive in the period 1978-1986. He reports that thickening of the ice sheet occurred in both the ablation and accumulation zone (order of magnitude 0.2 m/yr). Although there are doubts regarding the accuracy of the results (Douglas et al., 1990), this work shows the enormous potential of radar altimetry to monitor changes on the large ice sheets.

Most outlet glaciers for which observations exist (this is mainly in central and southern part of the west coast of Greenland) have retreated strongly over the last century (Weidick, 1984). As the retreat occurred in many regions, on a relatively short time scale (100 years), and in a period of significant warming in Greenland, increased ablation rates must be responsible for this. However, the large ablation zones of the inland ice must have suffered from this too. The implications for past sea level rise will be discussed shortly.

A few estimates have been made of the sensitivity of Greenland mass balance to climatic change. They are listed in Table 9.5. The method of Ambach and Kuhn (1989) is based on a new analysis of the EGIG data. In their approach, the mass and energy budget at the equilibrium line is expanded with a linear perturbation technique, allowing the calculation of the change in the equilibrium-line altitude (dELA) associated with small changes in temperature, precipitation and radiation. By extrapolating dELA to the entire ice sheet, an estimate can then be made of the change in ablation and accumulation area, and, by assigning mass-balance values, of the total ice mass budget. Bindschadler's (1985) calculation is based on the same mass-balance measurements, but a (minor) correction is made for a retreating ice margin. The value listed as Oerlemans et al. (1990) results from a straightforward sensitivity test with an energy balance model applied to four regions of the ice sheet. Braithwaite and Olesen (1990) have used an energy balance model to study their ablation measurements in southwest Greenland, and attempted to extrapolate the result to the entire ablation zone. There is a reasonable agreement between all those studies, but this is partly due to dependence of the input.

Above, only temperature has been considered as a climatic input parameter. In fact, changes in the seasonal cycle, in precipitation and cloud patterns have occurred and will occur in the future. The potential importance of such factors can be studied by sensitivity tests, and some results have been listed in Table 9.5. It has been suggested that even in the relatively warm climate of Greenland, snow accumulation may increase when temperature goes up (e.g., Reeh and Gundestrup, 1985). If annual precipitation would increase uniformly by 5% per degree warming, the "precipitation effect" can offset about 30% of the "temperature effect". Setting the annual precipitation proportional to the amount of precipitable water in a saturated atmospheric column [see Oerlemans and Van der Veen, 1984, p. 140] would imply, for mean conditions over the Greenland ice sheet, a 4% increase in precipitation for a 1°C warming. This leads to a best estimate of the sensitivity of 0.3 ± 0.2 mm yr^{-1} per degree C. The error bar is large because:

i) There is considerable uncertainty on how precipitation patterns over Greenland will change in a warmer climate.

ii) It is unknown whether iceberg calving from the outlet glaciers will increase due to increased basal water flow (Bindschadler, 1985). However, the ice possibly involved in rapid retreat of calving fronts is almost afloat, so the contribution to sea-level rise will be negligible. Consequent thinning of grounded ice further upstream is not likely to affect sea level within the next 100 years.

iii) It is unknown how factors like surface albedo and cloudiness will change.

With a record of mean summer temperatures, the sensitivity can be used to produce an estimate of Greenland's contribution to past sea level rise. As shown in Figure 9.5, the 1866-1980 summer temperature departures (relative to the 20-year average for the reference period 1866-1885) shows an overall warming of about 0.5°C. However, the decadal changes are pronounced, with a large warming of about 2°C occurring up to 1930-35 and a cooling trend thereafter. By summing the product between

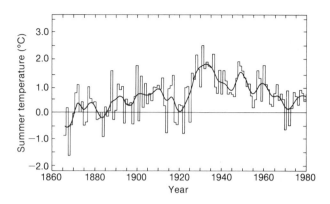

Figure 9.5: Summer (JJA) temperature (°C) as departures from reference period 1866-1885 averaged over Greenland. The smoothed curve is a moving 10-year filter.

the sensitivity value and the temperature departure for each year from 1880-1980, the 100-year contribution to sea level is estimated. Assuming initial conditions in equilibrium and a sensitivity of 0.3 ± 0.2 mm yr^{-1} per degree, the summation yields 23 ± 16 mm (or 0.23 ± 0.16 mm yr^{-1}). So the contribution from Greenland to past sea level rise appears to be somewhat less than that from glaciers and thermal expansion.

9.4.5 The Antarctic Ice Sheet.

The question of balance of the Antarctic ice sheet proves to be a very difficult one. From a physical point of view, regarding the very long time scale introduced by geodynamics and thermomechanical coupling in the ice sheet, it seems unlikely that the present ice sheet has adjusted completely to the last glacial-interglacial transition. A detailed modelling study by Huybrechts (1990), in which a glacial cycle of the Antarctic ice sheet is simulated on a 40km grid, suggests that the large-scale imbalance will not be more than a few percent of the annual mass turnover (corresponding to a rate of sea level change of less than 0.1 mm yr^{-1}). This does not exclude the possibility, however, that climate fluctuations with a shorter time scale have pushed the ice sheet out of balance. Also, there is increasing evidence that marine ice sheets, like the West Antarctic, could exhibit pulsating mass discharge which is not climate-related, but may have important consequences for sea level.

Budd and Smith (1985) made an assessment of the net balance by compiling a set of accumulation and ice velocity measurements (Table 9.6). The latter allow to make a rough estimate of the ice discharge from the main ice sheet across the grounding line, viz. 1879 x 10^{12} kg/yr. They find a number of about 2088 x 10^{12} kg/yr for the accumulation and estimate the net balance to be positive by

Table 9.6 *Antarctic mass balance (10^{12} kg/yr), † = without Antarctic Peninsula. Proper reference for SPRI (Scott Polar Research Institute) map and data: Drewry (1983).*

	Flux at grounding line	Surface balance (*grounded* ice)	Net
Budd and Smith (1985)	~1879	2088	0 to +418
Digitization SPRI *map* Huybrechts (1990)		2168	
Radok et al. (1986)		2158 1765 †	
Giovinetto and Bentley (1985)		1468 †	
Fortuin and Oerlemans (1990) [based on SPRI *data*]		1817	

209 x 10^{12} kg/yr. This would correspond to a rate of sea level change of about -0.6 mm yr^{-1}. Subsequent estimates of the total accumulation have produced lower values. Giovinetto and Bentley (1985) state that accumulation over the grounded part of the ice sheet is only 1468 x 10^{12}kg/yr. With the mean value for discharge from above, this yields a net balance of -411 x 10^{12} kg/yr. Fortuin and Oerlemans (1990) find, on the basis of a data set independently compiled from the archives of the Scott Polar Research Institute (SPRI), a mass gain at the surface of 1817 x 10^{12} kg/yr. With the discharge number from Budd and Smith, this then implies a net balance of -62 x 10^{12} kg/yr.

It must be stressed that the inference of ice mass discharge from a limited number of *surface* velocity measurements involves many uncertainties. The ratio of surface velocity to vertical mean velocity is such an uncertain factor. More seriously, outflow velocities vary dramatically from point to point, so lateral extrapolation and interpolation around the coast introduces very large errors. A comprehensive comparison of earlier estimates of the surface mass balance was given by Giovinetto and Bull (1987). Their discussion suggests that the total surface accumulation over grounded ice is not known to an accuracy better than 10%. When considering the net balance, this figure will be worse.

In conclusion, it is unknown whether the Antarctic ice sheet is currently in balance and whether it has been contributing to sea level rise over the last 100 years or not. A 20% imbalance of mass turnover cannot be detected in a definite way from present data.

Several methods exist to investigate how accumulation on the Antarctic ice sheet may change when temperature

changes. Analysis of the gas content in the deep Antarctic ice cores gives an indication of how accumulation varied between glacial and interglacial conditions (Lorius et al., 1984; Jouzel et al., 1989). In fact, it gives support to the view that accumulation on the interior is roughly proportional to the saturation mixing ratio of water vapour in the air above the inversion, as first suggested by Robin (1977). Another method involves regression analysis on measured temperatures and accumulation rates (Muszynski, 1985; Fortuin and Oerlemans, 1990). However, it is not so clear that a relation between accumulation and temperature based on spatial variation can be applied to climatic change. It is also possible to use precipitation rates as predicted by general circulation models of the atmosphere. Although the quality of these models has increased gradually, simulation of the climate of the polar regions still shows serious shortcomings (Schlesinger, 1990), and the results concerning glacier mass balance must be considered with much caution. So far, a systematic comparison between observed accumulation on the ice sheets and output from such models has not yet been published.

Table 9.7 lists a number of estimates of the change in Antarctic mass balance for a uniform warming of 1 degree C. Muszynski's estimate is the highest: a decrease of 0.38 mm/yr in sea level. The multiple regression analysis reported in Fortuin and Oerlemans (1990) yields a substantially lower value. In this analysis, which was based on a much larger newly compiled data set, a distinction was made between ice shelves, escarpment region and interior. Accumulation is strongly related to both temperature and latitude, parameters which also have a high mutual

Table 9.7 *Estimates of the change in Antarctic mass balance for a 1 °C warming. Δq_S represents saturation water vapour mixing ratio of air above the inversion.*

Source	Change in sea level (mm/yr)	Remarks
Muszynski (1985)	-0.38	Regression on 208 data points
Fortuin and Oerlemans (1990)	-0.139 (interior) -0.061 (escarpment) -0.200 (total)	Regression on 486 data points (only grounded ice)
Proportional to water vapour mixing ratio	-0.34	20 km grid over grounded ice

correlation. Taking this correlation out leads to a significantly weaker temperature dependence of the accumulation, but it can be argued that this approach is preferable when considering climate sensitivity. The value listed under "proportional to water vapour mixing ratio" was calculated by integrating over a 20km grid covering the entire ice sheet, with temperatures extrapolated from the data set used in the multiple regression mentioned above. The values thus obtained are rather close to the one suggested by Muszynski's work.

Support for the idea that higher temperatures will lead to significantly larger accumulation also comes from observations on the Antarctic Peninsula. Over the past 30 years temperature has gone up here by almost 2°C, whereas accumulation increased by as much as 25% in parallel with this (Peel and Mulvaney, 1988). Although this cannot be taken as proof of a causal relationship, it is in line with the sensitivity estimates listed in Table 9.7, which span a factor of two.

In summary, all quoted studies show an increase in accumulation with warming and thus a decrease in sea level. An ablation zone does not effectively exist in Antarctica, and a large warming would be required in order for ablation to influence mass balance.

9.4.6 Possible Instability of The West Antarctic Ice Sheet.
Most of the early attention to the issue of sea level rise and greenhouse warming was related to the stability of the West Antarctic ice sheet. Parts of this ice sheet are grounded far below sea level and may be very sensitive to small changes in sea level or melting rates at the base of adjacent ice shelves (e.g., Mercer, 1978; Thomas et al., 1979; Lingle, 1985; Van der Veen, 1986). In case of a climatic warming, such melting rates could increase and lead to disappearance of ice rises (places where the floating ice shelf runs aground). Reduced back stress on the main

ice sheet and larger ice velocities may result, with subsequent thinning of the grounded ice and grounding-line retreat.

It is hard to make quantitative statements about this mechanism. Several attempts have been made to model this ice sheet-shelf system and to study its sensitivity (Thomas et al., 1979; Lingle, 1985; Van der Veen, 1986, 1987; Budd et al., 1987). Van der Veen (1986), in a rather extensive study, concludes that the earlier estimates of the sensitivity of West Antarctica were too large. Budd et al. (1987) also give an extensive discussion on the response of the West Antarctic ice sheet to a climatic warming. Their considerations are based on a large number of numerical experiments with flow-band models. According to these experiments, very large ice-shelf thinning rates (10 to 100 times present values) would be required to cause rapid disintegration of the West Antarctic ice sheet. For a probably more realistic situation of a 50% increase in ice-shelf thinning rate for a one-degree warming (order of magnitude), the associated sea level rise would be about 0.1mm/yr for the coming decades.

Much of the drainage of the West Antarctic ice sheet goes through a number of fast flowing ice streams, the dynamics of which were not properly included in the modelling studies mentioned above. In recent years it has become clear from new observational studies (e.g., Bentley, 1987; Alley et al., 1987; MacAyeal, 1989) that those ice streams show much variability on a century and may be even decadal time scale. Although much of this variability is probably not related directly to climate change, it demonstrates the potential of this part of the ice sheet to react quickly to any change in boundary conditions. A comprehensive model of the ice streams and their interaction with the main ice body does not yet exist, unfortunately. Still, as argued by D.R. MacAyeal (abstract to the 1989-American Geophysical Union meeting on sea

Table 9.8 *Estimated contributions to sea-level rise over the last 100 years (in cm).*

	LOW	BEST ESTIMATE	HIGH
Thermal expansion	2	4	6
Glaciers/small ice caps	1.5	4	7
Greenland Ice Sheet	1	2.5	4
Antarctic Ice Sheet	-5	0	5
TOTAL	-0.5	10.5	22
OBSERVED	10	15	20

level change, unpublished), an extreme limit of the response of the West Antarctic ice sheet to greenhouse warming can be estimated. In his view the accelerated discharge of ice only occurs in the regions where sufficient sub-glacial sediments (the lubricant for the ice streams) is present. For a typical greenhouse warming scenario, the bulk of the increased mass outflow would occur between 100 and 200 years from now, and the actual projected West Antarctic contribution to sea level rise would be -10 cm after 100 yrs (increase in surface accumulation still dominating), +40 cm after 200 yrs, and +30 cm after 300 yrs (ice stream discharge stopped).

In summary, there is no firm evidence to suggest that the Antarctic ice sheet in general, or the West Antarctic ice sheet in particular, have contributed either positively or negatively to past sea level rise. On the whole, the sensitivity of Antarctica to climatic change is such that a future warming should lead to increased accumulation and thus a negative contribution to sea level change.

9.4.7 Other Possible Contributions

Sea level could also have been affected by net increases or decreases in surface and groundwater storage. In particular, groundwater depletion (through pumping) and drainage of swamps, soils and wetlands would contribute to a MSL rise. On the other hand, increases in surface storage capacity - especially large dams but also the combined effects of many small reservoirs and farm ponds - would detract from sea level.

Decreases in groundwater levels are commonly reported from all over the world from many different environments. This suggests that total groundwater storage volumes have been diminishing, particularly during the last 50 years. Data are meagre, however. One rough estimate (Meier, 1983; also see Robin, 1986) is that, globally, net depletion has amounted to about $2000km^3$ (equivalent to 0.55cm in sea level) during this century. Land drainage, particularly in Northwest Europe and North America over the last 100 years, has reduced soil and shallow groundwater storage over wide areas, but the actual amounts of water are difficult to estimate.

Substantial increases in surface storage have occurred since the 1930s. Newman and Fairbridge (1986) estimated that this has amounted to about $18750km^3$ (-5.2cm in sea level, using $362 \times 10^6 km^2$ for ocean area) over the period 1932-1982. Golubev (1983; Also see Robin, 1986), however, makes a much lower estimate, $5500km^3$ (-1.5cm in sea level).

Overall, the estimates appear too imprecise and the data insufficient, especially for groundwater changes, to be able to conclude much about the possible net effects on past sea level rise.

9.4.8 Synthesis

The estimated contributions to past sea level rise can now be summarised (Table 9.8). Assuming the contribution from Antarctica has been zero, the combined contributions from thermal expansion, mountain glaciers and the Greenland ice sheet over the last 100 years total 10.5cm. This is within the range of observed sea level rise (10 - 20cm), albeit at the lower end. The range of uncertainty is large: -0.5cm to 22cm.

Table 9.9 *Estimates of future global sea level rise (cm) (Modified from Raper et al., 1990)*

	CONTRIBUTING FACTORS				TOTAL RISE [a]		
	Thermal Expansion	Alpine	Greenland	Antarctica	Best Estimate	Range [f]	To (Year)
Gornitz (1982)	20	20 (Combined)			40		2050
Revelle (1983)	30	12	13		71 [b]		2080
Hoffman et al. (1983)	28 to 115	28 to 230 (Combined)				56 to 345	2100
						26 to 39	2025
PRB (1985)	c	10 to 30	10 to 30	-10 to 100		10 to 160	2100
Hoffman et al. (1986)	28 to 83	12 to 37	6 to 27	12 to 220		58 to 367	2100
						10 to 21	2025
Robin (1986) [d]	30 to 60 [d]	20±12 [d]	to +10 [d]	to -10 [d]	80 [i]	25 - 165 [i]	2080
Thomas (1986)	28 to 83	14 to 35	9 to 45	13 to 80	100	60 to 230	2100
Villach (1987) (Jaeger, 1988) [d]					30	-2 to 51	2025
Raper et al. (1990)	4 to 18	2 to 19	1 to 4	-2 to 3	21 [g]	5 to 44 [g]	2030
Oerlemans (1989)					20	0 to 40	2025
Van der Veen (1988) [h]	8 to 16	10 to 25	0 to 10	-5 to 0		28 to 66	2085

[a] - from the 1980s
[b] - total includes additional 17cm for trend extrapolation
[c] - not considered
[d] - for global warming of 3.5°C
[f] - extreme ranges, not always directly comparable
[g] - internally consistent synthesis of components
[h] - for a global warming of 2-4°C
[i] - estimated from global sea level and temperature change from 1880-1980 and global warming of 3.5±2.0°C for 1980-2080

9.5 How Might Sea Level Change in the Future?

Various estimates of future sea level rise are noted in Table 9.9. Such estimates are very difficult to compare because different time periods are chosen, and because assumptions regarding future greenhouse gas concentrations, changes in climate, response times, etc., are either different or not clearly stated. In general, most of the studies in Table 9.9 foresee a sea level rise of somewhere between 10cm and 30cm over the next four decades. This represents a rate of rise that is significantly faster than that experienced, on average, over the last 100 years.

Projections for the present assessment are made using the standard IPCC greenhouse gas forcing scenarios. These consist of a "Business-as-Usual" scenario, and three lower scenarios (B-D) in which greenhouse gas emissions are substantially reduced. Three projections are made for each scenario (12 projections in total) reflecting the high, low and best-estimate assumptions for each of the contributing factors, as described below.

9.5.1 *Methods and Assumptions*

Estimates of the thermal expansion effects are obtained using the upwelling-diffusion model of Wigley and Raper (1987) described in 9.4.1 and in Section 6. For each scenario, the model is run using a climate sensitivity (ΔT_{2x}) of 1.5°C, 2.5°C and 4.5°C for the low, best-estimate and high projections, respectively, with the diffusivity set to 0.63cm^{2}sec^{-1} and π set to 1 (see Section 6 for the justification of the choice of diffusivity and π values).

Concerning glaciers and small ice caps, significant warming may decrease the ice-covered area within a hundred years. Thus, in order to make realistic estimates of the glacier contribution, the changes in glacier area have to be taken into account. This is accomplished using a simple, global glacier melt model (Raper et al., 1990). The model contains three parameters that have to be prescribed: initial ice volume, a global-mean glacier response time and a representative glacier temperature sensitivity parameter. The parameter values were chosen to match estimated rates of glacier volume loss over the last 100 years. The model was run from 1861 to 2100 (implying that, at present, glaciers are in disequilibrium).

With respect to the Greenland and Antarctic ice sheets (including the West Antarctic ice sheet and the Antarctic Peninsula), the dynamic response can effectively be ignored for the time-scales considered here. The static changes in the surface mass balance can thus be represented by the sensitivity values discussed above, that is:

$\Delta h = 0.3 \pm 0.2$mm/yr per degree for the Greenland ice sheet

$\Delta h = -0.3 \pm 0.3$mm/yr per degree for the Antarctic ice sheet

Based on the latest results from transient runs of fully-coupled ocean-atmosphere GCM's (Stouffer et al., 1989), it was assumed that temperature changes were equivalent to the global mean, except in Greenland where temperature changes were enhanced by a factor of 1.5.

9.5.2 *Discussion*

The resultant projections of global sea level rise to the year 2100 are shown in Figures 9.6 and 9.7. Under the Business-as-Usual scenario, the best estimate is that, for the year 2030, global sea level would be 18cm higher than today. Given the stated range of uncertainty in the contributing factors, the rise could be as little as 8cm or as high as 29cm. By the year 2070, the projected range is 21-71cm with a best-estimate of 44cm, although it should be cautioned that projections this far into the future are fraught with many uncertainties, many of which are external to thermal expansion and land ice melting.

The major contributing factors to the sea level rise are thermal expansion of the oceans and glaciers and small ice caps. The minor contributions to sea level from the Greenland and Antarctic ice sheets are positive and negative, respectively (Table 9.10).

For scenarios B, C and D (Figure 9.7), the sets of projections are similar. This is because with low forcing scenarios, the temperature and sea level effects are more sensitive to ΔT_{2x} and the history of forcing change up to 1990 than to forcing change post-1990. The best-estimates for the year 2070 fall in the range 27-33cm, about one-third less than the Business-as-Usual case.

The fact that sea level continues to rise throughout the 21st century - even under scenarios of strict emission reductions demonstrates the strong effect of past changes in greenhouse gas concentrations on future climate and sea level. This is because of the lag effects introduced by the thermal inertia of the oceans and the continuing response of land ice to climate changes. In effect, this creates a very substantial sea level rise "commitment". This is illustrated

Table 9.10 *Factors contributing to sea level rise (cm), 1985 - 2030. "Business-as-Usual" Scenario - Best Estimate for 2030.*

	Thermal Expansion	Mountain Glaciers	Greenland	Antarctica	TOTAL
HIGH	14.9	10.3	3.7	0.0	28.9
BEST ESTIMATE	10.1	7.0	1.8	-0.6	18.3
LOW	6.8	2.3	0.5	-0.8	8.7

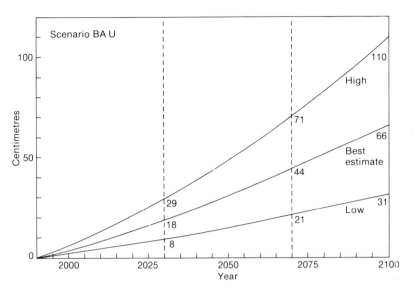

Figure 9.6: Global sea-level rise, 1990-2100, for Policy Scenario Business-as-Usual.

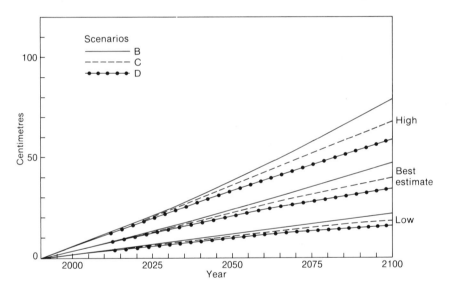

Figure 9.7: Global sea-level rise, 1990-2100, for Policy Scenarios B, C, D.

in Figure 9.8. Here, the IPCC "Business-as-Usual" Scenario of greenhouse forcing is imposed to the year 2030 with no further changes in forcing thereafter. Sea level, however, continues to rise at almost the same rate for the remainder of the century.

This section has been concerned primarily with *global* mean sea level rise. It should be borne in mind that sea level will not rise uniformly around the world. First, at any given coastal location sea level will be influenced by local and regional vertical land movements. In some circumstances, these are large and will mask climate related changes in ocean volume. Second, dynamic

processes in the ocean and atmospheric circulation will also cause sea level to change regionally. For example, a sensitivity study with a dynamic ocean model showed regional differences of up to a factor of two relative to the global-mean value (Mikolajewicz et al., 1990). Finally, changes in the frequency of extreme sea level events may be most important in their impact on coastal zones, but are currently difficult to quantify because of the uncertainties in regional predictions of climatic change.

In general, for the coming decades, the present best-estimate projection of sea level rise for the Business-as-

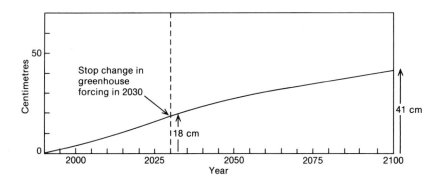

Figure 9.8: Commitment to sea level rise in the year 2030. The curve shows the sea level rise due to Business-as-Usual emissions to 2030, with the additional rise that would occur in the remainder of the century even if climate forcing was stabilised in 2030.

Usual case does not represent a major departure from those found in the most recent literature (Table 9.9).

9.6 Summary and Conclusions

This chapter has addressed three questions:

Has global-mean sea level been rising over the last 100 years?

What are the causal factors that could explain a past rise in sea level? and,

What increases in sea level can be expected in the future?

The array of data and methodological problems inherent in estimating the rate of past sea level change is large. The selection of data and its manipulation can make a difference of more than a factor of two in the global trend estimate. While recent analyses of MSL trends involve more refined means of data correction and analysis, they generally support, not alter, the broad conclusions of previous assessments. It is our judgement that:

Global sea level has been rising.

The average rate of rise over the last 100 years has been 1.0 - 2.0mm/yr.

There is no firm evidence of an acceleration in global MSL rise over this century (although there is some evidence that sea level rose faster in this century compared to the previous two centuries).

It appears that the past rise in sea level is due largely to thermal expansion of the oceans and increased melting of glaciers and the margins of the Greenland ice sheet. There is no firm basis for supposing that the Antarctic ice sheet has contributed either positively or negatively to past sea level change. In general, these findings support the conclusion, based on analyses of tide gauge records, that there has been a globally-coherent, secular rise in sea level, and that the causes are most likely related to climatic change.

Future changes in sea level were estimated for each of the IPCC forcing scenarios. For each scenario, three projections - best estimate, high and low - were made corresponding to the estimated range of uncertainty in each of the potential contributing factors, and in the climate sensitivity and resulting global warming predictions.

It is found that:

For the "Business-as-Usual" Scenario at year 2030, global-mean sea level is 8-29cm higher than today, with a best-estimate of 18cm. At the year 2070, the rise is 21-71cm, with a best-estimate of 44cm.

Most of the contribution is estimated to derive from thermal expansion of the oceans and the increased melting of mountain glaciers.

The Antarctic ice sheet contributes negatively to sea level due to increased accumulation associated with warming. Increased outflow of ice from the West Antarctic ice sheet is likely to be limited, but the uncertainty is large.

The Greenland ice sheet contributes positively to sea level rise, but part of the enhanced melting and runoff may be offset by increased snowfall in the higher parts, so the uncertainties are very large.

For the lower forcing scenarios (B,C and D), the sea level rise projections are similar, at least until the mid-21st century. On average these projections are approximately one-third lower than those of the "Business-as-Usual" Scenario.

Even with substantial decreases in the emissions of the major greenhouse gases, future increases in sea level are unavoidable - a sea level rise "commitment" - due to lags in the climate system.

In general, this review concludes that a rise of more than 1 metre over the next century is unlikely. Even so, the rate of rise implied by the Business-as-Usual best-estimate is 3-6 times faster than that experienced over the last 100 years. The prospect of such an increase in the rate of sea level rise should be of major concern to many low-lying coasts subject to permanent and temporary inundation, salt intrusion, cliff and beach erosion, and other deleterious effects.

References

Alley, R.B., D.D. Blankenship, S.T. Rooney and C.R. Bentley (1987): Till beneath ice stream B 4. A coupled ice-till flow model. *J. Geophys. Res*, **92**, 8931-8940.

Ambach, W. (1963): Untersuchungen sum Energieumsatz in der Ablationszone des Grönlandischen Inlandeises. *Meddelelser om Gronland*, **174(4)**, 311 pp.

Ambach, W. (1979): Zur Nettoeisablation in einem Höhenprofil am Grönlandischen Inlandeis. *Polarforschung*, **49**, 55-62.

Ambach, W. and M. Kuhn (1989): Altitudinal shift of the equilibrium line in Greenland calculated from heat balance characteristics. In: J. Oerlemans (ed.), *Glacier Fluctuations and Climatic Change*, Kluwer (Dordrecht), pp. 281-288.

Aubrey, D.G. (1985): Recent sea levels from tide gauges: problems and prognosis. In: *Glaciers, Ice Sheets and Sea Level: Effect of a CO2-induced climatic change*. DOE/ER/60235-1 (U.S. Department of Energy Carbon Dioxide Research Division, Washington) pp. 73-91.

Barnett, T.P. (1983): Recent changes in sea level and their possible causes. *Clim. Change*, **5**, 15-38.

Barnett, T.P. (1984): Estimation of "global" sea level change: a problem of uniqueness. *J. Geophys. Res*, **89**, 7980-7988.

Barnett, T.P. (1985): Long-term climatic change in observed physical properties of the oceans. In: *Detecting the Climatic Effects of Increasing Carbon Dioxide*, (Eds., M.C. MacCracken and F.M. Luther). U.S. DOE/ER-0235, pp. 91-107.

Barnett, T.P. (1988): Global sea level change. In: *NCPO, Climate variations over the past century and the greenhouse effect*. A report based on the First Climate Trends Workshop, 7-9 September 1988, Washington D.C. National Climate Program Office/NOAA, Rockville, Maryland.

Bauer, A. (1968): Nouvelle estimation du bilan de masse de l'Inlandsis du Groenland. *Deep Sea Res.*, **14**, 13-17.

Bader, H. (1961): *The Greenland Ice Sheet*. Cold Region Science and Engineering Report, I-B2.

Benson, C.S. (1962): *Stratigraphic studies in the snow and firn of the Greenland ice sheet*. SIPRE Res. Report no. 70.

Bentley, C.R. (1987): Antarctic ice streams: a review. *J. Geophys. Res.*, **92**, 8843-8858.

Bindschadler, R.A. (1985): Contribution of the Greenland Ice Cap to changing sea level: present and future. In: *Glaciers, Ice Sheets and Sea Level: Effects of a CO2-induced climatic change*. National Academy Press (Washington), pp. 258-266.

Björnsson, H., (1979): Nine glaciers in Iceland. *Jökull*, **29**, 74-80.

Bolin, B., B. Döös, J. Jäger and R.A. Warrick (Eds.) *The Greenhouse Effect, Climatic Change, and Ecosystems*, SCOPE 29. John Wiley and Sons (Chichester)

Braithwaite, R.G. and O.B. Oleson (1989): Ice ablation in West Greenland in relation to air temperature and global radiation. *Zeitschr. Gletscherk. Glazialgeol.*, **20**, 155-168.

Braithwaite, R.G. and O.B. Oleson (1989): Calculation of glacier ablation from air temperature, West Greenland. In: J. Oerlemans (ed.), *Glacier Fluctuations and Climatic Change*. Kluwer (Dordrecht), pp. 219-234.

Braithwaite, R.G. and O.B. Oleson (1990): Increased ablation at the margin of the Greenland ice sheet under a greenhouse effect climate. *Annals of Glaciology,* **14**, in press.

Budd, W.F., B.J. McInnes, D. Jenssen and I.N. Smith (1987): Modelling the response of the West Antarctic Ice Sheet to a climatic warming. In: *Dynamics of the West Antarctic Ice Sheet*, (Eds., C.J. van der Veen and J. Oerlemans), Reidel, pp. 321-358.

Budd, W.F. and I.N. Smith (1985): The state of balance of the Antarctic Ice Sheet, an updated assessment 1984. In: *Glaciers, Ice Sheets and Sea Level: Effects of a CO2-induced Climatic Change*. National Academy Press (Washington), pp. 172-177.

Carter, W.E., D.G. Aubrey, T. Baker, C. Boucher, C. LeProvost, D. Pugh, W.R. Peltier, M. Zumberge, R.H. Rapp, R.E. Schutz, K.O. Emery and D.B. Enfield (1989): *Geodetic Fixing of Tide Gauge Bench Marks*. Woods Hole Oceanographic Institution, Technical Report WHOI-89-31.

Douglas, B.C., R.E. Cheney, L. Miller, R.W. Agreen, W.E. Carter and D.S. Robertson (1990): Greenland ice sheet: is it growing or shrinking? *Science*, in press.

Drewry, D., (1983): *Antarctica, Glaciological and Geophysical Folio*, Scott Polar Research Institute, Cambridge, UK.

Ekman, M. (1988): The world's longest continued series of sea level observations, *Pure and App. Geophys.*, **127**, 73-77.

Flint, R.F. (1971): Glacial and Quaternary Geology. John Wiley, New York, 892pp.

Fortuin, J.P.F. and J. Oerlemans (1990): Parameterization of the annual surface temperature and mass balance of Antarctica. *Annals of Glaciology*, **14**, in press.

Giovinetto, M.B. and C.R. Bentley (1985): Surface balance in ice drainage systems of Antarctica. *Antarctic Journal of the United States*, **20**, 6-13.

Giovinetto, M.B. and C. Bull (1987): *Summary and analysis of surface mass balance compilations for Antarctica, 1960-1985*. Byrd Polar Research Center, Rep. no. 1, 90pp.

Golubev, G.N. (1983): Economic activity, water resources and the environment: a challenge for hydrology. *Hydrol. Sci. J.*, **28**, 57-75.

Gornitz, V. (1990): Mean sea level changes in the recent past. In: *Climate and Sea Level Change: Observations, Projections and Implications*, (Eds., R.A. Warrick and T.M.L. Wigley). Cambridge University Press, Cambridge (in press).

Gornitz, V. and A. Solow (1989): Observations of long-term tide-gauge records for indications of accelerated sea level rise, Workshop on Greenhouse-Gas-Induced Climatic Change, May 8-12, 1989, University of Massachusetts, Amherst, MA.

Gornitz, V. and S. Lebedeff (1987): Global sea level changes during the past century. In: *Sea-level fluctuation and coastal evolution*, (Eds., D. Nummedal, O.H. Pilkey and J.D. Howard), SEPM Special Publication No. 41.

Gornitz, V., S. Lebedeff and J. Hansen (1982): Global sea level trends in the past century. *Science, 215*, 1611-1614.

Greuell, W. (1989): *Glaciers and climate: energy balance studies and numerical modelling of the historical front variations of the Hintereisferner (Austria)*. Ph.D. Thesis. University of Utrecht (The Netherlands), 178 pp.

Grove, J.M. (1988): *The Little Ice Age*. Methuen (London).

Haeberli, W., H. Bösch, K. Scherler, G. Østrem and C.C. Wallén (1989): *World Glacier Inventory. Status 1988*. IAHS(ICSI)-UNEP-UNESCO.

Haeberli, W., P. Müller, P. Alean and H. Bösch (1989): Glacier changes following the little ice age - A survey of the international data basis and its perspectives. In: *Glacier Fluctuations and climatic change*, (Ed. J. Oerlemans), Reidel (Dordrecht), pp. 77-101.

Hoffert, M.I. and B.P. Flannery (1985): Model projections of the time-dependent response to increasing carbon dioxide. In: *Projecting the Climatic Effects of Increasing Carbon Dioxide*, (Eds., M.C. MacCracken and F.M. Luther), U.S. Department of Energy, Carbon Dioxide Research Division, pp. 149-190.

Hoffman, J.S., J.B. Wells and J.G. Titus (1986): Future global warming and sea level rise. In: *Iceland Coastal and River Symposium*, (Ed. G. Sigbjarnason), Reykjavik, National Energy Authority, 245-266.

Hoffman, J.S., D. Keyes and J.G. Titus (1983): *Projecting Future Sea Level Rise: Methodology, Estimates to the Year 2100, and Research Needs*, U.S. GPO 055-000-0236-3. GPO, Washington, D.C.

Huybrechts, Ph. (1990): The Antarctic ice sheet during the last glacial-interglacial cycle: a 3-D model experiment. *Annals of Glaciology, 14*, in press.

Jaeger, J. (1988): Developing policies for responding to climatic change: a summary of the discussions and recommendations of the workshops held in Villach 1987 and Bellagio 1987. WMO/TD-No.225.

Jouzel, J. and 9 others (1989): A comparison of deep Antarctic ice cores and their implications for climate between 65,000 and 15,000 years ago. *Quat. Res., 31*, 135-150.

Kasser, P. (1967): *Fluctuations of Glaciers, 1959-1965*. Vol. I. IASH(ICSI)-UNESCO.

Kasser, P. (1973): *Fluctuations of Glaciers, 1965-1970*. Vol. II. IASH(ICSI)-UNESCO.

Kasser, P. and W. Haeberli (eds) (1979): *Die Schweiz und ihre Gletscher. Von der Eiszeit bis zur Gegenwart*. Kümmerly und Frei, Bern.

Kostecka, J.M. and I.M. Whillans (1988): Mass balance along two transects of the west side of the Greenland ice sheet. *J. Glaciology, 34*, 31-39.

Kuhn, M. (1990): Contribution of glaciers to sea level rise. In: *Climate and Sea Level Change: Observations, Projections and Implications*, (Eds., R.A. Warrick and T.M.L. Wigley). Cambridge University Press, Cambridge (in press).

Lingle, C.S. (1985): A model of a polar ice stream and future sea level rise due to possible drastic retreat of the West Antarctic Ice Sheet. In: *Glaciers, Ice Sheets and Sea Level: Effects of a*

CO_2-induced Climatic Change. National Academy Press (Washington), pp. 317-330.

Lorius, C., D. Raynaud, J.R. Petit, J. Jouzel and L. Merlivat (1984): Late glacial maximum Holocene atmospheric and ice thickness changes from ice core studies. *Annals of Glaciology, 5*, 88-94.

MacAyeal, D.R. (1989): Ice-shelf response to ice-stream discharge fluctuations: III. The effects of ice-stream imbalance on the Ross Ice Shelf, Antarctica. *J. Glaciology, 35*, 38-42.

MacCracken, M.C. and F.M. Luther (Eds.) *Detecting the Climatic Effects of Increasing Carbon Dioxide*, U.S. DOE/ER-0235.

Meier, M.F. (1983): Snow and ice in a changing hydrological world. *Hydrol. Sci., 28*, 3-22.

Meier, M.F. (1984): Contribution of small glaciers to global sea level. *Science, 226*, 1418-1421.

Mercer, J.H. (1978): West Antarctic ice sheet and CO_2 greenhouse effect: a threat of disaster. *Nature 271*, 321-325.

Mikolajewicz, U., B. Santer and E. Maier-Reimer (1990): Ocean response to greenhouse warming. Max-Planck-Institut für Meteorologie, Report 49.

Müller, F. (1977): *Fluctuations of Glaciers, 1970-1975*. Vol. III. IASH(ICSI)-UNESCO.

Newman, W.S. and R.W. Fairbridge (1986): The management of sea level rise. *Nature, 320*, 319-321.

Oerlemans, J. (1989): A projection of future sea level. *Climatic Change, 15*, 151-174.

Oerlemans, J. (1990): A model for the surfacebalance of ice masses: Part I. Alpine glaciers. *Z. Gletscherk. Glazialgeol.*, submitted.

Oerlemans, J., R. van de Wal and L.A. Conrads (1990): A model for the surfacebalance of ice masses: Part II. Application to the Greenland ice sheet. *Z. Gletscherk. Glazialgeol.*, submitted.

Oerlemans, J. and C.J. van der Veen (1984): *Ice sheets and climate*. Reidel (Dordrecht), 217 pp.

Ohmura, A. (1987): New temperature distribution maps for Greenland. *Z. Gletscherk. Glazialgeol., 23*, 1-45.

Ohmura, A. and N. Reeh (1990): New precipitation and accumulation maps for Greenland. *J. Glaciology*, submitted.

Østrem, G., O. Liestøl and B. Wold (1977): Glaciological investigations at Nigardbreen, Norway. *Norsk Geogr. Tisddkr, 30*, 187-209.

Peel, D.A. and R. Mulvaney (1988): Air temperature and snow accumulation in the Antarctic Peninsula during the past 50 years. *Annals of Glaciology, 11*, 206-207.

Peltier, W.R. and A.M. Tushingham (1989): Global sea level rise and the greenhouse effect: might they be connected? *Science, 244*, 806-810.

Peltier, W.R. and A.M. Tushingham (1990): The influence of glacial isostatic adjustment on tide gauge measurements of secular sea level. *J. Geophys. Res.*, (in press).

Pirazzoli, P.A., D.R. Grant and P. Woodworth (1987): Trends of relative sea level change: past, present and future. XII International INQUA Congress, Ottawa, July 31-August 9, 1987, Special Session 18: "Global Change".

Polar Research Board (PRB) (1985): *Glaciers, Ice Sheets and Sea Level: Effect of a CO_2-induced Climatic Change*. Report of a Workshop held in Seattle, Washington, September 13-15, 1984. U.S. DOE/ER/60235-1.

Pugh, D.T., N.E. Spencer and P.L. Woodworth (1987): Data holdings of the Permanent Service for Mean Sea Level, Bidston, Birkenhead. Permanent SErvice for Mean Sea Level, 156pp.

Pugh, D.T. (1990): Improving sea level data. In: *Climate and Sea Level Change: Observations, Projections and Implications*, (Eds., R.A. Warrick and T.M.L. Wigley). Cambridge University Press, Cambridge (in press).

Radok, U., Barry, R.G., Jenssen, D., Keen, R.A., Kiladis, G.N. and McInnes, B. (1982): *Climatic and Physical Characteristics of the Greenland Ice Sheet*, CIRES, University of Colorado, Boulder.

Radok, U., T.J. Brown, D. Jenssen, I.N. Smith and W.F. Budd (1986): *On the Surging Potential of Polar Ice Streams. Part IV. Antarctic Ice Accumulation Basins and their main Discharge Regions*. CIRES, Boulder / University of Melbourne. Rep. DE/ER/60197-5.

Raper, S.C.B., R.A. Warrick and T.M.L. Wigley (1990): Global sea level rise: past and future. In: *Proceedings of the SCOPE Workshop on Rising Sea Level and Subsiding Coastal Areas*, Bangkok 1988, (Ed. J.D. Milliman), John Wiley and Sons (Chichester), in press.

Reeh, N. (1985): Greenland ice sheet mass balance and sea level change. In: *Glaciers, Ice Sheets and Sea Level: Effects of a CO_2-induced Climatic Change*. National Academy Press, (Washington), pp. 155-171.

Reeh, N. and N.S. Gundestrup (1985): Mass balance of the Greenland ice sheet at Dye 3. *J. Glaciology*, **31**, 198-200.

Revelle, R. (1983): Probable future changes in sea level resulting from increased atmospheric carbon dioxide. In: NAS *Changing Climate*, pp. 433-447. NAS (Washington D.C.).

Robin, G. de Q. (1977): Ice cores and climatic change. *Phil. Trans. R. Soc., London*, Series B, **280**, 143-168.

Robin, G. de Q. (1986): Changing the sea level. In: *The Greenhouse Effect, Climatic Change, and Ecosystems*, (Eds., B. Bolin, B. Döös, J. Jäger and R.A. Warrick). John Wiley and Sons (Chichester), pp. 323-359.

Roemmich, D. (1985): Sea level and the thermal variability of the ocean. In: *Glaciers, Ice Sheets and Sea Level: Effects of a CO_2-induced Climatic Change*. National Academy Press, (Washington), pp. 104-115.

Stouffer, R.J., S. Manabe and K. Bryan (1989): Interhemispheric asymmetry in climate response to a gradual increase of atmospheric CO_2. *Nature*, **342**, pp. 660-662.

Thomas, R.H., T.J.D. Sanderson and K.E. Rose (1979): Effects of a climatic warming on the West Antarctic Ice Sheet. *Nature*, **227**, 355-358.

Thomson, R.E. and S. Tabata (1987): Steric height trends of ocean station PAPA in the northeast Pacific Ocean. *Marine Geodesy*, **11**, 103-113.

Trupin, A. and J.Wahe, 1990: Spectroscopic analysis of global tide guage sea level data. *Geophys. J. Int.*, **100**, 441-454.

Van der Veen, C.J. (1986): *Ice Sheets, Atmospheric CO_2 and Sea Level*, Ph.D. Thesis, University of Utrecht (The Netherlands).

Van der Veen, C.J. (1987): Longitudinal stresses and basal sliding: a comparative study. In: *Dynamics of the West Antarctic Ice Sheet*, (Eds., C.J. van der Veen and J. Oerlemans), Reidel, 223-248.

Van der Veen, C.J. (1988): Projecting future sea level. *Surv. in Geophys.*, **9**, 389-418.

Vivian, R. (1975): *Les Glaciers des Alpes Occidentales*. Allier, Grenoble.

Weidick, A. (1984): Review of glacier changes in West Greenland. *Z. Gletscherk. Glazialgeol.*, **21**, 301-309.

Wigley, T.M.L. and S.C.B. Raper (1987): Thermal expansion of sea water associated with global warming. *Nature*, **330**, 127-131.

Wigley, T.M.L. and S.C.B. Raper (1990): Future changes in global-mean temperature and thermal-expansion-related sea level rise. In: *Climate and Sea Level Change: Observations, Projections and Implications*, (Eds., R.A. Warrick and T.M.L. Wigley), Cambridge University Press, Cambridge (in press).

Woodworth, P.L. (1990): A search for accelerations in records of European mean sea level. *International J. Climatology*, **10**, 129-143.

Zwally, H.J. (1989): Growth of Greenland ice sheet: Interpretation. *Science*, **246**, 1589-1591.

10

Effects on Ecosystems

J.M. MELILLO, T.V. CALLAGHAN, F.I. WOODWARD, E. SALATI, S.K. SINHA

Contributors:
J. Aber; V. Alexander; J. Anderson; A. Auclair; F. Bazzaz; A. Breymeyer; A. Clarke; C. Field; J.P. Grime; R. Gifford; J. Goudrian; R. Harris; I. Heaney; P. Holligan; P. Jarvis; L. Joyce; P. Levelle; S. Linder; A. Linkins; S. Long; A. Lugo, J. McCarthy, J. Morison; H. Nour; W. Oechel; M. Phillip; M. Ryan; D. Schimel; W. Schlesinger; G. Shaver; B. Strain; R. Waring; M. Williamson.

CONTENTS

EXECUTIVE SUMMARY

Ecosystem Metabolism and Climate Change

Photosynthesis, plant and microbial respiration tend to increase with increasing temperatures, but at higher temperatures respiration is often the more sensitive process. As a consequence, global warming may result in a period of net release of carbon from the land to the atmosphere. The magnitude of this release is uncertain. Factors that will influence the amount of carbon released include local patterns of climate change and the responses of the biota to simultaneous changes in soil moisture and atmospheric CO_2 concentration.

Increased soil water availability will tend to stimulate plant growth in dry ecosystems and increase carbon storage in cold and wet ecosystems like lowland tundra. A number of recent modelling studies have predicted that water stress will be a primary cause of tree death in the southern temperate forests of the Northern Hemisphere as climate changes. Forest death and replacement by grasslands would result in a net flux of carbon from the terrestrial biosphere to the atmosphere.

Increased atmospheric CO_2 has the potential to increase plant growth in a variety of ways: stimulation of photosynthesis; depression of respiration; relief of water and low light stresses; relief of nutrient stress by several mechanisms (greater nutrient use efficiency, increased nutrient uptake through root-microbial associations, increased symbiotic nitrogen fixation); and delay of senescence that prolongs the growing season. Some of the mechanisms that promote increased growth could be particularly important in arid/semi-arid and infertile areas. However, there is great uncertainty about whether or not these mechanisms operate for prolonged periods in natural ecosystems. For example, there are no field data from whole-ecosystem studies of forests that demonstrate a "CO_2 fertilization" effect. If elevated CO_2 does stimulate the growth of woody vegetation, this could lead to long-term net carbon storage in terrestial ecosystems.

Ecosystem Structure and Climate Change

Because species respond differently to climatic change, some will increase in abundance while others will decrease. Ecosystems will therefore change in structure. Over time some species may be displaced to higher latitudes or altitudes. Rare species with small ranges may be prone to local or even global extinction.

Warming rates are predicted to be rapid (0.3°C per decade) and there is great uncertainty about how species will respond to these rapid changes. Ecosystems of large stature such as forests may not be able to migrate fast enough to keep pace with climate change. In past times, species migrations were largely unaffected by human land use. Barriers to migration now exist (e.g., human settlements, highways, etc.). Therefore, inferences from previous migrations cannot be applied without caution to the present and future situations.

Human Activities, Ecosystem Changes and the Climate System

Human activities such as deforestation in the tropics and forest harvest and regrowth in mid-latitudes of the Northern Hemisphere are influencing the climate system by affecting greenhouse gas fluxes. Deforestation in the tropics is releasing 1.6 ± 1 Pg C annually to the atmosphere.

The net exchange of carbon between the land and the atmosphere due to forest harvest and regrowth in the mid-latitudes of the Northern Hemisphere is uncertain. These regrowing forests may be accumulating 1-2 Pg C annually. One analysis suggests that an equivalent amount of carbon is released back to the atmosphere through the burning and decay of previously harvested wood.

The issue of carbon storage in the mid-latitudes of the Northern Hemisphere is further complicated by the eutrophication of the region with nitrogen. Nitrogen in agricultural fertilizers and in acid rain may be promoting carbon storage at the rate of 0.5 - 1.0 Pg C annually, but there is considerable uncertainty in this estimate.

Reforestation as a Means of Managing Atmospheric CO_2

Reducing the atmospheric CO_2 concentration through an afforestation program would require the planting of a vast area of forest. Approximately 370×10^6 ha of temperate forest would have to be planted in order to accumulate 1 Pg C annually. This assumes a forest with an annual carbon accumulation rate of 2.7 t per hectare. The carbon accumulation would continue for almost a century. After that time, the forest would be mature and would not sequester more carbon.

Methane and Nitrous Oxide Fluxes

Microbial activity is the dominant source to the atmosphere of methane and nitrous oxide. Warmer and wetter soil conditions may lead to increased fluxes of these gases to the atmosphere. Changes in land use, and fertilizer and atmospheric inputs of nitrogen, also have the potential to affect methane and nitrous oxide fluxes.

Deforestation and Regional Hydrology

The conversion of large areas of tropical forest to grassland will likely change the hydrological regime of the region. Rainfall will be reduced and surface water flow will be affected.

Marine Ecosystems

Climate change will probably affect ocean circulation and mixing patterns. Circulation and mixing control nutrient availability to the oceans' microscopic plants (phytoplankton) and their access to solar radiation required for photosynthesis. Since nutrients are an important controller of net primary production in marine environments, production will be changed to the degree that upper ocean physical processes change in response to climate change. Different nutrient and mixing regimes are characterized by different plankton communities, which have wide ranging efficiencies of processing carbon, with important implications for long term ocean storage of organic carbon.

10.0 Introduction

On the basis of current evidence from climate modelling studies it appears that the change in globally averaged surface temperature due to doubling CO_2 probably lies in the range 1.5 to 4.5°C (Section 5). Temperature changes of this magnitude in the Earth's history have been associated with shifts in the geographic distribution of terrestrial biota. For example, the boreal forests of Canada extend well north of the current timber line during the Medieval Warm Epoch (800 to 1200 AD); a time when temperatures in that region were about 1°C warmer than today's. At the same time, farmers in Scandinavia grew cereal crops as far north as 65° latitude (Lamb, 1977). Evidence from the past suggests that the potential for ecosystem change in a warmer future is large (Warrick et al. 1986a).

A shift in the geographic distribution of terrestrial biota is a long-term (decades to centuries) response to climate change. Responses to a changing climate will also occur at other time-scales. In the short term (minutes to years), likely ecosystem responses include changes in the rates of processes such as photosynthesis and decomposition, and changes in the interactions between species such as those between plants and insect pests. In the intermediate term (years to decades), these changes in processes and interactions will lead to changes in community structure. For example, in a mixed forest type in the mid-latitude region, where both deciduous and coniferous tree species coexist, a warmer climate could lead to the loss of the conifers.

Some climate-induced changes of ecosystem structure and function are expected to feed back to the climate system. For instance, the warming of high latitude wetlands will almost certainly increase the production of CH_4 and its release to the atmosphere and this will accelerate warming.

10.1 Focus

In this section we consider two general issues, the effects of global change on ecosystems and the effects of ecosystem changes on the climate system. We center most of our discussion on process-level responses of ecosystems to global change. To understand many ecosystem responses to climate change, we consider them in the context of other components of global change such as increases in the atmospheric concentration of CO_2. We also consider the ecological consequences of tropical deforestation and the eutrophication of Northern Hemisphere areas with nitrogen in agriculture fertilizers and in acid precipitation as examples of ecosystem changes influencing climate systems. While the primary focus of this section is on terrestrial ecosystems, we end the section with a brief discussion of climate change and marine ecosystems.

10.2 Effects of Increased Atmospheric CO_2 and Climate Change on Terrestrial Ecosystems

Increases in atmospheric CO_2, warming, and changes in precipitation patterns all have the potential to affect terrestrial ecosystems in a variety of ways. Here we review some of the major effects and identify some of the ways that these three factors interact to influence ecosystems.

10.2.1 Plant and Ecosystem Responses to Elevated CO_2

Current climate models estimate that even if man-made emissions of CO_2 could be kept at present rates, atmospheric CO_2 would increase to about 450 ppmv by the year 2050, and to about 520 ppmv by the year 2100 (Section 1). Regardless of how the climate changes over this period, the Earth's biota will be living in a "CO_2-rich" environment. How will plants and ecosystems respond to elevated CO_2?

10.2.1.1 Plant responses
In this part of the report we will refer to two general groups of plants - "C_3" plants and "C_4" plants. These plant groups differ in a number of ways including certain aspects of the biochemical pathways they use in the photosynthesis process. Most of the Earth's plant biomass (about 95%) is accounted for by C_3 species, but a number of plants important to humans, such as maize, are C_4 species.

10.2.1.1.1 Carbon budget: Atmospheric CO_2 affects various components of a plant's carbon budget including photosynthesis, respiration, and biomass accumulation and allocation.

Photosynthesis - It has been shown many times that a doubling of CO_2 in the atmosphere will cause a short-term (minutes to hours) increase in photosynthesis (Kimball 1983, Gifford 1988). In some plants the increase is reduced after longer-term (weeks to months) exposure (Tissue and Oechel 1987, Fetcher et al. 1988, Sage et al. 1990). This reduction may occur because other factors such as low nutrient availability eventually limit CO_2 uptake.

Respiration - Two types of respiration are recognized in plants; one, known as "photorespiration," is intimately associated with photosynthesis, and the other, "dark respiration," includes all plant respiration except photo-respiration. Photorespiration of C_3 plants is greatly reduced at high CO_2. The pattern is not so clear for "dark" respiration. The published data on dark respiration rate per unit of dry weight or leaf area indicate increases in some cases (Oechel and Strain 1985) and decreases in others (Gifford et al. 1985).

Biomass accumulation - When grown at high CO_2 levels under favorable environmental conditions (e.g., favorable temperature, plentiful water and nutrients), C_3 plants almost always show increases in biomass accumulation.

The C_4 plants are less responsive to high CO_2 levels in terms of biomass accumulation, but nonetheless the response is generally positive. For both C_3 and C_4 plants, the response is very species-dependent and closely linked to environmental conditions (Mooney et al. 1990).

Allocation - Increases in CO_2 affect how plants allocate carbon among their various organs. Many studies indicate that with increasing atmospheric CO_2, plants allocate proportionally more carbon below ground than above ground, causing an increase in root to shoot ratios (Larigauderie et al. 1988, Curtis et al. 1990). High CO_2 can also increase the number of branches, tillers, flowers or fruits that a plant has (e.g., Curtis et al. 1989).

The ways in which other environmental factors interact with CO_2 to determine carbon allocation in plants is largely unknown. This is a serious gap in our knowledge and is a major stumbling block to the development of mechanistic, whole-plant models of carbon dynamics.

Tissue quality - Plant tissue quality can change with exposure to high CO_2. Changes in tissue quality include higher carbohydrate levels (Sionit et al. 1981) and, at least in one instance, higher levels of soluble phenolics and structural compounds (Melillo 1983). Nutrient concentrations are also often decreased (Curtis et al. 1990, see 10.2.1.1.3). These changes in tissue quality could have far-reaching consequences for herbivory, host-pathogen relationships, and soil processes such as decomposition and nutrient cycling. Much more work is needed in this area before we can make generalizations about the linkages between elevated CO_2, tissue chemistry and ecosystem effects.

10.2.1.1.2 Interactions between carbon dioxide and temperature: Temperature and CO_2 interact to affect photosynthesis and growth. Although the reactions are species specific, the general response for C_3 plants is that the optimum temperature increases for net photosynthesis. Idso and colleagues (1987) have suggested that plant growth response to elevated CO_2 seems to be greater at higher temperatures. If, however, temperature becomes extremely high, enzyme degradation will limit both photosynthesis and growth. Likewise, plants growing at low temperatures are not as responsive to elevated CO_2 for physiological reasons that lead to a feedback inhibition of photosynthesis.

10.2.1.1.3 Carbon dioxide and environmental stress: Elevated CO_2 can influence plant responses to limitations of water, light and nutrient availability and other environmental factors (Table 10.1).

Water stress - Water use can be affected by high CO_2. Short-term measurements show that increased CO_2 reduces water-loss (transpiration) rates per unit leaf area and increases water use efficiency (WUE), which is the ratio of

Table 10.1 *Relative effects of increased CO_2 on plant growth and yield: a tentative compilation[1] (from Warrick et al., 1986b)*

	C_3	C_4
Under non-stressed conditions	++	0 to +
Under environmental stress:		
Water (deficiency)	++	+
Light intensity (low)	+	+
Temperature (high)	++	0 to +
Temperature (low)	+	?
Mineral nutrients:	0 to +	0 to +
Nitrogen (deficiency)	+	+
Phosphorous (deficiency)	0?	0?
Potassium (deficiency)	?	?
Sodium (excess)	?	+

1   Sign of change relative to control CO_2 under similar environmental constraints
++ strongly positive
+   positive
0   no effect
?   not known or uncertain

photosynthesis to transpiration (Farquhar and Sharkey 1982). Increased WUE could lead to increased biomass accumulation for plants growing in arid environments.

The net effect of high CO_2 on total water use per unit land area under field conditions is less certain. This is because the increases in leaf area and root extension observed in high-CO_2 plants tend to increase total water use and may counteract the effect of low transpiration per unit leaf area. Gifford (1988) has concluded that for both physiological and meteorological reasons, high CO_2 concentration might exert little or no effect on regional evapotranspiration, but this issue is far from resolved.

Low light - Carbon dioxide enrichment can increase plant growth at low light intensity. In fact, the relative enhancement of growth at low light can even be greater than at high light (Gifford 1979). For some plants, however, the relative enhancement of growth by high CO_2 appears equal at low and high light (Sionit et al. 1982).

Nutrient stress - High CO_2 can increase plant growth in some situations of nutrient-stress. A number of C_3 plants growing under nitrogen-deficient conditions exhibited increased growth when the CO_2 concentration was doubled (Wong 1979, Sionit et al. 1981, Goudriaan and de Ruiter 1983). In these instances there was an increase in the nitrogen use efficiency (NUE); that is, the ratio of carbon gain to nitrogen used was increased.

10.2.1.1.4 Phenology and senescence:
Elevated CO_2 has been shown to influence the phenology and senescence of plants. Annual plants may develop more quickly under elevated CO_2, reaching full leaf area, biomass, and flower and fruit production sooner than plants at ambient CO_2 (Paez et al. 1984). Early leaf and seed production could shift the population dynamics and competitive relationships of plants growing under field conditions.

There is also evidence of delayed senescence of some species under elevated CO_2 (Hardy and Havelka 1975, Mooney et al. 1990). Delayed leaf senescence could extend the growing season and this could lead to increased biomass accumulation (Mooney et al. 1990). In ecosystems with cold climates, however, the growing season could also expose plants to frost damage (Oechel and Strain 1985).

10.2.1.2 Community and ecosystem responses to elevated carbon dioxide

10.2.1.2.1 Plant-plant interactions:
We can expect changes in the interactions of C_3 plants with elevated CO_2. As we noted earlier (see 10.2.1.1.1) the responses of C_3 plants to increased CO_2 are species dependent.

Some ecosystems such as temperate zone grasslands can contain a mixture of C_3 and C_4 plants. Elevated CO_2 could affect the competition between them. Based on what we know about the biochemistry and physiology of C_3 and C_4 species, we would expect that as the CO_2 concentration increases, the C_3 plants should do progressively better than the C_4 plants, unless there is water stress. A number of studies have shown just these results. For example, Bazzaz and Carlson (1984) studied the competition between C_3 and C_4 herbaceous plants grown under two moisture regimes and three levels of CO_2 (300, 600, and 1200 ppm). The C_3 species grew progressively more rapidly (was a better competitor) than the C_4 species as the CO_2 and moisture levels increased.

10.2.1.2.2 Interactions between plants and animals:
The effects of increased CO_2 on plant-animal interactions have received relatively little attention. Some work has been done on herbivory and the conclusion is that herbivory may be indirectly affected by high CO_2 concentrations. Several reports (Overdieck et al. 1984, Lincoln et al. 1984, 1986, Lincoln and Couvet 1989, Fajer et al. 1989) have indicated that rates of herbivory increase on plant tissues grown at high CO_2. These increases in herbivory appear to be related to changes in the tissue quality of plants exposed to elevated CO_2 (see 10.2.1.1.1). The increased herbivory could affect plant growth as well as feed back to ecosystem-level phenomena like nutrient cycling. Increased herbivory would be expected to accelerate nutrient cycling.

Linkages between the effects of elevated CO_2 on plant phenology and herbivory have been suggested (Oechel and Strain 1985), but to our knowledge no research has been carried out on this topic. One argument is that changes in the timing of herbivore feeding relative to plant phenology could affect productivity and competitive ability of the plants concerned.

10.2.1.2.3 Interaction between plants and microbes:
Elevated atmospheric CO_2 and climate change will probably have major effects on microbial symbionts of plants such as nitrogen-fixing bacteria and mycorrhizal fungi.

Symbiotic nitrogen fixing organisms have large requirements for energy provided as plant photosynthate. These organisms are primarily responsible for giving plants access to the large reservoir of nitrogen in the atmosphere by converting that gaseous nitrogen into organic nitrogen. For many ecosystems, high rates of productivity are linked to nitrogen fixation. Numerous experiments have shown that climatic variables and CO_2 concentration are important controllers of the relationship between plants and symbiotic nitrogen fixers. For example, a field experiment by Hardy and Havelka (1975) showed that over a nine-week period, plants grown with supplemental CO_2 exhibited a five-fold increase in nitrogen fixation rate over untreated controls.

Mycorrhizae are symbiotic associations between the host-plant root and a mycorrhizal fungus. As with symbiotic nitrogen fixers, the mycorrhizal fungi depend on plants for a supply of reduced carbon. Thus, climate and CO_2 changes that affect a plant's ability to fix atmospheric CO_2 have the potential to affect mycorrhizal functioning. Mycorrhizae may affect plant nutrition, especially phosphorus nutrition, and plant water relations. Luxmoore and co-workers (Luxmoore et al. 1986, Norby et al. 1986, O'Neill et al. 1987) have shown that mycorrhizal infection is enhanced by elevated CO_2, and the increased infection resulted in increased plant growth on nutrient-poor soils.

10.2.1.2.4 Decomposition:
Free-living soil micro-organisms are responsible for organic matter decay. Decay rate is a function of the chemical quality of the organic matter and environmental factors such as temperature and moisture (see 10.2.2.2.3). Earlier we noted that plants grown in elevated CO_2 have altered tissue chemistry such as higher carbon to nitrogen ratios. These changes in tissue chemistry could slow decomposition and possibly lead to plant nutrient stress.

10.2.1.2.5 Whole-ecosystem exposure to elevated carbon dioxide: Many of the direct effects of elevated CO_2 on plant growth have been observed in short-term studies in the laboratory. Serious questions have been raised about whether or not these phenomena actually occur in the field and if they do, whether they are long-term or only transient. Some answers to these questions may be gained from a review of two recent experiments on intact ecosystems - one a tussock tundra ecosystem in Alaska and the other a mid-latitude salt-marsh ecosystem in Maryland. These are the only whole-ecosystem experiments we know of in which the entire system has been subjected to doubled CO_2 concentrations for more than one growing season. Both experiments have been run for three years.

At the tundra site, CO_2 and temperature were controlled in greenhouses placed over intact field plots (Oechel and Riechers 1987, Tissue and Oechel 1987). Experimental treatments included ambient CO_2 and temperature conditions, elevated CO_2 (510 and 680 ppmv) and ambient temperature, and elevated CO_2 and temperature (680 ppmv CO_2, +4°C temperature above ambient).

At the salt-marsh site CO_2 was controlled through the use of open top chambers set over intact field plots (Drake et al. 1989). The experimental treatments included ambient CO_2 and temperature, and elevated CO_2 (ambient plus 340 ppm) and ambient temperature.

The tundra ecosystem is floristically diverse but is dominated by a sedge, while the marsh system is comprised largely of pure patches of two higher plants, a sedge and a grass. The tundra and salt marsh sedges are C_3 plants and the salt marsh grass is a C_4 plant. The plant and ecosystem responses of these two systems (Table 10.2) generally follow predictions based on the interactions of CO_2 and the other environmental factors discussed earlier.

Significant ecosystem-level effects were noted in both the tundra and the salt marsh. For the tundra plots exposed to elevated CO_2, there was a complete homeostatic adjustment of whole ecosystem carbon flux within three years, with the result being no change in net carbon storage in CO_2-treated plots relative to controls. However, the combination of elevated CO_2 and temperature rise resulted in an increase in net carbon storage that lasted for the three years of observations (Oechel and Riechers 1986, 1987).

The CO_2 treatment in pure stands of the C_3 marsh plant resulted in increased net carbon storage for the whole system. In the pure stands of C_4 grass, net carbon storage for the whole system was not increased.

One of the most important points that can be made about the comparison of the responses of the two ecosystems to elevated CO_2 is that the interactions among temperature, CO_2, and nutrient availability are key controlling factors. Temperature affects both plant photosynthetic response to elevated CO_2 and nutrient availability through organic matter decomposition (see 10.2.2.2.3). Both temperature

and nutrient availability exert control on the growth of plant organs where fixed carbon can be stored. In cold, low nutrient environments, the growth of storage organs is slow and this can lead to an "end product" inhibition of photosynthesis. If temperature is not limiting, but nutrients are, the increased allocation of fixed carbon to roots could result in more of the soil volume being "mined" to meet plant nutrient demand, thereby allowing the plant to utilize the CO_2 .

Have we correctly interpreted the interactions and do they operate in other terrestrial ecosystems? We do not know the answers to these questions. There have been no long-term studies of the responses of most of the world's ecosystems to elevated CO_2 or climate change. For example, we currently have no information about many of the responses of forests and other woody ecosystems to elevated CO_2. Some scientists have argued that limitations of water, nutrients and light, will prevent these ecosystems, especially unmanaged foresets, from showing significant responses to elevated levels of atmospheric CO_2 (e.g., Kramer, 1981). The responses of forests to increased CO_2 are very uncertain. A high research priority for the near future has to be a series of whole-ecosystem manipulations, including forest manipulations, in which key controlling factors such as CO_2, temperature, moisture and nutrient availability are varied.

10.2.1.3 Summary

Increased atmospheric CO_2 has the potential to alter ecosystem metabolism. Net primary production could be enhanced by increased CO_2 in a variety of ways including the following: stimulation of photosynthesis; depression of respiration; relief of water and low light stresses; relief of nutrient stress by several mechanisms (greater nutrient use efficiency, increased nutrient uptake through root-microbial associations, increased symbiotic nitrogen fixation); and delay of senescence that prolongs the growing season.

Elevated CO_2 could also lead to net carbon storage, especially if the growth of woody vegetation is stimulated and there is not an equal stimulation of decomposition by some other factor such as warming. At this time we have no evidence that elevated CO_2 has increased net carbon storage in natural ecosystems dominated by woody vegetation.

Increased CO_2 could also change species composition by affecting plant reproductive processes, competition, plant-animal interactions, and plant-microbe interactions.

Table 10.2 *Effects of doubling CO₂ on several plant and ecosystem properties and processes. In the arctic all species are C₃. Saltmarsh communities are mono-specific stands of the sedge Scirpus olneyi (C₃) and the grass Spartina patens (C₄). The symbols indicate the response to elevated compared to normal ambient CO₂ as an increase (+), decrease (-), no change (0), or no data as a blank.*

		ARCTIC	SALTMARSH C_3	C_4
I	**PLANT EFFECTS**			
A	Carbon exchange			
	Photosynthesis	0	+	0
	Acclimation of photosynthesis	+	0	0
	Plant respiration	0	-	-
	Decomposition of dead shoots		-	-
B	Growth			
	Shoot expansive growth	0	0	0
	Root biomass	-/0	+	0
	Number of shoots	+	+	0
	Size of shoots	0	0	0
	Root/shoot ratio	-/0	+	0
C	Tissue Composition			
	N tissue concentration	-	-	0
	Carbon/nitrogen	+	+	0
	Starch content	+		
	Tissue density/specific wt.	+	0	0
	Salt content	-		
D	Development/reproduction			
	Senescence	-	-	0
	Tillering	+	+	0
	Number of flowers	-	0	0
	Number of seeds/stem		0	0
	Sexual/asexual reproduction	-		
E	Water Use			
	Transpiration	0	-	-
	Water use efficiency	0	+	+
	Leaf tempertaure	0	+	+
	Leaf water potential		+	+
II	**ECOSYSTEM EFFECTS**			
	Evapotranspiration	0	-	-
	Net carbon storage	+/0	+	0
	Acclimation of NCE to CO2	+	0	0
	Net ecosystem respiration Species		-	-
	composition	+	+	0
	Water use	0	-	-
	Nitrogen content of canopy	-	0	0
	Soil enzyme activity	+/-		
	Soil solution nitrogen	-/0		

NCE = Net Carbon Exchange. (Moung et al., 1990)

10.2.2 Plant and Ecosystem Responses to Changes in Temperature and Moisture

Temperature and moisture are considered major controllers of plant and ecosystem processes. They exert a strong influence on birth, growth and death rates of plants. They also act as primary controllers of the biogeochemistry of ecosystems. In this section we review some of the aspects of these controls.

10.2.2.1 Plant responses to changes in temperature and moisture

10.2.2.1.1 *Carbon budget:* Temperature and moisture are important controllers of the carbon budgets of plants. These environmental factors directly influence basic processes such as photosynthesis and respiration.

Temperature - Photosynthesis and plant respiration respond differently to temperature. For example, gross photosynthesis of many mid-latitude plants ceases at temperatures just below 0°C (minimum) and well above 40°C (maximum), with optimum rates in the range of 20-35°C (Figure 10.1.b). Above 0°C the response of photosynthesis to temperature is initially rapid, but slows in the optimum range. In contrast, plant respiration rate tends to be slow below 20°C, but at higher temperatures it accelerates rapidly up to the temperature where the rate of respiration equals the rate of gross photosynthesis, and there can be no net assimilation of carbon (Figure 10.1.b). The response of net photosynthesis is broadly similar to that of overall growth (Figures 10.1.a and 10.1.b). These differences in the responses of photosynthesis and

respiration to temperature increases have been used to support the argument that global warming may result in a reduction in net carbon uptake by plants (Woodwell, 1987).

It must be noted that some plants do appear to have the ability to adjust to temperature changes. Cases have been documented where respiratory response to temperature is adjusted to stay within a limited range of rates in spite of the prevailing temperature (e.g., McNulty and Cummins, 1987).

Moisture - Water stress can decrease photosynthesis in a wide range of plants (e.g., Hsiao, 1973). As we discussed earlier (10.2.1.4), water stress can be alleviated, at least in the short term, when plants are exposed to increased atmospheric CO_2.

The effects of water stress on respiration are complex. A reduction in moisture supply will, in many cases, reduce plant growth and so reduce respiration associated with growth (Hanson and Hitz, 1982). If water stress is severe it will cause cellular damage and this may lead to increased rates of protein turnover with an associated increase in respiration.

10.2.2.1.2 *Phenology and senescence:* As well as influencing the rate of plant growth and metabolism, temperature and moisture can influence the timing of development and senescence. For example, the annual leaf canopy development and abscission in autumn-deciduous trees and shrubs are partially under temperature control. In the drier tropical ecosystems, moisture often functions as the primary controller of canopy development and leaf senescence (Long, 1990). Thus, depending on the ecosystem, either temperature or moisture can function as a primary determinant of the length of the growing season.

Canopy development in northern ecosystems often depends on winter temperatures. In the well documented case of the spruce *Picea sitchensis* (Cannell and Smith, 1986), bud-burst is only triggered after a winter period providing 140 days with a temperature of less than 5°C, followed by a spring period of warmer weather. If the winter temperatures increase, as predicted, so that there is a period shorter than 140 days when temperatures are less than 5°C, then bud-burst may be delayed or even fail to occur. This is the type of climate effect that will result in the migration of plant species to higher latitudes or higher elevations (see 10.2.2.3).

Not all species will respond to climate changes by migrating to higher latitudes or higher elevations. Field experiments have shown that species are able to evolve new phenological characteristics in a new climate. This has been observed to occur within 8 years of transplanting populations to a new climate (Woodward, 1990). The rapidity of this response was observed in an herbaceous perennial, which showed marked temporal changes in seed germination and dormancy in the new environment. The

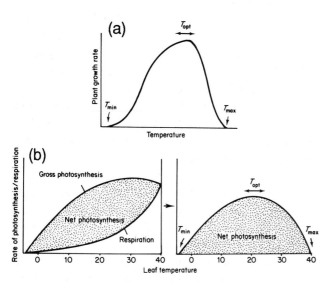

Figure 10.1: Schematic representation of plant responses to temperature. Panel (a) is the general response of plant growth, and panel (b) is the general response of photosythesis and respiration. T_{min} = a minimum temperature, T_{opt} = optimum temperature range, and T_{max} = a maximum temperature. (From Fitter and Hay, 1981).

universality of this response is unknown but it could prove to be a powerful agent for disrupting community synchrony, for initiating structural change and for enhancing the capacity of species to migrate.

10.2.2.2 Community and ecosystem responses
10.2.2.2.1 Plant community composition: Changes in climate will likely alter differentially the regeneration success, growth and mortality rates of plants. The resulting changes in competitiveness of species or species groups will affect community composition. Where species occur at their distributional limits, in transition zones, small changes in climate are likely to promote disproportionately large responses in the plant species. Changes in community structure will, therefore, be identified sooner in transitional zones between vegetation types than elsewhere.

10.2.2.2.2 Interactions between plants and animals: Of crucial importance in plant community functioning is the synchronous operation of the life cycles of interacting plants, animals and soil organisms. Complex synchronies are found in communities in which the life cycles of plants and pollinating and seed-dispersing animals must be closely linked. Changes in climate could disrupt these synchronies.

Climate-related stresses, such as drought stress, can make plants susceptible to insect attack. As an example, oak wilt disease in the USSR appears to be dependent on the decreased ability of the trees to resist leaf-eating insects during drought (Israel et al., 1983).

Warming may expand the overwintering ranges of some plant pests and this could prove a serious problem for agroecosystems. For example, in the United States the potato leafhopper, a serious pest on soybeans and other crops, currently overwinters only in a narrow band along the coast of the Gulf of Mexico. Warmer winter temperatures could cause a doubling or a tripling of its overwintering range. This would increase invasion populations and lead to greater insect density and increased plant damage (Smith and Tirpak, 1989).

10.2.2.2.3 Decomposition: Climate is an important controller of the decomposition of both surface litter and organic matter dispersed through the soil profile. The chemical composition of the decomposing material also influences decay rate.

Surface litter - The decomposition of surface litter is very clearly related to climatic factors, with rates generally increasing with increasing temperature and precipitation in well-drained sites. In poorly-drained sites, excessive moisture (waterlogging) can slow decay rates. Climatic control is most often quantified by a relationship with actual evapotranspiration (AET).

It has been known for some time that litters from different plant species decay at different rates under similar conditions (Minderman, 1968). This has been linked to differences in the "quality" of carbon. Litters with high lignin concentrations will decay more slowly than those with lower concentrations. The ratio of lignin to nitrogen has proven a good predictor of litter decay rate in temperate and boreal ecosystems (Melillo et al., 1982).

A number of attempts have been made to integrate the climate and chemical quality controls of litter decay. These models generally indicate that when AET is low the decomposition rates do not vary much with lignin concentrations, but as AET increases resource quality accounts for more of the variation in decomposition rates (Meentemeyer, 1978, 1984, Pastor and Post, 1986).

Soil organic matter - Temperature, moisture and soil texture are important controllers of soil organic matter decomposition. These factors assume more or less importance depending on the ecosystem. In tundra ecosystems, soil respiration can be limited by an excess of moisture (waterlogging) as well as by low temperatures.

Soil respiration in the well drained forests of the boreal and temperate zones are most often temperature limited (Van Cleve and Sprague, 1971, Bunnell et al., 1977). Soil respiration in these systems is rarely limited by moisture deficit (Anderson, 1973, Schlesinger, 1977, Moore 1984). In very dry, sub-tropical forests, lack of moisture can limit decay (Carlyle and U Ba, 1988).

Overall, litter decomposition and SOM accumulation in climax grasslands follow predictable climate- and soil texture-related patterns (Brady, 1974) which are amenable to the development of simulation models of carbon dynamics (Hunt, 1977, McGill et al., 1981, Van Veen and Paul, 1981, Parton et al., 1987). In dry grassland sites, soil respiration is primarily a function of soil moisture (Hunt, 1977, Warembourg and Paul, 1977, Orchard and Cook, 1983), although computer model simulations of the response of semi-arid soils of North America to warming suggest that higher soil temperatures will result in increased carbon losses (Schimel, private communication). In mesic grasslands, as in forests, temperature is the main determinant of carbon mineralization rates for the soil system. Fine textured soils, those rich in clays, are thought to render soil organic matter more stable through the mechanism of physical protection than are the coarse-textured sandy soils.

There is concern that global warming will accelerate the decomposition of surface litter and soil organic matter, especially at high latitudes of the Northern Hemisphere. In the Arctic tundra, there are about 160 Pg carbon stored in the soil (Schlesinger, 1984). Most of it is frozen in permafrost with only 20-40cm thawing in the summer. Even for the thawed material the cold, and in some cases wet, conditions preclude rapid decomposition. But what will happen if a CO_2 doubling is accompanied by a 4-8°C temperature increase?

The experiments of Billings and his colleagues (1982, 1983, 1984) suggest that wet sedge tundra will become a net source of carbon to the atmosphere, at least for a short time, if climate changes such that air temperature is increased and there is greater soil drainage. Billings argued that warmer temperatures and a lowered water table would result in greatly increased rates of soil respiration. Using Billing's data, Lashof (1989) calculated that a 4°C temperature rise would increase the net annual flux of carbon from the tundra to the atmosphere by 1 Pg.

The consequences of soil warming in northern forests are clear from an experiment conducted on a boreal forest in Alaska. Experimental heating of the surface soil, to 9°C above ambient temperature for three summers, was carried out in black spruce forest (Van Cleve et al., 1983). For the entire period there was a 20% reduction in the amount of organic carbon in the surface soil of the heated site as compared to no reduction in an adjacent control site. The increased decay of the soil organic matter resulted in an increase in nutrients available to plants. As a result of the more favourable temperature and nutrient regimes, foliage showed increased photosynthetic rates and significantly higher concentrations of nitrogen, phosphorus and other nutrients important to plant growth.

10.2.2.2.4 Models of ecosystem response to climate change: A family of population-based forest growth models (e.g., JABOWA, FORET, LINKAGES) has been used to consider the effects of climate change on forest composition, carbon storage capacity and geographic distribution. For example Pastor and Post (1988) ran a population-based forest growth model (LINKAGES) for several sites including a spruce-northern hardwood site in northeastern Minnesota. Forest growth was simulated on two soil types found in this region; one with low water-holding capacity and the other with high water-holding capacity. The simulations were initiated from bare plots with the seeds of trees commonly found in the area. For 200 years the sites were allowed to follow the forest dynamics appropriate to the current climate. The climate conditions were then changed linearly to reach a "2xCO2" climate over the next 100 years and then remained constant the final 200 years of the simulation.

On the soil with high water-holding capacity the spruce-northern hardwood forest was replaced by a more productive northern hardwood forest (Figure 10.2). The aboveground carbon mass in the mature post-climate-change forest was about 50% greater than in the pre-climate-change forest. The northern hardwood forest was more productive for two reasons. First, in the model, northern hardwoods have a faster growth rate and can attain a greater biomass than the spruce. Second, the

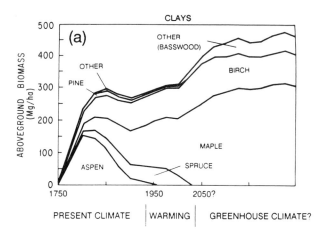

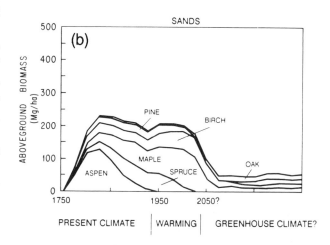

Figure 10.2: Predictions of biomass and species composition of Minnesota forests under climatic conditions predicted with CO_2 doubling. The predictions are based on a forest growth model (LINKAGES). Climate inputs were the same for the two runs, but panel (a) shows simulated forest growth on a soil with high water-holding capacity, and panel (b) shows simulated forest growth on a soil with low water-holding capacity (After Pastor and Post, 1988).

warmer climate, as well as the relatively easy-to-decompose litter of the hardwoods, increases nitrogen availability and this enhancement amplifies the effect of warming on productivity.

On the soil with the low water-holding capacity, the spruce-northern hardwood forest was replaced by a stunted pine-oak forest of much lower carbon storage capacity. At maturity, the oak-pine forest contained only 25% of carbon contained in the original spruce-northern hardwood forest.

In this example, temperature, plant-soil water relations, and nitrogen cycling all interacted to affect ecosystem structure and function. Changes in climate resulted in changes in forest composition, and depending on soil water relations, either an increase or a decrease in ecosystem carbon storage capacity.

Solomon (1986) used another population-based forest growth model (FORENA) for 21 locations in eastern North America with 72 species of trees available as seeds at all times. Initial soil conditions were the same at all sites. The simulations were all initiated from a bare plot and were allowed to follow the forest dynamics appropriate to the modern climate with undisturbed conditions for 400 years. After year 400 the climatic conditions were changed linearly to reach the "$2xCO_2$" climate in year 500 (see Solomon, 1986 for details).

At the end of 400 years the forests on the 21 sites had reached structural maturity and most contained the appropriate species mixes as judged by comparing them to what actually grows at the sites. The major effects of the changes in climate that resulted from a doubling of atmospheric CO_2 were as follows:

A slower growth of most deciduous tree species throughout much of their geographical range;

A dieback of the dominant trees, particularly in the transition between boreal/deciduous forests;

A reduction of carbon storage in the vegetaion in the southern two-thirds of the region, and a gain in the far north;

An invasion of the southern boreal forest by temperate deciduous trees that was delayed by the presence of the boreal species;

A shift in the general pattern of forest vegetation similar to the pattern obtained from the Holdridge map experiments (see 10.2.2.3.1).

The overall reduction in carbon stocks in the vegetation of forests of the eastern North America was estimated to be 10% (Figure 10.3). If the results are generalized to all temperate forests the annual flux from these systems to the atmosphere would be between 0.1 and 0.2 Pg C (Lashof, 1989).

Solomon (1986) discussed the possible effects of several important ecological processes that were not included in the model. For example, insects and other pathogens, as well as air pollutants, could enhance the mortality simulated by the models. Also, plant migration (and associated lag effects) could have a negative influence on forest productivity.

These forest growth models are useful tools for making preliminary evaluations of forest ecosystem responses to climate change. They do have a number of limitations and these have been reviewed recently (Smith and Tirpak, 1989). For example, major uncertainties exist regarding the kinds and rates of response of individual tree species to changes in the environment including the CO_2 increases. Efforts are currently under way to improve the physiological and soil process components of these models.

Climate change may lead to an increase in the frequency of extreme weather events such as tropical storms. Drier

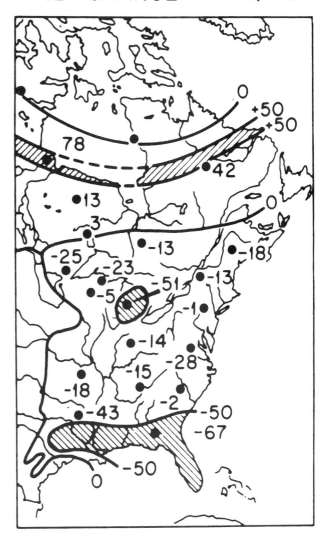

$2xCO_2-1xCO_2$ BIOMASS NET CHANGE = -11 t/ha

Figure 10.3: Carbon storage dynamics (in megagrams per hectare) simulated at 21 sites in eastern North America. Map shows differences above contemporary climate and $2xCO_2$ climate. Only carbon in above-ground biomass is represented. (From Solomon, 1986).

conditions in some regions may lead to increased fire frequency. Large disturbances such as severe storms and large fires can destroy vegetation, increase susceptibility of sites to erosion, change nutrient cycling rates and dramatically alter animal habitat. The effects of these large-scale disturbances are currently not considered by these forest growth models.

10.2.2.3 Large-scale migration of biota

One of the major consequences of climate change could be the migration of biota across the landscape. The migrations could have many effects including the release of large amounts of carbon to the atmosphere from dying and decaying forests. The releases could be large enough to further increase warming (Woodwell, 1987, Lashof, 1989). Both modelling studies and palaeoecological studies have been used to examine the relationships between climate change and forest migration.

10.2.2.3.1 Simulation of global scale response using vegetation-climate relationships: A very general approach to examining the possible responses of the world's ecosystem types to climate change is to use hypothetical relationships between climate and vegetation derived in present-day conditions, and to apply these to scenarios of changed climate. Emanuel et al. (1985a,b) employed a lifezone classification of Holdridge (1947, 1964). This system hypothesizes zonation of vegetation across gradients of average total annual precipitation, the ratio of potential evapotranspiration to precipitation and mean annual biotemperature (average annual temperature computed by setting all values below 0°C to zero). Using the Holdridge system, Emanuel and his co-workers (1985a,b) predicted a large shrinkage of the boreal forest

(by 37%) and tundra (by 32%) and expansion of grassland lifezones under warmer climates due to a CO_2 doubling. Because the temperature changes in the climate-change scenario used by Emanuel and his colleagues were small toward the equator, there were smaller changes in the tropical life zones. In this modelling exercise, precipitation was maintained at current levels for all areas.

There are several uncertainties attached to this type of assessment (Emanuel et al., 1985a), notably the selection of climate scenarios and low resolution of the data grid (0.5° x 0.5°). In addition, the response of ecosystems to factors such as CO_2 and the rate of climate change is not considered.

10.2.2.3.2 Palaeo-ecological evidence: The IPCC projections of climate change indicate a rapid rise in global temperature with an increase of about 0.3°C per decade over the next century (Section 5). Rapid increase in temperature may create problems for large-stature ecosystems such as forests.

The significance of projected rates of temperature change becomes clear when the consequent geographic shifts in isotherms are considered. For example, in mid-continental North America, each degree (°C) change in temperature corresponds to a distance of 100-125 km. If similar temperature-distance relationships are assumed in

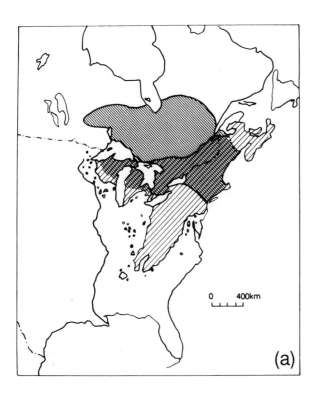

Figure 10.4: Present and future range of eastern hemlock (*Tsuga canadensis*) under two climate scenarios predicted by (a) Hansen et al., 1983 and (b) Manabe and Wetherald, 1987. Light diagonal shading is the present range, and dark diagonal shading the potential range with CO_2 doubling. Cross hatched area of overlap is where the trees are likely to be found 100 years from now (Davis, 1988).

the future, a 3°C rise in temperature would lead to a 300 to 375 km northward displacement in isotherms during the next century.

Based on the fossil pollen record, we know that the rate of movement of forest trees during the Holocene was generally 25 and 40 km per century (Davis, 1981; Huntley and Birks, 1983), with the fastest rate 200 km per century (Ritchie and MacDonald, 1986). With these rates of movement, most tree species would not be able to change their geographical distribution as fast as the projected shifts in suitable climate. Zaluski and Davis (unpublished data cited in Davis, 1988) provide an example based on the past rates of spread of eastern hemlock (*Tsuga canadensis*). Under two scenarios for future climate, this species would not be able to migrate fast enough to occupy much of its potential range 100 years from now (Figure 10.4).

Prehistorical species migrations were largely unaffected by human land use. In contrast, modern species migrations may be severely restricted by land use, for example the progression of the altitudinal treeline with global warming may be prevented by stock grazing. Also, because suitable habitats for many rare species characteristic of areas of low productivity are infrequent and fragmentary, being surrounded by very productive agricultural land, it is unlikely that these species will spread naturally even though climate change will increase their vigour and reproductive capacity. On the other hand, the migration of weedy species may be enhanced by land-use change, for example along corridors of dispersal formed by roads, railways, etc. and in open ground created by various forms of human disturbance.

Very little is known about migration rates under present or likely future climates. Models of migrations of invasive species based on diffusion (Williamson, 1989) or epidemic theory (Carter and Prince, 1981) suggest that the outcome of invasions is often unexpected and accurate prediction has rarely ever been achieved. The relationship between species migrations and climate is not simple and geographical barriers to dispersal may be locally important.

Quaternary pollen records also show that plant communities continuously change over time. Communities disassemble and new ones arise presumably because species respond to climatic variations according to unique sets of physiological and ecological requirements (Graham, 1986). The resulting new combinations of vegetation, climate and soils can change the spatial patterns of such fundamental processes as primary production (Pastor and Post, 1988). More subtle, but still important, relationships such as those evolved between host and pathogen may be disrupted by the stress of new conditions, resulting in increased frequency of epidemics (Leonard and Fry, 1986).

Past changes in plant associations and abundances have been so marked that some types of ecosystems have also been transient. For example, 20,000-30,000 years ago,

when the Earth's climate was about 5°C colder than present, large parts of North America, Europe and Asia were covered by herbaceous vegetation that did not resemble modern tundra, grassland or steppe. Thus a plant community that was dominant in the Northern Hemisphere for 10,000 years does not exist today. Some well-known modern ecosystems have had much shorter existences. As an example, old-growth Douglas-fir (*Pseudotsuga menziesii*) forests of the Pacific Northwest, renowned for their long-lived, massive trees and huge ecosystem carbon storage, are first recognized in the fossil record about 6000 years ago - representing only 5 to 10 generations of the dominant tree species. (Brubaker, personal communication).

Thus the dynamic record of the Earth's vegetation clearly demonstrates that ecosystems as well as communities may be short lived in the face of changing climate. Ecosystems represent a set of processes linking biota and their geochemical environment in a particular climate. Because natural variations in the Earth's climate have followed a relatively unique path dictated by changes in large-scale climatic controls, a variety of ecosystems come and go over time. To the extent that human influences will cause unique future climates, we should expect fundamental changes in current ecosystems.

10.2.2.4 Summary

Both photosynthesis and plant and microbial respiration tend to increase with increasing temperatures, but at higher temperatures respiration is often the more sensitive process (see 10.2.2.1.1). As a consequence, global warming may result in a period of net release of carbon from the land to the atmosphere (Woodwell, 1987). The magnitude of the release would depend on the magnitude and the seasonality of the temperature change and the responses of the various processes to that change. One estimate is that the annual carbon flux from the boreal zone associated with a 4°C global temperature change would be in the range of 0.5-2.0 Pg C (Lashof, 1989).

Besides being dependent on temperature, photosynthesis and plant and soil respiration can be influenced by soil water. Increased water availability will tend to stimulate plant growth in dry ecosystems and to increase carbon storage in cold and wet ecosystems like lowland tundra. A number of recent modelling studies have predicted that water stress will be a primary cause of forest death in the southern and central regions of North America as climate changes (e.g., Solomon 1986, Smith and Tirpak, 1989). Forest death has the potential for releasing large stores of carbon to the atmosphere.

A major consequence of climate change could be the migration of biota across the landscape. Communities will not migrate as units. Individual species will migrate at different rates depending upon a variety of species specific

Table 10.3 *Area coverage, plant carbon and net primary production for major terrestrial ecosystems according to Whittaker and Likens (1975); (2) Atjay et al. (1979); (3) Olson et al. (1983). The amount of carbon in soil is also shown following the classifications by (2) Atjay et al. (1979), and according to (4) Schlesinger (1977), based on the classification by Whittaker and Likens (1975). From Bolin, 1986.*

	Area 10^{12} m^2			Plant carbon Pg C			Primary Production Pg C yr^{-1}			Detritus, soil Pg C	
	(1/4)	(2)	(3)	(1)	(2)	(3)	(1)	(2)	(3)	(2)	(4)
(1) Tropical rain forest	17.0	10.3	12.0	344	193	164	16.8	10.5	9.3	82	
(2) Tropical seasonal forest	7.5	4.5	6.0	117	51	38	5.4	3.2	3.3	41	288
(3) Temperate forest	12.0	7.0	8.2	174	88	65	6.7	4.6	4.9	72	161
(4) Boreal forest	12.0	9.5	11.7	108	96	127	4.3	3.6	5.7	135	247
(5) Woodland, shrubland interupted woods	8.5	4.5	12.8	23	24	57	2.7	2.2	4.6	72	59
(6) Savannah	15.0	22.5	24.6	27	66	49	6.1	17.7	10.7	264	63
(7) Temperate grassland	9.0	12.5	6.7	6	9	11	2.4	4.4	2.6	295	170
(8) Tundra, alpine	8.0	9.5	13.6	2	6	13	0.5	0.9	1.8	121	163
(9) Desert, semidesert	18.0	21.0	13.0	6	7	5	0.7	1.3	0.9	168	104
(10) Extreme desert	24.0	24.5	20.4	0	1	0	0.0	0.1	0.5	23	4
(11) Cultivated land	14.0	16.0	15.9	6	3	22	4.1	6.8	12.1	128	111
(12) Swamps, marshes and coastal land	2.0	2.0	2.5	14	12	7	2.7	3.3	3.6	225	145
(13) Bogs and peatland		1.5	0.4		3	1		0.7	0.2		
(14) Lakes and streams	2.0	2.0	3.2	0	0	1	0.4	0.4	0.4	0	0
(15) Human areas		2.0			1			0.2		10	
TOTAL:	149.0	149.3	151.1	827	560	560	52.8	59.9	60.6	1636	1515

characteristics as well as environmental factors such as natural and human-caused barriers to dispersal. Predictions of migration rates are very uncertain. It is possible that many tree species will not be able to change their geographical distribution as fast as the projected shifts in suitable climate and extinctions may occur.

10.3 The Effects of Terrestrial Ecosystem Changes on the Climate System

In this part of the report we consider how a variety of changes in terrestrial ecosystems can affect the climate system. We begin by considering the types of ecosystem changes that will affect net carbon storage in ecosystems.

10.3.1 Carbon Cycling in Terrestrial Ecosystems

Terrestrial ecosystems contain about 2000 Pg of carbon (Table 10.3), almost three times the amount held in the atmosphere. Each year about 5% of the terrestrial carbon stock is exchanged with the atmosphere. Plants take up about 100 Pg C from the atmosphere through the process of

photosynthesis and release 40 Pg C through respiration. The difference between total photosynthesis and plant respiration, 60 Pg C, is called net primary production (NPP). At the global scale, tropical rain forest ecosystems are the most productive and desert ecosystems are the least productive (Table 10.3).

In an unperturbed world, NPP is approximately balanced by the release of carbon from soils to the atmosphere through microbial respiration. This carbon balance can be changed considerably by direct human impact (landuse changes, particularly deforestation) and by other changes in the environment such as the composition of the atmosphere as well as by changes in climate as discussed earlier (10.2.2.4).

10.3.1.1 Deforestation in the Tropics

The current IPCC estimate for the annual net release of carbon from the land to the atmosphere due to deforestation and related land use in the tropics is 1.6 ± 1.0 Pg (Section 1). The large uncertainty associated with this number is related to the fact that we have poor knowledge of the rate of deforestation, of the carbon stocks of the forests being cleared, and of the dynamics of carbon loss through soil decay processes following clearing (Houghton, 1990).

10.3.1.2 *Forest regrowth in the mid-latitudes of the Northern Hemisphere*

Several analyses suggest that the forest harvest and regrowth cycle is such that at this time the regrowing forests in the mid-latitudes of the Northern Hemisphere are accumulating carbon at the rate of 1-2 Pg annually (Armentano and Ralston, 1980, Johnson and Sharpe, 1983, Melillo et al., 1988). What is not clear at the present time is the rate of oxidation of the cut wood. One analysis (Melillo et al., 1988) suggests that the carbon in the cut wood is being returned to the atmosphere through decay and burning at about the rate that the regrowing forests are accumulating carbon. This subject deserves further attention especially in light of the recent modelling study of the global carbon cycle that suggests that mid-latitude ecosystems of the Northern Hemisphere are accumulating carbon at the rate of 2.0 to 3.4 Pg C annually (Tans et al, 1990).

10.3.1.3 *Eutrophication and toxification in the mid-latitudes of the Northern Hemisphere*

The issue of carbon storage in the mid-latitudes of the Northern Hemisphere is further complicated by the increased availability of nutrients such as nitrogen from agricultural fertilizers and from combustion of fossil fuel. This increased nitrogen availability may result in net carbon storage in plants and soils (Melillo and Gosz, 1983, Peterson and Melillo, 1985). In the mid to late 1980's, between 70 and 75 Tg N per year have been applied to agricultural fields, mostly in the mid-latitudes of the Northern Hemisphere (Eichner, 1990). In addition, about 18 Tg N have been deposited each year on the forests of this region in acid precipitation (Melillo et al., 1989). The addition of this nitrogen to agricultural plots and forests could be causing an accumulation of between 0.5 and 1.0 Pg carbon in soils and woody vegetation (Melillo, private communication). However, it should be noted that the greater availability of nitrogen is, at least for the forests, associated with increasing levels of pollutants which could reduce plant growth. We are uncertain about the net effects on the terrestrial carbon balance of eutrophication and toxification of mid-latitude ecosystems of the Northern Hemisphere.

10.3.2 *Reforestation as a Means of Managing Atmospheric CO_2*

It is sometimes suggested that we should reverse the long-term trend of land clearing and grow new forests to absorb the excess CO_2. Is this suggestion a reasonable one? What rates of carbon uptake and storage can we expect? How much land area would be needed to make a difference? How long would the uptake continue?

Recently, Jarvis (1989) estimated that a rapidly growing forest in the temperate zone can accumulate a mean of 2.7tC/ha annually for almost 100 years. To accumulate 1 Pg C per annum, the equivalent of about 17% of current annual fossil fuel CO_2 emissions, the new growing forest would have to occupy 370×10^6 ha. This is an area equivalent to about one half the size of the Amazon Basin.

The accumulation will not be linear. Early on it will be rapid, but as the forest matures, the accumulation will slow down until the annual respiration rate of the forest is about equal to carbon uptake. If the forest is left unmanaged, the trees will eventually die at different times and be replaced naturally; over a large area net uptake of CO_2 by the vegetation from the atmosphere and return of CO_2 by organic matter decomposition to the atmosphere will be in balance and the forest will no longer act as a net carbon sink.

Houghton (1990) has made a similar set of calculations for the tropics. He first made an estimate of land available in the tropics for afforestation. Counting only lands that supported forests in the past, and that are not currently used for either crops or settlement, he concluded that 500×10^6 ha of tropical lands could be reforested. He also reasoned that an additional 365×10^6 ha could be reforested if the current area occupied by the crops and fallow of shifting cultivation was reduced to the area required to produce similar yields under low-input continuous cultivation. The total area, 865×10^6 ha is larger (by about 25%) than the Amazon Basin.

Since much of the land that would be involved in the reforestation is probably degraded (low in plant nutrients, especially phosphorus), the average rate of carbon accumulation was estimated by Houghton to be about 1.7tC/ha. Given these estimates of available land and carbon accumulation rate, Houghton calculated that about 1.5 Pg C might be withdrawn from the atmosphere each year over the next century (150 Pg C total). After that time the new forest, if left untouched, would be in steady-state with respect to carbon exchange with the atmosphere. Houghton noted that the estimate is optimistic because it fails to consider ownership of the land, the expense of reforestation, or the ability of degraded lands to grow trees, even at the modest rates used in his calculations.

Although simple, these calculations are important because they illustrate the enormity of the task facing us if we decide to manage atmospheric CO_2 by afforestation.

10.3.3 *Methane and Nitrous Oxide Fluxes*

The trace gas concentrations of the Earth's atmosphere have been increasing for more than a century (Section 1). Microbial activity is the dominant source to the atmosphere of two of these gases, methane (CH_4) and nitrous oxide

(N$_2$O). Climatic factors play an important part in controlling the rates at which these gases are produced. In addition, atmospheric inputs of nitrogen and sulphur to ecosystems may affect CH$_4$ production and consumption. Nitrogen additions to ecosystems as well as forest clearing can stimulate N$_2$O production. The controls on the fluxes of these two gases are discussed below.

10.3.3.1 Methane

Bacteria known as "methanogens" produce methane. These organisms can only metabolize and live in the strict absence of oxygen. Methanogenic ecosystems usually are aquatic, such as swamps, marshes, fens, paddies, lakes, tundra, and so on, where oxygen-deficient zones develop due to O$_2$ consumption by respiration and limitation of O$_2$ diffusion from the atmosphere. Other ecosystems such as wet meadows, potholes, and inundation zones, may be methane producers for at least part of the year, but may support consumption of atmospheric CH$_4$ when they dry up. Most of the aerobic soils do not support CH$_4$ production and emission; quite the contrary, they seem to function as CH$_4$ consumption sites (Seiler and Conrad, 1987, Steudler et al., 1989). The guts of termites and the rumens of cattle are also anoxic methanogenic environments. An overview of the relative importance of various sources and sinks for CH$_4$ are given in Section 1 (Table 1.2).

A number of factors besides the oxygen status of the environment control CH$_4$ production. These factors include: temperature, the availability of carbon substrate, pH and electron acceptors.

Most methanogenic bacteria have temperature optima of 30-40°C. Thus, if the supply of organic matter is not limiting, increasing the temperature generally will stimulate CH$_4$ production in most methanogenic environments. It is estimated that a temperature increase of 10°C will lead to a 2.5-3.5 fold increase in CH$_4$ production (Conrad et al., 1987). This assumes no change in soil moisture; that is, the soils remain waterlogged.

Warming in high latitude ecosystems such as wet tundra and the boreal wetlands will almost certainly lead to increased CH$_4$ emissions from these ecosystems which currently account for the release of about 40 Tg of CH$_4$ to the atmosphere annually. Based on the simple relationship between temperature and methanogenic activity described above, a 4°C rise in temperature in the soils of the high latitude systems could lead to a 45% to 65% increase in methane release from these systems. If warming is accompanied by drying, then there may ultimately be a reduction of CH$_4$ release to the atmosphere since the soils would become oxygen filled, a condition unsuitable for the methanogens. Under drier conditions the carbon-rich northern soils would become net sources of CO$_2$ until a new equilibrium was reached between carbon inputs from plants and CO$_2$ loss associated with decomposition (see 10.2.2.2.3).

Sediments or anaerobic soils rich in organic matter are often active in CH$_4$ production. The addition of organic substrates to methanogenic sites generally stimulates CH$_4$ production, provided that temperature and pH are not major limiting factors. Most methanogenic bacteria grow optimally in a narrow range around pH 7 (Conrad and Schutz, 1988).

Bicarbonate functions as the electron acceptor during the process of methanogenesis. Electron acceptors such as nitrate and sulfate in general are preferred over bicarbonate and thus inhibit CH$_4$ production. So, any input of oxidized compounds such as sulfate or nitrate in fertilizers or acid rain will reduce CH$_4$ production. The significance of this reduction for the global CH$_4$ budget is unclear at this time, but it is not likely to be large.

The major sink for CH$_4$ is reaction with OH in the troposphere. Soils also function as a sink for CH$_4$, with the magnitude of the sink being in the range of 30 $\pm$ 15 Tg CH$_4$ per year. Recent evidence indicates that the magnitude of the soil sink for CH$_4$ is controlled by temperature, moisture and some aspects of the nitrogen cycle in the soil (Steudler et al., 1989). Warmer temperatures increase CH$_4$ uptake by aerobic soils, while high soil moisture and high rates of nitrogen turnover in soils reduce CH$_4$ uptake. The argument has been made that the eutrophication of the soils of the mid-latitudes of the temperate region, with 18 Tg N per year associated with acid rain, may have reduced CH$_4$ uptake by soils enough to have contributed to the atmospheric increase of CH$_4$ (Melillo et al., 1989). The logic here is that either a reduction in a CH$_4$ sink or an increase in a CH$_4$ source will lead to an increase in atmospheric CH$_4$ content since the gas has such a long lifetime in the atmosphere. While this nitrogen-methane interaction is an interesting example of a subtle impact of industrialisation on the global environment, the importance of this mechanism is not yet established.

10.3.3.2 Nitrous oxide

Our current understanding of the global budget of N$_2$O is reviewed in Section 1 (Table 1.4). The budget is largely controlled by microbial processes.

Nitrous oxide can be produced by four processes: denitrification, nitrification, assimilatory nitrate reduction and chemodenitrification. Of these processes, denitrification in aerobic soils is probably the most important source of N$_2$O (Matson and Vitousek, 1990). Denitrification is defined as the dissimilatory reduction of oxides of nitrogen to produce N$_2$O and N$_2$ by a diverse group of bacteria.

The cellular controllers of denitrification are O$_2$, nitrate (NO$_3$) and carbon. Moisture has an indirect effect on denitrification by influencing the O_2 content of soil. If other conditions are appropriate, then temperature becomes an important controller of denitrification rate.

Tiedje (1988) has suggested that the controllers of denitrification vary among habitats. He indicated that oxygen availability is the dominant factor limiting denitrification in habitats that are exposed to the atmosphere, such as soil, while in dominantly anaerobic habitats, such as sediments, NO_3 is the most important cellular controller. When anaerobic zones occur in soils, then NO_3 availability, or carbon availability, as controlled by wetting and drying or freezing and thawing cycles, can become a critical factor in controlling denitrification. In fertilized soils, carbon is commonly second in importance after O_2, whereas in unfertilized soils, is often second in importance.

Denitrifiers in the natural environment are capable of producing either N_2O or N_2 as end products. Numerous factors have been reported to affect the proportion of N_2O produced relative to N_2 in denitrifying cells and soils. If the availability of oxidant (N-oxide) greatly exceeds the availability of reductant (most commonly carbon), then the oxidant may be incompletely utilized; that is, N-oxides will not be completely reduced to N_2. In other words, the dominant product of denitrification may be N_2O in systems where, at least for a time, nitrate supply is high and carbon supply is low, but not excessively so (Firestone and Davidson, 1989).

A wetter climate may lead to increased N_2O production since soil moisture influences O_2 availability to microbes and low oxygen tension is a precondition of denitrification. Warming will accelerate N_2O production by accelerating the nitrogen cycle in the soil and making more nitrate available to the denitrifiers.

In the tropics, conversion of forest to pasture has been shown to increase N_2O flux (Luizao et al., 1990). Nitrous oxide emissions also increase immediately after forest harvest in the temperate zone (Bowden and Bormann, 1986). In both of these cases, ecosystem disturbance has resulted in warmer and wetter soils that are cycling nitrogen more rapidly than pre-disturbance soils.

10.3.4 Ecosystem Change and Regional Hydrologic Cycles

Vegetation affects continental and regional hydrology by influencing the processes of evapotranspiration and surface runoff. Inclusion of vegetation in general circulation model simulations influences continental rainfall and other climate paramaters (e.g., Sellers, 1987, Sato et al., 1989). At the regional scale, Pielke and co-workers have shown that vegetation influences boundary layer growth and dynamics, and that vegetation mosaics affect small-scale and meso-scale circulations, convective activity and rainfall (Avissar and Pielke, 1989).

At the continental scale, several simulations have been done on the climatic impact of complete deforestation of the Amazon Basin (Henderson-Sellers and Gornitz, 1984; Wilson, 1984; Dickinson and Henderson-Sellers, 1988; Lean and Warrilow, 1989; Nobre et al., 1990). The Amazon Basin contains about half of the world's tropical rainforests and plays a significant role in the climate of that region. It is estimated that approximately half of the local rainfall is derived from local evapotranspiration (Salati et al., 1978). The remainder is derived from moisture advected from the surrounding oceans. A major modification of the forest cover could therefore have a significant climatic impact. Reduced evapotranspiration and a general reduction in rainfall, although by variable amounts, was found in most simulations.

The studies by Lean and Warrilow (1989) and Nobre et al. (1990) show reductions of about 20% in rainfall in simulations in which forest was replaced by grassland. Lean and Warrilow showed that albedo and roughness changes contributed almost equally to the rainfall reduction. Nobre et al. suggest that the switch to a more seasonal rainfall regime, which they obtained, would prevent forest recovery.

Recall that these simulations were "all or nothing"; forest or grassland. The consequences of intermediate changes (partial land conversion) have not yet been studied but, given the nonlinearity of the coupled vegetation-climate system, may not be intermediate. The consequences of fragmentation of continental ecosystems and partial conversion will need to be studied using mesoscale simulations as well as by using general circulation models, and both must be pursued.

Consequences of altered ecosystem-atmosphere exchange on runoff, resulting from changing vegetation, have not been much studied but are of major importance. Appropriate regional scale simulations for runoff routing and river flow are just now being developed (e.g., Vorosmarty et al., 1989) and will be crucial tools in evaluating the surface hydrology component of the hydrologic cycle, and its interactions with ecosystems and climate.

The central issue for upcoming research in this area is to analyze the feedbacks between vegetation and climate to understand how the effects of climate forcing will be modulated by vegetation response. This question has two components. First, the biophysical coupling between vegetation and the atmosphere must be understood and quantified to allow simulation of evapotranspiration, albedo and roughness as a function of vegetation attributes. Second, the response of vegetation to climate forcing must be understood and quantified so that in our models vegetation structure and physiology can change as climate changes. Coupling these two allows for feedback; this must then be analyzed on a regional basis to determine the potential for positive and negative feedbacks between ecosystems and the atmosphere. This effort must include global and regional simulations and large-scale field studies for parameterization and validation.

10.3.5 Summary

The effects of ecosystem change on the net carbon balance of terrestrial ecosystems is potentially large but it is uncertain. Deforestation is clearly a source of carbon to the atmosphere of 1.6 ± 1.0 Pg annually. The net carbon balance of mid-latitude ecosystems of the temperate zone is uncertain. The combination of forest regrowth and eutrophication of these systems with nitrogen appears to be causing them to function as carbon sinks in the range of 1.5 to 3.0 Pg annually. The combined effects of wood burning and decay of cut wood and the toxification of mid-latitude ecosystems with a variety of pollutants may be offsetting the carbon uptake in these systems. The magnitude of net carbon storage in mid-latitude terrestrial ecosystems of the Northern Hemisphere is very uncertain.

The management of atmospheric CO_2 by global-scale reforestation will be an enormous task. To remove just 1 Pg C annually by this mechanism would require a very large area.

Warmer and wetter soil conditions have the potential for increasing the fluxes of CH_4 and N_2O from terrestrial ecosystems to the atmosphere. If the northern wetland soils are warmed by 4°C, the CH_4 flux to the atmosphere could increase by as much as 36 Tg annually. If the soil moisture decreases in northern ecosystems, CH_4 production would be decreased while CO_2 flux would increase.

At the regional scale, changes in vegetation structure such as deforestation, have the potential to alter the hydrological cycle. Both precipitation and surface runoff can be affected. We do not yet have models to predict the affects of complex changes in land use patterns on regional hydrology.

10.4 Marine Ecosytems and Climate Change

10.4.1 Climate Change and Community Response

Based on the record of the past, there is little doubt that global warming will result in different distributions of marine planktonic organisms than those of today (CLIMAP Project, 1976). If the ocean warming were to be simply and positively correlated with latitude, the expansion of habitat in a poleward direction, which has occurred during the Holocene, would continue. But since the rate of change is expected to be very rapid (see Section 5), questions immediately arise regarding the potential of the biota to accommodate to these rates of change.

Changes in temperature and precipitation will have an influence on the circulation of surface waters and on the mixing of deep water with surface water. This mixing

exchanges water to great depth in a few places in the ocean such as the North Atlantic. Changes in circulation and/or a restriction of the mixing could reduce ocean productivity.

The palaeo-record for global temperature patterns during recent ice age cycles and the output of general circulation models that simulate increases in global temperatures resulting from higher concentrations of radiatively active gases in the troposphere both indicate that the warmer the planet the less the meridional gradient in temperature. Temperatures in polar regions could warm 2-3 times the global mean warming associated with a doubling of the equivalent of preindustrial atmospheric CO_2 concentrations.

At high latitudes, warming would result in diminished temporal and spatial extent of sea-ice; some models even predict an ice-free Arctic. A significant reduction in the extent and persistence of sea-ice in either polar region would have profound consequences for marine ecosystems. Sea-ice itself is a critical habitat for Arctic marine plankton (Clarke, 1988). The underneath of the ice and interstices in the ice are highly productive habitats for plankton. The low light in winter limits primary production in the water column, and the relatively high concentrations of algae living on and actually in the ice, are an important source of food for herbivores both while sea-ice is in place and when it breaks up in the spring. The quantitative importance of sea-ice in the high latitude marine ecosystem is now well established, with important food web implications for fish, seabirds, and marine mammals (Gulliksen and Lonne, 1989). During the spring melting the resultant freshwater lens is also believed to be critical in the life cycle of many pelagic polar species. The most intense aggregations or "blooms" of plankton occur at the ice margin, evidently in response to the density stratification resulting from the overlying lens of fresh water, and perhaps also in response to the release of nutrient materials that had accumulated in the ice. Although total global primary production for a polar region might not be very different with or without the ice edge "blooms", such transient peaks in plankton abundance can be critical in the life cycle of certain higher trophic level organisms, and are most definitely of importance in terms of the flux of carbon to deep water and the ocean floor.

Certain marine animals, mammals and birds in particular, have life history strategies that reflect adaptation to sea ice. An extreme case may be the polar bear, which shares ancestors with the grizzly bear, and would not exist today had it not been for the reliability of the ice habitat for hunting its primary prey, the seal.

In sub-polar and temperate waters the effects of global warming on the plankton habitat in near-surface waters are at present unpredictable. Physical, chemical, and biological conditions in these regions are inherently variable on both seasonal and annual bases. This results primarily from the variability in atmospheric forcing, including surface heat exchange with the atmosphere and associated thermal stratification and destratification in the upper ocean, seasonal storm events and associated mixing, and cloud cover. These processes and properties determine the mass upward flux of nutrients from the deep ocean, the residence time of plankton in the upper sunlit region of the water column, and the availability of light at the surface of the sea. These seasonal variations in conditions give rise to seasonality in primary production with attendant "blooms" and corresponding high rates of organic carbon flux to the deep-sea (Honjo, 1984).

Long-term oceanographic studies in several regions demonstrate high correlations between the abundance and productivity of marine ecosystems and atmospheric conditions. A clear example of this is the 1950-80 period in the vicinity of the United Kingdom. During this period there was a decline in the abundance of plankton that has been correlated with changes in wind strength and direction (Dickson et al., 1988). Temperature changes per se were small and poorly correlated with the plankton changes, and it is judged that changes in wind-driven mixing in the upper ocean was the primary factor contributing to the decline in plankton abundance. From studies such as this it seems likely that the type of climate changes being forecast in association with increased radiative forcing in the troposphere will have significant effects on plankton abundance and productivity.

In addition, the effect of intensity and frequency of mixing events on plankton assemblages depends on the physiology and structure of the plankton. Diatoms, for example, are phytoplankton that typically dominate in cold nutrient-rich waters, such as those that are seasonally well mixed. Because they have high sinking rates, diatoms require a turbulent mixed layer in order to remain successful constituents of the plankton community (Smayda, 1970). Diatoms are the preferred food for many organisms in the marine food web, and when replaced by other types of phytoplankton, fish productivity can be dramatically reduced (Barber and Chavez, 1983). The dense cell encasement of the diatom helps to explain why these phytoplankton are an important contributor to the flux of organic carbon to the deep-sea.

Along-shore winds contribute to the mixing of deep water with surface water in many coastal waters and across the equatorial Pacific. The direction, intensity, duration, and frequency of these wind events determine the extent and timing of the mixing events. Because this process, which is typically highly seasonal, is very important in stimulating the primary production processes that lie at the base of the food webs for many species, it can be anticipated that the global climate change will affect higher trophic levels, including fish, in these regions.

10.4.2 Interaction Between the Land and the Ocean

Climate change has the potential to change the rate of delivery of materials from the land to the ocean and such changes could affect the biological component of ocean ecosystems. For example, climate change could lead to an increase in wind erosion and the delivery of fine particles by aeolian transport from the continents to the surface ocean. Depending on their composition, such particles could have either a biostimulatory (Martin and Fitzwater, 1987) or a biotoxic effect on oceanic productivity and other marine processes. Similarly, climate change could increase water-driven erosion and the amount of material transported to the world's coastal oceans by river systems. Again, the effects will be dependent on the nature of the terrestrial material.

10.4.3 Interactions Between the Ocean and the Atmosphere

Climate-induced changes in ocean ecology are of importance in relation both to the sustainability and management of living resources and to biogeochemical feedback on the climate system. Several atmospheric feedback processes are well-defined in General Circulation Models, but those relating to ocean productivity are not yet sufficiently well understood to be included. The three main effects of global warming that are expected to operate on ocean plankton (as mentioned above) would all tend to decrease the ocean uptake of CO_2, i.e., a positive feedback.

Ocean waters are currently a major source of dimethylsulphide (DMS) to the atmosphere. The oxidized products of this DMS, which is produced by plankton, may increase cloudiness through nucleation on sulphate aerosols in the troposphere (Charlson et al., 1987), and may increase albedo in the stratosphere from sulphate aerosols (Ryaboshapko, 1983). One group of plankton, the Coccolithophorids, are apparently a major source of DMS, and their bloom processes would most likely respond, although in uncertain ways, to changes in ocean-atmosphere exchanges resulting from climate change.

While neither the direction nor magnitude of many of these effects is known with certainty, changes from present-day values are expected to be greatest in mid- to high-latitude ocean regions. There is evidence for major changes in the functioning of North Atlantic ecosystems in the transitions between glacial and inter-glacial periods (Broecker and Denton, 1990), supporting the view that changes in ocean plankton, once initiated, may enhance the rate of climate change (the 'plankton multiplier') until a new near-equilibrium is reached.

10.4.4 The Carbon System and the Biological Pump

The oceans are by far the largest active reservoir of carbon. Recent estimates of the total amount of dissolved inorganic carbon in the sea establish its range as between 34,000 and

38,000 Pg carbon. Only a small fraction is CO_2 (mole fraction 0.5 percent); the bicarbonate ion with a mole fraction of 90 percent and the carbonate ion with a mole fraction of just under 10 percent are the dominant forms of dissolved inorganic carbon. The dissolved organic carbon pool has been reported to be similar in size to the pool of terrestrial soil carbon, but recent data suggest that it may in fact be considerably larger.

Although the oceans are the largest active reservoirs of carbon and cover 70 percent of the globe, the total marine biomass is only about 3 x Pg carbon (though such estimates are uncertain at best), or just over 0.5 percent of the carbon stored in terrestrial vegetation. On the other hand, the total primary production is 30 to 40 x Pg carbon/yr. A portion of this production results in a sink for atmospheric CO2, primarily through the sinking of particulate carbon. As a consequence of this "biological pump," the concentration of dissolved inorganic carbon is not uniform with depth; the concentration in surface waters is 10 to 15 percent less than that in deeper waters. There is a corresponding depletion of phosphorus and nitrogen in surface waters, even in areas of intense upwelling, as a result of biological uptake and loss of detrital material.

The fate of this material depends, in part, upon its chemical characteristics. If it is in the form of organic tissue, then it is oxidized at intermediate depths, which results in an oxygen minimum and a carbon, nitrogen, and phosphorus maximum. If it is carbonate, it dissolves below the lysocline, raising both alkalinity and the concentration of carbon, at depths where the high pressure increases the solubility of calcium carbonate.

Thus the "biological pump" lowers the partial pressure of CO_2 in surface waters and enhances the partial pressure in waters not in contact with the atmosphere. The functioning of the biological pump involves the supply of nutrients to surface waters, food web dynamics, and sinking losses of particulates to the deep sea. It may be expected to respond both to changes in the strength of the overall thermohaline circulation and to variations in the abundance of nutrients, primarily nitrogen and phosphorus.

A portion of the nutrient flux to the surface returns to the deep sea unused by the biota, carried along by the return flow of waters in downwelling systems at high latitude. It is important to define the physical, chemical, and biological processes that regulate the concentration of organic nutrients in descending water masses, the flux of so-called preformed nutrients. The concentration of preformed nutrients may be expected to reflect physical processes, and it can be influenced also by biological activity to the extent that this activity can result in packaging of carbon, nitrogen, and phosphate in fecal material that can fall to the deep, providing a path for transfer of nutrients from the surface to the deep, independent of the physical processes such as those responsible for the formation of deep water in high latitudes.

10.4.5 Summary

Climate change can affect the productivity and the storage of organic carbon in marine ecosystems. The community composition of marine ecosystems will also be affected. Details of these effects cannot be predicted at the present time. There is also the possibility that the net exchange of trace gases (e.g., organic sulphur gases) between the oceans and the atmosphere could be altered but this, too, is uncertain.

References

Anderson, J.M., 1973: Carbon dioxide evolution from two deciduous woodland soils. *Journal of Applied Ecology* **10**:361-378..

Armentano, T.V. and E.S. Menges, 1986: Patterns of change in the carbon balance of organic-soil wetlands of the temperate zone. *Journal of Ecology* **74**:755-774.

Atjay, G.L., P. Ketner and P. Duvigneaud, 1979: Teresstrial primary production and phytomass, pp. 129-182. In: B. Bolin, E. Degens, S. Kempe and P. Ketner (eds.), *The Global Carbon Cycle*, SCOPE 13. Wiley, Chichester.

Avissar, R., and R.A. Pielke, 1989: A parameterization of heterogeneous land surfaces for atmospheric numerical models and its impact on regional meteorology. *American Meteorological Society* **117**-2113.

Barber, R.T. and F.P. Chavez, 1983: Biological consequences of El Niño. *Science* **222**:1203-1210.

Bazzaz, F.A. and R.W. Carlson, 1984: The response of plants to elevated CO_2. I. Competition among an assemblage of annuals at different levels of soil moisture. *Oecologia* (Berlin) **62**:196-198.

Billings, W.D., J.O. Luken, D.A. Mortensen and K.M. Petersen, 1982: Arctic tundra: A sink or source for atmospheric carbon dioxide in a changing environment? *Oecologia* (Berlin) **53**:7-11.

Billings, W.D., J.O. Luken, D.A. Mortensen and K.M. Peterson, 1983: Increasing atmospheric carbon dioxide: possible effects on arctic tundra. *Oecologia* (Berl.) **58**:286-289.

Billings, W.D., K.M. Peterson, J.D. Luken and D.A. Mortensen, 1984: Interaction of increasing atmospheric carbon dioxide and soil nitrogen on the carbon balance of tundra microcosms. *Oecologia* **65**:26-29.

Bolin, B., 1986: How much CO_2 will remain in the atmosphere? pp. 93-155. In: B. Bolin, B. Doos, J. Jager and R. Warrick (eds.), *The Greenhouse Effect, Climate Change, and Ecosystems*, SCOPE 29. John Wiley and Sons, London.

Bowden, W.B., and F.H. Boorman, 1986: Transport and loss of nitrous oxide in soil water after forest clear-cutting. *Science* **233**;867-869.

Brady, N.C., 1974: The nature and properties of soils. MacMillans, New York.

Broecker, W.S., and G.H. Denton, 1990: What drives glacial cycles? *Scientific American* **262**:49-56.

Bunnell, F., D.E.N. Tait, P.W. Flanagan and K.Van Cleve, 1977: Microbial respiration and substrate weight loss. I. A general model of the influences of abiotic variables. *Soil Biology and Biochemistry* **9**:33-40.

Cannell, M.G.R., and R.I. Smith, 1986: Climatic warming, spring budburst and frost damage on trees. *J. Appl. Ecol.* **23**:177-191

Carlyle, J.C. and T.U. Ba, 1988: Abiotic controls of soil respiration beneath an eighteen-year-old Pinus radiata stand in Southeast Australia. *Journal of Ecology* **76**:654-662.

Charlson, R.J., J.E. Lovelock, M.O. Andreae and S.G. Warren, 1987: Oceanic phytoplankton, atmospheric sulfur, cloud albedo and climate. *Nature* **326**:655-661.

Carter, K.N., and S.D. Prince, 1981: Epidemic models used to explain biogeographical distribution limits. *Nature* **293**:644-645.

Charlson, R.J., J.E. Lovelock, M.O. Andreae and S.G. Warren, 1987: Oceanic phytoplankton, atmospheric sulfur, cloud albedo and climate. *Nature* **326**:655-661.

Clarke, A., 1988: Seasonality in the Antarctic marine environment. *Comp. Biochem. Physiol.* **90B**:461-473.

CLIMAP Project, 1976: The surface of the ice-age earth. *Science*: **191**:1131-1137.

Conrad, R., and H. Schutz, 1988: Methods of studying methanogenic bacteria andmethanogenic activities in aquatic environments. Pp. 301-343 in: B. Austin (ed.) *Methods in Aquatic Bacteriology*. Chichester:Wiley.

Conrad, R., H. Schutz and M. Babbel, 1987: Temperature limitation of hydrogen turnover and methanogenesis in anoxic paddy soil. *FEMS Microbiol. Ecol.* **45**:281-289.

Curtis, P.S., B.G. Drake, P.W. Leadly, W. Arp and D. Whigham, 1989: Growth and senescence of plant communities exposed to elevated CO_2 concentrations on an estuarine marsh. *Oecolgia* (Berlin) **78**:20-26.

Curtis, P., L.M. Balduman, B.G. Drake and D.F. Whigham, 1990: The effects of elevated atmospheric CO_2 on belowground processes in C_3 and C_4 estuarine marsh communities. *Ecology*, in press.

Davis, M.B., 1981: Quaternary history and the stability of forest communities, pp. 132-153. In: D.C. West, H.H. Shugart and D.B. Botkin (eds.), *Forest Succession*. Springer-Verlag, New York.

Davis, M.B., 1981: Quaternary history and the stability of forest communities. Pp. 132-153 in: D.C. West, H.H. Shugart and D.B. Botkin (eds.) *Forest Succession*. Springer-Verlag, New York.

Davis, M.B., 1988: Ecological Systems and Dynamics. Pp. 69-106 in: *Toward an Understanding of Global Change*. National Academy Press, Washington, DC.

Dickinson, R.E. and A. Henderson-Sellers, 1988: Modelling tropical deforestation: A study of GCM land-surface parameterizations. *Quart. J. R. Met. Soc.* **114**:439-462.

Dickson, R.R., P.M. Kelly, J.M. Colebrook, W.C. Wooster and D.H. Cushing, 1988: North winds and production in the eastern North Atlantic. *J. Plankt. Res.* **10**:151-169.

Drake, B.G., P.W. Leadley, W. Arp, P.S. Curtis and D. Whigham, 1989: The effect of elevated atmospheric CO_2 on C_3 and C_4 vegetation on Chesapeake Bay. Proceedings of the Symposium on the Physiological Ecology of Aquatic Plants. Aarhus, Denmark.

Eichner, M., 1990: Nitrous oxide emissions from fertilized soils: summary of available data available. *Journal of Environmental Quality* **19**:272-280.

Emanuel, W.R., H.H. Shugart and M.P. Stevensohn, 1985a: Climate change and the broad-scale distribution of terrestrial ecosystem complexes. *Climatic Change* **7**:29-43.

Emanuel, W.R., H.H. Shugart and M.P. Stevenson, 1985b: Response to comment : Climatic change and the broad-scale distribution of terrestrial ecosystem complexes. *Climatic Change* **7**:457-460.

Fajer, E.D., M.D. Bowers and F.A. Bazzazz, 1989: The effects of enriched carbon dioxide atmospheres on plant-insect herbivore interactions. *Science*. **243**:1198-1200.

Farquhar, G.D. and T.D. Sharkey, 1982: Stomatal conductance and photosynthesis. *Annual Reviews of Plant Physiology* **33**:317-345.

Fetcher, N., C.H. Jaeger, B.R. Strain and N. Sionit, 1988: Long-term elevation of atmospheric CO_2 concentration and the carbon exchange rates of saplings of Pinus taeda and Liquid amber styraciflua L. *Tree Physiology* **4**:255-262.

Firestone, M.K., and E.A. Davidson, 1989: Microbial basis for NO and N_2O production and consumption in soil, pp. 7-22. In: M.O. Andreae and D.S. Schimel (eds.). *Exchange of Trace Gases betweeen Terrestrial Ecosystems and the Atmosphere*.

Fitter, A.H., and R.K.M. Hay, 1981: Environmental Physiology of Plants. Academic Press, London. pp. 355.

Gifford, R.M., 1979: Growth and yield of CO_2-enriched wheat under water-limited conditions. *Aust. J. Plant Physiol.* **6**:367-378.

Gifford, R.M., 1990: The effect of CO_2 and climate change on photosynthesis, net primary production and net ecosystem production. *Ecological Applications* (submitted).

Gifford, R.M., H. Lambers and J.I.L. Morison, 1985: Respiration of crop species under CO_2 enrichment. *Physiologia Plantarum* **63**:351-356.

Goudriaan, J., and H.E. de Ruiter, 1983: Plant response to CO_2 enrichment, at two levels of nitrogen and phosphorous supply. 1. Dry matter, leaf area and development.. *Neth. J. Agric. Sci.* **31**:157-169.

Graham, R.W., 1986: Response of mammalian communities to environmental changes during the late Quaternary. Pp. 300-313 in: J. Diamond and T.J. Case (eds.) *Community Ecology*. Harper & Row, New York.

Gulliksen, B., and O.J. Lonne, 1989: Distribution, abundance and ecological importance of marine symnpagic fauna in the Arctic. *Rapp. P. v. Reun. Cons. Int. Explor. Mer.* **188**:133-138.

Hanson, J. , G. Russell, D. Rind, P. Stone, A. Lacis, S. Lebedeff, R. Ruedy and L. Travis, 1983: Efficient three-dimensional global models for climate studies: Models I and II. April *Monthly Weather Review* **3(4)**:609-662

Hanson, A.G., and W.D. Hitz, 1982: Metabolic responses of mesophytes to plant water deficits. *Ann. Rev. Plant Physiol.* **33**:163-203.

Hardy, R.W.F. and U.D. Havelka, 1975: Photosynthate as a major factor limiting N_2 fixation by field grown legumes with emphasis on soybeans. Pp. 421-439 in: R. S. Nutman (ed.) *Symbiotic Nitrogen Fixation in Plants*, Cambridge University Press, London.

Heal, O.W., P.W. Flanagan, D.D. French and S.F. MacLean, 1981: Decomposition and accumulation of organic matter. Pp. 587-633 in: L. C. Bliss, O. W. Heal and J. J. Moore (eds.) *Tundra Ecosystems: A Comparative Analysis*. Cambridge University Press, Cambridge.

Hincklenton, P.R. and P.A. Jolliffe, 1980: Alterations in the physiology of CO_2 exchange in tomato plants grown in CO_2-enriched atmospheres. *Canadian Journal of Botany* **58**:2181-2189.

Henderson-Sellers, A., and V. Gornitz, 1984: Possible climatic impacts of land cover transformations, with particular emphasis on tropical deforestation. Climatic Change. **6**:231-258.

Holdridge, L.R., 1947: Determination of world plant formations from simple climatic data. *Science* **105**:367-368.

Holdridge, L.R., 1964: Life Zone Ecology. Tropical Science Center, San Jose, Costa Rica.

Honjo, S., 1984: The study of ocean fluxes in time and space by bottom tethered sediment trap arrays. GOFFS Proceedings of a Workshop, National Academy Press, Washington, D.C.

Houghon. R., 1990: The global effects of tropical deforestation. *Environmental Science and Technology*. **24**:414-422.

Hsaio, T., 1973: Plant responses to water stress. *Annual Reviews of Plant Physiology* **24**:519-570.

Hunt, H.W., 1977: A simulation model for decomposition in grasslands. *Ecology* **58**:469-484.

Huntley, B. and H.K.B. Birks, 1983: *An atlas of past and present pollen maps of Europe: 0-13,000 years ago*. Cambridge University Press.

Idso, S.B., B.A. Kimball, M.G. Anderson and J.R. Mauney, 1987: Effects of atmospheric CO_2 enrichment on plant growth: the interactive role of air temperature. Agriculture, *Ecosystems and the Environment* **20**:1-10.

Israel, Yu A., L.M. Filipova, L.M. Insarov, G.E. Semenov and F.N. Semeniski, 1983: The background monitoring and analysis of the global change in biotic states. Problems of Ecological Monitoring and Ecosystem Modelling. **IV**:4-15.

Jarvis, P.G., 1989: Atmospheric carbon dioxide and forests. *Phil. Trans. R. Soc.* London B **324**:369-392.

Johnson, W.C., and C.M. Sharpe, 1983: The ratio of total to merchantable forest biomass and its application to the global carbon budget. *Canadian Journal of Forest Research*. **13**:372-383.

Kimball, B.A., 1983: Carbon-dioxide and agricultural yield: An assemblage and analysis of 430 prior observations. *Agronomy Journal* **75**:779-788

Kramer, P.J., 1981: Carbon dioxide concentration, photosynthesis, and dry matter production. *BioScience* **31**:29-33.

Lamb, H.H., 1977: Climate: Present, Past and Future. Vol. 2, London, Methuen. pp. 835

Larigauderie, A., D.W. Hilbert and W.C. Oechel, 1988: Interaction between high CO_2 concentrations and multiple environmental stresses in Bromus mollis. *Oecologia* **77**:544-549.

Lashof, D.A., 1989: The dynamic greenhouse: feedback processes that may influence future concentrations of atmospheric trace gases and climatic change. *Climatic Change* **14**:213-242.

Lean, J., and D.A. Warrilow, 1989: Simulation of the regional impact of Amazon deforestation. *Nature* **342**:411-413.

Leonard, K.J. and W.E. Fry (eds.). 1986. *Plant Disease Epidemiology, Population Dynamics and Management*. Macmillan, New York. 372 pp.

Lincoln, D.E., N. Sionit and B.R. Strain, 1984: Growth and feeding response of Pseudoplusia includens (Lepidoptera:Noctuidae) to host plants grown in controlled carbon-dioxide atmospheres. *Environmental Entomology* **13**:1527-1530.

Lincoln, D.E., and D. Couvet and N. Sionit, 1986: Response of an insect herbivore to host plants grown in carbon dioxide enriched atmospheres. *Oecologia* (Berlin) **69**:556-560.

Lincoln, D.E., and D. Couvet, 1989: The effects of carbon supply on allocation to allelochemicalas and caterpillar consumption of peppermint. *Oecologia* (Berlin) **78**:112-114..

Long, S.P., 1990: Primary production in grasslands and coniferous forests in relation to climate change: an overview of the information available for modelling change in this process. *Ecological Applications* (In press).

Luizao, F., P. Matson, G. Livingston, R. Luizao and P. Vitousek, 1990: Nitrous oxide flux following tropical land clearing. *Global Biochemical Cycles*. in press.

Luxmoore, R.J., E.G. O'Neill, J.M. Ellis and H.H. Rogers, 1986: Nutrient uptake and growth responses of Virginia pine to elevated atmospheric CO_2. *J. Environ. Qual.* **15**:244-251.

Manabe, S., and R.T. Wetherald, 1987: Large-scale changes in soil wetness induced by an increase in carbon-dioxide. *Journal of Atmospheric Sciences*. **44**:1211-1235.

Martin, J.H. and S. Fitzwater, 1987: Iron deficiency limits phytoplankton in the northeast Pacific subarctic. *Nature* **331**:341-343.

Matson, P.A., and P.M. Vitousek, 1990: Ecosystem approaches for the development of a global nitrous oxide budget. *Bioscience*, in press.

McGill, W.B., C.A. Campbell, J.F. Doormaar, E.A. Paul and D.W. Anderson, 1981: PHOENIX - A model of the dynamics of carbon and nitrogen in grassland soils. In: F.E. Clark and T. Rosswall (eds.) *Terrestrial Nitrogen Cycles*. Ecological Bulletins, Stockholm.

McNulty, A.K., and W. Cummins, 1987: The relationship between respiration and temperature in leaves of the arctic plant Saxifraga cornua. *Plant, Cell and Environment* **10**:319-325.

Meentemeyer, V., 1978: Macroclimate and lignin control of litter decomposition rates. *Ecology* **59**:465-472.

Meentemeyer, V., 1984: The geography of organic matter decomposition rates. *Annals of the Association of American Geographers* **74**:551-560.

Melillo, J.M., 1983: Will increases in atmospheric CO_2 concentrations effect decay processes? pp. 10-11 Annual Report, The Ecosystems Center, Marine Biological Laboratory, Woods Hole, MA.

Melillo, J.M., J.D. Aber and J.F. Muratore, 1982: The influence of substrate quality of leaf litter decay in a northern hardwood forest. *Ecology* **63**:621-626.

Melillo, J.M., and J.R. Gosz, 1983: Interactions of biogeochemical cycles in forest ecosystems, pp. 177-222. In: B. Bolin and R.B. Cook (eds.), *The Major Biogeochemical Cycles and their Interactions*. John Wiley and Sons, New York.

Melillo, J.M., J.R. Fruci, R.A. Houghton, B. Moore III and D.L. Skole, 1988: Land-use change in the Soviet Union between 1850 and 1980: causes of a net release of CO_2 to the atmosphere. *Tellus* **40B**:116-128.

Melillo, J.M., P.A. Steudler, J.D. Aber and R.D. Bowden, 1989: Atmospheric deposition and nutrient cycling, pp. 263-280. In: M. O. Andreae and D. S. Schimel (eds.), *Exchange of Trace Gases between Terrestrial Ecosystems and the Atmosphere*. Dahlem Workshop Report (LS) 47. John Wiley and Sons, New York.

Minderman, G., 1968: Addition, decomposition and accumulation of organic matter in forests. *Journal of Ecology* **56**:355-362.

Mooney, H.A., B.G. Drake, R.J. Luxmoore, W.C. Oechel and L.F. Pitelka, 1990: How will terrestrial ecosystems interact with the changing CO_2 concentration of the atmosphere and anticipated climate change? *Bioscience*, in press.

Moore, T.R., 1984: Litter decomposition in a sub-arctic, spruce-lichen woodland in eastern Canada. *Ecology* **65**:299-308.

Nobre, C, J. Shukla and P. Sellers, 1990: Amazon deforestation and climate change. *Science* **247**:1322-1325

Norby, R.J., E.G. O'Neill and R.G. Luxmoore, 1986: Effects of atmospheric CO_2 enrichment on the growth and mineral nutrition of Quercus alba seedlings in nutrient poor soil. *Plant Physiology* **82**:83-89.

O'Neill, E.G., R.J. Luxmoore and R.J. Norby, 1987: Elevated atmospheric CO_2 effects on seedling growth, nutrient uptake, and rhizosphere bacterial populations of Liriodendron tulipifera L. *Plant and Soil* **104**:3-11.

Oberbauer, S.F., N. Sionit, S.J. Hastings and W.C. Oechel, 1986: Effects of CO_2 enrichment and nutrition on growth, photosynthesis and nutrient concentration of Alaskan tundra plant species. *Can. J. Bot.* **64**:2993-2998.

Oechel, W.C. and B.R. Strain, 1985: Native species responses. Chapter 5, pp. 118-154 in: B. R. Strain and J. D. Cure (eds.) *Direct Effects of Carbon Dioxide on Vegetation*, State-of-the-Art Report. U. S. Department of Energy, Office of Basic Energy Sciences, Carbon Dioxide Research Division, Washington, DC.

Oechel, W.C. and G.H. Riechers, 1986: Impacts of increasing CO_2 on natural vegetation, particularly tundra. In: C. Rosenzweig and R. Dickinson (eds.) *Climate-Vegetation Interactions*. OIES-UCAR Report OIES-2, Boulder, CO.

Oechel, W.C., and G.I. Riechers, 1987: Response of a Tundra Ecosystem to Elevated Atmospheric Carbon Dioxide. U.S. Department of Energy, Washington, DC.

Olson, J.S., J.A. Watts and L.J Allison, 1983: Carbon in Live Vegetation of major World Ecosystems . United States Department of Energy, TR004 pp. 164.

Orchard, V.A. and F.J. Cook, 1983: Relationships between soil respiration and soil moisture. *Soil Biology and Biochemistry* **15**:447-453.

Orchard, V.A., and F.J. Cook, 1983: Relationships between soil respiration and soil moisture. *Soil Biology and Biochemistry* **15**:447-453.

Overdieck, D., D. Bossemeyer and H. Lieth, 1984: Long-term effects of an increased CO_2 concentration level on terrestrial plants in model-ecosystems. I. Phytomass production and competition of Trifolium repens L. and Lolium perenne L. *Progress in Biometeorology* **3**:344-352.

Paez, A., H. Hollmers and B.R. Strain, 1984; CO_2 enrichment and water interaction on growth on two tomato cultivars. *Journal of Agricultural Science* **102**:687-693.

Parton, W.J., D.S. Schimel, C.V. Cole and D.S. Ojima, 1987; Analysis of factors controlling soil organic matter levels in Great Plains grasslands. *Soil Science Society of America Journal* **51**:1173-1179.

Pastor, J. and W.M. Post, 1986: Influence of climate, soil moisture and succession on forest soil carbon and nutrient cycles. *Biogeochemistry* **2**:3-27.

Pastor, J. and W.M. Post, 1988: Response of northern forests to CO_2-induced climate change. *Nature* **334**:55-58.

Peterson, B.J., and J.M. Melillo, 1985: The potential storage of carbon caused by eutrophication of the biosphere. *Tellus* **37B**:117-127.

Ritchie, J.C. and G.M. MacDonald, 1986: The patterns of post-glacial spread of white spruce. *Journal of Biogeography* **13**:527-540.

Ryaboshapko, A.G., 1983: The atmospheric sulphur cycle. Pp. 203-296 in: M. V. Ivanov and J. R. Freney (eds.) *The Global Biogeochemical Sulphur Cycle*. John Wiley & Sons, New York.

Salati, E., J. Marques and L.C.B. Molion, 1978: Origem e distribuicao das Chuvas na Amazonia. *Interciencia* **3**:200-206.

Sage, R.F., T.D. Sharkey and J.R. Seeman, 1989: Acclimation of photosynthesis to elevcated CO_2 in five C_3 species.. *Plant Physiology* **89**:590-596.

Sato, N., P.J. Sellers, D.A. Randall, E.K. Schneider, J Shukla, J.L. Kinter III, Y-T. Hou and E. Albertazzi, 1989: Effects of implementing the simple biosphere model in a general circulation model. *Journal of the Atmospheric Sciences* **46**:2757-2769.

Schlesinger, W. H., 1977: Carbon balance in terrestrial detritus. *Ann. Rev. Ecol. Syst.* **8**:51-81.

Schlesinger, W. H., 1984: Soil organic matter: a source of atmospheric CO_2, pp. 111-127. In: G. M. Woodwell (ed.), *The role of Terrestrial Vegetation in the Global Carbon Cycle, Methods of Appraising Changes*, SCOPE 23. John Wiley and Sons, Chichester.

Seiler, W. and R. Conrad, 1987: Contribution of tropical ecosystems to the global budgets of trace gases, especially CH_4, H_2, CO and N_2O. Pp. 133-162 in: R. E. Dickinson (ed.) *The Geophysiology of Amazonia. Vegetation and Climate Interactions*. Wiley, New York.

Sellers, P.J., 1987: Modelling effects of vegetation on climate, pp. 133-162. In: R.E. Dickinson (ed.), *The Geophysiology of Amazonia*, Wiley and Sons.

Smayda, T.J., 1970: The suspension and sinking of phytoplankton in the sea. *Oceanography, Marine Biology Annual Review* **8**:353-414.

Sionit, N., D.A. Mortensen, B.R. Strain and H. Hellmers, 1981: Growth response of wheat to CO_2 enrichment and different levels of mineral nutrition. *Agronomy Journal* **73**:1023-1027.

Sionit, N., H. Hellmers and B.R. Strain, 1982:. Interaction of atmospheric CO_2 enrichment and irradiance on plant growth. *Agronomy Journal* **74**:721-725.

Smayda, T.J., 1970: The suspension and sinking of phytoplankton in the sea. *Oceanography*, Marine Biology Annual Review **8**:353-414.

Smith, J.B. and D Turpak (eds.), 1989: The potential effect of global change on the United States. U.S. Enviromental Protection Agency, 413pp.

Solomon, A.M., 1986: Transient response of forests to CO_2-induced climate change: simulation modelling experiments in eastern North America. *Oecologia* (Berlin) **68**:567-579.

Steudler, P.A., R.D. Bowden, J.M. Melillo and J.D. Aber, 1989: Influence of nitrogen fertilization on methane uptake in temperate forest soils. *Nature* **341**:314-316.

Tans, P.P., I.Y. Fung and T. Taakahashi, 1990: Observational constraints on the global atmospheric CO_2 budget. *Science* **247**:1431-1438.

Tiedje, J.M., 1988: Ecology of denitrification and dissimilatory nitrate reduction to ammonium. Pp. 179-244 in: J. B. Zehnder (ed.) *Biology of Anaerobic Microorganisms*. Wiley, New York.

Tissue, D.T. and W.C. Oechel, 1987: Response of Eriophorum vaginatum to elevated CO_2 and temperature in the Alaskan arctic tundra. *Ecology* **68**(2):401-410.

Van Cleve, K. and D. Sprague, 1971: Respiration rates in the forest floor of birch and aspen stands in interior Alaska. *Arctic and Alpine Research* **3**:17-26.

Van Cleve, K., L. Oliver, R. Schlentner, L.A. Viereik and C.T. Dyrness, 1983: Productivity and nutrient cycling in taiga forest ecosystems. *Canadian Journal of Forest Resources* **13**:747-766.

Van Veen, J.A. and E.A. Paul, 1981: Organic carbon dynamics in grassland soils. I. Background information and computer simulation. *Canadian Journal of Soil Science* **61**:185-201.

Vorosmarty, C.J., B. Moore III, A.L. Grace, M.P. Gildea, J.M. Melillo, B.J. Peterson, E.B. Rastetter and P.A. Steudler, 1989: Continental scale models of water balance and fluvial transport: An appilication to South America. *Global Biogeochemical Cycles* **3**:241-265.

Warembourg, F.R. and E.A. Paul, 1977: Seasonal transfers of assimilated ^{14}C in grassland: Plant production and turnover, soil and plant respiration. *Soil Biology and Biochemistry* **9**:295-301.

Warrick, R.A., H.H. Shugart, M.Ja. Antonovsky, J.R. Tarrant and C.J. Tucker, 1986a: The effects of incerased CO_2 and climate change on terrestial ecosystems, pp. 363-392. In: B. Bolin, B. Doos, J. Jager and R. Warrick (eds), *The Greenhouse Effect, Climate Change, and Ecosystems*, SCOPE 29. John Wiley and Sons, London.

Warrick, R.A., R.M. Gifford and M.L. Parry, 1986b: CO_2, climatic change and agriculture, pp. 393-473. In: B. Bolin, B. Doos, J. Jager and R. Warrick (eds), *The Greenhouse Effect, Climate Change, and Ecosystems*, SCOPE 29. John Wiley and Sons, London.

Whittaker, R.H., and G.E. Likens, 1975: The biosphere and man, pp. 305-328. In: H. Leith and R.H.Whittaker (eds.), Primary Productivity of the Biosphere. *Ecol. Studies* **14**, Springer-Verlag. Berlin, Heidelberg, New York.

Williamson, M., 1989: MAthematical models of invasion, pp. 329-360. In: J.A. Drake, H.A. Mooney, F. di Castri, R.H. Groves, F.J. Kruger, M. Rejmanek and M. Williamson (eds.), *Biological Invasions: A Global Perspective*, SCOPE 37. Wiley, Chichester.

Wilson, M.F., 1984: The construction and use of land surface information in a general circulation model. Ph.D. thesis, University of Liverpool, UK.

Wong, S.C., 1979: Elevated atmospheric partial pressures of CO_2 and plant growth: I. Interactions of nitrogen nutrition and photosynthetic capacity in C_3 and C_4 species. *Oecologia* (Berlin) **44**:68-74.

Woodward, F.I., 1990: The impact of low temperatures in controlling the geographical distribution of plants. *Phil. Trans. R. Soc.* **326**:585-593.

Woodwell, G.M., 1987: Forests and climate: surprises in store. *Oceanus* **29**:71-75.

11

Narrowing the Uncertainties: A Scientific Action Plan for Improved Prediction of Global Climate Change

G. MCBEAN, J. MCCARTHY

Contributors:
K. Browning; P. Morel; I. Rasool.

CONTENTS

EXECUTIVE SUMMARY

The IPCC has the responsibility for assessing both the state of scientific knowledge of climate and climatic changes due to human influences. The World Climate Research Programme (WCRP), sponsored by the World Meteorological Organization (WMO) and the International Council for Scientific Unions (ICSU), and the International Geosphere-Biosphere Programme (IGBP), sponsored by ICSU, together constitute the international framework of the quest for scientific understanding of climate and global change.

The scientific strategy to achieve effective prediction of the behaviour of the climate system must be based on a combination of process studies, observation and modelling. Sections 1 to 10 of this report identified several areas of scientific uncertainty and shortcomings. To narrow these uncertainties, substantial scientific activities need to be undertaken. The following 5 areas are considered the most critical:

1) control of the greenhouse gases by the Earth system;
2) control of radiation by clouds;
3) precipitation and evaporation;
4) ocean transport and storage of heat; and
5) ecosystem processes.

Within the WCRP, the Global Energy and Water Cycle Experiment (GEWEX) is addressing (2) and (3), while the World Ocean Circulation Experiment (WOCE) is concerned with (4) and parts of (1). Two core activities of the IGBP, the Joint Global Ocean Flux Study (JGOFS) and the International Global Atmospheric Chemistry Programme (IGAC) are designed to investigate the control of greenhouse gases by the oceanic and terrestrial biospheres, while the Biospheric Aspects of the Hydrological Cycle (BAHC) is a complement to GEWEX that also addresses (3). An additional core project of the IGBP focuses on Global Change and Terrestrial Ecosystems (GCTE). Both the WCRP and IGBP have other essential core activities, such as the Tropical Oceans - Global Atmosphere (TOGA) Programme and the study of Past Global Changes (PAGES), that contribute to these efforts to reduce uncertainties in climate predictions.

Narrowing the uncertainties in future climate change predictions requires strongly enhanced national participation in these internationally coordinated programmes.

This will require increased commitments to the endeavours of the WCRP and the IGBP. These programmes are the result of many years of planning and they represent consensus statements of the international science community regarding the maturity of the fundamentals that underpin these projects and the readiness of the community to commit to these timely endeavours.

In order to proceed with this agenda, all nations must reaffirm their commitment to observe and document the fundamental aspects of the climate system and the changes occurring within it, including:

1) improvement of the global atmosphere and land surfaces observing system. The World Weather Watch and Global Atmospheric Watch need to be fully implemented and augmented by: improved atmospheric sounders, radiometers and wind observations; active sensors for wind and rain; vegetation sensors; and an improved commitment to quality control and archival of all data;

2) development of a global ocean and ice observing system. Satellite observations of ocean surface temperature, wind and topography, sea-ice concentration and colour, operational upper-ocean heat and freshwater monitoring, and systematic sea-level and deep-ocean measurements are required;

3) establishment of a comprehensive system for climate monitoring. It is essential that existing networks (WWW, GAW, IGOSS, GEMS, GSLS) be maintained and, where appropriate, enhanced. Special attention needs to be given to calibration and quality control, documentation and international coordination and data exchange.

The analysis and interpretation of the observational data will require the understanding arising from the projects of the WCRP and the IGBP, and will involve the use of more refined climate models. The next generation of predictive models will require additional computing resources in order to incorporate the more sophisticated understanding of the climate system arising from this research effort. This effort will lead to predictions that have higher spatial resolution than can be attained at this time.

The time scales for narrowing the uncertainties must be measured in terms of several years to more than a decade. Advances must await the conduct of several major experiments, many of which will be about a decade in duration, and the development of new technologies for space-based observation and numerical computation. It is essential that government funding agencies recognize the magnitude of both the financial and human resources needed to undertake these research programmes and make the necessary commitments.

11.1 Introduction

In order to deal with the issues posed by increased atmospheric greenhouse gas concentrations and to prepare human societies for the impacts of climate change, climate predictions must become more reliable and precise. Present shortcomings include:

Significant uncertainty, by a range of three, regarding the sensitivity of the global average temperature and mean sea-level to the increase in greenhouse gases;

Even larger uncertainties regarding regional climatic impacts, such that current climate change predictions have little meaning for any particular location;

Uncertainty in the timing of the expected climate change;

Uncertainty in the natural variations.

To overcome these shortcomings, substantial improvements are required in scientific understanding which will depend on the creative efforts of individual scientists and groups. Nevertheless the scale of the task demands international coordination and strong national participation.

The IPCC has responsibility for assessing the current state of scientific knowledge of climate and climatic changes due to human influences. The World Meteorological Organization (WMO) and the International Council of Scientific Unions (ICSU) established in 1980 the World Climate Research Programme (1), to promote scientific research on physical climate processes and to develop a capability for predicting climate variations. Several major internationally coordinated climate research projects organized by the WCRP are now underway. The Intergovernmental Oceanographic Commission (IOC) assists with the oceanographic component. Furthermore, ICSU established in 1986 the International Geosphere-Biosphere Programme (2) to study the interactive physical, chemical and biological processes responsible for change in the Earth system, especially those which are most susceptible to change, on time scales of decades to centuries. The IGBP with its emphasis on biogeochemical aspects and the WCRP with its emphasis on physical aspects, together constitute the international framework of the quest for the scientific understanding of global change. This report deals with climate change but it must be stressed that climate change is but one of a wide range of environmental issues that are confronting the world. Many of these issues are linked and scientific study of one issue will frequently aid in understanding others.

(1) The World Climate Research Programme (WCRP) is jointly sponsored by the World Meteorological Organization and the International Council of Scientific Unions. The main goals of the WCRP are to determine to what extent transient climatic variations are predictable and to lay the scientific foundation for predicting the response of the Earth's climate to natural or man-made influences. The main components of the WCRP are the numerical experimentation programme to develop improved models of the Earth's climate, the Global Energy and Water Cycle Experiment (GEWEX), the Tropical Ocean and the Global Atmosphere (TOGA) Programme, and the World Ocean Circulation Experiment (WOCE). Each programme includes a range of projects to study specific aspects or physical processes of the Earth system. An example is the International Satellite Cloud Climatology Project, to determine the quantitative effect of clouds on the Earth radiation balance and climate.

(2) The International Geosphere-Biosphere Programme (IGBP) is an inter-disciplinary research initiative of the International Council of Scientific Unions, to describe and understand the interactive physical, chemical and biological processes that regulate the total Earth system, the unique environment that it provides for life, the changes that are occurring and the manner in which changes are influenced by human actions. A central objective of the IGBP is to establish the scientific basis for quantitative assessments of changes in the Earth's biogeochemical cycles, including those which control the concentration of carbon dioxide and other chemicals in the atmosphere.

11.2 Problem Areas and Scientific Responses

To achieve effective prediction of the behaviour of the climate system, we must recognize that this system is influenced by a complex array of interacting physical, chemical and biological processes. The scientific strategy to address these processes must include both observation and modelling. We must be able to understand the mechanisms responsible for past and present variations and to incorporate these mechanisms into suitable models of the natural system. The models can then be run forward in time to simulate the evolution of the climate system. Such a programme includes three essential steps:

Analysis of observational data, often obtained from incomplete and indirect measurements, to produce coherent information and understanding;

Application of observational information and understanding to construct and validate time-dependent mathematical models of natural processes;

Running such models forward to produce predictions that can (and must) be tested against observations to determine their "skill" or reliability over relatively short time-periods.

Sections 1 to 10 of this assessment have identified several areas of scientific uncertainty. The following 5 areas are considered the most critical.

11.2.1 Control of the Greenhouse Gases by the Earth System

Greenhouse gases in the atmosphere, such as carbon dioxide and methane, are part of vast natural cycles. For some greenhouse gases, the current rates of release which are directly attributable to human activities are small percentages of large natural fluxes between the atmosphere, the ocean and terrestrial ecosystems while for others, human activities result in dominant emissions. The atmospheric carbon content is a very small fraction of existing reservoirs of carbon in ocean waters and sediments. Relatively minor adjustments in the world ocean circulation and chemistry, or in the life cycle of terrestrial vegetation, could significantly affect the amount of CO_2 or CH_4 in the atmosphere, even were anthropogenic emissions to be stabilized. In particular, global warming is likely to decrease the absorption of carbon dioxide by sea water and lead to widespread melting of methane gas hydrates in and under the permafrost and also release CH_4. Conversely, positive changes in the biogenic storage of carbon in the ocean could increase the oceanic CO_2 uptake and ameliorate the greenhouse effect.

Current knowledge of oceanic and terrestrial biogeochemical processes is not yet sufficient to account quantitatively for exchanges between the atmosphere, ocean and land vegetation (Section 1). The international Joint Global Ocean Flux Study (JGOFS), a core project of IGBP, has been designed to investigate the oceanic biogeochemical processes relating to the cycle of carbon in the ocean and to assess the capacity of the ocean for absorbing CO_2 (3). A central question being addressed relates to the role of the ocean and its circulation (see description of the World Ocean Circulation Experiment in Section 11.2.4) in the uptake of CO_2 produced from the burning of fossil fuels. This uptake occurs via both physical and biological processes. Neither is well quantified on a global scale, and the regulation of the biological processes is at present only poorly understood. In particular, the biogeochemical processes responsible for the long-term

storage of a portion of the total primary production cannot at this time be resolved sufficiently in time and space to say how they might be affected by climate change. The first component of JGOFS, relating to process studies, began with a pilot study in the North Atlantic in 1989. Two time-series of measurements have been initiated at stations in the vicinity of Bermuda and Hawaii. JGOFS will result in an order of magnitude improvement in the precision of the assessment of the ocean's role in sequestering CO_2 from the atmosphere.

(3) The Joint Global Flux Study (JGOFS), a core project of the IGBP, is organized by ICSU's Scientific Committee on Oceanic Research. Its major goal is to determine and understand on a global scale the time-varying fluxes of carbon and associated biogenic elements in the ocean and to evaluate the related exchanges with the atmosphere, the sea floor, and continental boundaries. This project has three major components: 1) a sequence of studies to elucidate the connections between various biogeochemical processes and distributions; 2) a global-scale survey and long time-series of measurements to improve the basic description of the carbon cycle; and 3) a modelling effort to identify critical processes and variables, to construct basin and global-scale fields from observed parameters and to predict future states of the ocean. The next generation of satellites will provide ocean colour data for the global assessment of temporal and spatial patterns in primary productivity, and the WOCE hydrographic survey will provide the ship opportunities and ancillary physical and chemical data required to generate the first global inventory of ocean carbon.

Another core project, the International Global Atmospheric Chemistry Programme (IGAC), is being designed to investigate the interactions between atmospheric chemistry and the terrestrial biosphere, particularly fluxes of carbon dioxide, methane and nitrogen oxides (4).

11.2.2 Control of Radiation by Clouds

Radiation is the primary energy source of the climate system and the principal heat input to the oceans (Section 2). These fluxes are very sensitive to the amount, distribution and optical properties of water and ice clouds (Sections 3 and 5) and are central to the problem of greenhouse heating. Aerosols also influence the net radiative flux at the surface. The net heat flux to the ocean determines the rate of ocean warming and volume expansion, which is likely to be the largest contribution to the global rise of the mean sea-level during the next century, if the Greenland and Antarctic ice sheets are neither gaining nor losing mass (Section 9).

(4) The International Global Atmospheric Chemistry Programme (IGAC), a core project of the IGBP, is jointly organized with ICSU's Commission on Atmospheric Chemistry and Global Pollution. Its goal is to document and understand the processes regulating biogeochemical interactions between the terrestrial and marine components of the biosphere and the atmosphere, and their role in climate. It consists of several research projects, which address natural variability and anthropogenic perturbations in the composition of the atmosphere over terrestrial tropical, polar, and mid-latitude regions as well as over the oceans. Other research efforts will address, through observations and modelling, the global distribution of chemically and radiatively important species (including emission rates and other processes governing their abundances) and the role of these substances in cloud condensation. The results will yield a marked improvement in our understanding of the processes responsible for regulating the abundance of atmospheric constituents that are of relevance to climate.

For the past two decades, the Earth radiation budget at the top of the atmosphere has been measured from satellites. The most recent radiometric measurement (Earth Radiation Budget Experiment) discriminates between cloudy and cloud-free areas, thus providing direct determination of the net energetic effect of clouds in the present climate. This information is not enough, however, to distinguish between the effect of different types of clouds at different altitudes. The ongoing International Satellite Cloud Climatology Project of WCRP (started in 1983) is working to assemble global statistics of the distribution and properties of different cloud types. In order to quantify the interannual variability of cloud systems, ten or more years of data are required. Because of the importance of changes in cloudiness in the radiation budget, these measurements must be continued. Process studies will be important in the study of feedbacks such as cloud-radiation interaction and its dependence on cloud water content, particle size and altitude.

Further modelling and observational research will nevertheless be necessary to achieve accurate representation in climate models of the role of clouds and radiation. The WCRP's Global Energy and Water Cycle Experiment (GEWEX), discussed more fully in Section 11.2.3, has as one of its objectives the more precise quantitative deduction of all energy fluxes within the atmosphere and at the air-sea and air-land interfaces (5).

(5) The Global Energy and Water Cycle Experiment (GEWEX) is a programme launched by the WCRP to observe, understand and model the hydrological cycle and energy fluxes in the atmosphere, on the land surface and in the upper ocean. The Programme will investigate the variations of the global hydrological regime and their impact on atmospheric and oceanic dynamics, as well as variations in regional hydrological processes and water resources, and their response to change in the environment such as the increase of greenhouse gases. The GEWEX Programme has several components. It incorporates a major atmospheric modelling and analysis component requiring a substantial increase in computer capabilities, because climate models with high spatial resolution are needed to achieve realistic simulations of regional climates. GEWEX will provide an order of magnitude improvement in the ability to model global precipitation and evaporation, as well as accurate assessment of the sensitivity of the atmospheric radiation and clouds to climate changes. Because of the complex interactions of clouds and radiation, the GEWEX Programme will cooperate with the TOGA Programme and other projects in studies of cloud processes. In addition, there will be studies to improve the extraction of atmospheric and land-surface information from satellite data. The GEWEX Programme also includes a series of land-surface experiments to develop understanding and parameterizations of evaporation and heat exchanges from inhomogeneous, vegetated surfaces. Advancement of hydrological models and their integration into climate models is another objective of the GEWEX Programme. The IGBP Core Project on the Biospheric Aspects of the Hydrological Cycle (BAHC) will deal with the complementary problem of resolving the role of the biosphere and land-surface processes in this context and develop methods to implement the interaction of the biosphere with the physical Earth system in global models.

Achieving the objectives of GEWEX will require the development of major new instruments to be flown on the next generation of multi-disciplinary satellite platforms in polar orbit or on the International Space Station and co-orbiting platforms (Earth Observing System, EOS). For this reason, the main GEWEX observing period must be timed to start with the launch of these satellite systems (expected in 1997 to 2000). The Experiment will last approximately five years and scientific interpretation and application of the results will take several further years.

11.2.3 Precipitation and Evaporation

The condensation of water is the main energy source of the atmospheric heat engine and the transport of water vapour by the atmospheric circulation is a key process in the redistribution of the sun's energy in the Earth system. Water vapour is also an important greenhouse gas. The vertical distribution of latent heating in precipitating clouds has a large effect on the large-scale circulation of the atmosphere. Precipitating clouds also play an important role in the general circulation through their effect on vertical transport of heat, moisture and momentum. The inflow of fresh water at high latitudes is a major factor in determining sea-water buoyancy, which forces the ocean circulation. The rates of accumulation of snow and the ablation of ice in the Antarctic and Greenland ice sheets are important sources of uncertainty for sea-level rise during the next century (Section 9). Changes in the hydrological regime, precipitation and evaporation, and consequent change in soil moisture and the availability of fresh water resources, are the most serious potential consequences of impending climate change in terms of its effect on man.

Unfortunately, present quantitative knowledge of the large-scale water budget is still very poor. For example, it has not yet been possible to measure or deduce from existing measurements either global precipitation or global evaporation (Section 7). About one third of the water run-off from continents to the ocean takes place as flow in small ungauged coastal rivers or underground aquifers. Much improved quantitative assessments of these components of the global water cycle are essential to achieve accurate predictions of future water resources in a changed climate. Values of precipitation and evaporation at sea can be inferred only roughly from general circulation models or satellite observations.

To address these and the other problems mentioned in Section 11.2.2, WCRP has launched the GEWEX programme (5) and specific projects aimed at the collection and analysis of observations available at present, such as the Global Precipitation Climatology Project (started in 1988) and the Global Run-off Data Project (1980). The Global Precipitation Climatology Project has undertaken to combine all available rain-gauge measurements and meteorological data with satellite observations of rain-clouds to produce the first global climatological record of monthly-mean total precipitation, including over the oceans. These projects require increased support by operational meteorological and hydrological agencies to upgrade the worldwide collection and exchange of essential ground-based measurements of rain, snow and river flow.

An ensemble of modelling and field studies of atmospheric and hydrological processes has already been initiated. The most significant achievements towards this objective are hydrological-atmospheric field studies aiming to close the energy and water budget of a land parcel of size commensurate with the spatial resolution of a general circulation model. A series of regional-scale (10 to 100 km) field studies (Hydrological-Atmospheric Pilot Experiments, HAPEX; First ISLSCP Field Experiment and others) in different major ecosystems was started in 1986 and future experiments will continue to the end of the century. Cooperation between WCRP and IGBP is being pursued to take into account the role of biological processes in evapotranspiration from terrestrial vegetation and, conversely, the effect of climate change on terrestrial ecosystems. Hydrological models of a continent-size river basin, driven by daily precipitation and evaporation estimated from analysed meteorological fields, are now being developed and will be applied during a continental-scale project in the mid-1990s. Other programmes, such as the Coupled Ocean-Atmosphere Response Experiment of the TOGA Programme, are important for determining the physical processes within mesoscale convective cloud systems.

11.2.4 Ocean Transport and Storage of Heat

The ocean plays a major role in the climate system through its storage and transport of heat. The response time of the upper ocean is relatively short (months to years) compared to the deeper ocean. It is now recognized that the largest portion of the interannual variability of the climate system is linked to the tropical oceans. For this reason, the Tropical Ocean - Global Atmosphere (TOGA) Programme (6) was originated and it is maintaining an intensive observational programme from 1985 to 1995. In addition, the TOGA Coupled Ocean-Atmosphere Response Experiment (COARE) is planned for the mid-1990s, to investigate the coupling between the warm western tropical Pacific Ocean, through cloud dynamics, with the high atmosphere. With the TOGA observational array of buoys, current meters, and ships, it has been possible to investigate the evolution of the tropical ocean and to initiate experimental forecasts of the El Niño phenomenon.

---

(6) The WCRP's Tropical Ocean - Global Atmosphere (TOGA) Programme is aimed at understanding and eventually predicting how the evolution of the tropical oceans interacts with and causes global-scale variability of the atmosphere. The El Niño - Southern Oscillation (ENSO) is the major cause of interannual variability of the climate system. TOGA began its observing period in 1985 and will continue until 1995. A large array of special oceanographic and atmospheric measurements have been deployed. The TOGA Programme also includes process studies and model development. One special activity is the investigation of monsoon dynamics.

If the atmosphere and upper-ocean alone were responding to the increase in greenhouse heating and the cloud-radiation feedback operated according to current knowledge, then the surface of the Earth would already be 1 to 2°C warmer than the temperatures of the nineteenth century. The response of the Earth's climate to increased greenhouse heating is being reduced by the thermal inertia of the ocean, determined by the largely unknown rate of penetration of heat into the upper 1000 metres. Deep-ocean warming results mainly from water sinking at high latitudes, frequently in the presence of sea ice, and subsequent circulation in the ocean. Quantitative modelling of the global ocean circulation is essential to determine the timing of global warming. The WCRP World Ocean Circulation Experiment (7) will provide the understanding and observations of the global ocean to enable the development of these ocean models.

---

(7) The World Ocean Circulation Experiment (WOCE) of the WCRP is a worldwide oceanographic programme to describe the oceanic circulation at all depths and on the global domain, during a five-year period (1990-1995). The primary goal of WOCE is to develop global ocean models for the prediction of climate change and to collect data sets necessary to test those models. Over the five years of the programme, there will be an intensification of the effort to determine air-sea fluxes globally by combining marine meteorological and satellite data, an upper-ocean measurements programme to determine the annual and inter-annual oceanic response to atmospheric forcing, and a programme of high quality, hydrographic and chemical tracer observations, surface and under-water drifters, current meter arrays, tide gauges and satellite altimetry to determine the basic features of the deep-ocean circulation. WOCE will provide adequate determination of global heat and fresh water fluxes in the bulk and at the surface of the ocean.

Three new satellite missions in support of WOCE are in the final stages of preparation or being planned:

The US-French TOPEX-POSEIDON precision altimetry mission, to measure the ocean surface topography with unprecedented accuracy for the purpose of determining ocean circulation (to be launched in 1992);

The European Space Agency ERS-1 and ERS-2 missions to measure wave height and ocean surface topography, wind stress and temperature at the surface of the ocean (to be launched in 1991 and 1994);

The US-Japanese ADEOS project to provide, in particular, more complete observations of the wind stress over the global ocean (to be launched in 1994).

Oceanographic agencies and institutions are joining forces to deploy the research vessel fleet needed to implement the WOCE hydrographic and geochemical surveys at sea, which call for 25 ship years. The concentration of a number of chemicals and isotopes will be measured throughout the ocean. By studying the distribution of substances that entered the environment at different times over the last century, geochemists and ocean modellers can estimate the time scales associated with the slow renewal and circulation of the ocean's deep water masses.

WOCE systems are to be activated for five years only. Maintaining systematic observations of essential oceanographic quantities after WOCE, to monitor the changes taking place in the ocean and to define the state of the global ocean circulation from which further dynamical predictions could be made, is a requirement which remains to be put into effect (Section 11.3.2).

11.2.5 Ecosystem Processes

As indicated above, both terrestrial and marine ecosystems are important as sources and sinks of biogenic gases that have radiative properties in the atmosphere. In addition, terrestrial ecosystems play an important role in the exchange of moisture and therefore energy between the land surface and overlying atmosphere. Thus predictions of climate, driven by anthropogenic increases in the atmospheric content of greenhouse gases, must take into consideration the likelihood that feedback from ecosystems will be altered by climate change itself. Temporal and spatial patterns in temperature, precipitation, and nutrient deposition (including extreme events) will directly affect soils, plant productivity, vegetation structure, and community composition. This influence is not limited to natural ecosystems; it also imposes regional constraints on agriculture and forestry. Large scale displacement of particular ecosystems will affect the climate system by altering local surface roughness and albedo.

A more quantitative understanding of the function of ecosystems in climate is important. In particular, research is needed to provide better global observations of the nature and extent of vegetation and soils. In addition, *in situ* studies must be scaled up to investigate the physiological and ecological processes that configure an ecosystem response to climate change. The interpretation of such data requires a modelling approach that also includes the ability of community constituents to migrate away from a unsuitable climate regime. The full use of this information in climate predictions requires higher resolution models (50 km) capable of simulating extreme conditions and the full range of the seasonal cycle. The IGBP is engaged in planning new studies of climate and land use changes on ecosystems, and their attendant feedbacks on climate. Partly to address these ecosystem dimensions of climate regulation, the IGBP has established a core project on Global Change and Terrestrial Ecosystems (GCTE) (8).

(8) Global Change and Terrestrial Ecosystems (GCTE) is a core project of the IGBP, aimed at developing the capacity to predict the effects of changes in climate, atmospheric composition and land-use practices on terrestrial ecosystems. This capacity is required both because the ecosystem changes are of direct importance to humans, and because they will have a feedback effect on further evapotranspiration, albedo and surface roughness. The project has two main foci: Ecosystem "Physiology" - the exchanges of energy and materials, and their distribution and storage - and Ecosystem Structure - the changes in species (functional type) composition and physiognomic structure, on the patch, landscape and regional (continental) scales. The project is based on close integration of experimentation and modelling. It consists of seven core activities, each of which is made up of a number of particular tasks, which include such topics as elevated CO_2 effects on ecosystem functioning, changes in the biogeochemical cycling of C, N, P and S, soil dynamics, vegetation dynamics and changes in spatial patterns.

11.3 Requirements for Narrowing Uncertainties in Future Climate Change

Climate research can provide a valuable service to society by providing the means for detection of future climate change, quantitative prediction of the timing and rate of the expected global changes and assessments of probable regional effects. In order to achieve these goals, it is necessary to develop a comprehensive Global Earth Observing System, to develop improved climate models and to acquire new scientific knowledge. As shown in Figure 11.1, there is an essential symbiosis between observations and modelling. Observations are required for long-term climate monitoring, especially for detection of climate change, and for local process studies. Modelling is required to support process studies as well as to provide the vehicle for climate predictions. Observations and modelling are brought together to develop an under-standing of the components of the climate system. At present, modelling and data assimilation studies are limited by availability of computer time; large increases in computer resources are a major requirement. A second major limitation on the advancement of climate research is the shortage of highly-trained scientists.

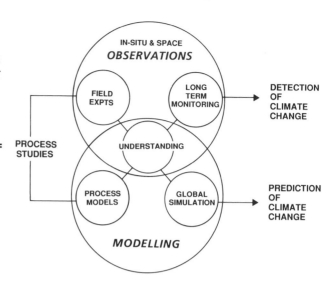

Figure 11.1: The symbiosis between observations and modelling.

11.3.1 Improvement of the Global Atmosphere and Land Surfaces Observing System

Basic information on climatic processes, climate variations and systematic trends originates from the operational meteorological observing systems of the World Weather Watch, complemented by various operational hydrological networks and environmental measurements of the Global Atmospheric Watch (9). In addition to the **maintenance** of the basic meteorological systems, specific **improvements** are needed, in particular:

a) improved infrared and microwave atmospheric sounding instruments on meteorological satellites, to obtain more accurate temperature/moisture information with better vertical resolution in the troposphere (e.g., high spectral resolution infrared spectrometer/radiometer);

b) improved tropospheric wind observations from geostationary satellites (cloud-drift winds) and platforms in low Earth-orbit (Doppler wind lidar);

c) rain-radars and passive multiple-frequency imaging radiometers in the microwave spectrum, to estimate rainfall over the whole globe; and

d) radiometers to determine more accurately the Earth's radiation budget.

Temporal and spatial patterns in key vegetation properties can, when calibrated and interpreted in the context of ground-truth data, be efficiently studied with satellite sensors. These data are critical in detecting regional shifts in ecosystem form and function in response to climate and land use changes. Sustaining these activities at levels appropriate to the study of climate change requires:

(9) Several international organizations operate major observational networks that provide information that needs to be included in a comprehensive system for climate monitoring. For climate monitoring purposes, it is essential that each of these networks be maintained and, where appropriate, enhanced. The important characteristics of climate monitoring systems include: long-term continuity; consistency of calibration; quality control; documentation of techniques; and international availability of data. The major networks are: the World Weather Watch and Global Atmosphere Watch of WMO; the Integrated Global Ocean Services System of WMO and the Intergovernmental Oceanographic Commission; the Global Environmental Monitoring System of UNEP; and the Global Sea-Level Service of IOC.

e) improved commitment to the acquisition and archival of AVHRR data for determining rates of change (for example, those arising from desertification and land clearing), shifts in seasonal vegetation cycles, and rates of plant production needed for carbon cycling modelling; and

f) improved high spectral resolution sensors, like those being defined for EOS, to determine vegetation characteristics (including physical and chemical properties) needed as input for global biogeochemical models.

11.3.2 Development of a Global Ocean and Ice Observing System

From the perspective of global climate change, systematic ongoing observations of the global ocean are needed for several purposes. The key to predicting the rate of change of the global system is to be found in observations of the ocean circulation and heat storage. Predictions of future climate change will eventually be carried out starting from the observed state of the combined atmosphere and ocean system. An example of this approach is the forecast of El Niño and other tropical climate disturbances by the WCRP Tropical Ocean and Global Atmosphere (TOGA) Programme.

A comprehensive ocean and ice observing system requires:

a) satellite observations of the ocean surface temperature, wind and topography, sea-ice concentration and chlorophyll content (ocean colour), and of the topography of the Antarctic and Greenland ice sheets, by an international array of space platforms in suitable orbits around the Earth;

b) an international operational upper-ocean monitoring programme, to determine the time and space dependent distribution of heat and fresh water in upper ocean layers, seasonal variations and long-term trends; and

c) an international programme of systematic sea-level and deep-ocean measurements, at suitable time and space intervals, to determine the state of the ocean circulation, ocean volume and transport of heat.

11.3.3 Establishment of a Comprehensive System for Climate Monitoring

In the previous two sections (11.3.1 and 11.3.2), improvements to the global atmosphere and land surface observing system and development of a global ocean and ice observing system were discussed. These observing systems must be coupled with existing observing systems to establish a comprehensive system for climate monitoring. It must be recognized that the requirements for climate monitoring are different from those for weather prediction. Failure to recognize this in the past means that there are a number of uncertainties which have been introduced into long-term climatological time series. A long-term commitment is now needed by the world's national weather services to monitor climate variations and change. Changes and improvements in observational networks should be introduced in a way which will lead to continuous, consistent long-term data sets sufficiently accurate to document changes and variations in climate (9). Some observing systems have and will be established for research purposes. These research systems are usually of limited duration and area and may have different emphases from a climate monitoring system (Figure 11.1). However, it is important that these observations also be integrated with the information from long-term monitoring systems.

Satellite observations are not yet of long enough duration to document climate variations. In order for these data to be most useful, it is very important that they be analysed and interpreted with existing *in situ* data. High priority should be given to the blending or integration of spaced-based and *in-situ* data sets in such a way as to build upon the strengths of each type of data. Examples of ongoing projects that have used mixed satellite data sets, careful cross calibration and coordinated data processing to produce global data sets are the WCRP's International Satellite Cloud Climatology Project (ISCCP) and the Global Precipitation Climatology Project (GPCP). For some applications, it is necessary to assimilate the observations into operational numerical weather prediction models for which purpose the observations need to be available in near real time. All satellite instruments should be calibrated both prior to launch and in flight. For those instruments now in operation, techniques should be developed to prevent or detect instrument drift. In planning new satellite observations for long-term monitoring, special attention should be paid to continuity of calibration and processing, archival and access to the data.

It is, further, imperative that there be strengthened international agreements and procedures for international exchange of existing basic climatological data (e.g., rain-gauge measurements and/or meteorological satellite data). National data centres must make available, through free exchange, to the world climate community, data sets collected in their countries. Existing CLIMAT and international data exchange (World Weather Records) must include essential variables which are absent at present. For most climate variables, the spatial and temporal resolution of the exchanged data is inadequate, precluding world-wide analyses of extremes. New methods of international exchange may be needed. For a worldwide ocean monitoring system, there are requirements for streng-thening international agreements to facilitate standard temperature, salinity and velocity measurements by all vessels and oceanographic drifting platforms within national Exclusive Economic Zones. Agreements among scientists and research institutions for international sharing of ocean-ographic data need to be strengthened.

There remains considerable data in manuscript form, which will prove valuable in producing more definitive analyses of climate change and variability. These data should be documented in computer compatible form. These and proxy data, including palaeo-climate data, are needed to reconstruct variations in past climates. In many cases, spatial and temporal resolution needs to be improved. Palaeodata can be particularly valuable in testing hypotheses regarding the mechanisms that have in the past linked physical and biogeochemical aspects of climate change. The IGBP has established a core project spec-ifically for studying Past Global Changes (PAGES) (10).

The WMO has undertaken to strengthen its activities to monitor the chemical composition and other characteristics of the atmosphere, away from pollution sources, and to incorporate these activities into a coordinated Global Atmosphere Watch (GAW) programme. The main objective of GAW is to establish a global monitoring network of about 20 observatories, complemented by regional stations whenever possible.

In addition to the above-mentioned needs to improve available instruments and methods to observe various climate parameters the development of a comprehensive system for monitoring climate change around the world is likely to be required under a convention on climate change. There are many international bodies involved in already existing monitoring activities such as the World Climate Research Programme, the World Climate Programme, the WMO Commission for Climatology, the Global Atmosphere Watch, the UNESCO Man and Biosphere, the Intergovernmental Oceanographic Commission and the

(10) The IGBP core project on Past Global Changes (PAGES) coordinates and integrates existing national and international palaeo-projects and implements new activities in order to obtain information on the pre-industrial variations of the Earth system and the baseline on which human impacts are superimposed. Typical research tasks are the separation of anthropogenic and corresponding responses and the documentation of possible internally forced processes. Of particular interest are the deconvolution of long-term climatic changes over a glacial cycle of ecosystems to the warming at the end of the last glaciation and changes in the atmospheric content of CO_2 and CH_4 throughout the last glacial cycle, and during periods of abrupt climatic change. All this information will help to evaluate both climatic and biogeochemical cycle models.

UNEP Global Environmental Monitoring System. The future overall system for monitoring climate change and its effects requires coordination of activities of these and other organizations.

11.3.4 Development of Climate Models

Improved prediction of climate change depends on the development of climate models, which is the objective of the climate modelling programme of the WCRP. Atmospheric general circulation models are based on numerical models used with considerable success by the national weather services to predict weather several days ahead. However, when adapted for climate prediction, these models need to be extended in several ways, all of which place heavy demands on computer resources. Extending the period of integration from days to decades has already been achieved, but so far only at the expense of poor spatial resolution and/or inadequate representation of the interaction between the atmosphere and the oceans. In addition, some feedback mechanisms in the climate system depend on biological-chemical interactions, the proper understanding of which are key tasks for the IGBP.

The oceans are represented in most climate models in a very simplified way that does not properly simulate the ocean's ability to absorb heat and hence retard (and to some extent alter the pattern of) global warming. These (so-called equilibrium) models provide an estimate of eventual climate changes but not the rate at which these will take place. In order to predict the evolution of climate realistically, it is necessary to develop further a new generation of models in which the atmosphere and oceans (and sea-ice) are fully coupled and in which the circulation of the oceans, as well as the atmosphere, is explicitly computed. The oceans have important eddies occurring on small scales and fine spatial resolution is essential for their explicit representation in a realistic manner.

Finer resolution than used at present is also required for the atmospheric component if regional variations of climate are to be predicted. Present-day climate models do not have sufficient resolution to represent in a meaningful way the climate of specific regions as small as, for example, the majority of individual nations. The implementation of finer resolution models will require significant advances in computer capability such as can be expected within the next 5 to 10 years. Improved parameterization of cloud and other processes will also need to be incorporated. To develop these parameterizations, very-fine mesh models will need to be developed, covering domains large enough to embrace entire mesoscale convective cloud systems, and appropriate field experiments will need to be carried out in different regions to provide input data and validation.

Models are required not only for prediction and process studies, but also for analysing the inevitably incomplete observational data sets to reconstruct and monitor climate change. Appropriate 4-dimensional data assimilation techniques exist for observations of atmospheric temperature and wind, but substantial improvements over the course of the next decade are needed to extend these methods for assimilating other parameters such as precipitation. Similar techniques will also be required for the analysis of ocean observations.

The development of advanced geosphere-biosphere models will be an important task for the IGBP. Such models are needed to introduce, as dynamic variables, the biological source and sink terms into descriptions of the way in which greenhouse gases will vary with climate change. That will become feasible with supercomputers available early in the 21st century. Such models will simulate detailed events with sufficient spatial and temporal resolution to permit explicit treatment of the strong non-linear interactions between physics, chemistry and biology that occur on small scales.

Model validation is a prerequisite to reducing uncertainties. The interaction between observational data and numerical modelling is a continuing process. It is essential for the development and testing of these models, for their operation and for their validation and eventual application to prediction. Confidence in models depend on comparisons with observations. Modellers must identify data sets needed, their temporal and spatial scales of measurement and the required accuracy and simultaneity of observations. The observational system must be designed to provide such measurements and the data analysis and information system must be able to transform these measurements into usable parameters.

11.3.5 International Research on Climate and Global Change

Carrying out this global multi-disciplinary programme requires unprecedented co-operation of the world scientific community and task-sharing by research institutions and responsible national administrations, with long-term commitments of support. Inasmuch as global climate change is recognized as a problem concerning all nations, establishing an effective global climate monitoring and prediction system must also be recognized as a responsibility to be shared by all nations. For the physical aspects of the climate system, research on the clouds and radiation, the global water cycle and the oceans should be given highest priority at this time. Within the World Climate Research Programme, this includes the Global Energy and Water Cycle Experiment (GEWEX) and World Ocean Circulation Experiment (WOCE). Resource commitments are required as a matter of urgency. Regarding biogeochemical aspects of the climate system, research on the role of the ocean and the terrestrial biosphere as sources and sinks of greenhouse gases should be given highest priority. Within the International Geosphere-Biosphere Programme, this includes the Joint Global Ocean Flux Study (JGOFS) and the International Global Atmospheric Chemistry Programme (IGAC). Internationally coordinated research on the role of ecosystems in climate, including a biospheric contribution to GEWEX through an IGBP core project on BAHC, is now being organized. These studies will be an obvious complement to climate studies and be important in laying the sound foundation for assessing the effects of climate change on ecosystem functions and subsequent feedbacks to the climate system.

As noted earlier, there are strong interactions between climate change and other global environmental issues. A research programme on climate change will provide valuable insight into these issues and as we gain understanding of the functioning of the total Earth system, research programmes will need to be modified and reframed in terms of a research programme aimed at supporting global environmental management in all its aspects.

It is recognized that there are important scientific questions that are being addressed by national and other research programmes independently of these international programmes. The emphasis in this report has been given to those scientific questions that require large experimental and observing systems, and therefore require international coordination of efforts.

11.3.6 Time-scales for Narrowing the Uncertainty

New concepts and scientific methods cannot mature overnight and the lead time for significant progress in knowledge must be estimated in terms of years or even a decade, as for any major new technical development.

Although progress will occur incrementally in all fields, it is instructive to identify times when the pace of progress will be intensified as a result of specific activities some of which are shown for illustrative purposes in Figure 11.2. Where the oceans are concerned, substantial advances can be expected after the mid-1990s as a result of the WOCE and JGOFS field programmes. On the same time scale, model developments will lead to better treatment of the oceans in coupled atmosphere-ocean general circulation models and hence to more realistic predictions of the rate of change of climate. For the atmosphere, clouds introduce the greatest uncertainty: some progress will occur during the 1990s as a result of the ISCCP and continued Earth radiation measurements from space. The Global Energy and Water Cycle Experiment, to start in the late 1990s, will provide the first comprehensive data set for marked improvement in the treatment of clouds and radiation in climate models. A proper treatment of clouds may require the development of more powerful satellite instruments such as three-dimensional cloud radars after the turn of the century. A similar time scale applies to the representation of precipitation and evaporation, and associated transport processes. The series of HAPEXs will continue through the mid-1990s and the Continental Scale Study of a large

river basin will follow. Advanced space instruments (Figure 11.3) involving a heavy investment in active sensors (lasers and radars) will begin producing important results at the end of the 1990s as part of the GEWEX. This is also the time when we shall begin to see global models incorporating chemical and biospheric aspects and advanced four-dimensional data assimilation techniques Improved understanding of the sources and sinks of the major greenhouse gases will only come following completion of JGOFS (for the marine component) and IGAC (for the terrestrial component and the atmospheric transformations) at the end of the century. Although still lacking some of these improvements, a major advance in the ability to predict the regional differences in climate change is expected to take place in the late 1990s, with the implementation of higher resolution models of the atmosphere. A broad summary of the time-scales for narrowing the uncertainties is given in Figure 11.4.

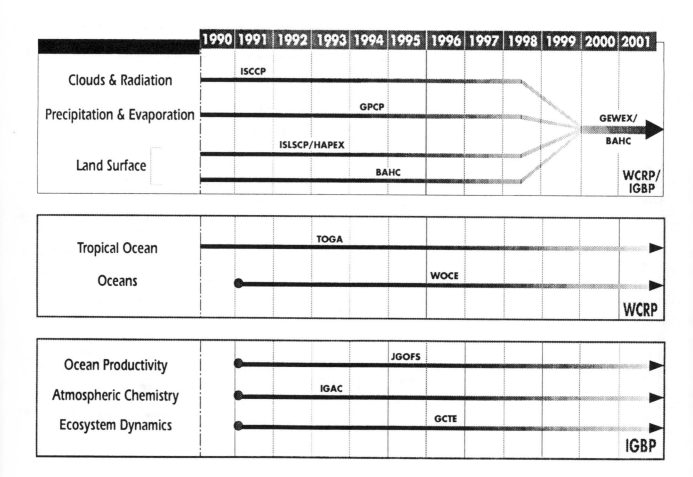

Figure 11.2: Schematic time schedules of major scientific programmes, as referred to in text.

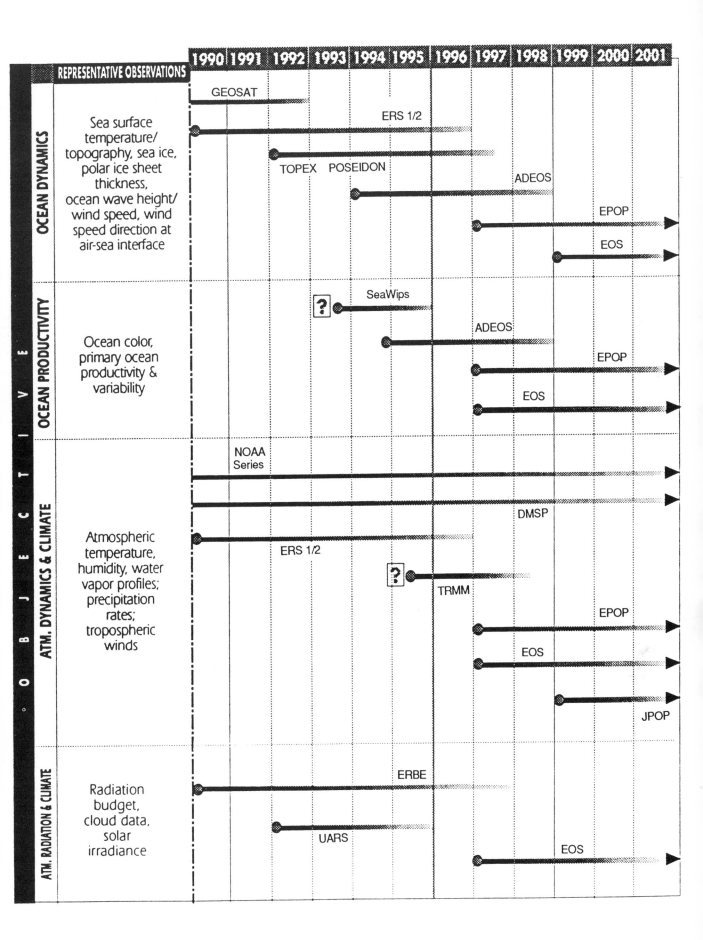

Figure 11.3: Illustrative examples of important satellite measurements.

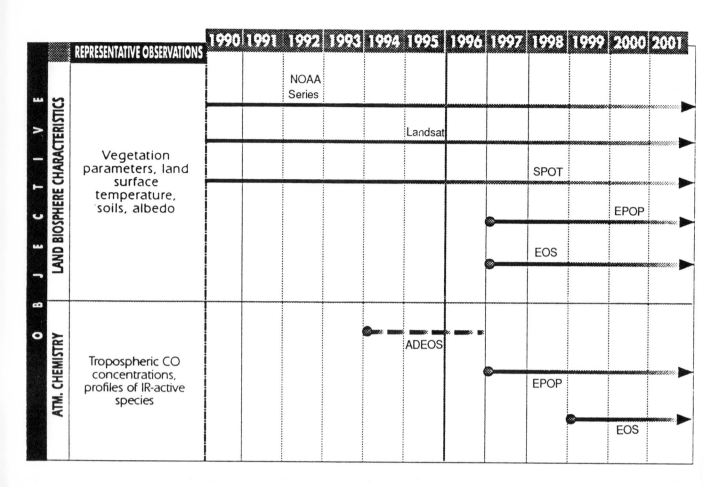

Figure 11.3: (Continued)

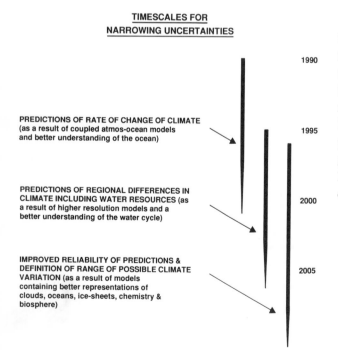

For these objectives to be met a long term commitment to observation and modelling activities must be maintained for at least the next decade. It is essential that government funding agencies recognize the magnitude of both the financial and human resources needed to undertake this research programme and the need to plan for appropriately long and sustained scientific efforts. Understanding and predicting the Earth's climate system must surely be the greatest scientific challenge yet to be faced by humankind. It is a worthy banner behind which the nations of the world can unite.

Figure 11.4: Timescales for narrowing uncertainties.

ANNEX

Climatic consequences of emissions

Model calculations contributed by:

C. Bruhl; E. Byutner; R.G. Derwent; I. Enting; J. Goudriaan; K. Hasselmann;
M. Heimann; I. Isaksen; C. Johnson; I. Karol; D. Kinnison; A. Kiselev; K. Kutz;
T-H. Peng; M. Prather; S.C.B. Raper; K.P. Shine; U. Siegenthaler; F. Stordal;
A. Thompson; D. Tirpak; R.A. Warrick; T.M.L. Wigley; D.J. Wuebbles.

Co-ordinators:
G.J. Jenkins; R.G. Derwent.

A.1 Introduction

Modelling studies have been undertaken by a number of research groups to investigate the climate consequences of several man-made emission scenarios. The first category of emission scenarios is that generated by IPCC Working Group III, which represents a broad range of possible controls to limit the emissions of greenhouse gases; these we refer to as **policy scenarios.** The second category of scenarios is generated by Working Group I to illustrate the way in which the atmosphere and climate would respond to changes in emissions; these we refer to as **science scenarios.** Many of the results have already been displayed in the appropriate sections of this report; they are brought together here to allow the complete emissions-climate pathway to be seen. The exploration of the climate consequences of both categories of emissions scenarios involved the sequence of modelling studies illustrated in Figure A.1.

A.1.1 Policy Scenarios

Four policy scenarios have been developed by Working Group III; they are described in Appendix 1 of this report.

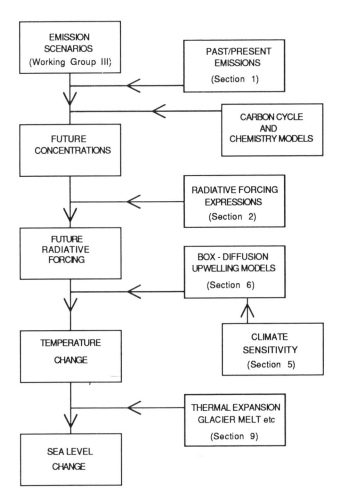

Figure A.1: Sequence of modelling studies.

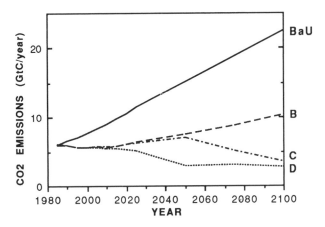

Figure A.2(a): Emissions of carbon dioxide (as an example) in the four "policy" scenarios generated by IPCC Working Group III.

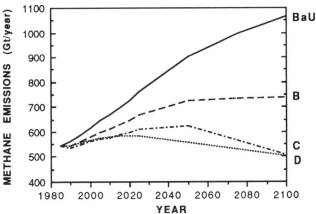

Figure A.2(b): Emissions of methane (as an example) in the four "policy" scenarios generated by IPCC Working Group III.

The first approximates to a Business-as-Usual (BaU) case. The other three incorporate a progressive penetration of controls on greenhouse gas emissions, and in this report are labelled Scenario B, C and D. (The BaU case had earlier been referred to as Scenario A).

Each scenario includes emissions of the main greenhouse gases, and other gases (such as NOx and CO) which influence their concentrations. The emissions of carbon dioxide and methane are shown in Figure A.2 as examples. For further information on the background to, and method of generation of, these policy scenarios, see Expert Group on Emissions Scenarios (1990).

A.1.2 Science Scenarios

The chain of processes in the atmosphere and other components of the climate system which lead from emissions to climate change (typified in this case by global temperature and sea level rise) can be illustrated by using a small number of cases where emissions are changed in some hypothetical manner, often exaggerated for clarity. The following science scenarios were selected by Working Group I; changes described apply equally to all the greenhouse gases:

S1:   Constant emissions at 1990 levels of all gases from 1990 onwards;

S2:   Reduce all emissions by 50% in 1990, and hold constant thereafter;

S3:   Decrease emissions at 2% per year compound from 1990;

S4:   Increase emissions at 2% per year compound until 2010, then decrease at 2% per year compound.

In each case the 1990 emissions are assumed to be those given in the Business-as-Usual policy scenario.

A.2   Pre-1990 Greenhouse Gas Concentrations

Concentrations of the greenhouse gases prior to 1989 are taken as those observed directly in the atmosphere or in ice-cores, as discussed in Section 1. The year 1765 was chosen as the pre-industrial baseline for greenhouse gas concentrations (and, hence, for man-made forcing and global mean temperature). 1990 concentrations were calculated by making a small extrapolation from most recent observations and trends.

A.3   Future Greenhouse Gas Concentrations

A number of models which contain representations of atmospheric chemistry and the carbon cycle have been used to make projections of future atmospheric concentrations of the greenhouse gases from the man-made emissions scenarios. The results from each of the models are not shown explicitly. For each scenario and gas, a single best-estimate was made of the concentration projection, and these are shown in Figure A.3 for the policy scenarios and A.4 - A.5 for the science scenarios. In most cases these best-estimates were simple means of a number of model results.

A quantitative understanding of the relationship between trace gas emissions and tropospheric concentrations requires a description in three dimensions of atmospheric dynamics, atmospheric chemistry and sources and sinks. The problem is complex and demands the use of large supercomputers. Hence, the results described in this Annex have been obtained with models which contain a number of simplifying assumptions. The models give a range of future concentrations depending on the assumptions adopted. Tropospheric models together with their representations of

atmospheric transport and chemistry have not been subject to comparison and evaluation in the same manner as the stratospheric assessment models. Consequently, there are no favoured approaches or recommended models.

In many cases, comparison of different model results reveal many shortcomings in the emissions scenarios themselves. Each emission scenario starts from 1985 and extends to 2100. Much is known about trace gas sources and sinks from these and other modelling studies of the 1985 atmosphere. The current emissions provided by Working Group III (WGIII) have not always been harmonised with previous model studies of the life cycles of these trace gases. Each set of model studies has therefore been adjusted in some way to accommodate the 1985 situation where the emissions are in some way inadequate.

Emissions of some gases, in particular man-made and natural hydrocarbons, were not projected by WGIII; each modelling group has therefore made its own assumptions about these, and this is an additional cause of differences between model results.

In the case of **nitrous oxide**, its atmospheric burden and its rate of increase have been well established from observations and the models have been set up to reproduce these observations. The lifetime calculated within each model determines the 1985 emissions required to support the current atmospheric burden and its rate of increase. There is a narrow range of emissions which satisfies this balance and it is model dependent; the WGIII emissions do not in general fall within this range, and modelling groups have therefore adopted one of two approaches to correct for this.

The WGIII emissions, taken together with current atmospheric concentration and its rate of increase, yield a lifetime of 167 years, somewhat longer than that generally accepted (see Section 1.5). Models with shorter lifetimes than 167 years will not reproduce current rate of increase of concentration, and some modellers have added further nitrous oxide emissions in order to accommodate this; this has led to a divergence between model results.

The 1985 atmospheric concentration of **methane** is well known, and its rate of increase shows that sources and sinks are not currently in balance. Not all aspects of these methane sources have yet been adequately quantified, neither has the sink through its reaction with OH. Whilst the accuracies in sources and sinks are adequate in budget studies to confirm that tropospheric OH can indeed account for the observed methane behaviour, the lack of accuracy has led to problems in the concentration calculations described here.

The atmospheric burden of methane and its rate of increase have been well established from observations and the models have been set up to reproduce them. The OH radical concentration distribution calculated within each model determines the 1985 emissions required to support

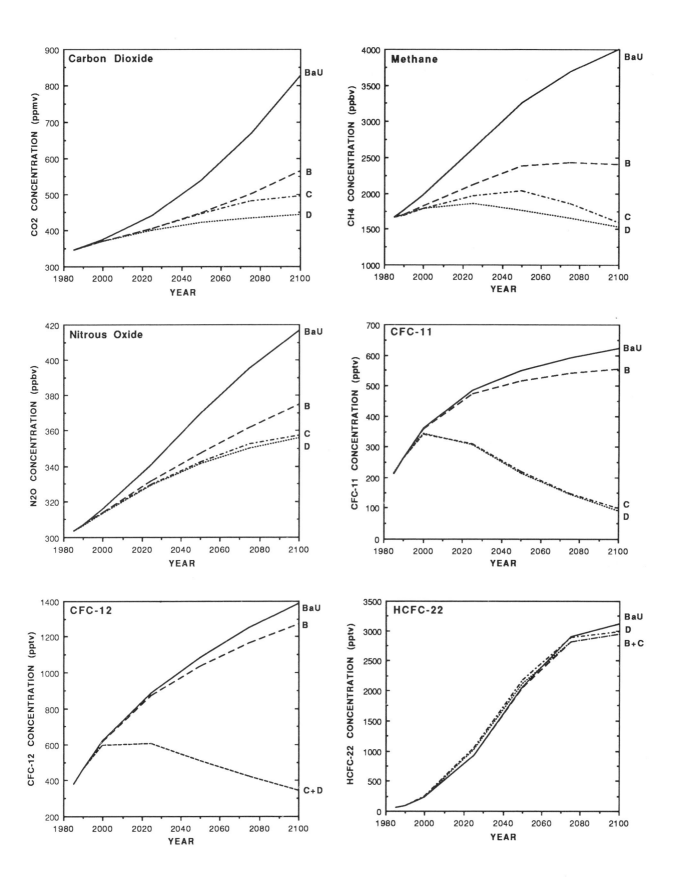

Figure A.3: Concentrations of the main man-made greenhouse gases resulting from the four IPCC WGIII "policy" emissions scenarios.

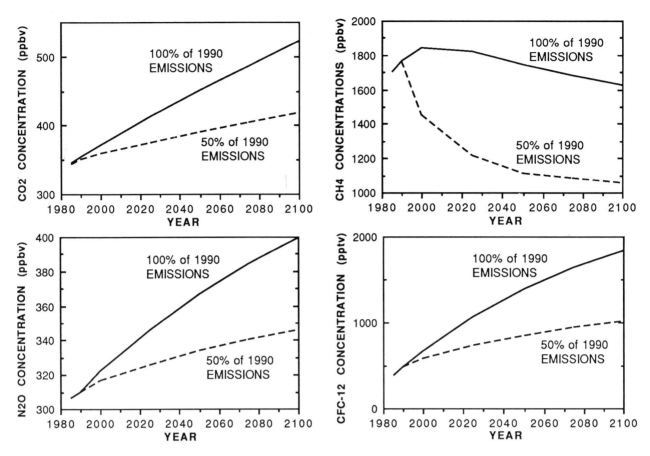

Figure A.4: Concentrations of carbon dioxide, methane, nitrous oxide and CFC-12 resulting from continuing emissions at 100% of 1990 levels, and emissions at 50% of 1990 levels.

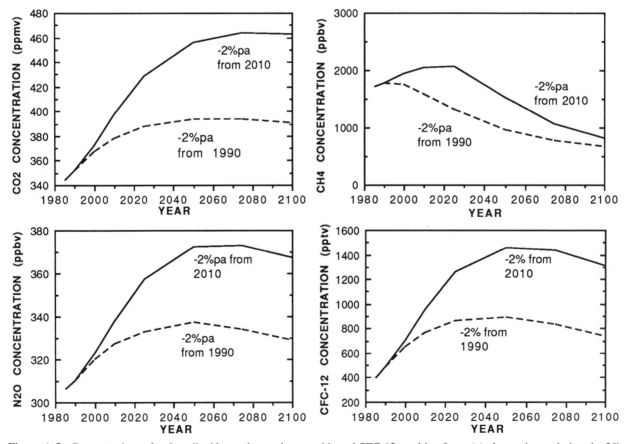

Figure A.5: Concentrations of carbon dioxide, methane, nitrous oxide and CFC-12 resulting from (a) decreasing emissions by 2% pa from 1990, and (b) increasing emissions by 2%pa until 2010, followed by decreasing emissions at 2%pa.

the current atmospheric burden and its rate of increase. There is a narrow, model dependent, range of methane, carbon monoxide and oxides of nitrogen emissions which satisfies this balance. The 1985 emissions provided in the scenarios do not fall within the required range of all the models, and modellers have therefore adopted one of two approaches, causing the part of divergence in model results. Some models have used the methane, carbon monoxide and NO_x emission scenarios exactly as provided and scaled up all their results to make the 1985 methane concentrations agree with observations. Other models have added extra methane, carbon monoxide or NO_x to make the 1985 methane concentrations agree with observations and maintained the extra injection throughout each calculation.

Differences in results can also arise because each model calculates a different scenario- and time-dependent tropospheric OH distribution. One model includes a feedback between composition changes to temperature changes to relative humidity changes back to OH radical concentrations. Almost all the models include the complex interaction between the future methane, carbon monoxide, ozone and nitrogen oxides concentrations on future OH radical concentrations.

CFCs 11 and 12 both have well quantified sources and stratospheric photolytic sinks. The relatively small differences between model calculations is due to differences in model transport and assumed or calculated atmospheric lifetimes. Such differences are similar to those reported in stratospheric ozone assessments (e.g., WMO, 1989)

Although **HCFC-22** sources are also all man-made its lifetime is determined, not by stratospheric photolysis but by tropospheric OH oxidation. However, the temperature dependence of the oxidation reaction is so large that virtually all of the atmospheric removal occurs in low latitudes in the lower region of the troposphere and in the upper stratosphere. The models used for this assessment generally have different 1985 tropospheric OH radical distributions and the different model formulations lead to different future OH distributions depending on the methane, carbon monoxide and nitrogen oxide emissions. In addition, one of the models employed includes an additional feedback whereby future global warming leads to increased humidities and hence increased tropospheric OH radical concentrations. Longer lifetimes imply greater tropospheric build-up of HCFC-22 by the year 2100.

Several modelling groups calculated future concentrations of tropospheric and stratospheric **ozone**, but because there was a wide divergence in the results, and because the relationship between concentration and forcing is not well established, the effects of ozone have not yet been included in the climate response.

A.4 Past and Future Radiative Forcing

The relationships between atmospheric concentration and radiative forcing derived in Section 2 were applied to the concentration histories of the greenhouse gases described

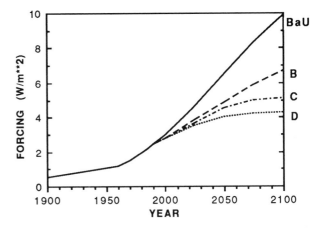

Figure A.6: Radiative forcing calculated from the four "policy" emissions scenarios.

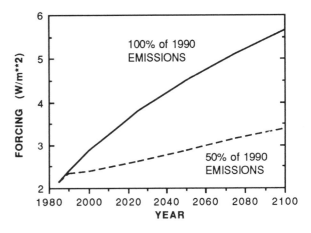

Figure A.7: Radiative forcing calculated to arise from continuing emissions of **all** man-made greenhouse gases at 100% of 1990 levels and 50% of 1990 levels.

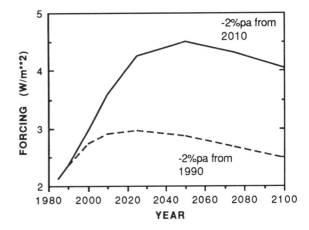

Figure A.8: Radiative forcing calculated to arise from (a) decreasing emissions of **all** man-made greenhouse gases by 2% pa from 1990, and (b) increasing emissions of **all** gases by 2%pa until 2010, followed by decreasing emissions at 2%pa.

above, and to the estimates of future concentrations from the four policy scenarios. The combined historical and future radiative forcing is illustrated for the four policy scenarios in Figure A.6, and those from the science scenarios in Figures A.7 and A.8.

A.5 Estimates of Future Global Mean Temperature

Ideally, the complete climate effects of the emission scenarios would be investigated using a comprehensive coupled atmosphere-ocean General Circulation Model, the results from which (using simple concentration increases) have been discussed in Section 6.4. Such model runs are prohibitively expensive and time consuming. Instead, estimates of the change in global mean surface air temperature due to man-made forcing (both historical and projected) were made using a box-diffusion-upwelling model of the type discussed in Section 6.6.

These models have a number of prescribed parameters (mixed-layer depth, upwelling rate, etc.) which are set to the optimum values discussed in Section 6. For each scenario, three values of climate sensitivity (the equilibrium temperature rise due to a doubling of carbon dioxide concentrations) are employed, as described in Section 5; 1.5°C, 2.5°C and 4.5°C. Results are given for each of these climate sensitivities, indicated as "high", "best estimate" and "low" in the figures.

Temperature rise estimates for the four policy scenarios are shown in Figure A.9, and those from the science scenarios are given in Figure A.10 and A.11 (best-estimate values only).

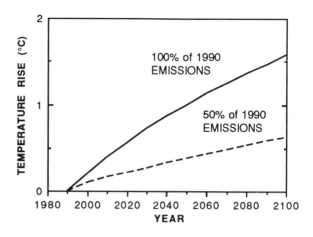

Figure A.10: Temperature rise calculated from continuing emissions of **all** man-made greenhouse gases at 100% of 1990 levels and 50% of 1990 levels.

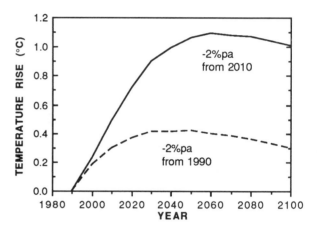

Figure A.11: Temperature rise calculated from (a) decreasing emissions of **all** man-made greenhouse gases by 2% pa from 1990, and (b) increasing emissions of **all** gases by 2%pa until 2010, followed by decreasing emissions at 2%pa.

A.6 Estimates of Future Global Mean Sea Level Rise

Box diffusion models are also used to estimate the sea level rise from the forcing projections; the thermal expansion part of future sea level rise is calculated directly by these models. The models also contain expressions for the contributions to sea level change from glacier and land ice melting, and changes in the mass-balance of the Greenland and Antarctic ice sheets.

Sea level changes estimated from the four policy scenarios are shown in Figure A.12, and for the science scenarios in Figure A.13 and A.14. Again, "high", "best-estimate" and "low" curves are shown for the policy scenarios, corresponding to the same climate sensitivities as used in the temperature rise estimates.

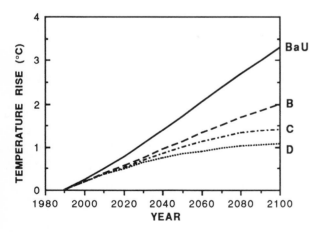

Figure A.9: Temperature rise calculated using a box-diffusion-upwelling model, due to the four IPCC WGIII "policy" emissions scenarios. Only the best-estimate value (corresponding to a climate sensitivity of 2.5°C) is shown.

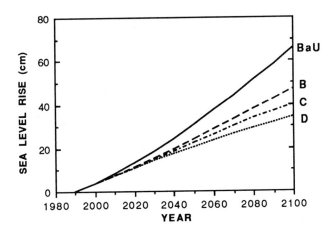

Figure A.12: Sea level rise due to the four IPCC WGIII "policy" emissions scenarios. Only the best-estimate value (corresponding to a climate sensitivity of 2.5°C) is shown.

A.7 Emissions to Sea Level Rise Pathway for Science Scenarios

In order to illustrate the timescales for adjustment of the climate to changes in emissions, the full pathway between emission change, through concentration change (using carbon dioxide as an example), radiative forcing, temperature rise and sea level rise is illustrated for each of the science scenarios in Figures A.15 to A.18.

References

Expert Group on Emissions Scenarios, IPCC Working Group III, Draft Report on Emissions Scenarios, February 1990.

WMO, 1989: Scientific Assessment of stratospheric ozone: 1989, Global Ozone Research and Monitoring Project, Report *20*, Geneva.

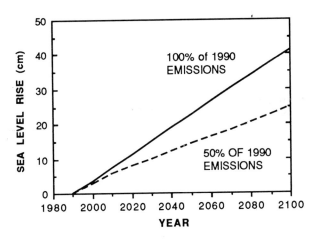

Figure A.13: Sea level rise calculated from continuing emissions of **all** man-made greenhouse gases at 100% of 1990 levels and 50% of 1990 levels.

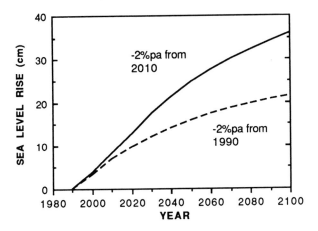

Figure A.14: Sea level rise calculated from (a) decreasing emissions of **all** man-made greenhouse gases by 2% pa from 1990, and (b) increasing emissions of **all** gases by 2%pa until 2010, followed by decreasing emissions at 2%pa.

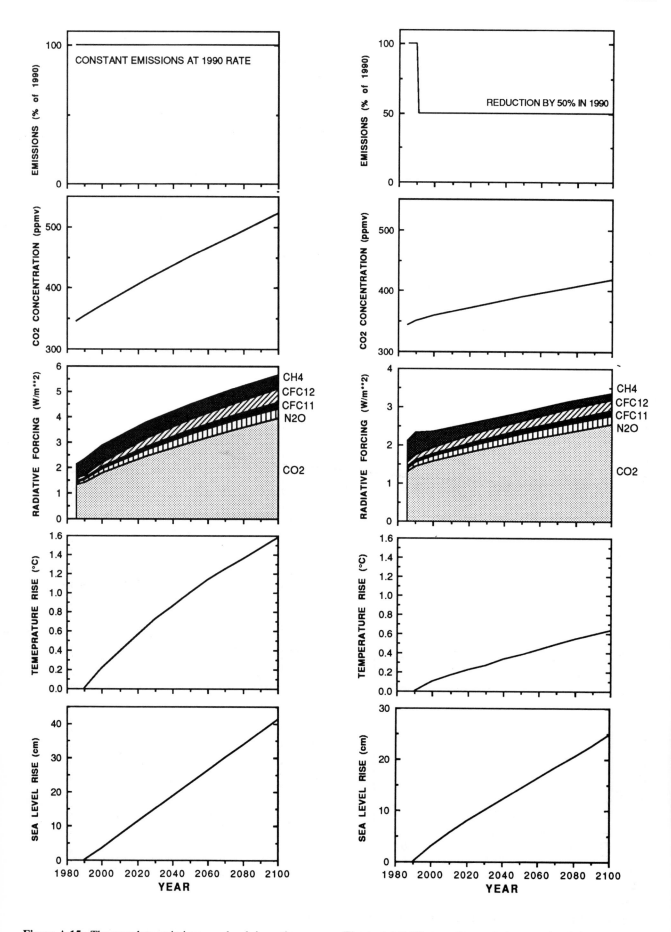

Figure A.15: The complete emissions- sea level rise pathway shown in the case of continuing emissions at 1990 levels.

Figure A.16: The complete emissions- sea level rise pathway shown for emissions continuing at 50% of 1990 levels.

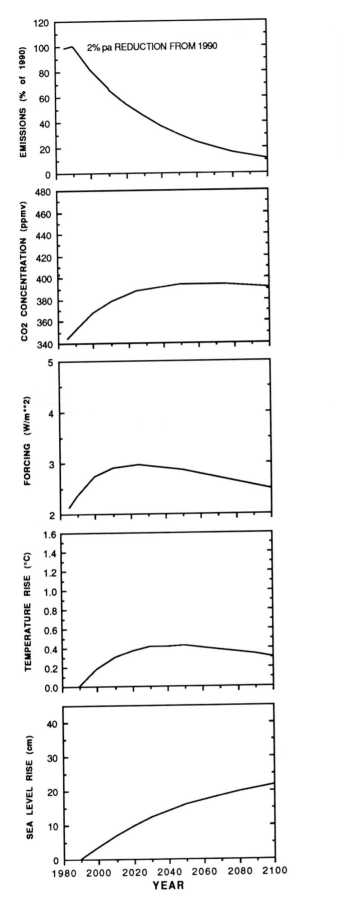

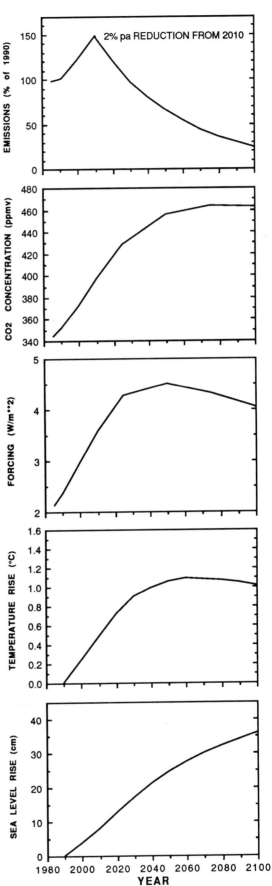

Figure A.17: The complete emissions- sea level rise pathway shown due to decreasing emissions of **all** man-made greenhouse gases by 2%pa from 1990.

Figure A.18: The complete emissions- sea level rise pathway shown due to increasing emissions of **all** gases by 2%pa until 2010, followed by decreasing emissions at 2%pa.

Appendix 1

EMISSIONS SCENARIOS FROM THE RESPONSE STRATEGIES WORKING GROUP OF
THE INTERGOVERNMENTAL PANEL ON CLIMATE CHANGE

The Steering Group of the Response Strategies Working Group (Working Group III) requested the USA and The Netherlands to develop emissions scenarios for evaluation by the IPCC Working Group I. The scenarios cover the emissions of carbon dioxide (CO_2), methane (CH_4), nitrous oxide (N_2O), chlorofluorocarbons (CFCs), carbon monoxide (CO) and nitrogen oxides (NO_x) from present up to the year 2100. Growth of the economy and population was taken as common for all scenarios. Population was assumed to approach 10.5 billion in the second half of the next century. Economic growth was assumed to be 2-3% annually in the coming decade in the OECD countries, and 3-5 % in the Eastern European and developing countries. The economic growth levels were assumed to decrease thereafter. In order to reach the required targets, levels of technological development and environmental controls were varied.

In the **Business-as-Usual Scenario** (Scenario A) the energy supply is coal intensive and on the demand side only modest efficiency increases are achieved. Carbon monoxide controls are modest, deforestation continues until the tropical forests are depleted and agricultural emissions of methane and nitrous oxide are uncontrolled.

For CFCs the Montreal Protocol is implemented albeit with only partial participation. Note that the aggregation of national projections by IPCC Working Group III gives higher emissions (10-20%) of carbon dioxide and methane by 2025.

In **Scenario B** the energy supply mix shifts towards lower carbon fuels, notably natural gas. Large efficiency increases are achieved. Carbon monoxide controls are stringent, deforestation is reversed and the Montreal Protocol implemented with full participation.

In **Scenario C** a shift towards renewables and nuclear energy takes place in the second half of next century. CFCs are now phased out and agricultural emissions limited.

For **Scenario D** a shift to renewables and nuclear in the first half of the next century reduces the emissions of carbon dioxide, initially more or less stabilizing emissions in the industrialized countries. The scenario shows that stringent controls in industrialized countries combined with moderated growth of emissions in developing countries could stabilize atmospheric concentrations. Carbon dioxide emissions are reduced to 50% of 1985 levels by the middle of the next century.

Appendix 2

ORGANISATION OF IPCC AND WORKING GROUP I

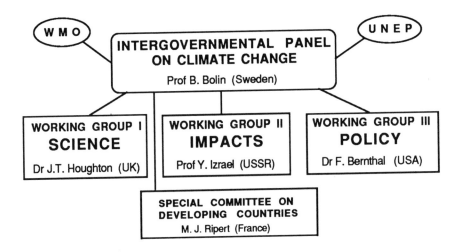

IPCC:

Chairman:	Professor B. Bolin *(Sweden)*
Vice Chairman:	Dr A. Al Gain *(Saudi Arabia)*
Rapporteur:	Dr J.A. Adejokun *(Nigeria)*
Secretary:	Dr N. Sundararaman *(WMO)*

Working Group I:

Chairman:	Dr J.T. Houghton *(United Kingdom)*
Vice Chairmen:	Dr M. Seck *(Senegal)*
	Dr A.D. Moura *(Brazil)*

Core Team at the UK Meteorological Office:

Co-ordinator:	Dr G.J. Jenkins
Technical Editor:	Mr J.J .Ephraums
Assistant Technical Editor:	Miss S.K. Varney
Computing:	Mrs A. Foreman
Sub-Editor:	Mr A. Gilchrist
Visiting Scientists:	Dr R.T. Watson *(NASA, USA)*
	Dr R.J. Haarsma *(KNMI, Netherlands)*
	Prof H-X. Cao *(SMA, PRC)*
	Dr T. Callaghan *(NERC, UK)*

Appendix 3

CONTRIBUTORS TO IPCC WG1 REPORT

SECTION 1

Lead Authors:

R .T. Watson	NASA Headquarters, USA
H. Rodhe	University of Stockholm, Sweden
H. Oeschger	Physics Institute, University of Bern, Switzerland
U. Siegenthaler	Physics Institute, University of Bern, Switzerland

Contributors:

M. Andreae	Max Planck Institute für Chemie, FRG
R. Charlson	University of Washington, USA
R. Cicerone	University of California, USA
J. Coakley	Oregon State University, USA
R. G. Derwent	Harwell Laboratory, UK
J. Elkins	NOAA Environmental Research Laboratories, USA
F. Fehsenfeld	NOAA Aeronomy Laboratory, USA
P. Fraser	CSIRO Division of Atmospheric Research, Australia
R. Gammon	University of Washington, USA
H. Grassl	Max Planck Institut für Meteorologie, FRG
R. Harriss	University of New Hampshire, USA
M. Heimann	Max Planck Institut für Meteorologie, FRG
R.A. Houghton	Woods Hole Research Centre, USA
V. Kirchhoff	Atmospheric and Space Science, INEP, Brazil
G. Kohlmaier	Institut für Physikalische und Theoretisch Chemie, FRG
S. Lal	Physical Research Laboratory, India
P. Liss	University of East Anglia, UK
J. Logan	Harvard University, USA
R.J. Luxmoore	Oak Ridge National Laboratory, USA
L. Merlivat	University of Paris, France
K. Minami	National Institute of Agro-Environmental Sciences, Japan
G. Pearman	CSIRO Division of Atmospheric Research, Australia
S. Penkett	University of East Anglia, UK
D. Raynaud	Laboratoire de Glaciologie et Geophysique de l'Environment, CNRS, France
E. Sanhueza	Max Planck Institute für Chemie, FRG
P. Simon	Institute for Space Aeronomy, Belgium
W. Su	Research Centre for Eco-Environmental Science, Academy of Sciences, China
B. Svensson	University of Agricultural Sciences, Sweden

A. Thompson	NASA Goddard Space Flight Center, USA
P. Vitousek	Stanford University, USA
M. Whitfield	Plymouth Marine Laboratory, UK
P. Winkler	German Weather Service, FRG
S. Wofsy	Harvard University, USA

SECTION 2

Lead Authors:

K.P. Shine	University of Reading, UK
R.G. Derwent	Harwell Laboratory, UK
D.J. Wuebbles	Lawrence Livermore National Laboratory, USA
J-J. Morcrette	ECMWF, UK

Contributors:

A.J. Apling	Global Atmosphere Division, Department of the Environment, UK
J.P. Blanchet	Atmospheric Enivironment Service, Canada
R. Charlson	University of Washington, USA
D. Crommelynck	Royal Meteorological Institute, Belgium
H. Grassl	Max Planck Institut für Meteorologie, FRG
N. Husson	Laboratoire de Meteorologie Dynamique, CNRS, France
G.J. Jenkins	Meteorological Office, UK
I. Karol	Main Geophysical Observatory, USSR
M.D. King	NASA Goddard Laboratory of the Atmosphere
V. Ramanathan	University of Chicago, USA
H. Rodhe	University of Stockholm, Sweden
G. Thomas	University of British Columbia, Canada
G-Y Shi	Institute of Atmospheric Physics, Academia Sinica, China
T. Yamanouchi	National Institute of Polar Research, Japan
W-C Wang	State University of New York, USA
T.M.L. Wigley	Climatic Research Unit, University of East Anglia, UK

SECTION 3

Lead Authors:

| U. Cubasch | Max Planck Institut für Meteorologie, FRG |
| R.D. Cess | State University of New York, USA |

Contributors:

F. Bretherton	University of Wisconsin, USA
H. Cattle	Meteorological Office, UK
J.T. Houghton	Meteorological Office, UK
J.F.B. Mitchell	Meteorological Office, UK
D. Randall	Colorado State University, USA
E. Roeckner	Max Planck Institut für Meteorologie, FRG
J. D. Woods	National Environment Research Council, UK
T. Yamanouchi	National Institute of Polar Research, Japan

SECTION 4

Lead Authors:

W.L. Gates	Lawrence Livermore National Laboratory, USA
P.R. Rowntree	Meteorological Office, UK
Q.-C. Zeng	Institute of Atmospheric Physics, Academy of Sciences, China

Appendix 3

Contributors:

P.A. Arkin	NOAA Climate Analysis Center, USA
A. Baede	KNMI, The Netherlands
L. Bengtsson	ECMWF, UK
A. Berger	Institute d'Astronomie et de Geophysique, Belgium
C. Blondin	Direction de la Meteorologie Nationale, France
G.J. Boer	Canadian Climate Center, Canada
K. Bryan	NOAA Geophysical Fluid Dynamics Laboratory, USA
R.E. Dickinson	National Center for Atmospheric Research, USA
S. Grotch	Lawrence Livermore National Laboratory, USA
D. Harvey	University of Toronto, Canada
E.O. Holopainen	University of Helsinki, Finland
R. Jenne	National Center for Atmospheric Research, USA
J.E. Kutzbach	University of Wisconsin, USA
H. Le Treut	Laboratoire de Meteorologie Dynamique du CNRS, France
P. Lemke	Max Planck Institut für Meteorologie, FRG
B. McAvaney	Bureau of Meteorology Research Centre, Australia
G.A. Meehl	National Center for Atmospheric Research, USA
P. Morel	WMO, Switzerland
T.N. Palmer	ECMWF, UK
L. P. Prahm	Danish Meteorological Institute, Denmark
S.H. Schneider	National Center for Atmospheric Research, USA
K.P. Shine	University of Reading, UK
I.H. Simmonds	University of Melbourne, Australia
J.E. Walsh	University of Illinois, USA
R.T. Wetherald	NOAA Geophysical Fluid Dynamics Laboratory, USA
J. Willebrand	Institut für Meerskunde der Universitat Kiel, FRG

SECTION 5

Lead Authors:

J. F. B. Mitchell	Meteorological Office, UK
S. Manabe	NOAA Geophysical Fluid Dynamics Laboratory, USA
V. Meleshko	Main Geophysical Observatory, USSR
T. Tokioka	Meteorological Research Institute, Japan

Contributors:

A. Baede	KNMI, The Netherlands
A. Berger	Institute d'Astronomie et de Geophysique, Belgium
G. Boer	Atmospheric Environment Service, Canada
M. Budyko	State Hydrological Institute, USSR
V. Canuto	NASA Goddard Insitute for Space Studies, USA
H-X Cao	State Meteorological Administration, China
R.E. Dickinson	National Center forAtmospheric Research, USA
H. Ellsaesser	Lawrence Livermore National Laboratory, USA
S. Grotch	Lawrence Livermore National Laboratory, USA
R.J. Haarsma	KNMI, The Netherlands
A. Hecht	Environmental Protection Agency Headquarters, USA
B. Hunt	CSIRO, Australia
B. Huntley	University of Durham, UK
B. Keshavamurthy	Physical Research Laboratory, India
R. Koerner	Geological Survey of Canada
C. Lorius	Laboratoire de Glaciologie, CNRS, France
M. MacCracken	Lawrence Livermore National Laboratory, USA
G. Meehl	National Center for Atmospheric Research, USA

E. Oladipo	Ahmadu Bello University, Nigeria
E. Perrott	Oxford University, UK
A.B. Pittock	CSIRO, Australia
L. P. Prahm	Danish Meteorological Institute, Denmark
D. Randall	Colorado State University, USA
P.R. Rowntree	Meteorological Office, UK
M.E. Schlesinger	University of Illinois, USA
S.H. Schneider	National Center for Atmospheric Research, USA
C. Senior	Meteorological Office, UK
N. Shackleton	University of Cambridge, UK
W.J. Shuttleworth	Institute of Hydrology, UK
R. Stouffer	NOAA Geophysical Fluid Dynamics Laboratory, USA
F. Street-Perrott	University of Oxford, UK
A. Velichko	State Hydrological Institute, USSR
K. Vinnikov	State Hydrological Institute, USSR
R.T. Wetherald	NOAA Geophysical Fluid Dynamics Laboratory, USA

SECTION 6

Lead Authors:

F.P. Bretherton	University of Wisconsin, USA
K. Bryan	NOAA Geophysical Fluid Dynamics Laboratory, USA
J.D. Woods	Natural Environment Research Council, UK

Contributors:

J. Hansen	NASA Goddard Institute of Space Studies, USA
M. Hoffert	New York University, USA
X. Jiang	University of Illinois, USA
S. Manabe	NOAA Geophysical Fluid Dynamics Laboratory, USA
G. Meehl	National Center for Atmospheric Research, USA
S.C.B. Raper	Climatic Research Unit, University of East Anglia, UK
D. Rind	NASA Goddard Insitute for Space Studies, USA
M.E. Schlesinger	University of Illinois, USA
R. Stouffer	NOAA Geophysical Fluid Dynamics Laboratory, USA
T. Volk	University of New York, USA
T.M.L. Wigley	Climatic Research Unit, University of East Anglia, UK

SECTION 7

Lead Authors:

C.K. Folland	Meteorological Office, UK
T.R. Karl	NOAA National Climate Data Centre, USA
K.Ya. Vinnikov	State Hydrological Institute, USSR

Contributors:

J.K. Angell	NOAA Air Resources Laboratory, USA
P.A. Arkin	NOAA Climate Analysis Center, USA
R.G. Barry	University of Colorado, USA
R.S. Bradley	University of Massachussetts, USA
D.L. Cadet	Institut National des Sciences de l'Universe, France
M. Chelliah	NOAA Climate Analysis Center, USA
M. Coughlan	Bureau of Meteorology, Australia
B. Dahlstrom	Meteorological and Hydrological Institute, Sweden
H.F. Diaz	NOAA Environmental Research Laboratories, USA
H. Flohn	Meteorologisches Institut, FRG
C. Fu	Institute of Atmospheric Physics, Academy of Sciences, China

P.Ya Groisman	State Hydrological Institute, USSR
A. Gruber	NOAA/NESDIS, USA
S. Hastenrath	University of Wisconsin, USA
A. Henderson-Sellers	Macquarie University, Australia
K. Higuchi	Atmospheric Environment Service, Canada
P.D. Jones	Climatic Research Unit, University of East Anglia, UK
J. Knox	Atmospheric Environment Service, Canada
G. Kukla	Lamont-Doherty Geological Laboratory, USA
S. Levitus	NOAA Geophysical Fluid Dynamics Laboratory, USA
X. Lin	State Meteorological Administration, China
N. Nicholls	Bureau of Meteorology, Australia
B.S. Nyenzi	Directorate of Meteorology, Tanzania
J.S. Oguntoyinbo	University of Ibadan, Nigeria
G.B. Pant	Institute of Tropical Meteorology, India
D.E. Parker	Meteorological Office, UK
A.B. Pittock	CSIRO, Australia
R.W. Reynolds	NOAA Climate Analysis Center, USA
C.F. Ropelewski	NOAA Climate Analysis Center, USA
C.D. Schönwiese	Institut für Meteorologie und Geophysics, FRG
B. Sevruk	University of Zurich, Switzerland
A.R. Solow	Woods Hole Oceanographic Institution, USA
K.E. Trenberth	National Center for Atmospheric Research, USA
P. Wadhams	Scott Polar Research Institute, UK
W.C. Wang	State University of New York, Albany, USA
S. Woodruff	NOAA Environmental Research Laboratories, USA
T. Yasunari	Institute of Geoscience, Japan
Z. Zeng	Atmospheric and Environmental Research Inc., USA and China

SECTION 8

Lead Authors:

T.M.L. Wigley	Climatic Research Unit, University of East Anglia, UK
T.P. Barnett	University of California, USA

Contributors:

T.L. Bell	NASA Goddard Space Flight Center, USA
P. Bloomfield	North Carolina State University, USA
D. Brillinger	University of California, USA
W. Degefu	National Meteorological Services Agency, Ethiopia
Duzheng Ye	Institute of Atmospheric Physics, Academy of Sciences, China
S. Gadgil	Institute of Science, India
G.S. Golitsyn	Institute of Atmospheric Physics, USSR
J.E. Hansen	NASA Goddard Institute for Space Studies, USA
K. Hasselmann	Max Planck Institut für Meteorologie, FRG
Y. Hayashi	NOAA Geophysical Fluid Dynamics Laboratory, USA
P.D. Jones	Climatic Research Unit, University of East Anglia, UK
D.J. Karoly	Monash University, Australia
R.W. Katz	National Center for Atmospheric Research, USA
M.C. MacCracken	Lawrence Livermore National Laboratory, USA
R.L. Madden	National Center for Atmospheric Research, USA
S. Manabe	NOAA Geophysical Fluid Dynamics Laboratory, USA
J.F.B. Mitchell	Meteorological Office, UK
A.D. Moura	Instituto de Pesquisas Espaciais, Brazil
C. Nobre	Instituto de Pesquisas Espaciais, Brazil
L.J. Ogallo	University of Nairobi, Kenya

E.O. Oladipo	Ahmadu Bello University, Nigeria
D.E. Parker	Meteorological Office, UK
S.C.B. Raper	Climatic Research Unit, University of East Anglia, UK
A.B. Pittock	CSIRO, Australia
B.D. Santer	Max Planck Institut für Meteorologie, FRG
M.E. Schlesinger	University of Illinois, USA
C.-D. Schönwiese	Institute für Meteorologie und Geophysics, FRG
C.J.E. Schuurmans	KNMI, The Netherlands
A. Solow	Woods Hole Oceanographic Institute, USA
K.E. Trenberth	National Center for Atmospheric Research, USA
K.Ya. Vinnikov	State Hydrological Institute, USSR
W.M. Washington	National Center for Atmospheric Research, USA
D. Ye	Institute of Atmospheric Physics, Academy of Sciences, China
T. Yasunari	University of Tsukuba, Japan
F.W. Zwiers	Canadian Climate Center, Canada

SECTION 9

Lead Authors:

R.A. Warrick	Climatic Research Unit, University of East Anglia, UK
J. Oerlemans	Institute of Meteorology and Oceanography, The Netherlands

Contributors:

P. Beaumont	St David's University College, UK
R.J. Braithwaite	Geological Survey of Greenland, Denmark
D.J. Drewry	British Antarctic Survey, UK
V. Gornitz	NASA Goddard Institute for Space Studies, USA
J.M. Grove	University of Cambridge, UK
W. Haeberli	Versuchsanstalt für Wasserbau, Switzerland
A. Higashi	International Christian University, Japan
J.C. Leiva	Instituto Argentino de Nivologia y Glaciologia, Argentina
C.S. Lingle	NASA Goddard Institute for Space Studies, USA
C. Lorius	Laboratoire de Glaciologie, CNRS, France
S.C.B. Raper	Climatic Research Unit, University of East Anglia, UK
B. Wold	Water Resources and Energy Administration, Norway
P.L. Woodworth	Proudman Oceanic Laboratory, UK

SECTION 10

Lead Authors:

J. Melillo	Woods Hole Marine Laboratory, USA
T. V. Callaghan	Institute of Terrestrial Ecology, UK
F. I. Woodward	University of Cambridge, UK
E. Salati	Universidade do Estado de Sao Paulo, Brazil
S. K. Sinha	Agriculture Research Institute, India

Contributors:

H. Abdel Nour	General National Forests Corporation, Sudan
J. Aber	University of New Hampshire, USA
V. Alexander	University of Alaska, USA
J. Anderson	Hatherly Laboratories, UK
A. Auclair	Quebec, Canada
F. Bazzaz	Harvard University, USA
A. Breymeyer	Institute of Geography and Spatial Organization, Poland
A. Clarke	British Antarctic Survey, UK
C. Field	Stanford University, USA

J.P. Grime	University of Sheffield, UK
R. Gifford	CSIRO, Australia
J. Goudrian	Agricultural University, The Netherlands
R. Harriss	University of New Hampshire, USA
I. Heany	Institute of Freshwater Ecology, Windermere Laboratory, UK
P. Holligan	Plymouth Marine Laboratory, UK
P. Jarvis	University of Edinburgh, UK
L. Joyce	Woods Hole Research Center, USA
P. Levelle	Laboratoire d'Ecologie de l'Ecole Normale, CNRS, France
S. Linder	University of Agricultural Sciences, Sweden
A. Linkins	Clarkson University, USA
S. Long	University of Essex, UK
A. Lugo	Institute of Tropical Forestry, USA
J. McCarthy	Harvard University, USA
J. Morison	University of Reading, Uk
W. Oechel	San Diego State University, USA
M. Phillip	Institute of Plant Ecology, Denmark
M. Ryan	Woods Hole Research Center, USA
D. Schimel	Colorado State University, USA
W. Schlesinger	Duke University, USA
G. Shaver	Woods Hole Research Center, USA
B. Strain	Duke University, USA
R. Waring	Oregon State University, USA
M. Williamson	York University, UK

SECTION 11

Lead Authors:

G. McBean	University of British Columbia, Canada
J. McCarthy	Harvard University, USA

Contributors:

K. Browning	Meteorological Office, UK
P. Morel	WMO, Switzerland
I. Rasool	OSSA, NASA, USA

ANNEX

Coordinators:

G.J.Jenkins	Meteorological Office, UK
R.G.Derwent	Harwell Laboratory, UK

Model calculations contributed by:

C. Bruhl	Max Plank Institute fur Chemie, FRG
E. Byutner	Main Geophysical Observatory, USSR
R.G.Derwent	Harwell Laboratory, UK
I. Enting	CSIRO, Australia
J.Goudriaan	Wageningen Agricultural University, Netherlands
K.Hasselmann	Max Planck Institute fur Meteorologie, FRG
M.Heimann	Max Planck Institute fur Meteorologie, FRG
I.S.A Isaksen	University of Oslo, Norway
C.E. Johnson	Harwell Laboratory, UK
I. Karol	Main Geophysical Observatory, USSR
D.Kinnison	Lawrence Livermore National Laboratory, USA
A.A.Kiselev	Main Geophysical Observatory, USSR
K.Kurz	Max Planck Institute fur Meteorologie, FRG

T.-H. Peng	Oak Ridge National Laboratory, USA
M.J. Prather	Goddard Institute for Space Studies, USA
S.C.B.Raper	University of East Anglia, UK
K.P.Shine	University of Reading, UK
U. Siegenthaler	University of Bern, Switzerland
F. Stordal	University of Oslo, Norway
A.M. Thompson	Goddard Space Flight Center, USA
D.Tirpak	Environmental Protection Agency, USA
R.A.Warrick	University of East Anglia, UK
T.M.L.Wigley	University of East Anglia, UK
D.J.Wuebbles	Lawrence Livermore National Laboratory, USA

Appendix 4

REVIEWERS OF IPCC WGI REPORT

The persons named below all contributed to the peer review of the IPCC Working Group I Report. Whilst every attempt was made by the Lead Authors to incorporate their comments, in some cases these formed a minority opinion which could not be reconciled with the larger concensus. Therefore, there may be persons below who still have points of disagreement with areas of the Report.

AUSTRALIA

R. Allan	CSIRO (Division of Atmospheric Research)
I. Allison	Australian Antarctic Division
G. Ayers	CSIRO (Division of Atmospheric Research)
B. Pittock	CSIRO
W. Bouma	CSIRO (Division of Atmospheric Research)
B. Bourke	Australian Bureau of Meteorology
B. Budd	Melbourne University
T. Denmead	CSIRO (Centre for Environmental Mechanics)
M. Dix	CSIRO (Division of Atmospheric Research)
I. Enting	CSIRO (Division of Atmospheric Research)
J. Evans	CSIRO (Division of Atmospheric Research)
S. Faragher	CSIRO (Division of Atmospheric Research)
R. Francey	CSIRO (Division of Atmospheric Research)
P. Fraser	CSIRO (Division of Atmospheric Research)
J. Frederiksen	CSIRO (Division of Atmospheric Research)
I. Galbally	CSIRO (Division of Atmospheric Research)
J. Garratt	CSIRO (Division of Atmospheric Research)
D. Gauntlett	Australian Bureau of Meteorology
R. Gifford	CSIRO (Division of Plant Industry)
S. Godfrey	CSIRO (Division of Oceanography)
H. Gordon	CSIRO (Division of Atmospheric Research)
J. Gras	CSIRO (Division of Atmospheric Research)
C. Griffith	Department of Arts, Sport, Environment, Tourism & Territories
A. Henderson-Sellers	Macquarie University
K. Hennesey	CSIRO (Division of Atmospheric Research)
T. Hirst	CSIRO (Division of Atmospheric Research)
R. Hughes	CSIRO (Division of Atmospheric Research)
B. Hunt	CSIRO (Division of Atmospheric Research)
J. Kalma	CSIRO (Division of Water)
D. Karoly	Monash University

H. Kenway	Department of Prime Minister & Cabinet
M. Manton	Australian Bureau of Meteorology
B. McAveney	Australian Bureau of Meteorology
A. McEwan	CSIRO (Division of Oceanography)
J. McGregor	CSIRO (Division of Atmospheric Research)
W. J. McG Tegart	ASTEC
I. Noble	Australian National University
G.Pearman	CSIRO (Division of Atmospheric Research)
M. Platt	CSIRO (Division of Atmospheric Research)
P. Price	Australian Bureau of Meteorology
P. Quilty	Australian Antarctic Division
M. Raupach	CSIRO (Centre for Environmental Mechanics)
B. Ryan	CSIRO (Division of Atmospheric Research)
L. Rikus	Australian Bureau of Meteorology
I. Simmonds	Melbourne University
N. Smith	Australian Bureau of Meteorology
N. Streten	Australian Bureau of Meteorology
L. Tomlin	Department of Industry, Technology & Commerce
G. B. Tucker	CSIRO (Division of Atmospheric Research)
M. Voice	Australian Bureau of Meteorology
I. Watterson	CSIRO (Division of Atmospheric Research)
P. Whetton	CSIRO (Division of Atmospheric Research)
M. Williams	Monash University
S. Wilson	Cape Grim Baseline Air Pollution Station
K H Wyrhol	University of Western Australia
J. Zillman	Australian Bureau of Meteorology

AUSTRIA

R. Christ	Federal Ministry of Environment, Youth and Family
H. Hojensky	Federal Environmental Agency
H. Kolb	Institut für Meteorologie und Geophysik
O. Preining	Institut für Experimentalphysik

BELGIUM

| A Berger | Instute d'Astronomie et de Geophysique |

BRAZIL

| C. Nobre | Instituto de Pesquisas Espaciais |

CANADA

M. Berry	Canadian Climate Centre
G. J. Boer	Canadian Climate Centre
R. Daley	Canadian Climate Centre
B. Goodison	Canadian Climate Centre
D. Harvey	University of Toronto (Department of Geography)
H. Hengeveld	Canadian Climate Centre
G. McBean	University of British Columbia
L. Mysak	Canadian Climate Centre
W.R. Peltier	Canadian Climate Centre
R W. Stewart	
F. Zwiers	Canadian Climate Centre

CHINA

Chong-guang Yuan	Institute of Atmospheric Physics
Cong-bin Fus	Institute of Atmospheric Physics
Ding Yihui	Academy of Meteorological Sciences
Du-zheng Ye	Institute of Atmospheric Physics
Guang-yu Shi	Institute of Atmospheric Physics
Ming-xing Wang	Institute of Atmospheric Physics
Qing-cun Zeng	Institute of Atmospheric Physics
Shi-yan Tao	Institute of Atmospheric Physics
Xue-hong Zhang	Institute of Atmospheric Physics
Zhou Xiuji	Academy of Meteorological Sciences
Zhao Zongci	Institute of Climate

DENMARK

P. Frich	Danish Meteorological Institute
K. Frydendahl	Danish Meteorological Institute
L. Laursen	Danish Meteorological Institute
L.P. Prahm	Danish Meteorological Institute
A. Wiin-Nielsen	Geophysical Institute (University of Copenhagen)

EGYPT

A.M. Mehanna	Meteorological Authority
A.L. Shaaban	Meteorological Authority

ETHIOPIA

W. Degufu	National Meteorological Services Agency
T. Haile	National Meteorological Services Agency

FEDERAL REPUBLIC OF GERMANY

P. Crutzen	SC IGBP
K. Hasselmann	Max Planck Institut für Meteorologie
B. Santer	Max Planck Institut für Meteorologie
C.D. Schönwiese	Institut für Meteorologir und Geophysik der Universitat Goethe
J. Willebrand	Institut für Meerskunde der Universitat Kiel

FINLAND

E. Holopainen	University of Helsinki

FRANCE

A. Alexiou	CCO, UNESCO
D. Cariolle	EERM/Centre National de Recherches Meteorologiques
C. Lorius	Laboratoire de Glaciologie, CNRS

GERMAN DEMOCRATIC REPUBLIC

S. Dyck	Dresden Technical University
H. Lass	Institute for Marine Research
D. Spaenkuch	Meteorological Service of GDR

INDIA

S. Gadgil	Indian Institute of Sciences

ITALY

M. Conte	Italian Meteorological Service (Climate Unit)
M. Olacino	National Research Council (Institute of Atmospheric Physics)
S. Yanni	Italian Meteorological Service

JAPAN
H. Akimoto	National Institute for Environmental Studies
H. Muramatsu	Disaster Prevention Research Institute, Kyoto University
T. Nakazawa	Meteorological Research Institute
A. Noda	Meteorological Research Institute
M. Okada	Meteorological Research Institute
Y. Sugimur	Meteorological Research Institute
R. Yamamoto	Laboratory for Climatic Research, Kyoto University
T. Yasunari	University of Tsukuba, Institute of Geoscience
M. Yoshino	University of Tsukuba, Institute of Geoscience

KENYA
I.J. Ogallo	University of Nairobi, Meteorology Department

NETHERLANDS
A. Baede	Royal Netherlands Meteorological Institute (KNMI)
T. Buishand	Royal Netherlands Meteorological Institute (KNMI)
J. de Ronde	Rijkswaterstaat, Department of Transport and Public Works
W. de Ruijter	Institute for Meteorological & Oceanography, University of Utrecht
A. Kattenberg	Royal Netherlands Meteorological Institute (KNMI)
J. Rozema	University of Amsterdam
C. Schuurmans	Royal Netherlands Meteorological Institute (KNMI)
R. J. Swart	Institute for Public Health and the Environment (RIVM)
H. Tennekes	Royal Netherlands Meteorological Institute (KNMI)
R. Van Dorland	Royal Netherlands Meteorological Institute (KNMI)
A. van Ulden	Royal Netherlands Meteorological Institute (KNMI)

NEW ZEALAND
J. S. Hickman	New Zealand Meteorological Service
W. A. Laing	DSIR Fruit and Trees
D. C. Lowe	DSIR Institute of Nuclear Sciences
M. R. Manning	DSIR Institute of Nuclear Sciences
A. B. Mullan	New Zealand Meteorological Service

NIGERIA
E.O. Oladipo	University of Ahmadu Bello, Geography Department

SAUDI ARABIA
J. C. McCain	Research Institute (King Fahd University of Petroleum & Minerals)
N. I. Tawfiq	Meteorological & Environmental Protection Administration

SENEGAL
E.S. Diop	COMARAF
M. Seck	Ministere de l'Equipement

SOVIET UNION
M. Budyko	Main Geophysical Observatory
G.S. Golitsyn	Academy of Sciences, Institute of Atmospheric Physics
I. L. Karol	Main Geophysical Observatory
V. Meleshko	Main Geophysical Observatory

SWEDEN
B. Bolin	University of Stockholm
B. Dahlstrom	Swedish Meteorological & Hydrological Institute

T. R. Gerholm	University of Stockholm
A. Johannsson	University of Stockholm
E. Kallen	University of Stockholm
H. Sundqvist	University of Stockholm
G. Walin	Goteborg University (Department of Oceanography)

SWITZERLAND

M. Beniston	ERCOFTAC - Ecole Polytechnique de Lausanne
H. C. Davies	Laboratory of Atmospheric Physics
J. Fuhrer	Swiss Federal Research Station
F. Gassmann	Paul Scherrer Institute
K. Kelts	Swiss Academy of Sciences
H. R. Luthi	Swiss Federal Office of Energy
P. Morel	WMO
A. Ohmura	Swiss Federal Institute of Technology
H. Oeschger	Physikalisches Institute, University of Bern

THAILAND

| P. Patvivatsiri | Meteorological Department, Bangkok |

UNITED KINGDOM

B. Hoskins	University of Reading
B. Huntley	University of Durham
P.D. Jones	University of East Anglia, Climatic Research Unit
P. Killworth	Hooke Institute
G. Needler	WOCE, Institute of Oceanographic Sciences
N. J. Shackleton	University of Cambridge
K.P. Shine	University of Reading
T. M. L. Wigley	University of East Anglia, Climatic Research Unit
J. D. Woods	Natural Environment Research Council

UNITED STATES OF AMERICA

J.D. Aber	University of New Hampshire
J. Angell	NOAA
A. Arking	NASA/GSFC
D.J. Baker	Joint Oceanographic Institutes Inc
T.C. Bell	NASA/GSFC
C.R. Bentley	University of Wisconsin at Madison
F.P. Bretherton	SSEC, University of Wisconsin
P Brewer	Pacific Marine Laboratory
W.E. Carter	National Geodetic Survey
T. Charlock	NAS/LARC
R. Cicerone	University of California at Irvine
N. Cobb	
R. Dahlman	
R.E. Dickinson	National Centre for Atmospheric Research
R.K. Dixon	U.S. EPA
J. Dutton	Pennsylvania State University
J.A. Eddy	UCAR
R. Ellingson	University of Maryland
H.W. Ellsaesser	Lawrence Livermore National Laboratory
E. Ferguson	NOAA
J. Firor	National Centre for Atmospheric Research
J. Fishman	NAS/LARC

B. Flannery	Exxon Research and Engineering Company
M. Ghil	University of California, Los Angeles
I. Goklany	
D. Goodrich	NASA Office of Climatic and Atmospheric Research
T.E. Graedel	Bell Telephone Laboratories
S. Grotch	Lawrence Livermore National Laboratory
K.Hanson	NOAA
D.J. Jacob	Harvard University
C.Y.J. Kao	Los Alamos National Laboratory
Y.J. Kaufman	NASA/GSFC
J.T. Kiehl	National Centre for Atmospheric Research
J. Mahlman	NOAA
T. Malone	St Joseph's College
G. MacDonald	MITRE Corporation
M. McFarland	DuPont
G.A. Meehl	National Centre for Atmospheric Research
G. North	Texas A&M University
J. O'Brien	CCCO
M. Oppenheimer	Environmental Defence Fund
R. Perhac	Electric Power Institute
A.M. Perry	Retired
D.Randall	Colorado State University
P. Risser	Universtiy of New Mexico
A. Robock	University of Maryland
D.R. Rodenhuis	U.S. Department of Commerce
W. Rossow	NASA/GISS
D. Schimel	Colorado State University
W.H. Schlesinger	Duke University
S.H. Schneider	National Centre for Atmospheric Research
J. Shukla	University of Maryland
J. Sigmon	Environmental Protection Agency
J. Smagorinsky	
D.W. Stahle	Universtiy of Arkansas
J. Steele	Woods Hole Oceanographic Institute
G. Stevens	Colorado State University
P.H. Stone	Massachusetts Institute of Technology
N.D. Sze	AER Inc.
J. Trabalka	Oak Ridge National Laboratory
K.Trenberth	National Centre for Atmospheric Research
K. van Cleve	Universtiy of Alaska
T.H. Vonder Haar	Colorado State University
J.E. Walsh	University of Illinois
D.E. Ward	USDA Forest Service
R.T. Watson	NASA HQ, Washington
T. Webb	Brown University
G. Weller	Universtiy of Alaska
R.T. Wetherald	Princeton University (GFDL)
R.S. Williams Jr.	U.S. Geological Survey
F.B. Wood	U.S. Congress
B. Worrest	Environmental Protection Agency

Appendix 5

ACRONYMS: INSTITUTIONS

AERE	Atomic Energy Research Establishment, Harwell, UK
CCC	Canadian Climate Centre, Downsview, Ontario, Canada
CNRS	Centre National de Recherches Meteorologiques, France
CRU	Climatic Research Unit, University of East Anglia, UK
CSIRO	Commonwealth Scientific & Industrial Research Organisation, Australia
EPA	Environmental Protection Agency, Washington, USA
GFDL	Geophysical Fluid Dynamics Laboratory, Princeton, USA
GISS	Goddard Institute of Space Sciences, New York, USA
ICSU	International Council of Scientific Unions
IPCC	Intergovernmental Panel on Climate Change
MGO	Main Geophysical Laboratory, Leningrad,USSR
MPI	Max Planck Institut, FRG
MRI	Meteorological Research Institute, Japan
NASA	National Aeronautics and Space Administration, USA
NATO	North Atlantic Treaty Organisation
NCAR	National Center for Atmospheric Research, Boulder, USA
NOAA	National Oceanic and Atmospheric Administration, USA
OSU	Oregon State University, USA
SCOPE	Scientific Committee On Problems of the Environment
UKDOE	Department of the Environment, UK
UKMO	Meteorological Office, Bracknell, UK
UNEP	United Nations Environment Programme
USDOE	Department of the Energy, USA
WMO	World Meteorological Organisation

Appendix 6

ACRONYMS: PROGRAMMES & MISCELLANEOUS

PROGRAMMES:

CLIMAP	Climatic Applications Project (WMO)
COADS	Comprehensive Ocean Air Data Set
GAW	Global Atmospheric Watch
ERBE	Earth Radiation Budget Experiment
ERS	Earth Resources Satellite
GEWEX	Global Energy and Water Cycle Experiment
GMCC	Geophysical Monitoring of Climatic Change
ICRCCM	Intercomparison of Radiation Codes in Climate Models
IGAC	International Global Atmospheric Chemistry Programme
IGBP	International Geosphere-Biosphere Programme
ISCCP	International Satellite Cloud Climatology Project
JGOFS	Joint Global Ocean Flux Study
SAGE	Stratospheric Aerosol and Gas Experiment
TOGA	Tropical Ocean and Global Atmosphere
WCRP	World Climate Research Programme
WOCE	World Ocean Circulation Experiment

MISCELLANEOUS

3-D	Three dimensional
AET	Actual Evapotranspiration
AGCM	Atmosphere General Circulation Model
AVHRR	Advanced Very High Resolution Radiometer
BaU	Business-as-Usual
BP	Boiling Point
CCN	Cloud Condensation Nuclei
CRF	Cloud Radiative Forcing
CW	Cloud Water
EBM	Energy Balance Model
EKE	Eddy Kinetic Energy
ENSO	El Niño Southern Oscillation
FC	Fixed Cloud
GCM	General Circulation Model
GP	Gross Prediction
GPP	Gross Primary Production
GWP	Global Warming Potential

LWC	Liquid Water Content
MCA	Moist Convective Adjustment
MSL	Mean Sea Level
NH	Northern Hemisphere
NPP	Net Primary Production
ODP	Ozone Depletion Potential
OGCM	Ocean General Circulation Model
OLR	Outgoing Longwave Radiation
PBL	Planetary Boundary Layer
PC	Penetrative Convection
RCM	Radiative Convective Model
RH	Relative Humidity
SH	Southern Hemisphere
SST	Sea Surface Temperature
TOA	Top of the Atmosphere
WUE	Water Use Efficiency
UV	Ultraviolet

Appendix 7

UNITS

SI (Systeme Internationale) Units:

Physical Quantity	Name of Unit	Symbol
length	meter	m
mass	kilogram	kg
time	second	s
thermodynamic temperature	kelvin	K
amount of substance	mole	mol

Fraction	Prefix	Symbol	Multiple	Prefix	Symbol
10^{-1}	deci	d	10	deka	da
10^{-2}	centi	c	10^2	hecto	h
10^{-3}	milli	m	10^3	kilo	k
10^{-6}	micro	μ	10^6	mega	M
10^{-9}	nano	n	10^9	giga	G
10^{-12}	pico	p	10^{12}	tera	T
10^{-15}	femto	f	10^{15}	peta	P
10^{-18}	atto	a			

Special Names and Symbols for Certain SI-Derived Units:

Physical Quantity	Name of SI Unit	Symbol for SI Unit	Definition of Unit
force	newton	N	$kg\ m\ s^{-2}$
pressure	pascal	Pa	$kg\ m^{-1}s^{-2}(=Nm^{-2})$
energy	joule	J	$kg\ m^2\ s^{-2}$
power	watt	W	$kg\ m^2s^{-3}(=Js^{-1})$
frequency	hertz	Hz	s^{-1}(cycle per second)

Decimal Fractions and Multiples of SI Units Having Special Names:

Physical Quantity	Name of Unit	Symbol for Unit	Definition of Unit
length	ångstrom	Å	$10^{-10}\ m = 10^{-8}cm$
length	micrometer	μm	$10^{-6}m = μm$
area	hectare	ha	$10^4\ m^2$
force	dyne	dyn	$10^{-5}\ N$
pressure	bar	bar	$10^5\ N\ m^{-2}$
pressure	millibar	mb	$1hPa$
weight	ton	t	$10^3\ Kg$

Non- SI Units:

°C   degrees Celsius (0°C = 273K approximately)
                Temperature differences are also given in °C (=K) rather than the
                more correct form of "Celsius degrees".

ppmv   parts per million (10^6)by volume
ppbv   parts per billion (10^9) by volume
pptv   parts per trillion (10^{12}) by volume

bp   (years) before present
kpb   thousands of years before present
mbp   millions of years before present

The units of mass adopted in this report are generally those which have come
into common usage, and have deliberately not been harmonised, e.g.,

GtC   gigatonnes of carbon (1 GtC = 3.7 Gt carbon dioxide)
MtN   megatonnes of nitrogen
TgS   teragrams of sulphur

Appendix 8

CHEMICAL SYMBOLS

O	atomic oxygen	HO_2	hydroperoxyl
O_2	molecular oxygen	NO_y	total active nitrogen
N	atomic nitrogen	pCO_2	partial pressure CO_2
N_2	molecular nitrogen	CFC	chlorofluorocarbon
H	hydrogen	DMS	Di Methyl Sulphide
Cl	chlorine		
Cl_2	molecular chlorine		
Br	bromine	CFC-11	$CFCl_3$
F	fluorine		(trichlorofluoromethane)
CH_4	methane	CFC-12	CF_2Cl_2
N_2O	nitrous oxide		(dichloro-difluoromethane)
NO	nitric oxide	CFC-13	CF_3Cl
NO_2	nitrogen dioxide	CFC-113	$C_2F_3Cl_3$
CO	carbon monoxide		(trichloro-trifluoroethane)
CO_2	carbon dioxide	CFC-114	$C_2F_4Cl_2$
H_2O	water		(dichloro-tetrafluoroethane)
CH_3Cl	methyl chloride	CFC-115	C_2F_5Cl
CH_3Br	methyl bromide		(chloropentafluoroethane)
O_3	ozone	HCFC-22	CHF_2Cl
OH	hydroxyl		(chlorodifluoromethane)
CCl_4	carbon tetrachloride	HCFC-123	$C_2HF_3Cl_2$ ($CHCl_2CF_3$)
NMHC	non-methane hydrocarbons	HCFC-124	$CHFClCF_3$
NO_x	nitrogen oxide	HFC-125	CHF_2CF_3
CH_3CCl_3	methyl chloroform	HCFC-132b	$C_2H_2F_2Cl_2$
HNO_3	nitric acid	HFC-134a	$C_2H_2F_4$ (CH_2FCF_3)
PAN: $CH_3CO_3NO_2$	peroxyacetylnitrate	HCFC-141b	CH_3CFCl_2
NO_3	nitrate radical	HCFC-142b	CH_3CF_2Cl
SO_2	sulphur dioxide	HFC-143a	CH_3CF_3
COS	carbonyl sulphide	HFC-152a	CH_3CHF_2
H_2S	dimethylsulphide		
BrO	bromine monoxide	HALON 1211	CF_2BrCl ($CBrClF_2$)
HCl	hydrochloric acid		(bromodichloromethane)
C_2H_6	ethane	HALON 1301	CF_3Br ($CBrF_3$)
ClO_2	chlorine dioxide		(bromotrifluoromethane)
$ClONO_2$	chlorine nitrate	HALON 2402	$C_2F_4Br_2$
HOCl	hypochlorous acid		(dibromo-tetrafluoroethane)
H_2O_2	hydrogen peroxide		

This is the Report of Working Group 1 of the
Intergovernmental Panel on Climate Change, which was set
up jointly by the World Meteorological Organization and
United Nations Environment Programme in 1988. It is an
assessment of how human activities may be changing the
Earth's climate through the Greenhouse Effect – potentially
the greatest global environmental challenge facing mankind.
The topics covered by this assessment include:

- Changes in greenhouse gases in the atmosphere
- The global climate system and how it is modelled
- Computer predictions of climate change
- Observed climate change over the last century
- Detection of climate change due to human activities
- Changes in sea level due to global warming
- The response of ecosystems to climate change
- Research required to narrow uncertainties

Several hundred international scientists participated in the
preparation and review of this assessment, making it the most
authoritative and strongly supported statement on climate
change that has been made by the scientific community so
far. The information presented here is of the highest quality.
It has been designed to provide a common guide to
policymakers worldwide, and will now form a solid scientific
foundation upon which forthcoming negotiations on the
response to climate change will be based. This assessment is,
therefore, an essential reference for all who are concerned
with climate change and its consequences.

WMO

UNEP

CAMBRIDGE UNIVERSITY PRESS

ISBN 0-521-40720-6

9 780521 407205